LENK'S VIDEO HANDBOOK

Other McGraw-Hill Books of Interest

Bartlett • CABLE COMMUNICATIONS

Benson • AUDIO ENGINEERING HANDBOOK

Benson • TELEVISION ENGINEERING HANDBOOK

Benson and Whitaker • TELEVISION AND AUDIO HANDBOOK

Coombs • PRINTED CIRCUITS HANDBOOK, 4/E

Croft and Summers • AMERICAN ELECTRICIANS' HANDBOOK

Fink and Beaty • STANDARD HANDBOOK FOR ELECTRICAL ENGINEERS

Fink and Christiansen • ELECTRONICS ENGINEERS' HANDBOOK, 4/E

Harper • HANDBOOK OF ELECTRONIC PACKAGING AND INTERCONNECTION, 2/E

Johnson • ANTENNA ENGINEERING HANDBOOK, 3/E

Lenk • LENK'S AUDIO HANDBOOK

Mee and Daniel • MAGNETIC RECORDING HANDBOOK, 2/E

Mee and Daniel • MAGNETIC RECORDING TECHNOLOGY, 2/E

Sherman • CD-ROM HANDBOOK, 2/E

Whitaker • ELECTRONIC DISPLAYS

Williams and Taylor • ELECTRONIC FILTER DESIGN HANDBOOK, 2/E

To order or receive additional information on these or any other McGraw-Hill titles, in the United States please call 1-800-822-8158. In other countries, contact your local McGraw-Hill representative. KEY=WM16XXA

LENK'S VIDEO HANDBOOK

Operation and Troubleshooting

John D. Lenk

Second Edition

McGraw-Hill

New York   San Francisco   Washington, D.C.   Auckland   Bogotá
Caracas   Lisbon   London   Madrid   Mexico City   Milan
Montreal   New Delhi   San Juan   Singapore
Sydney   Tokyo   Toronto

Library of Congress Cataloging-in-Publication Data

Lenk, John D.
　　Lenk's video handbook : operation and troubleshooting / John D.
Lenk. — 2nd ed.
　　　　p.　　cm.
　　Includes index.
　　ISBN 0-07-037616-6.
　　1. Television—Repairing.　2. Video tape recorders—Maintenance
and repair.　3. Videodisc players—Maintenance and repair.　4. Home
video systems—Maintenance and repair.　I. Title.
TK6642.L457　　　1997
621.388′337—dc20　　　　　　　　　　　　　　　　96-34523
　　　　　　　　　　　　　　　　　　　　　　　　　　　　CIP

McGraw-Hill

A Division of *The McGraw·Hill Companies*

1 2 3 4 5 6 7 8 9 0　DOC/DOC　9 0 1 0 9 8 7 6

ISBN 0-07-037616-6

*The sponsoring editor for this book was Stephen S. Chapman, the editing
supervisor was Peggy Lamb, and the production supervisor was Donald F.
Schmidt. It was set in Times Roman by Terry Leaden of McGraw-Hill's
Professional Book Group composition unit.*

Printed and bound by R. R. Donnelley & Sons Company.

 This book is printed on recycled, acid-free paper containing a minimum of 50% recycled, de-inked fiber.

McGraw-Hill books are available at special quantity discounts to use as premiums
and sales promotions, or for use in corporate training programs. For more information, please write to the Director of Special Sales, McGraw-Hill, 11 West 19th
Street, New York, NY 10011. Or contact your local bookstore.

Greetings from the Villa Buttercup!
To my wonderful wife, Irene,
thank you for being by my side all these years!
To my lovely family, Karen, Tom, Brandon, Justin, and Michael.
And to our Lambie and Suzzie, be happy wherever you are!

To my special readers: May good fortune find your doorway,
bringing you good health and happy things.
Thank you for buying my books!

And special thanks to Steve Chapman, Stephen Fitzgerald, Leslie
Wenger, Ted Nardin, Mike Hays, Patrick Hansard, Lisa Schrager,
Mary Murray, Carol Wilson, Florence Trimble, Janet Gomolson,
Gemma Velten, Fran Minerva, Jane Stark, Ann Wilson, Andrew
Yoder (best-selling author), and Robert McGraw of McGraw-Hill
for making me an international bestseller again!

This is book number 86.
Abundance!

CONTENTS

Chapter 3. Basic Video Color Circuits 3.1

Chapter 4. Digital Video Circuits 4.1

Chapter 5. VHS Video Circuits 5.1

Chapter 6. Beta Video Circuits 6.1

PREFACE

As in the case of the first edition, this new and improved second edition provides you with how-it-works and how-to-service-it information that can be put to immediate use on any video equipment. With more than 50 percent new material added, the second edition truly becomes a "something for everyone" video handbook.

If you are a beginner or student, with no knowledge of video troubleshooting, read the first chapter, which describes the basic troubleshooting approach for any type of video equipment or system.

If you do not understand troubleshooting for a particular type of video equipment (TV, monitor, digital TV, VCR, camcorder, or video-disc player), there are chapters that describe the specific troubleshooting approach for that type of equipment.

Even if you are an experienced technician or field-service engineer, the step-by-step examples throughout the book (based on actual troubleshooting problems) serve to fill the gap between the theory of video and practical how-to knowledge required for troubleshooting modern video equipment.

This book concentrates on *troubleshooting approaches*, rather than trying to provide detailed troubleshooting for all models of all video equipment (an impossible task). This is done by providing a separate chapter for a cross section of video devices. In each chapter, you will find (1) an introduction that describes the equipment's purpose, (2) circuit descriptions or circuit theory (not generally available in the manufacturer's service manuals), (3) sample test and adjustment procedures (including a full discussion of any special test equipment required), and (4) a logical troubleshooting approach for the video circuits (based on manufacturers' recommendations).

Chapter 1 provides a summary or refresher of basic troubleshooting techniques for all types of video equipment. It concentrates on basic troubleshooting requirements that are imposed by complex and simple video equipment that is solid-state, digital, IC, microprocessor-based, or whatever.

Chapter 2 is devoted to the basics of video circuits, concentrating on TV/monitors (the most familiar video devices in consumer electronics). The second edition includes new circuit information on VIF/SIF, vertical/horizontal sweep, sync separator, high voltage, vertical oscillator, horizontal oscillator, system control, and CRT drive.

Chapter 3 is devoted to the basics of video color and provides coverage similar to that of Chapter 2. The second edition includes new circuit information on video/chroma processing, comb filters, Y/C processing, RGB drives, and localizing color troubles.

Chapter 4 is devoted to video circuits that use some form of digital pulses. Such circuits include frequency synthesis (FS) tuning, PLL, PSC, on-screen display (OSD), infrared (IR) remote control, and digital TV (picture in picture). The second edition includes new circuit information on tuning memory, AFT, FM traps, channel surfing, PinP processing, and PinP troubleshooting.

Chapter 5 is devoted to the video circuits found in VHS VCRs. The chapter also covers other portions of the VCRs that are directly related to the video circuits (such as the system control and servo system for the tape-drive mechanism). In addition to providing VHS-circuit operation (including HQ, S-VHS, and VHS-C), the chapter includes troubleshooting, adjustment, and service notes for all types of VHS video circuits.

Chapter 6 is devoted to the video circuits found in Beta VCRs. The chapter also covers other portions of Beta VCRs that are directly related to the video circuits (such as system control and servo system for the tape-drive mechanism).

Chapter 7 is devoted to the video circuits found in video cameras and camcorders. Because a camcorder is essentially a video camera combined with a VCR, the examples here are based on the circuits found in camcorders (both VHS and 8 mm), including Newvicon, Saticon, MOS, and CCD video-camera circuits. The second edition includes new circuit information on CCD cameras (timing generators, drives, PG, sync generator, output demodulation, AGC/chroma separation, white balance, clamp, blanking, gamma correction, white clip, pedestal, iris processing, AGC, AWB, RGB, chroma/luma, encoding/decoding, and matrix).

Chapter 8 describes the differences between the 8-mm format and the VHS camcorder formats discussed in Chapter 7. The second edition includes new circuit information on 8-mm servo, automatic track finding (ATF), tape counter, capstan servo, free-speed compensation, capstan motor drive, and error correction.

Chapter 9 (an entirely new chapter in the second edition) is devoted to video-disc players, and includes basic theory, troubleshooting, adjustment, and service notes. Some of the subjects covered are CDV, laser-discs, DVD, optics, track formats, laser safety, mute circuits, spindle-motor, system control, laser control, servo, focus, tracking and slider, tilt, disc signal processing, video processing, digital memory, and analog and digital audio.

John D. Lenk

ACKNOWLEDGMENTS

Many professionals have contributed to this book. I gratefully acknowledge the tremendous effort needed to produce this book. Such a comprehensive work is impossible for one person, and I thank all who contributed, both directly and indirectly.

I give special thanks to the following: Karen Allen of the Benjamin Group for Toshiba, John Taylor of Zenith, and the staff at Mitsubishi, Sony, Hitachi, Sanyo, Thomson Consumer Electronics (RCA), and Philips Consumer Electronics.

I also wish to thank Joseph A. Labok of Los Angeles Valley College for help and encouragement throughout the years.

And a very special thanks to Steve Chapman, Stephen Fitzgerald, Leslie Wenger, Ted Nardin, Mike Hays, Patrick Hansard, Lisa Schrager, Mary Murray, Carol Wilson, Peter Mellis, Judy Kessler, Florence Trimble, Janet Gomolson, Gemma Velten, Fran Minerva, Jane Stark, Melanie Holscher, Fred Perkins, Robert McGraw, Barbara McCann, Kimberly Martin, Jane Schmidt, Tracy Baer, Judith Reiss, Charles Love, Betty Crawford, Jeanne Myers, Peggy Lamb, Thomas Kowalczyk, Don Schmidt, Jaclyn Boone, Katherine Brown, Jennifer Priest, Kathy Greene, Donna Namorato, Monika Macezinskas, Susan Kagey, Bob Ostrander, Lori Flaherty, Stacy Spurlock, Midge Haramis, B. J. Peterson, and Andrew Yoder (best-selling author) of McGraw-Hill's Professional Publishing for having that much confidence in me.

And to Irene, my wife, business manager and Super Agent, I extend my thanks. Without her help this book could not have been written.

CHAPTER 1

UNDERSTANDING AND TROUBLESHOOTING VIDEO CIRCUITS

This chapter provides a summary or refresher of basic troubleshooting techniques for all types of video equipment. No matter what circuit is involved, a logical approach is needed to find and correct any fault. Certain troubleshooting requirements are imposed by complex and simple video equipment, whether it is solid state, digital, IC, microprocessor-based, or whatever. For example, with any video equipment or system, you must know how the equipment works under normal conditions. This means that you must study the equipment to find out how each circuit normally works. Of course, studying the equipment can mean several different things.

If the equipment is military or industrial, there is an instruction manual that provides all sorts of information (theory of operation, operating and test procedures, adjustment procedures, parts lists and location diagrams, etc.). In some very complex video systems, there are several manuals (service, overhaul, operation, parts catalog, etc.).

If the video device is of the consumer-electronics or home-entertainment type (TV or monitor, VCR, camcorder, etc.), datasheets, or fact sheets, are usually available. Although these sheets do not include the elaborate descriptions found in technical manuals, they do contain condensed data (schematics, test and adjustment procedures, waveforms, voltage and resistance data, etc.) which are adequate for the type of video equipment. In the absence of a datasheet, most home-entertainment equipment is provided with a schematic diagram that shows some waveforms and voltage and resistance data, as applicable.

No matter what information is available, study the data thoroughly before attempting to troubleshoot the video device. In rare cases, there is no information to study. If you have ever serviced equipment under these conditions, you will realize the value of technical data and of studying the data thoroughly. You will also realize the value of studying the author's many troubleshooting books, which describe both operational theory and test and adjustment procedures in boring detail.

You must know the function of all controls and adjustments and how to operate them. Often, conditions that appear to be serious defects are the result of operator trouble. Also, as equipment ages, some adjustments may become critical. In any event, you must be capable of operating the equipment to do a good service job. (Besides, it makes a bad impression on the customer if you cannot find the on-off switch, especially after the second service call.)

With simple video equipment, the operating procedures are obvious or are standardized (such as the operating procedures for TV sets). However, if you are hopelessly lost, try to get the Owner's Manual, which usually describes operation and special precautions for the equipment. (Use the pretext of wanting to "look up the serial number" to lay hands on the manual.)

You must know how to use test equipment and how to perform test and adjustment procedures. You are in real trouble in any troubleshooting situation if you cannot use electronic test equipment effectively. Remember that some video equipment requires special test instruments (such as lightboxes, light meters, adjustment charts, and light sources for camcorders). Fortunately, most video troubleshooting procedures can be performed with three basic instruments: the meter (VOM and/or digital meter), the oscilloscope (which we shall call a *scope*), and a signal source (RF generator, sweep generator, NTSC color generator, etc.).

You must be able to perform a checkout procedure on the repaired equipment. No matter how simple the repair, this step should not be omitted. Sometimes, one problem is the result of another. If both are not cured, the problem will still be there. A classic example of this is an intermittent short or arc between two points (say, between two pins on the edge connector of a PC board). Assume that the short burns out a resistor on the PC board or burns out an entire IC module.

To make an adequate checkout, it is again necessary to understand the equipment and the operating procedure (usually found in the technical manual or datasheet). Generally, if the video device performs all the operating functions in the proper sequence, it can be considered as repaired and ready for use. Also, a checkout of the equipment after repair may point out the need for readjustment of controls.

You must know how to use tools to repair a problem once it has been located. Most video repairs can be made with basic tools (soldering tools, pliers, screwdrivers, diagonal cutters, etc.). However, special techniques must be used for certain equipment and circuits. Repair of PC boards and removal and replacement of IC modules are typical examples. Remember that if the video equipment contains mechanical components (such as VCRs and camcorders), metric tools are required.

Finally, you must be able to logically analyze the information about malfunctioning equipment and apply a systematic, logical procedure to find the problem. In short, you must be able to think. The information to be analyzed may be the equipment's performance (the appearance of the picture on a TV screen) or indications taken from test equipment (voltage and resistance measurements or waveforms). Either way, it is the analysis of the information that makes for logical, efficient troubleshooting.

1.1 TYPICAL TROUBLESHOOTING SEQUENCE FOR VIDEO EQUIPMENT

There are four basic steps in the troubleshooting sequence: (1) determine the symptoms of failure, (2) localize the trouble to a complete functional unit or module, (3) isolate the trouble to a circuit (or circuit board) within the module, and (4) locate the specific trouble. On very simple video equipment, or on equipment where there is only one functional unit (TV, VCR, camcorder, etc.), step 2 can be omitted. Before going into the details of the steps, let us consider what is done by each step.

1.1.1 Determining Failure Symptoms in Video Circuits

Determining symptoms means that you must know what the equipment is supposed to do when operating normally and, more important, that you must be able to recognize when the normal job is not being done. Everyone knows what a TV set is supposed to do, but no one knows how well each set is to perform (and has performed in the past) under all operating conditions (with a given antenna, lead-in, cable, location, etc.).

All TV sets have operating controls and two built-in aids for evaluating performance (the loudspeaker and the picture tube). Using the normal and abnormal symptoms produced by the loudspeaker and picture tube, you must analyze the symptoms to answer the questions, "How well is this set performing and where in the set could there be trouble that produces these symptoms?"

The determining-symptoms step does not mean that you charge into the set with screwdriver and soldering tool, nor does the step mean that test equipment should be used extensively. Instead, the step means that you make a visual check, noting both normal and abnormal performance indications. The step also means that you operate the controls to gain further information. At the end of the determining-symptoms step, you definitely know that something is wrong and have a fair idea of what is wrong, but you probably do not know just what area of the equipment is faulty. This is established in the next step of troubleshooting.

1.1.2 Localizing Trouble to a Functional Video Unit or Module

Most video equipment can be subdivided into units or areas which have a definite purpose or function. The term *function* is used in video troubleshooting to denote an operation in a specific area of the equipment. For example, in a basic black and white TV, the functions can be divided into RF, IF, audio, video, picture tube, and power supply.

To localize the trouble systematically and logically, you must have a knowledge of the function units of the video device and must correlate all the symptoms previously determined. So the first consideration in localizing the trouble to a functional unit is a valid estimate of which area might be causing the indicated symptoms. Initially, several technically accurate possibilities may be considered as the probable trouble area.

As a classic (oversimplified) example, if both picture and sound are poor on a TV set, the trouble might be in either the RF or IF stages, since these functional areas are common to both picture and sound reproduction. On the other hand, if the picture is good but the sound is poor, the trouble is probably in the audio stages, since these functional areas apply only to sound.

Using Diagrams.   Video troubleshooting involves (or should involve) the extensive use of diagrams. Such diagrams may include a *functional block diagram* and always includes *schematic diagrams. Practical wiring diagrams,* such as found in military-style service literature, are almost never available in video service. However, typical video literature does include *printed wiring diagrams* (on those boards where individual components can be replaced).

The block diagram, such as shown in Fig. 1.1, shows the functional relationship of all major sections or units in the equipment (a black and white TV set, in this case). The block diagram is thus the most logical source of information when localizing trouble to a functioning unit or section. Unfortunately, not all video service literature includes a block diagram (or the block diagram shows very little useful detail). It may be necessary to use only schematic diagrams.

Schematic diagrams show the functional relationship of all parts in the equipment. Such parts include transistors, ICs, transformers, capacitors, resistors, diodes, etc. Generally, the schematic presents too much information (not directly related to the specific symptoms noted) to be of maximum value during the localizing step.

In a typical video-equipment schematic, the decisions being made regarding the probable trouble area may become lost among all the details of the schematic. However, the schematic is very useful in later stages of troubleshooting (and when a block diagram is not available). In comparing the block diagram and the schematic

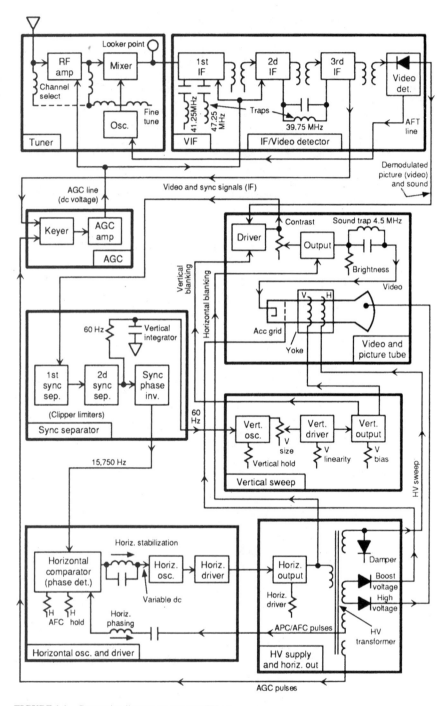

FIGURE 1.1   Composite discrete-component TV set.

during the localize step, note that each transistor or IC shown on the schematic is usually represented as a block on the block diagram.

In some video service literature, the physical relationship of parts is given on *component-location diagrams* or *photos* (also called parts-placement diagrams or photos). These location or placement diagrams often supplement the printed wiring diagrams and do not necessarily show all parts. Instead, the diagrams concentrate on major parts such as ICs, transistors, diodes, and adjustment points or controls. For this reason, location and placement diagrams are the least useful when localizing trouble and are most useful when locating specific parts during other phases of troubleshooting.

To sum up, it is logical to use a block diagram instead of a schematic or location diagram when you want to make a valid estimate about the probable trouble areas (or circuits). The use of a block diagram also permits you to use a troubleshooting technique known as *bracketing* (or *good input-bad output*). If the block diagram includes major test points, as it may in some well-prepared service literature, the block also permits you to use test equipment to help narrow down the probable trouble cause. However, test equipment is used more extensively during the isolation step of troubleshooting.

1.1.3 Isolating Troubles to a Video Circuit

After the trouble is localized to a single functional area, the next step is to isolate the trouble to a circuit in the faulty area. To do this, you concentrate on circuits in the area that could cause the trouble and ignore the remaining circuits. The isolating step involves the use of test equipment such as meters, scopes, and signal generators for *signal tracing* and *signal substitution* in the suspected faulty area. By making valid educated estimates and properly using the applicable diagrams, bracketing techniques, signal tracing, and signal substitution, you can systematically and logically isolate the trouble to a single defective circuit.

Repair techniques or tools to make necessary repairs to the equipment are not used until after the specific trouble is located and verified. That is, you still do not charge into the equipment with solder tools and screwdriver at this point. Instead, you are now trying to isolate the trouble to a specific defective circuit so that, once the trouble is located, it can be repaired.

1.1.4 Locating the Specific Trouble in Video Equipment

Although this troubleshooting step refers only to locating the specific trouble, the step also includes a final analysis or review of the complete procedure and the use of repair techniques to remedy the problem. This final analysis permits you to determine whether some other malfunction affected the part or whether the part located is the actual cause of trouble.

Inspection using the senses—sight, smell, hearing, and touch—is very useful in trying to locate the problem. This inspection is usually performed first, to gather information that may lead more quickly to the defective part. (The inspection is often referred to as a *visual inspection* in service literature, although it involves all the senses.) Among other things to look for during visual inspection are burned, charred, or overheated parts, arcing in the circuit, and burned-out parts.

In video equipment where access to the circuits is relatively easy, a rapid visual inspection should be performed first. Then the active device—transistor or IC—can be checked. A visual inspection is always recommended as the first step for all equipment. A possible exception is equipment in which access to the majority of circuits is very difficult but where certain parts (usually active devices such as transistors and ICs or possibly complete PC boards) can easily be removed and tested (or replaced).

The next step in locating the specific problem in typical video equipment is the use of a scope to observe *waveforms* and a meter to measure voltages. The scope can also be used as a substitute meter to measure voltages. Of course, a meter is best when making *resistance* and *continuity* checks to pinpoint a defective part.

Note that in most present-day video service literature the voltages (and possibly the resistances) are often given on the schematic diagram, but this information may also appear in chart form (following the military style). No matter what information is given and what form the information takes, you must be able to use test equipment to make the measurements. That is why we stress test equipment and procedures throughout this book.

After the trouble is located, you should make a final analysis of the complete troubleshooting procedure to verify the trouble. Then you can repair it and check out the equipment for proper operation.

1.1.5   Developing a Systematic Troubleshooting Procedure for All Video Equipment

The development of a systematic and logical troubleshooting procedure requires (1) a logical approach to the problem, (2) knowledge of the equipment, (3) the interpretation of test information, and (4) the use of information gained in each step.

Some technicians feel that knowledge of the equipment involves remembering past failures as well as the location of all test points, all adjustment procedures, and so on. This approach may be good in troubleshooting only one type of video equipment but has little value in developing a basic troubleshooting approach.

It is true that recalling past equipment failures may be helpful, but you should not expect that the same trouble is the cause of a given symptom in every case. In any video equipment, many trouble areas may show about the same symptom indications.

Also, you should never rely only on your memory of adjustment procedures, test-point locations, and so on, in dealing with any troubleshooting problem. This is one of the functions of service literature, which contains diagrams and information on the equipment. In the remaining chapters of this book, we discuss the specific use of service literature and the type of information to be found in it. The important point is that you should learn to become a systematic, logical troubleshooter, not a memory expert.

1.1.6   Relationships among Video Troubleshooting Steps

Thus far, we have established the overall troubleshooting approach. Now let us make sure that you understand how each troubleshooting step fits together with others by analyzing their relationships.

Assume that you are troubleshooting a VCR, you are well into the fourth step (locate), and you find nothing wrong with the circuit that is supposedly causing the problem. That is, all waveforms, voltage measurements, and resistance measurements are normal. What is your next step? You might assume that nothing is wrong, that the problem is "customer or operator trouble." This is poor judgment. First, there must be something wrong with the VCR since some abnormal symptoms were recognized (during at least one of the first three steps) before you got to the locate step.

Never assume anything when troubleshooting. Either the equipment is working properly or it is not (first step). Either observations and measurements are made or they are not made (second and third steps). You must draw the right conclusions from the observations, measurements, and other factual evidence, or you must *repeat the troubleshooting procedure.*

There are those who believe that repeating the troubleshooting procedure means starting all over from the first step. Some service literature recommends this since it is possible for anyone, even an experienced technician, to make mistakes. When performed logically and systematically, the troubleshooting procedure keeps mistakes to a minimum. However, voltage and resistance measurements may be interpreted erroneously, waveform observations or bracketing may be performed incorrectly, or many other mistakes may occur through simple oversight.

Despite such recommendations by other service-literature writers, this author contends that "repeat the troubleshooting procedure" means *retrace* your steps, one at a time, until you find the place where you went wrong. Perhaps a previous voltage or resistance measurement was interpreted erroneously in the locate step, or perhaps bracketing was incorrectly performed in the isolate step. The cause must be determined logically and systematically by taking a *return path* to the point at which you went astray.

1.2 PRACTICAL VIDEO TROUBLESHOOTING SEQUENCE

Now that we have established a basic troubleshooting sequence, let us apply this approach to some practical troubleshooting situations. The remainder of this chapter is devoted to examples of how the sequence can be used in troubleshooting a basic black and white TV set.

Note that these examples are very generalized and do not apply to any specific TV model. They are given here to establish how the troubleshooting sequence and basic techniques (bracketing, waveform measurement, voltage measurement, etc.) are used in practical troubleshooting (whether the video device is a discrete component, IC, or plug-in module).

1.2.1 Some Classic Video Trouble Symptoms

Table 1.1 lists the most common troubles for a TV set. The troubles are grouped into functional areas (or circuits) of the set. These areas, or circuit groups, correspond to those in the block diagram shown earlier in Fig. 1.1. Note that the TV set is essentially a discrete-component device (containing no ICs or plug-in modules). There are millions of such sets in use, even though the trend is toward IC and plug-in board design.

Some of the symptoms listed in Table 1.1 point to only one area of the TV set as a *probable cause* of trouble. For example, if there is no vertical sweep (the picture-tube screen shows only a horizontal line) but other set functions are normal (good sound, for example), the trouble is probably in the vertical sweep circuits. On the other hand, if both sound and picture are weak or poor, the trouble *could be* in the RF tuner, in the IF and video detector, or possibly in the driver of the video-amplifier and picture-tube circuits (in this particular set).

In the discussions of the localize, isolate, and locate steps that follow, we give examples of how the symptoms can be used as the first step in pinpointing trouble to an area, to a circuit within the area, and finally to a part within the circuit. Before going into these steps, let us follow the basic troubleshooting sequence we have established.

TABLE 1.1   Common Troubles in a Discrete-Component TV Set

RF tuner
No picture, no sound
Poor picture, poor sound
Hum bars or hum distortion
Picture smearing and sound separated from picture
Ghosts
Picture pulling
Intermittent problems
Problems on some (but not all) channels

IF amplifier and video detector
No picture, no sound
Poor picture, poor sound
Hum bars or hum distortion
Picture smearing, pulling, or overloading
Intermittent problems

Video amplifier and picture tube
Intermittent problems
Retrace lines in picture
Sound in picture
Poor picture quality
Contrast problems
No picture, sound normal
No picture, no sound
No raster (dark screen)

AGC
Poor picture
No picture, no sound

Sound IF and audio
Poor or weak sound
No sound

1.3 PRACTICAL VIDEO TROUBLE-SYMPTOM DETERMINATION

It is not difficult to realize that there definitely is trouble when video equipment does not operate. For example, there obviously is trouble when a TV set is plugged in and turned on and there is no sound, picture, or pilot light (or other front-panel indication, such as an LED turn-on). A different problem arises when the TV set is still operating but is not doing a good job. Assume that the picture and sound are present but that the picture is weak and that there is a buzz in the sound.

Another problem in determining trouble symptoms is improper use of the equipment by the operator. In complex equipment, operators are usually trained and checked out on the equipment. The opposite is true of home-entertainment or consumer video equipment used by the general public. However, no matter what equipment is involved, it is always possible for an operator (or customer) to report a "prob-

TABLE 1.1 Common Troubles in a Discrete-Component TV Set (*Continued*)

Sync separator
Picture pulling
No horizontal sync
No vertical sync
No sync (horizontal or vertical)

High-voltage power supply and horizontal output
Nonlinear horizontal display
Foldback or foldover
Narrow picture
Picture overscan
Dark screen

Low-voltage power supply
Picture pulling and excessive vertical height
Distorted sound and no raster
No sound, no picture raster, transformer buzz
No sound and no picture raster

lem" that is actually the result of improper operation. For these reasons, you must first determine the signs of failure, regardless of the extent of malfunction, caused by either the equipment or the customer. This means that you must know how the equipment operates normally and how to operate the equipment controls.

1.3.1 Video Symptom Recognition

Symptom recognition is the art of identifying *abnormal* and *normal* signs of operation in electronic equipment. A true trouble symptom is an *undesired change* in equipment performance or a *deviation* from the standard. For example, the normal TV picture is a clear, properly contrasted representation of an actual scene. The picture should be centered within the vertical and horizontal boundaries of the scene. If the picture suddenly begins to roll vertically, you would recognize this as a trouble symptom because the condition does not correspond to the normal performance that is expected.

Now assume that the picture is weak, caused perhaps by a poor broadcast signal in the area or a defective antenna (or a poor cable picture). If the RF and IF stages of the TV set do not have sufficient gain to produce a good picture under these conditions, you could mistake this for a trouble symptom. A poor picture (for this particular model of TV operating under these conditions) is not abnormal operation nor is it an undesired change. Thus, the condition is not a true trouble symptom and should be so recognized.

1.3.2 Equipment Failure versus Degraded Performance

Equipment failure means that either the entire piece of equipment or some functional part of it is not operating properly. For example, the total absence of a picture on the screen of a TV set (when all controls are properly set) is a form of equipment failure, even though there may be sound (background noise) from the speaker.

Degraded performance is present whenever the equipment is working but is not presenting normal performance. For example, the presence of snow on the screen is degraded performance, since the set has not yet failed, but the performance is abnormal.

Example of Equipment Failure and Degraded Performance.   Generally, the terms *equipment failure* and *degraded performance* apply to the equipment's built-in indicators (TV screen, loudspeakers, etc.). However, with some equipment it may be necessary to use test equipment to distinguish between the two conditions. For example, assume that the circuit in question is a simple, single-stage video amplifier, such as shown in Fig. 1.2. Such circuits are often found between ICs in video equipment and provide both amplification and buffering of the signals passing from one IC to the other.

Using the service literature, or your knowledge of similar circuits, you know that if a video signal is applied to the transistor base (input), an amplified video signal should appear at the collector (output). The amount of amplification depends on circuit gain. Assume that the input is 1 V and that the gain is 10. Let us analyze the four possible waveforms shown in Fig. 1.2.

Waveform A is an example of equipment (or circuit) failure. With a normal video signal at the base (input) there is no waveform at the collector (output), indicating that the circuit has failed to perform completely. (The next troubleshooting step is substitution of parts and/or voltage and resistance checks, discussed in following sections of this chapter.)

Waveform B shows neither equipment failure nor degraded performance. That is, the collector signal (output) is substantially the same as the base (input) but about 10 times greater in amplitude. This is the normal indication, and no trouble exists.

Waveform C is an example of degraded performance. There is a video signal at the

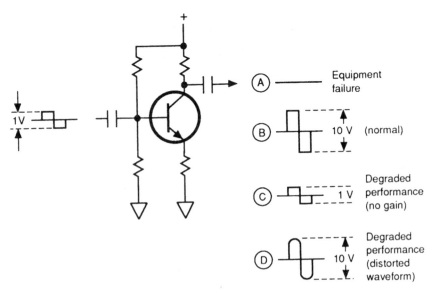

FIGURE 1.2   Single-stage video amplifier showing examples of normal and degraded performance and failure.

collector (output), so the video circuit is performing part of the normal function. However, the output waveform is not much greater in amplitude than the input waveform. Thus the video amplifier is producing little or no gain (in any event, the gain is well below the desired factor of 10).

Waveform D is also an example of degraded performance. There is a video signal at the output, and the signal has the desired gain of 10. However, the output waveform is distorted (the square wave input is badly rounded at the output, always a problem for any video circuit).

Note that the processes used here to get the waveforms are examples of signal substitution and signal tracing (mentioned in Sec. 1.1.3). Both techniques are discussed in great detail throughout this book.

1.3.3 Evaluation of Video Trouble Symptoms

Symptoms evaluation is the process of obtaining more detailed descriptions of the trouble symptoms. The recognition of the original trouble symptoms does not in itself provide enough information for you to decide on the probable cause or causes of the trouble because many faults produce similar trouble symptoms.

To evaluate a trouble symptom, you generally must operate controls associated with the symptom and apply your knowledge of electronic circuits, supplemented with information gained from the service literature. Of course, the mere adjustment of operating controls is not the complete story of symptom evaluation. However, the discovery of an incorrect setting can be considered part of the overall symptom evaluation process.

Example of Evaluating Symptoms.   When the screen of a TV set is not on (no raster), there obviously is trouble. The trouble could be caused by the brightness control being turned down (assuming that the power cord is plugged in and that the on-off switch is set to on). However, the same symptoms can be produced by a burned-out picture tube (filament open) or by a failure of the high-voltage power supply, among *many other* possible causes. Think of all the time you may save if you check the operating controls first, before you attack the set with soldering tool and pickax.

As mentioned earlier, to do a truly first-rate job of determining trouble symptoms, you must have a complete and thorough knowledge of the normal operating characteristics of the set. Your knowledge helps you decide whether the set is doing the job for which it was designed. In most video service literature, this is called "knowing your equipment."

In addition to knowing the set, you must be able to properly operate all the controls to determine the symptom—to decide on normal or abnormal performance. If the trouble is cleared by manipulating the controls, the trouble analysis may or may not stop at this point. By using your knowledge of the set, *you should find the reason* the specific control adjustment removed the apparent trouble.

1.3.4 Practical versus Theoretical Symptom Evaluation

There are two approaches to troubleshooting that often result in failure: the extreme theoretical and extreme practical.

The extreme theoretical approach is typified by those who feel that all troubleshooting can be done "on paper." An example of this is an engineer who insisted (to the author) that the only information required in a technical manual for a computer of the engineer's design was a complete set of logic equations. The engineer insisted that if the technician knows the logic, any problem can be solved. The author pointed out to no avail that it would be quite helpful if the technician also knew such facts as the location of the operating controls (particularly the on-off switch), the operating sequence, how to make internal adjustments, and so on.

The extreme practical approach is typified by the old-time TV repairpeople who replaced every vacuum tube in the set, one by one, until the problem was cleared. If this did not solve the problem, they replaced the filter capacitor (which they called a condenser). If the trouble was still present, they put the set on the shelf, tagged it a "tough dog," and waited until they had enough time to replace each and every part, if necessary.

The same approach is sometimes carried over into more complex video equipment, where components are mounted on plug-in circuit boards (CBs) or cards. The plug-ins can be replaced one by one until the trouble is cleared. In fact, some technical manuals recommend this procedure. If the plug-ins are sealed and are to be replaced as a unit (individual parts are not available), this approach has some merit. But even under these conditions, the practical, parts-replacement approach *without any trouble analysis* can be a waste of time. (For one reason, there are often front-panel controls and indicators that are not mounted on the boards.)

Even where plug-in replacement is recommended as an early step in troubleshooting, it is your knowledge of how the equipment works (together with a run-through of the operating sequence) that can often pinpoint the plug-in that contains the trouble. To sum up, a good troubleshooter must have an effective combination of theory and practice.

1.4   PRACTICAL TROUBLE LOCALIZATION (TO A FUNCTIONAL VIDEO CIRCUIT GROUP)

Localizing trouble means that you must determine which of the functional circuit groups in the equipment are actually at fault. You do this by systematically checking each area selected until the faulty one is found. If none of the functional circuit groups show improper performance, you must take a return path and recheck the symptom information (and observe more information, if possible). Several circuits could be causing the trouble, and the localize step narrows the list to those in one functional area, as indicated by a particular block on a block diagram.

The problem of trouble localization is simplified when a block diagram and a list of trouble symptoms are available for the equipment being serviced. Table 1.1 lists trouble symptoms for a TV set and is based on the block diagram of Fig. 1.1. Remember that this information applies to a *typical* or composite discrete-component TV. However, the general arrangements can be applied to many similar TV sets. Thus the illustrations serve as a universal starting point for trouble localization.

1.4.1   Bracketing in Video Circuits

Bracketing makes use of the block diagram or schematic to localize the trouble to a functional circuit group. (Bracketing is also known as the *good-input–bad-output technique.*) With bracketing, you can systematically narrow down the trouble to a circuit group and then to a faulty circuit. Symptom analysis and/or a signal-tracing test are used in conjunction with, or are a part of, bracketing.

Bracketing starts by placing brackets (at the good input and bad output) on the block or schematic. Bracketing can be done mentally or the brackets can be physically marked with a pencil, whichever is most effective for you. No matter what system is used, with the brackets properly positioned, you know that trouble exists *somewhere between the two brackets.*

The technique involves moving the brackets, one at a time (either the good input or the bad output), and then making tests to find if the trouble is within the newly bracketed area. The process continues until the brackets localize a circuit group.

The most important factor in bracketing is to find where the brackets should be moved in the elimination process. This is determined from your deductions based on your knowledge of the equipment and on the symptoms. All moves of the brackets should be aimed at localizing the trouble with a minimum of tests.

1.4.2 Bracketing Examples (TV and Monitor Bracketing)

Bracketing may be used with or without actual measurement of voltage or signals. That is, sometimes localization can be made on the basis of symptom evaluation alone. In practical troubleshooting both symptom evaluation and tests are usually required, often simultaneously. The following examples show how bracketing is used in both cases.

Assume that you are servicing a discrete-component TV set similar to the one shown in Fig. 1.1 and there is no vertical sync. That is, all functions are normal, but the picture rolls vertically. You could start by placing a good-input bracket at the input to the vertical sweep circuits and a bad-output bracket at the vertical coils of the deflection yoke, as shown in Fig. 1.3. However, from a practical standpoint, your first move should be adjustment of the vertical hold control. (Note that Fig. 1.3 is similar to the vertical sweep section shown in Fig. 1.1.)

If the trouble is not cleared by adjustment of the vertical hold control, confirm the good-input bracket by measuring the input sync pulses. Make this measurement at the input to the vertical oscillator rather than at the sync separator, as shown in Fig. 1.3. It is possible that the line between the sync separator and the vertical oscillator is open or partially shorted. (A complete short would probably cause failure of the sync-separator circuits and could affect the horizontal circuits as well.)

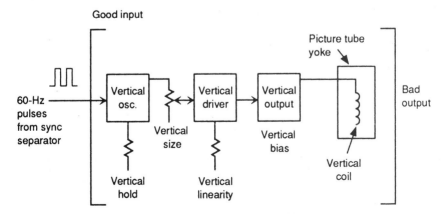

FIGURE 1.3 Vertical sweep circuit showing example of bracketing.

If the vertical sync pulses are normal at the input of the vertical oscillator but there is no vertical sync (even with adjustment of vertical hold), you have localized the trouble to a specific video circuit in the vertical sweep group, as discussed in Sec. 1.5.

1.4.3 Localization to More Than One Video Circuit Group

Several factors should be considered when you have localized trouble to more than one video circuit group. As a guide, if you can run a test that eliminates several circuits, or circuit groups, make that test first (before making a test that eliminates only one circuit). This requires an examination of the diagrams (block and/or schematic) and a knowledge of how the set operates. The decision also requires that you apply logic.

Test point accessibility is a prime factor to consider. A test point can be a special jack located at an accessible spot (say at the top of the chassis). The jack (or possibly a terminal) is electrically connected (directly or by a switch) to some important operating voltage or signal path. At the other extreme, a test point can be any point where wires join or where parts are connected together.

Another factor (although definitely not the most important) is your past experience and history of repeated failures. Past experience with identical or similar sets and related trouble symptoms, as well as the probability of failure based on records of repeated failures, should have some bearing on the choice of a first test point. However, all circuit groups related to the trouble symptom should be tested, no matter how much experience you may have had with the equipment. Of course, experience may help you decide which group to test first.

Those who have had any practical experience in troubleshooting know that the steps of a localization sequence rarely proceed in textbook fashion. Just as true is the fact that many troubles listed in the service literature may never occur in the equipment you are servicing. These troubles are included in the literature (in chart form or as *troubleshooting trees*) to guide you and are not meant to be hard and fast rules.

In many cases, it may be necessary to modify your troubleshooting procedure. The physical arrangement of the equipment may pose special troubleshooting problems. Also, special knowledge gained from experience with similar sets may simplify the task of localizing the trouble.

1.4.4 Modifying the Localization Procedure for Replaceable Video Modules

When the majority of components are located on replaceable modules, it is usually necessary to modify the localization procedure. The trend in present-day video equipment is toward the use of replaceable modules, such as circuit boards that either plug in or that require only a few soldered connections. In such video equipment, it is possible to replace each module or board, in turn, until the trouble is cleared.

For example, if replacement of a video-amplifier module restores normal operation, the defect is in that module. This conclusion can be confirmed by reinserting the suspected defective module. Although this confirmation process is not part of theoretical troubleshooting, it is a good practical check, particularly in the case of a plug-in module. Often, a trouble symptom of this sort may be the result of *dirty contacts* between the plug-in and the main-frame connector (and can be cured with an eraser).

Some service literature recommends that tests be made before all modules are arbitrarily replaced, usually because the modules are not necessarily arranged according to functional area. Thus there is no direct relationship between the trouble symptom and the module. In such cases, always follow the service literature recommendations.

Of course, if modules are not readily available in the field, you must make tests to localize the trouble to a module (so that you can order the right module, for example). Also, operating controls are not usually found on replaceable modules and must be tested separately.

1.4.5 Obvious and Ambiguous Trouble Symptoms During Localization

Some trouble symptoms lead to an obvious localization process, while others are not quite so obvious. The following are some examples of this using the TV set in Fig. 1.1 and the troubles listed in Table 1.1.

Obvious Trouble Symptoms

If the problem is one of sound only, with a good picture, the sound IF and audio circuit group (not shown) is suspect.

If there is good sound and a good picture raster but a poor picture, start localization at the video amplifier and picture-tube circuit group.

If there is a problem on only one channel (in picture, sound, or both) the RF tuner is the likely suspect.

If there is no sound and no raster, start localization by checking the input and output of the low-voltage power supply (not shown).

If the raster is present but there is no sound and no picture on the raster, start localization by checking the signal at the output of the IF and video detector and at the output of the driver in the video amplifier and picture-tube circuits. Start at the same point if the trouble symptom is a poor picture and poor sound (or it points to circuits common to both sound and picture).

If horizontal sync is good, together with good sound but there is a vertical problem (lack of sync, insufficient height, poor linearity, etc.), look for problems in the vertical sweep circuits.

If there is good horizontal and vertical sync but the picture is narrow or there is obvious distortion, start with the high-voltage supply and horizontal output circuit group (probably with the horizontal output stage).

If both horizontal and vertical sync appear to be abnormal, start by checking the input and both outputs of the sync separator.

Ambiguous Trouble Symptoms. *If the driver output is abnormal but the IF output is good,* you have localized trouble to the video driver. On the other hand, *if the IF output is abnormal,* the trouble can be in the IF and video detector, RF tuner, or automatic gain control (AGC) circuits.

To eliminate the AGC circuits as a trouble suspect, apply a fixed dc voltage to the AGC line and check operation of the set. (This is known as *clamping* the AGC line.) If operation is normal with the AGC line clamped, but not when the clamp is removed, you have localized trouble to the AGC circuits.

Note that AGC circuit problems often produce ambiguous symptoms and are therefore difficult to localize. This is because a keyed AGC circuit, shown in Fig. 1.1 (and found in millions of discrete-component TV sets), requires two inputs (pulses from the horizontal section and IF signals from the IF section). For example, if the IF amplifiers are defective, the AGC circuits do not receive proper IF signals and do not produce a proper dc voltage to the AGC line. In turn, lack of AGC voltage can cause the IF amplifiers to operate improperly. Conversely, if the AGC circuits are defective, the IF amplifiers do not receive proper AGC voltages and do not deliver a proper IF signal to the AGC circuits.

Here is a simple guide that can be applied to most AGC circuits. If clamping the AGC line eliminates the problem, the trouble is probably localized to the AGC circuits. Start by checking both inputs (pulses and IF) to the AGC circuits.

If the AGC line is clamped and the IF and video-detector output is not normal, check the signal at the IF section input. (This input is at the first IF amplifier, which is also the RF tuner output.)

Most TV sets have a readily accessible test point at the IF-input and RF-output junction, often known as the *looker point.* If the signal at the looker point is abnormal, the trouble is localized to the RF tuner. If the looker point signal is normal, but the video detector output is abnormal, you have localized trouble to the IF and video-detector section.

Horizontal circuits often produce ambiguous symptoms in TV-set operation. If the horizontal oscillator or driver fail completely, there are no pulses to the high-voltage supply and horizontal output circuits. As a result, there is no high voltage to the picture tube (and no picture raster).

This same symptom can be caused by failure of the high-voltage transformer, the high-voltage output stage, or the picture tube itself. (In some sets, the high-voltage and output circuit group also supplies voltages to the picture-tube accelerator grids. If these voltages are absent, there is no raster.) Also, in some sets, failure of the video amplifier and picture-tube circuits can cut off the picture tube.

If you have good sound but there is no raster (or the screen is completely dark), there are two practical courses of localization for virtually any TV set. First, if convenient, check the high voltage to the picture tube as well as the voltages to all picture-tube elements (accelerator, grids, heater, cathode, etc.).

If it is not convenient to check the picture-tube voltages, check the input to the high-voltage and horizontal output group (which is also the output of the horizontal oscillator and driver circuit group, in sets similar to those of Fig. 1.1). This second choice of localization must also be followed if you find the picture-tube voltages normal.

If the signal at the horizontal oscillator and driver output is normal, the high-voltage and horizontal-output group is suspect. If the signal is abnormal, the horizontal oscillator and driver is the likely problem area. As discussed, failure in the horizontal circuits can also affect the AGC (no pulses to the AGC results in no AGC output).

Of course, if you have good sound and a good raster but there is *abnormal horizontal sync (or distortion and/or picture pulling),* start with the horizontal oscillator and driver group. Check both input pulses from the sync separator and AFC feedback pulses from the horizontal output.

1.5 PRACTICAL TROUBLE ISOLATION (TO A VIDEO CIRCUIT)

There is extensive testing and the use of diagrams in the isolation step. However, there are exceptions to this rule.

1.5.1 Isolating Trouble in IC and Plug-In Video Equipment

Trouble can be isolated to the IC or plug-in module input-output, but not to the circuits (or individual parts) of the IC, in any video equipment. No further isolation is necessary, since parts within the IC cannot be replaced on an individual basis.

This same condition is true for some video equipment where groups of circuits are mounted on sealed, replaceable boards or cards. (Note that not all modules or boards are sealed; many have replaceable parts.) Also remember that both IC's and individual parts are often combined on plug-in cards and boards (for example, where the entire IF and video-detector is replaced by an IC). All parts of the circuit, except transformers, are included in the IC. Such modules may plug in but often require some solder connections.

1.5.2 Using Diagrams in the Isolation Process

As in the localization process, a block diagram is the most convenient tool for isolation. Of course if you do not have a block, you must use the schematic diagram to narrow down the trouble area with logical tests and decisions.

The isolation process is much easier if you can recognize *circuit groups,* as well as individual circuits. If you can subdivide (mentally or otherwise) the schematic into circuit groups, rather than individual circuits, you can isolate the group by a single test at the input or output of the group. For example, in Fig. 1.1, all circuits after the video detector, and ahead of the vertical and horizontal circuits, can be considered as part of the sync-separator circuit group. This group has one input (from the video detector) and two outputs (60 and 15,750 Hz). If either output is abnormal, with a good input, the trouble is isolated to the sync-separator circuit group.

During the isolation process, you are interested in three bits of information: the *signal path* (or paths), the *signal form* (waveform, amplitude, frequency, etc.), and the *operating and adjustment controls* in the various circuits along the signal paths. If you know what signals are supposed to go where, and how the signals are affected by controls, you can isolate trouble quickly in virtually any video equipment.

1.5.3 Comparison of Video Signals or Waveforms

Both *signal tracing* and *signal injection,* or *substitution,* are used in the isolation step. With either technique, the isolation step involves *comparing actual video signals* or waveforms produced along the paths of the circuits against the signals or waveforms given in the service literature. If you find a signal or waveform that is absent or abnormal, you have isolated the problem to a circuit group (and preferably to a circuit in that group).

In the basic isolation process, you compare *input and output* of circuit groups (or circuits) in the signal paths. For a typical solid-state circuit, the input is at the base (or at the gate in case of a FET), while the output is at the collector or emitter (or at the drain or source for FETs). For a typical solid-state circuit group, the input is at the *first base* (or gate) in the signal path, whereas the output is at the *last collector or emitter* (or drain or source) in the *same path.*

Input-output relationships are not always easy to find on IC video equipment. For example, the IC is usually shown as a box, with the pins identified only by number. If you are lucky there will be an arrow pointing to an input pin or away from an output pin. If you are very fortunate, all of the pins will be identified by function (possibly on the schematic, in chart form, or not at all).

Remember that the physical locations of circuit groups within the video device almost never have any relation to the representation on diagrams (block and/or schematic). You must consult part-placement diagrams or photos to find the physical location. Also remember that you should not discard any information gained from the previous steps (symptoms and localization) during the isolation step. This information may permit you to identify those circuit groups (or circuits) that are probable trouble sources.

1.5.4 Video Signal Tracing and/or Signal Substitution

Both signal-tracing and signal-substitution (or signal-injection) techniques are used frequently in troubleshooting all types of video equipment. The choice between tracing and substitution depends on the test equipment used. For example, some generators designed for TV service have outputs that simulate signals found in all major signal paths of the set (RF, IF, sound IF, video pulses, audio, alignment patterns, stereo-TV signals, color signals, etc.). If you have such a generator, signal injection is the logical choice, since you can test all of the circuit groups on an individual basis (independent of other circuit groups). However, it is possible to troubleshoot TV sets with signal tracing alone (and this technique is recommended by many technicians).

Signal tracing is done by examining the signals at test points with a monitoring device such as a scope or meter. In signal tracing, the probe of the monitoring device is moved from point to point, with a signal applied at a fixed point. The applied signal can be generated from an external device, or the normal signal associated with the equipment can be used (such as using the regular broadcast signal to trace through a TV set).

Signal substitution is done by injection of an artificial signal (from a signal generator, sweep generator, etc.) into a circuit or circuit group (or to the complete video equipment) to check performance. In signal injection, the injected signal is moved from point to point, with an indicating or monitoring device remaining fixed at one point. The monitoring can be done with external test equipment (or the picture tube, in the case of a TV set or monitor).

Signal tracing and substitution are often used simultaneously in troubleshooting video equipment. For example, when troubleshooting the RF tuner and IF sections of a TV set (particularly a nondigital TV set), it is common practice to inject a sweep signal at the input and monitor the output with a scope (this technique is described in Chap. 2).

1.5.5 Using the Half-Split Technique in Video Circuits

The half-split technique is based on the idea of simultaneous elimination of the maximum number of circuit groups or circuits with each test. This saves both time and effort. In using half-split, you place brackets at the good- and bad-output points in the normal manner and study the symptoms. Unless the symptoms point definitely to one circuit or circuit group which might be the trouble source, the most practical place to make the first test is at a convenient test point *halfway between the brackets.*

Using an oversimplified example, assume that you are troubleshooting a TV set similar to the one in Fig. 1.1 and you have localized the problem to the IF and video-detector circuits (you get a good signal at the looker point and a bad signal at the video detector). Using half-split, the first test in the group should be between the second and third IF stages.

If the signal is good, you have eliminated two stages out of a possible four (the first and second IFs are good, and the problem is in the third IF or video detector). If the signal is bad, you have again eliminated two stages out of four (the problem is in the first or second IF stage). The next logical step is to monitor between the first and second IF stages (again using the half-split).

1.6 LOCATING A SPECIFIC TROUBLE IN VIDEO EQUIPMENT

The final step of troubleshooting—locating the specific trouble—requires testing of the various branches of the faulty circuit to find the defective part. The proper performance of the locate step enables you to find the cause of trouble, repair it, and return the equipment to normal operation. A follow-up to this step is to record the trouble so that future troubles may be easier to locate. Such a "history" on a certain model of video equipment may point out consistent failures which could be caused by a design error.

1.6.1 Locating Trouble in Plug-In Video Modules

As discussed in Sec. 1.4.4, the trend in present-day video equipment is toward IC and replaceable-module design. Because of this, some technicians assume that it is not necessary to trace specific troubles to individual parts. They assume that all troubles can be cured by module replacement. Some technicians are trained that way. The assumption is not always true.

Although the use of replaceable modules often minimizes the number of steps required in troubleshooting, it is still necessary to check circuit branches to parts outside the module (as is discussed throughout this book). Front-panel operating controls and indicators are a good example of this, since such controls are not located in the replaceable modules. Instead, the controls are connected to the terminals of an IC, circuit board, or plug-in.

1.6.2 Inspection Using the Senses

After the trouble is isolated to a circuit, the first step in locating the trouble is to perform a preliminary inspection using the senses. For example, burned or charred parts can often be spotted by sight or smell.

Overheated parts, such as hot transistor cases, can be located quickly by touch. (If one case is much hotter than the others, give this transistor special attention.) The sense of hearing can be used to listen for high-voltage arcing between wires, or wires and the main frame, for "cooking" or overloaded or overheated transformers or for hum (or lack of hum), whichever is the case. Although all of the senses are used, the procedure is referred to most frequently as a "visual inspection" in service literature.

1.6.3 Testing to Locate Faulty Parts in Video Circuits

Many transistors, ICs, and diodes are not easily replaced (unlike the ancient vacuum tube). Thus the old-time procedure of replacing tubes at the first sign of trouble has not carried over into present-day video equipment. Instead, the circuits are analyzed by testing to locate faulty parts.

Active Video Devices. For testing and troubleshooting purposes, transistors, ICs, and diodes may be considered active devices in any video equipment. Because of their key positions in the circuit, these devices are a convenient point from which to evaluate operation of the entire circuit (through waveform, voltage, and resistance tests). Making such tests at the terminals or pins of the active device often results in locating the trouble quickly.

Video Waveform Testing. Usually, the first step in video circuit testing is to analyze the output waveform of the circuit or active device (typically, the collector, or possibly the emitter, or a transistor). Of course, in some circuits (such as power supplies) there is no output waveform (only a dc voltage). (In a few video circuits, there is no waveform of any significance.)

In addition to checking for the presence of waveforms on a scope, the waveform must be analyzed in detail to check the amplitude, duration, phase, and/or shape. As discussed throughout this book, a careful analysis of waveforms can often pinpoint the most likely branch of a circuit that is defective.

Transistor and Diode Testers in Video Work. It is possible to test transistors and diodes in circuit, using in-circuit testers. These testers are usually quite good for transistors and diodes used at lower frequencies, particularly in the audio range. However, most in-circuit testers do not show the high-frequency characteristics. (The same is true for out-of-circuit testers.) For example, it is quite possible for a transistor to perform well in the audio range but be totally inadequate for video work.

Video Voltage Testing. After waveform analysis and/or in-circuit tests, the next logical step is to measure voltages at the active-device terminals or leads. Always pay particular attention to those terminals that show an abnormal waveform. These are the terminals most likely to show abnormal voltage.

Relative Voltages. It is often necessary to troubleshoot video circuits without the benefit of adequate voltage and resistance information. This can be done using the schematic diagram to make a logical analysis of the relative voltages at the transistor terminals. For example, with an NPN transistor, the base must be positive in relation to the emitter if there is to be emitter-collector current flow. That is, the emitter-base junction is forward-biased when the base is more positive (or less negative) than the emitter. (The approach of using relative voltages in video troubleshooting is discussed further in this chapter.)

Resistance Measurements. After waveform and voltage measurements are made, it is often helpful to make resistance measurements at the same point on the active device (or at other points in the circuit), particularly where abnormal waveforms and/or voltages are found. Suspected parts often can be checked by a resistance measurement, or a continuity check can be made to find point-to-point resistance of the suspected branch.

Considerable care must be used when making resistance measurements in video

circuits. The junctions of transistors act like diodes. When biased with the right polarity (by the ohmmeter battery), the diodes conduct and produce false resistance readings. (This is discussed further in Sec. 1.6.8.)

Current Measurements.   In rare cases, the current in a particular circuit branch can
be measured directly with a meter. However, it is usually simpler and more practical
to measure the voltage and resistance of a circuit and then calculate the current.

1.6.4 Video Waveform and Signal Measurements

When testing to locate trouble, the waveform measurements are made with the circuit in
operation and usually with an input signal (or signals) applied. The signals can be from an
external generator, or you can use alternative sources (such as a TV broadcast signal).

If you use the waveform reproductions found in the service literature, follow all of
the notes and precautions described in the literature. Usually, the literature specifies
the position of operating controls, typical input signal amplitudes, and so on.

If you are servicing a TV or monitor, note that most TV waveforms are measured
with the scope sweep setting at 30 or 7875 Hz. The 30-Hz frequency is one-half the
vertical frequency of 60 Hz. Similarly, the 7875 Hz is one-half the horizontal frequency of 15,750 Hz. With these sweep settings, *two waveforms* are displayed when checking either horizontal or vertical sweep signals.

Figure 1.4 shows waveform reproductions found in typical video service literature

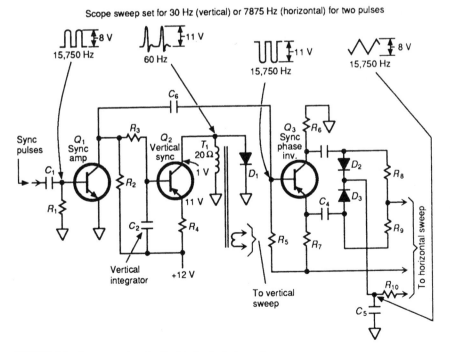

FIGURE 1.4   Sync-separator circuits showing typical waveforms.

(the sync-separator circuits of a discrete-component TV set, in this case). The approximate shape of the waveform is given on the schematic (rather than in a separate chart), together with the voltage amplitude. There may or may not be a note on the schematic indicating the approximate frequency of the waveform.

Note that only the horizontal sync pulses (15,750 Hz) are shown at the input to the sync amplifier Q_1 (at the base). As a practical matter, it is reasonable to assume that there are vertical sync pulses at the same point. Thus, you could monitor the Q_1 base with a scope and expect to see two waveforms representing the vertical sync pulses (if the scope is set to a 30-Hz sweep).

Relationship of Waveforms to Video Trouble Symptoms. Complete failure of a circuit usually results in the absence of a waveform. A poorly performing circuit usually produces an abnormal or distorted waveform. Also, exact waveforms are not always critical in all circuits. The representations of waveforms given in video literature are usually approximations of the actual waveform (unless photos are used). Also, the same waveform does not always appear exactly the same when measured with different scopes.

1.6.5 Video Voltage Measurements

When testing to locate trouble, make the voltage measurements with the circuit in operation but (usually) with no signals applied. If you are using the voltage information found in service literature, follow all notes and precautions. Usually, the information specifies the position of operating controls, typical input voltages, and so on.

In some video literature, the voltages are given on the schematic, together with the waveforms, as shown in Fig. 1.4. In other video literature, the voltage information is presented in chart form. Either system is quite accurate, but both systems require that you find the actual physical location of the terminals where the voltages are to be measured. (In some military-style literature, voltage and resistance charts are arranged to simulate the actual physical relationship or position of the terminals.)

Because of the safety practice of setting a voltmeter to the highest scale before making measurements, the terminals having the highest voltage should be checked first. Note that in the video circuits of Fig. 1.4 (Q_2 and Q_3), the collector is grounded, and the emitter has the highest voltage. (The order in which you check voltages makes little difference when using an *autoranging voltmeter*. However, it is good practice to check the highest voltages first, to establish the habit.) Then the terminals having less voltage should be checked in descending order.

If you have had any practical experience in troubleshooting, you know that voltage (as well as resistance and waveform) measurements are seldom identical to those listed in literature (because of resistance tolerances and meter and scope accuracy). However, it is essential that the measurements be within tolerances specified in the literature (particularly in the case of critical circuits).

Generally, the most important factors to consider in voltage-waveform-resistance measurement accuracy are the *symptoms and the output signal.* If there is no output, expect a fairly large variation of measurements in the trouble area. Trouble which results in critical performance that is just out of tolerance may cause only a slight change in circuit measurements.

1.6.6   Video Resistance Measurements

Resistance measurements must be made with no power applied. However, in some cases, various operating controls must be in certain positions to produce resistance readings similar to those found in the service literature. This is particularly true of controls that have variable resistances.

Always observe any notes or precautions found in the literature. In any circuit, always make sure that the filter capacitors are discharged before making resistance measurements. After all safety precautions and notes have been observed, measure the resistance from the terminals to the active device to the chassis (or ground) or between any two points that are connected by wiring.

In some video literature, resistance information is given on the schematic, together with the waveforms and voltage data. In other literature, the resistance information is presented in chart form. Do not be surprised if you find video literature with little or no resistance data. Quite often, the only resistances given are the values of resistors and the dc resistances of the coils and transformers.

The reasoning for the omission of resistance values from various terminals has some merit. If there is a condition in any active-device terminal circuit that produces an abnormal resistance (say, an open resistor or a resistor that has changed drastically in value), the voltage at that terminal is abnormal. If such an abnormal voltage reading is found, it is necessary to check out each resistance in the terminal circuit on an individual basis.

Because of the *shunting effect* of other parts connected in parallel, the resistance of an individual part or circuit may be difficult to check. In such cases, it is necessary to disconnect one terminal of the part being tested from the rest of the circuit. This leaves the part open at one end, and the value of resistance measured is that of the part only.

1.6.7   Duplicating Video Waveform, Voltage, and Resistance Measurements

If you are responsible for service of one type or model of video equipment, it is strongly recommended that you duplicate all waveform, voltage, and resistance measurements found in the service literature with your own test equipment. This should be done with video equipment that is known to be good and operating properly.

With a good set of duplicate measurements, you can spot even slight variations during troubleshooting. Always make the initial measurements with test equipment normally used in troubleshooting. If more than one set of test equipment is used, make the initial measurements with all available test equipment, and record any variations.

1.6.8   Using Schematics in Video Troubleshooting (Examples)

Regardless of the type of trouble symptom, the actual fault can eventually be traced to one or more of the circuit parts (transistors, ICs, diodes, etc.). The waveform-voltage-resistance checks then indicate which branch within a circuit is at fault. You must locate the particular part that is causing the trouble in the branch.

For this, you must be able to read a schematic diagram. These diagrams show what is inside the blocks on a block diagram and provide the final picture of equipment operation. Quite often, you must troubleshoot video equipment with only a schematic. If you are fortunate, the schematic will give some voltages and waveforms.

The following example shows how the schematic diagram is used to locate a fault within a circuit (sync separator of a discrete-component TV). Although this example involves only a selected circuit, the same basic troubleshooting principles apply to all video circuits.

Assume that the circuit in Fig. 1.4 is being serviced. The reasoning that led to this particular circuit is as follows. The symptom is a complete lack of vertical sync. That is, the TV picture rolls vertically and cannot be controlled by the vertical hold. All other functions, including the horizontal sync, are normal. This localizes the trouble to either the vertical sweep circuits or to the sync separator circuits (Table 1.1).

Isolating Trouble to the Sync Separator. You isolate the trouble to the sync separator by means of waveform measurement. Your first waveform measurement is at the collector of Q_2, which is (simultaneously) the vertical output of the sync separator and the input to the vertical sweep circuits (Fig. 1.1). The waveform is absent (or abnormal), and you place a bad-output bracket at the collector of Q_2.

All of the remaining waveforms shown in Fig. 1.4 are normal. Thus, you can assume that Q_1 is functioning normally (Q_1 is passing the horizontal sync pulses) and should pass the vertical sync pulses. However, it is better not to assume anything. Instead, confirm that there are vertical sync pulses at both the input (base) and output (collector) of Q_1. Note that neither of these pulses is shown in Fig. 1.4 (which is quite typical for video service literature).

Confirming Vertical Pulses. To overcome the lack of information, set the scope sweep to 30 Hz, and measure whatever waveforms appear at the base and collector of Q_1. If you find two pulses, they are the vertical sync pulses. The amplitude of these pulses should be about equal to the horizontal pulse amplitude (7-V base, 10-V collector). Assume that this is the case in making your measurement. Now, it is reasonable to assume that Q_1 is *probably* good and that you have a good input to Q_2.

Locating the Fault in the Q_2 Circuit. With the trouble isolated to Q_2, you must now locate the specific fault in the Q_2 circuit, using voltage and resistance measurements. The parts involved are R_2, R_3, R_4, C_2, Q_2, D_1, and T_1. Note that only two voltages are given for the terminals of the active device Q_2 (1-V collector, 11-V emitter). This indicates a 1-V drop across the T_1 winding, yet the normal waveform shows a pulse of about 10 V in amplitude. This indicates that Q_2 is normally biased to (or beyond) cutoff and is switched on by pulses from the junction of R_3 and C_2 (which form the vertical integrator, as discussed in Chap. 2).

To bias a PNP transistor to cutoff, the base must be less negative (or more positive) than the emitter. Since a positive 12-V supply is used, the base of Q_2 must be more positive than the emitter (11 V) but less than 12 V. Thus, if you find some voltage in the 11- to 12-V range from the base of Q_2 to ground, the voltage is probably correct and R_3 is probably good.

Clearing the Vertical Integrator. To clear the vertical integrator R_3 and C_2 from any suspicion, measure the waveforms at the collector of Q_1 and the base of Q_2, using scope sweeps of 30 and 7875 Hz. If R_3 and C_2 are doing their jobs, waveforms should appear at the base of Q_2 only when the 30-Hz sweep is used. If no waveforms appear, or if both the horizontal and vertical waveforms are found at the base of Q_2, R_3, and C_2 are suspect.

Checking Q_2. Now assume that there is a good pulse at the base of Q_2, and you have thus narrowed the problem down to Q_2, R_4, D_1, or T_1. At this point, you could check Q_2 by substitution or with an in-circuit transistor tester, if convenient. However, you will probably do better to make voltage measurements at all terminals of Q_2. This pinpoints any obvious part failures. For example, if R_4 or T_1 is open, all voltages are abnormal. If R_4 is shorted (resistor leads shorted), the emitter is at the supply voltage (12 V) instead of at 11 V.

Ambiguous Measurements. Remember that voltage measurements alone may not solve the problem. For example, if D_1 is shorted or leaking badly, it is still possible to get a near-normal voltage reading at the Q_2 collector, but a poor pulse waveform. Also, voltage measurements in pulse circuits are sometimes ambiguous. The dc voltage is given as 1 V, yet the pulse (or ac) voltage is 10 V. If the meter is set to measure dc, the pulses may increase the average dc voltage (in some meters) so that an abnormal reading appears, even though the circuit is functioning normally.

If D_1 is open, the dc voltage can be normal, but the pulse waveform is abnormal. D_1 functions to remove any negative-going pulses at the collector of Q_2. As another example, assume that the primary winding of T_1 has partially shorted turns. This produces a near-normal dc voltage but drastically reduces the pulse output to the vertical sweep circuits.

Resistance Measurements. Unless you pinpoint the problem with waveform and voltage measurements, your last step is to make resistance measurements. As shown in Fig. 1.4, no resistance values are given for any of the Q_2 terminals. So there is no point in measuring from the terminals to ground (as is standard practice in troubleshooting for most military-style equipment). Instead, you must check the resistance of each part on an individual basis.

Accurate Resistance Readings. To get accurate resistance readings of individual parts, you must disconnect one lead from the remainder of the circuit. If you do not, the effect of solid-state devices in the circuit can further confuse the troubleshooting process. For example, assume that you measure the T_1 primary resistance by connecting an ohmmeter across the winding. If the ohmmeter leads are connected so that the positive terminal of the ohmmeter battery is connected to the D_1 anode, D_1 is forward-biased, and the ohmmeter reads the combined D_1 and T_1 resistances.

This problem can be eliminated by *reversing the ohmmeter* leads and measuring the resistance both ways. If there is a difference in the resistance values with the leads reversed, check the schematic for possible forward-bias conditions in *diode and transistor* junctions of the associated circuit. Whenever practical, simply disconnect one lead of the part being measured. The problem of making resistance measurements during troubleshooting is discussed in Sec. 1.9.1.

1.6.9 Internal Adjustments during Video Troubleshooting

Adjustment of controls (both internal adjustment and operating controls) can affect circuit conditions. This may lead to false conclusions during troubleshooting. For example, the bias on the base of vertical output transistor Q_1 (in Fig. 1.5, which shows the vertical drive circuit of a typical discrete-component TV) is set by vertical bias potentiometer R_7. In turn, the value of the bias determines the portion of the sweep used by Q_1. Thus, the sweep voltages applied to the vertical-deflection yoke are set, in part, by the R_7. That is, adjustment of R_7 affects both height and linearity of the vertical sweep.

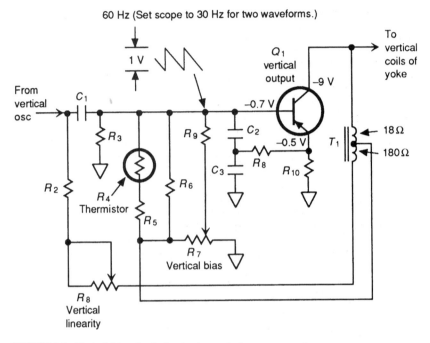

FIGURE 1.5   Vertical drive circuit showing internal adjustment controls.

However, the main purpose of R_7 is to compensate when a major part in the circuit (such as the output transistor, output transformer, or vertical-deflection yoke) is replaced. R_7 usually does not require adjustment during troubleshooting. On the other hand, the vertical linearity control R_8 has a considerable effect on the linearity of the vertical sweep and often requires adjustment during the troubleshooting process.

These two extremes often lead some technicians to one of two unwise courses of action. First, the technician may launch into a complete alignment procedure (or whatever internal adjustments are available) once the trouble has been isolated to a circuit. No internal control (no matter how inaccessible) is left untouched. The technician reasons that it is easier to make adjustments than to replace parts. While such a procedure eliminates improper adjustment as a possible fault, it can also create more trouble than is repaired. Indiscriminate internal adjustment is the technician's version of operator trouble.

At the other extreme, a technician may replace board after board, when a simple screwdriver adjustment will repair the problem. This usually means that the technician simply does not know how to perform the adjustment procedure or does not know what the control does in the circuit. Either way, a study of the service literature should resolve the problem.

To take the middle ground, do not make any internal adjustments during the troubleshooting procedure until trouble has been isolated to a circuit and then only when the trouble symptom or test results indicate possible maladjustment.

For example, assume that the vertical oscillator of a TV set is provided with a back-panel adjustment control (vertical hold) that sets the frequency of oscillation.

(Most discrete-component TV sets are provided with a vertical hold control.) If measurements show that the vertical oscillator is off frequency (not at 60 Hz), it is logical to adjust the vertical hold control. However, if waveform measurements show only a very low output (but it is on frequency), adjustment of the vertical hold control during troubleshooting could be confusing (and could cause further problems).

An exception to this rule is when the service literature recommends alignment or adjustment as part of the troubleshooting procedure. Generally, alignment or adjustment is checked after test and repair are performed. This assures that the repair (board replacement) procedure has not upset circuit adjustments.

1.6.10 Video Troubles Resulting from More Than One Fault

A review of all symptoms and test information obtained thus far helps you to verify the component (board, IC, or part) as the sole trouble or to isolate the faulty component. This is true whether the malfunction is caused by the isolated component or by some entirely unrelated cause.

If the isolated component can produce all of the symptoms, it is logical to assume that you have found the sole cause of trouble. If not, you must use your knowledge of the circuit to find what other component can produce all the symptoms.

The failure of one component often results in abnormal voltages or current that can damage other parts. Trouble is often isolated to a faulty part, which is the result of an original trouble, rather than the source of trouble. For example, assume that the troubleshooting procedure isolates a transistor as the cause of trouble. The transistor is burned out by excessive current. The problem is to find how the excessive current was produced. There are several possibilities.

There can be an extremely large input signal which overdrives the transistor. This shows a fault somewhere in the circuits ahead of the input connection. (Many ICs have internal circuits to prevent this condition.) Power surges can also cause the component to burn out. (Power-line surges are a common problem in digital ICs, as many computer owners already know.)

Thermal runaway can produce a burned-out transistor. (This condition can also affect some ICs but is not as common as with transistors.) Thermal runaway occurs when current heats the transistor, causing a further increase in current, resulting in more heat (because of the nature of many solid-state materials). This runaway condition continues until the heat-dissipation capabilities of the transistor are exceeded. *Bias-stabilization circuits* are generally included in most well-designed solid-state video equipment to prevent thermal runaway.

Of course, a burned-out transistor can be caused by something simple like a short, or it is possible for a component to simply "go bad." No matter what the cause, your job is to find the trouble, verify the cause, and repair the trouble as efficiently as possible.

1.6.11 Repairing Video Troubles

In a strict sense, repairing the trouble is not part of the troubleshooting process. However, repair is an important part of the total effort involved in getting the equipment back into operation. Repairs must be made before the equipment can be checked out and made ready for operation. Here are some important points to consider.

Never replace a component if it fails a second time unless you are sure that the cause of trouble is eliminated.

Always use exact replacements (except in emergency conditions). Never install a replacement component that has characteristics or a rating inferior to those of the original.

Always consult the service literature for any information on parts. Read all notes on the schematics. Many parts are critical, some to performance and some to safety.

Install replacement components in the same physical location as the original. In video circuits, changing the location of components (different lead lengths and so on) may cause the circuit to malfunction.

1.6.12 Operational Checkout of Video Equipment

Once the repairs are complete, make an operational check to verify that the equipment is free of faults and is performing properly again. Operate the equipment through all the operating modes. When the operational check is complete and the equipment is "certified" (by you and/or the customer) to be operating normally, make a brief record of the symptoms, faulty parts, and remedy for future reference. This is particularly helpful if you must troubleshoot similar equipment.

1.6.13 Safety Precautions in Video Troubleshooting

It is assumed that you are familiar with all of the standard safety precautions for electronic troubleshooting. Such precautions include checking for leakage that can cause "hot" chassis and metal covers, handling electrostatically sensitive (ES) devices, replacing leadless components, repairing PC boards, and so on. Brief summaries of these precautions are included in the appropriate chapters and in the remaining sections of this chapter. Compare the summaries with what you find in the service literature. (Most video service literature includes all necessary safety precautions for a particular type of equipment.)

1.7 VIDEO TROUBLESHOOTING NOTES

The following notes summarize practical suggestions for troubleshooting all types of video equipment.

Transient voltages: Be sure that power to the equipment is turned off or that the line cord is removed when making in-circuit test or repairs (except for voltage measurements and/or signal tracing, of course). Components can be damaged from the transient voltages developed when components are changed (plugging in a new board, for example). Remember that certain circuits may be "live" even with the power switch set to off (such as with an instant-on TV set).

Disconnected parts: Do not operate the equipment with any parts, such as loudspeakers or picture-tube yokes, disconnected. This can result in damage to transistors and/or ICs.

Sparks and voltage arcs: Use a meter and a high-voltage probe to measure the second-anode potential of a picture tube in any video equipment. Do not arc the second-anode lead to the chassis for a spark test. This can damage some transistors and ICs.

Intermittent conditions: If you run into an intermittent condition and find no fault using routine checks, try tapping (not pounding) the components. If this does not work, try rapid heating and cooling. (A small portable hair dryer and a spray-type circuit cooler make good heating and cooling sources.) Apply heat first, then cool. The quick change in temperature normally causes an intermittently defective component to go bad permanently (the component opens or shorts). Never hold a heated soldering tool directly on a component case.

Operating control settings: If transistor or IC pin appears to have a short, check the setting of any operating or adjustment controls associated with the circuit. For example, the gain control between video amplifier stages can show what appears to be a short to ground (from a transistor element or IC pin) simply because the control is set to zero or minimum.

Making a record of gain readings: If you must service any particular make or model of video device, record the transistor/IC gain readings of a unit that is working properly for future reference. Compare these gain readings with the values listed in service literature.

Shunting capacitors: Do not shunt suspected capacitors with known-good capacitors when troubleshooting video equipment. This technique is good only if the suspected capacitor is open and can cause damage to other circuits (because of the voltage surge).

Test connections: Many metal-case transistors (and a few ICs) have their cases tied to the collector (or to some point within the IC). Avoid using the case as a test point unless you are certain as to what point or circuit element is connected to the case. Avoid clipping onto some of the subminiature resistors used in certain video equipment. Any subminiature component can break if handled roughly.

Injecting signals: Make sure that there is a blocking capacitor in the signal-generator output when injecting a signal into a video circuit. If there is not, connect a capacitor between the generator output and the point of signal injection (transistor base, IC pin, etc.).

1.8   MEASURING VIDEO VOLTAGES IN CIRCUIT

One of the basic troubleshooting techniques for video equipment is to measure the voltages at all pins of the ICs and at all elements (base, emitter, and collector) of the transistors. This tells you instantly if any voltages are absent or abnormal and provides a good starting point for troubleshooting.

1.8.1   Basic Video Transistor Connections

Figure 1.6 shows the basic connections for both PNP and NPN transistor video circuits (with capacitors removed for clarity). The purpose of this figure is to establish normal video transistor relationships. With a normal pattern established, it is relatively simple to find an abnormal condition.

In most video circuits, the emitter-base junction is forward-biased to get current

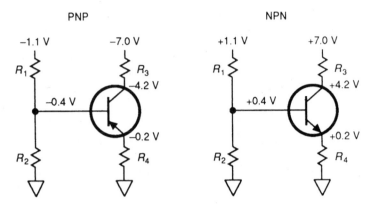

FIGURE 1.6 Basic connections for transistors in video circuits.

flow. In a PNP, this means that the base is made more negative (or less positive) than the emitter. Under these conditions, the emitter-base junction draws current and causes electron flow from the collector to the emitter. In an NPN, the base is made more positive (or less negative) than the emitter to cause current flow from emitter to collector.

1.8.2 Rules for Practical Analysis of Video Transistor Voltages

The following general rules can be applied to practical analysis of transistor voltages in troubleshooting video circuits:

1. The middle letters in PNP and NPN always apply to the base.
2. The first two letters in PNP and NPN refer to the *relative bias* polarities of the emitter with respect to either the base or the collector. For example, the letters PN (in PNP) show that the emitter is positive in relation to both the base and emitter. The letters NP (in NPN) show that the emitter is negative in relation to both the base and collector.
3. The collector-base junction is always reverse-biased.
4. The emitter-base junction is usually forward-biased in video circuits. There are exceptions where a transistor is completely cut off until video pulses are applied.
5. A base input voltage that aids or increases the forward bias also increases the emitter and collector currents.
6. A base input voltage that opposes or decreases the forward bias also decreases the emitter and collector currents.
7. The dc electron flow is always against the direction of the arrow on the emitter.
8. If electron flow is into the emitter, electron flow is out from the collector.
9. If electron flow is out from the emitter, electron flow is into the collector.

Using these basic rules, normal video transistor voltages can be summed up this way:

1. For an NPN transistor in a video circuit, the base is positive, the emitter is not quite as positive, and the collector is far more positive.

2. For a PNP transistor in a video circuit, the base is negative, the emitter is not quite as negative, and the collector is far more negative.

1.8.3 Practical Measurement of Video Transistor Voltages

There are two schools of thought on how to measure video transistor voltages. The most common method is to measure from a *common or ground to the element*. Video service literature generally specifies transistor voltages this way. For example, all of the voltages for the PNP in Fig. 1.6 are negative with respect to ground. (The positive test lead of the meter is connected to ground, and the negative test lead is connected to the elements, in turn.) This method is sometimes confusing to those not familiar with transistors, since the rules appear to be broken. (In a PNP transistor, some elements should be positive, but all elements are negative.) However, the rules still apply.

In the case of the PNP in Fig. 1.6, the emitter is at -0.2 V, whereas the base is at -0.4 V. The base is *more negative* than the emitter. Thus the *emitter is positive in relation to the base,* and the base-emitter junction is forward-biased (normal). On the other hand, the base is at -0.4 V, whereas the collector is at -4.2 V. The *base is less negative than the collector.* As a result, the *base is positive with respect to the collector,* and the base-collector junction is reverse-biased (normal).

Some troubleshooters prefer to measure transistor voltages from *element to element (between electrodes)* and note the *difference in voltages*. For example, in the circuit in Fig. 1.6, a 0.2-V differential exists between base and emitter. The element-to-element method of measuring transistor voltages quickly establishes forward-to-reverse bias.

1.9 TROUBLESHOOTING WITH VIDEO TRANSISTOR VOLTAGES

This section presents an example of how voltages measured at the elements of a transistor can be used to analyze failure in video circuits.

Assume that an NPN transistor circuit is measured and that the voltages found are similar to those shown in Fig. 1.7a. Except in one case, these voltages show a defect. It is obvious that Q_1 is not forward-biased because the base is less positive than the emitter (reverse bias for an NPN). The only video circuit in which this might be normal is one that requires a large *trigger voltage* or *video pulse* (positive in this case) to turn Q_1 on.

The first troubleshooting clue in Fig. 1.7a is that the collector voltage is almost as large as the collector source at R_3. This means that very little current is flowing through R_3 in the collector-emitter circuit. Q_1 could be defective. However, the trouble is more likely to be one of bias. The emitter voltage depends mostly on the current through R_4. Unless the value of R_4 has changed substantially (this is unusual), the problem is incorrect base bias.

The next step in this case is to measure the bias-source voltage at R_1. If the voltage is at 0.7 V, instead of the required 2 V (as shown in Fig. 1.7b), the problem is obvious; the external bias voltage is incorrect. The condition should show up as a defect in the power supply and should appear as an incorrect voltage in other circuits.

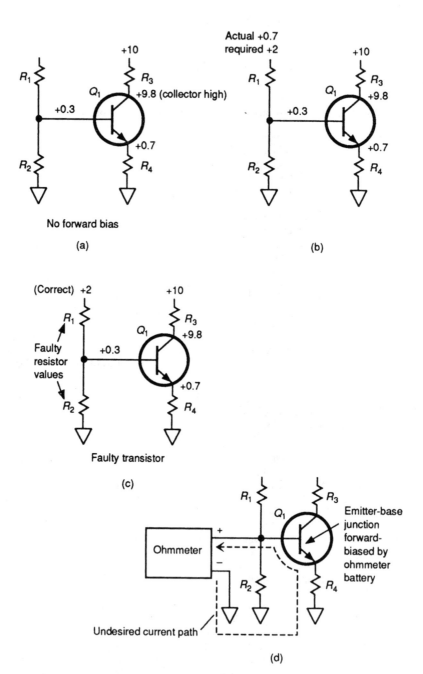

FIGURE 1.7 Troubleshooting with transistor voltages.

If the source voltage is correct, as shown in Fig. 1.7c, the trouble is probably a defective R_1 or R_2 (or Q_1).

The next step is to remove all voltage from the video device and measure the resistance of R_1 and R_2. If either value is incorrect, the corresponding resistor must be replaced. If both values are good, it is reasonable to check the value of R_4. However, it is more likely that Q_1 is defective. This can be established by test and/or replacement.

1.9.1 Practical In-Circuit Resistance Measurements

Be careful when measuring resistance values in transistor circuits with the resistors still connected. One reason for this is that the voltage produced by the meter could damage some transistors. More important, there is a chance for error because transistor junctions pass current in one direction. This can complete a circuit through other resistors and produce false indications.

For example, assume that a meter is connected across R_2 with the negative terminal of the meter (internal battery) connected to ground, as shown in Fig. 1.7d. Because R_4 is also connected to ground, the negative terminal is connected to the end of R_4. Further, because the positive terminal is connected to the base of Q_1, the base-emitter junction is forward-biased and there is current flow. In effect, R_4 is in parallel with R_2, and the meter reading is incorrect. To prevent the problem, *disconnect either end of* R$_2$, *and/or reverse the meter leads before making the measurement.*

1.10 TROUBLESHOOTING VIDEO TRANSISTORS IN CIRCUIT

The forward-bias characteristics of transistors can be used to troubleshoot video circuits without removing the transistor from the circuit. There are two basic methods. Silicon transistors normally have a voltage differential of about 0.4 to 0.8 V between emitter and base. Germanium transistors have a voltage differential of about 0.2 to 0.4 V. The polarities of voltages at the emitter and base depend on the type of transistor (NPN or PNP).

The voltage differential between emitter and base acts as a forward bias for the transistor. Sufficient differential or forward bias turns the transistor on, resulting in a corresponding amount of emitter-collector current flow. Removal of the voltage differential, or an insufficient differential, produces the opposite results, cutting the transistor off (no emitter-collector flow or very little flow). Now let us see how these characteristics can be used to troubleshoot video transistors in circuit.

1.10.1 Troubleshooting Video Circuits by Removal of Forward Bias

Figure 1.8a shows the test connections for an in-circuit test by removal of forward bias. First, measure the emitter-collector differential voltage under normal circuit conditions. Then short the emitter-base junction and note any change in emitter-collector differential.

If Q_1 is operating, the removal of forward bias causes the emitter-collector current flow to stop, and the emitter-collector voltage differential increases (the collector voltage rises to or near the supply value). For example, assume that the supply voltage is 12 V and that the differential between collector and emitter is 6 V when Q_1 is operat-

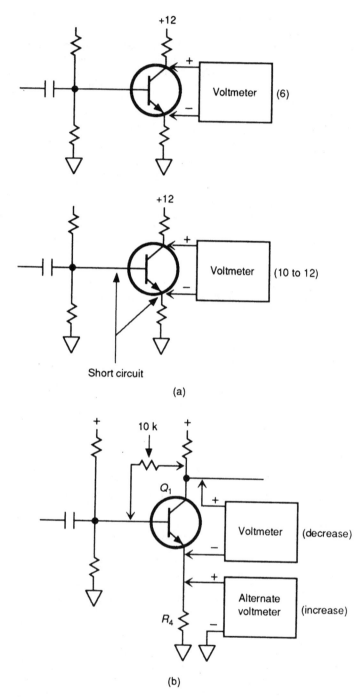

FIGURE 1.8   In-circuit transistor tests.

ing normally (no short). When the emitter-base junction is shorted, the emitter-collector voltage differential should rise to (or near) 12 V.

1.10.2 Troubleshooting Video Circuits by Application of Forward Bias

Figure 1.8*b* shows the test connections for test by application of forward bias. First, measure the emitter-collector differential under normal circuit conditions (or measure the voltage across R_4, as shown). Next, connect a 10-k resistor between the collector and base, and note any change in emitter-collector differential (or any change in voltage across R_4).

If Q_1 is operating, the application of forward bias causes the emitter-collector current flow to start (or increase), and the emitter-collector voltage differential decreases (or the voltage across R_4 increases).

1.10.3 Go/No-Go Tests in Video Circuits

The tests in Fig. 1.8 show that the transistor is operating on a go/no-go basis, which is usually sufficient for most video applications. However, the tests do not show transistor gain or leakage and do not establish operation of the transistor at the high frequencies often found in video equipment. For these reasons, some troubleshooters reason that the only satisfactory test of a transistor is *in-circuit operation*. If a transistor does not perform the intended function in a given circuit, the transistor must be replaced. So the most logical method of test is replacement.

1.11 USING TRANSISTOR TESTERS IN VIDEO TROUBLESHOOTING

Transistors can be tested in or out of circuit using commercial testers. The use of such testers in video troubleshooting is generally a matter of opinion. At best, such testers show the gain and leakage of transistors at low frequencies, under one set of conditions (fixed voltage, current, etc.). In the author's opinion, transistors used for high-frequency or switching applications in video circuits are best tested by substitution (in circuit) or with special test equipment (out of circuit).

1.12 TESTING VIDEO TRANSISTORS OUT OF CIRCUIT

There are four basic tests required for transistors in practical video troubleshooting: gain, leakage, breakdown, and switching. All of these tests are best made with a scope using appropriate adapters (curve tracers, switching-characteristic checkers, etc.). However, it is possible to test a transistor with an ohmmeter. These simple ohmmeter tests show if the transistor has leakage and produces gain. As discussed, the only true test of a transistor is in the circuit.

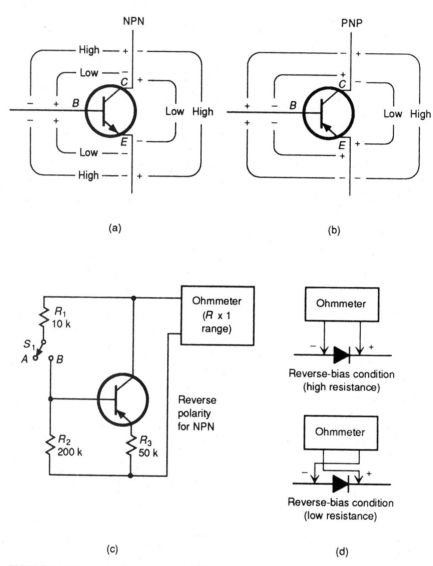

FIGURE 1.9 Out-of-circuit transistor and diode tests.

1.12.1 Testing Video Transistor Leakage with an Ohmmeter

With this method, transistors are considered as two diodes connected back to back. Each diode should show low forward resistance and high reverse resistance. These resistances can be measured with an ohmmeter as shown in Fig. 1.9a and b. Use the same ohmmeter range for each pair of measurement. Avoid using the $R \times 1$ range of an ohmmeter with high internal-battery voltage.

If the reverse resistance reading is low but not shorted, the transistor is leaking. If both forward and reverse readings are very low or show a short, the transistor is shorted. If both forward and reverse readings are very high, the transistor is open. If the forward and reverse readings are the same or nearly equal, the transistor is defective.

Actual resistance values depend on the meter range and battery voltage. *The ratio of forward-to-reverse resistance* is the best indicator of transistor leakage. Almost any transistor used in video circuits shows a ratio of at least 30:1. Many video transistors show ratios of 100:1 or greater.

1.12.2 Testing Video Transistor Gain with an Ohmmeter

Normally, there is little or no current flow between the emitter and collector of a transistor until the base-emitter junction is forward-biased. This fact can be used to make a basic gain test of a transistor using an ohmmeter. The test circuit is shown in Fig. 1.9c. In this test, the $R \times 1$ range is used. Any internal-battery voltage can be used, provided that the maximum collector-emitter breakdown voltage is not exceeded. In position A of switch S_1, there is no voltage applied to the base, and the base-emitter junction is not forward-biased. Under these conditions, the meter should show a high resistance. When switch S_1 is set to B, the base-emitter circuit is forward-biased (by the voltage across R_1 and R_2), and current flows in the emitter-collector circuit. This is indicated by a lower-resistance reading on the ohmmeter.

1.13 TESTING VIDEO-CIRCUIT DIODES OUT OF CIRCUIT

Again, the most practical test of a diode in video circuits is substitution. However, the forward and reverse resistances of diodes can be measured out of circuit as shown in Fig. 1.9d. If resistance is low in the reverse direction, the diode is probably leaking. If resistance is high in both directions, the diode is probably open. A low resistance in both directions usually shows a shorted diode.

It is possible for a defective video-circuit diode to show some difference in forward and reverse resistances. The main concern is the *ratio of forward-to-reverse resistance* (often known as the *front-to-back* or *back-to-front ratio*). The actual ratio depends on the type of diode. A typical small-signal diode in video circuits has a ratio of several hundred to one, whereas a power-supply rectifier diode can get by with ratios of 10 to 1.

1.14 TESTING AND HANDLING MISCELLANEOUS VIDEO COMPONENTS

Video components are best tested by monitoring the input and output waveforms during in-circuit operation. This is done as part of the troubleshooting procedure (as discussed throughout the remainder of this book). The waveforms are compared to those in the service literature.

Most video components can also be tested out of circuit using a scope with appropriate adapters. There are generally no satisfactory quick tests for the components (except for the transistors and diodes as discussed). Even if you devise a good quick test, you prove very little from a practical troubleshooting standpoint. Generally, sub-stitution is the best and ultimate test.

1.14.1 Handling ES Devices

If it becomes necessary to remove an ES device for test during the troubleshooting sequence of a video circuit, it must be treated with care. The service literature for most video equipment with ES devices (IGFETs, MOSFETs, etc.) often includes instruc-tions for handling the devices.

In circuit, an ES device is just as rugged as any comparable component. Out of cir-cuit, the ES device is subject to damage from *static charges* when handled. Unless otherwise directed by the service literature, use the following procedure when han-dling any ES device:

1. First turn off the power.

2. When the ES device is to be removed from a video equipment for test or replace-ment, your body should be at the same potential as the equipment. This can be done by placing one hand on the chassis before removing the component.

3. Before connecting the ES device to an external test circuit, put the hand holding the device against the tester and connect the tester lead to the ES-device lead (typi-cally, to the source lead for a MOSFET or IGFET). Make certain that the chassis, panel, and any other point you touch on the equipment or tester is at ground poten-tial *before touching* the point. In obsolete or defective equipment, the chassis or panel may be above or below ground.

4. When handling an ES device, the leads of the device must be shorted together. Generally, this is done in shipment by a *shorting ring* or a piece of metal foil or possibly a piece of wire. Connect the tester leads to the ES device. Then remove the shorting ring, wire, or foil. Leave the shorting device in place if the ES device is not to be tested immediately.

5. When soldering or unsoldering an ES device, the soldering tool tip must be at ground potential (no static charge). It may be convenient to collect a clip lead from the barrel of the soldering tool to the tester. The use of soldering guns is not rec-ommended for most ES devices.

6. Always remove power from the circuit before inserting or removing an ES device or a plug-in module containing an ES device. The voltage transients developed when terminals are separated may damage the ES device. This same caution applies to video circuits with conventional transistors and ICs. However, the chances of damage are greater with ES devices.

1.15 TROUBLESHOOTING VIDEO ICS

There is some difference of opinion about testing ICs in or out of circuit during video troubleshooting. An in-circuit test is the most convenient because the power source is available, and you need not unsolder the IC (which can be a real job). Of course, you must first measure the power-source voltages (plus, minus, ground, etc.) applied at the IC terminals to make sure that voltages are available and correct. If any of the volt-ages are absent or abnormal, this is a good starting point for troubleshooting. If the IC is digital or a microprocessor, reset voltages and clock signals must also be checked (as discussed in Chap. 4).

With all of the basic voltages and signals established, the in-circuit IC is tested by applying the appropriate input and monitoring the output. In some cases, it is not necessary to inject an input because the normal input is supplied by the circuits ahead of the IC. One drawback to testing a video IC in circuit is that the circuits before (input) and after (output) the IC may be defective. This can lead you to think that the IC is bad. For example, assume that the IC is used as the IF and video-detector stages of a TV set. To test such an IC, you inject a signal at the IC input (typically, from the TV tuner, Fig. 1.1) and monitor the video-detector output signal). Now assume that the IC output terminal is connected to a short circuit. There is no output indicated, even though the IC and the input signals are good. Of course, this shows up as an incorrect resistance reading (if resistance measurements are made).

Out-of-circuit tests for ICs have two obvious disadvantages: You must remove the IC, and you must supply the required power. However, if you test a suspected IC after removal and find that the IC is operating properly out of circuit, it is logical to assume that there is trouble in the circuits connected to the IC. This is very convenient to know *before you install a replacement IC.*

1.15.1 Video IC Power-Source Measurements

Although the test procedures for video ICs are essentially the same as those used for solid-state video circuits of the same type, measurement of the power-source voltages applied to the IC are not identical. Some ICs require connection to both a positive and a negative power source. This is particularly true when the IC contains an op amp or similar balanced circuits. Also, in some older equipment where the ICs are in metal cases, the case may be "hot." However, in most present-day video equipment, the ICs have plastic or other nonconducting cases.

1.16 EFFECTS OF CAPACITORS IN VIDEO TROUBLESHOOTING

During the troubleshooting process, suspected capacitors can be removed from the circuit and tested on bridge-type checkers. This establishes that the capacitor value is correct. With a correct value, it is reasonable to assume that the capacitor is not open, shorted, or leaking. From another standpoint, if the capacitor shows no shorts, opens, or leakage, it is also fair to say that the capacitor is good. So, from a practical troubleshooting standpoint, a simple test that checks for shorts, opens, or leakage is usually sufficient.

There are two basic methods for a quick check of capacitors during video troubleshooting: one with circuit voltages and one with an ohmmeter.

1.16.1 Checking Capacitors with Video-Circuit Voltages

As shown in Fig. 1.10a, this method involves disconnecting one lead of the capacitor (the ground or cold end) and connecting a voltmeter between the disconnected lead and ground. In a good capacitor, there should be a momentary voltage indication (or surge) as the capacitor charges up to the voltage at the hot end.

If the voltage indication remains high, the capacitor is probably shorted. If the voltage indication is steady but not necessarily high, the capacitor is probably leaking. If there is no voltage indication whatsoever, the capacitor is probably open.

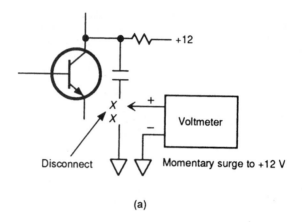

(a)

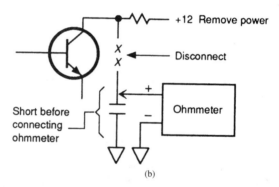

(b)

FIGURE 1.10   Video-circuit capacitor tests.

1.16.2   Checking Video-Circuit Capacitors with an Ohmmeter

As shown in Fig. 1.10b, this method involves disconnecting one lead of the capacitor (usually the hot end) and connecting an ohmmeter across the capacitor. Make certain that all power is removed from the circuit. As a precaution, short across the capacitor to make sure that no charge is retained (after the power is removed). In a good capacitor, there should be a momentary resistance indication (or surge) as the capacitor charges up to the voltage of the ohmmeter battery.

    If the resistance indication is near zero and remains so, the capacitor is probably shorted. If the resistance indication is steady at some high value, the capacitor is probably leaking. If there is no resistance indication (or surge) whatsoever, the capacitor is probably open.

1.16.3 Functions of Capacitors in Video Circuits

The functions of capacitors should be considered when troubleshooting video circuits. Consider the following example. The emitter resistor in a video circuit (such as the R_{10} in Fig. 1.5) is used to stabilize Q_1 gain and prevent thermal runaway (as discussed in Sec. 1.6.10). With R_{10} in the circuit, any increase in Q_1 collector current produces a greater drop in voltage across R_{10}. When all other factors remain the same, the change in Q_1 emitter voltage reduces the base-emitter forward-bias differential, thus tending to reduce Q_1 collector current.

When circuit stability is more important than gain (as is usually the case in video circuits), the emitter resistor is not bypassed. When signal gain must be high, the emitter resistance is bypassed to permit passage of the signal. If a video circuit does use emitter bypass (rare), and the bypass capacitor is open, stage gain is reduced drastically, although the transistor voltages remain substantially the same.

1.16.4 Low-Gain Symptoms in Video Circuits

If there is a low-gain symptom in any video amplifier (with emitter bypass) and the voltages appear normal, check the bypass capacitor. This can be done using either method shown in Fig. 1.10.

1.16.5 Coupling Capacitors in Video Circuits

The effects of defective coupling capacitors should be considered when troubleshooting video circuits. Consider the following examples.

Electrolytic Capacitors: Some video circuits use electrolytic capacitors for both coupling and bypass, particularly at low frequencies. From a troubleshooting standpoint, electrolytics tend to have more leakage than other capacitors (ceramic, etc.). However, good-quality electrolytics should have leakage of less than 10 μA at normal video-circuit voltages.

Defects in Coupling Capacitors: The function of C_6 in Fig. 1.4 is to pass signals from Q_1 to Q_3. If C_6 is shorted or badly leaking, the voltage from Q_1 (collector) is applied to the base of Q_3. This forward-biases Q_3, causing heavy current flow and possible burnout of Q_3. In any event, Q_3 is driven into saturation, and stage gain is reduced (and/or the output from Q_3 is abnormal).

If C_6 is open, there is little or no change in the voltage at Q_1 and Q_3, but the signal from Q_1 does not appear at the base of Q_3. From a troubleshooting standpoint, if C_6 is suspected (of shorts, leakage, or opens), try substitution (or the tests shown in Fig. 1.10).

1.16.6 Defects in Video Decoupling or Bypass Capacitors

The function of C_5 in Fig. 1.4 is to pass video signals to ground and thus provide a return path for the signals. If C_5 is shorted or leaking badly, the output from Q_3 is drastically reduced (or totally absent). If C_5 is open, there is little or no change in Q_3 voltage, but the signal waveform is abnormal (generally distorted). Either way, the most practical test of a suspected C_5 is substitution.

CHAPTER 2

BASIC VIDEO CIRCUITS (BLACK AND WHITE TVS AND MONITORS)

This chapter is devoted to the basics of video circuits. A black and white TV set is used as an example, because this is the most familiar video device in consumer electronics. The chapter describes functions, operation, circuit theory, test and alignment procedures, and a practical troubleshooting approach for the circuits involved.

2.1 BASIC BLACK AND WHITE TV-BROADCAST SYSTEM

As shown in Fig. 2.1, the basic TV-broadcast system consists of a transmitter capable of producing both AM and FM signals, a television camera, and a microphone. TV program sound (or audio) is transmitted through the microphone, which modulates the FM portion of the transmitter. The picture (or video) portion of the program is broadcast by means of the AM transmitter.

The TV broadcast channels are about 6 MHz wide, with the sound (FM) carrier at a frequency 4.5 MHz higher than the picture (AM) carrier. For example, on VHF Channel 7, the picture (AM) is transmitted at a frequency of 175.25 MHz, with the sound (FM) transmitted at 179.75 MHz. Channel 7 occupies the band of frequencies between 174 and 180 MHz.

The sound portion of the TV signal uses conventional FM broadcast principles, which need not be described here. However, the picture portion of the signal is unique to TV in that both picture information and synchronizing pulses are transmitted on the AM carrier.

Figure 2.1 also shows the relationship between the picture tube in the TV set and the station-camera picture tube. Both tubes have an electron beam, which is emitted by the tube cathode and strikes the tube surface. Both tubes have horizontal and vertical sweep systems that deflect the beam so as to produce a rectangular screen (or raster) on the tube surface. The vertical sweep is at a rate of 60 Hz, with the horizontal sweep at 15,750 Hz.

The electron beam of the station-camera picture tube is modulated by the amount of light that strikes the camera tube screen. In turn, the transmitted AM signal is modulated by the electron beam. The amplitude of the transmitted AM signal is determined by the intensity of the light at any given instant. The position of the electron

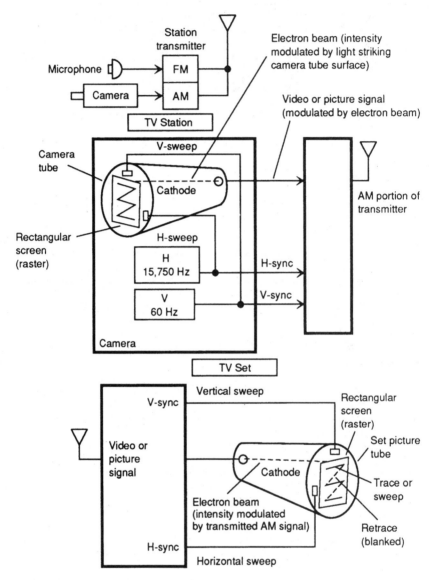

FIGURE 2.1   Basic TV broadcast system.

beam at a given instant is set by the horizontal and vertical sweep circuits, which, in turn, are triggered by pulses in the camera. These same pulses are transmitted on the AM carrier and act as synchronizing pulses (sync pulses) for the TV set horizontal and vertical sweep circuits.

The electron beam of the TV set tube is modulated by the transmitted AM signals so as to "paint" a picture on the tube screen. The amplitude of the AM signal determines the intensity of the light produced on the TV screen at any instant. For example, if the camera sees an increase in light, the electron beams in both tubes (camera and receiver) are increased, and the TV set tube shows an increase in light.

The TV set tube horizontal and vertical deflection systems are triggered by the horizontal and vertical synchronizing pulses transmitted on the AM portion of the TV broadcast signal. The electron beam of the TV set follows the beam in the camera tube. Both beams are at the same corresponding spot at the same instant.

Assume that the camera is focused on a white card with a black numeral 3 at the center, as shown in Fig. 2.2. As the camera tube electron beam is swept across the surface, light is reflected from the card onto the camera-tube surface. When the beam passes across the white-card background portion of the reflected light, the beam intensity is maximum. Beam intensity drops to minimum when the beam is at any portion of the light reflected from the (black) numeral 3. This varying electron beam modulates the transmitted AM carrier.

The electron beam starts at the top of the camera-tube screen, sweeps across to one side, and is blanked during the return to the other side (during retrace), until the beam finally reaches the bottom of the screen. The beam is then blanked and returned to the screen top. If the camera is focused on the numeral 3 pattern, the camera translates the entire pattern, line by line, into a picture signal (a voltage that varies in amplitude with intensity of the reflected light).

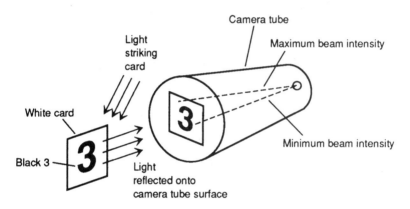

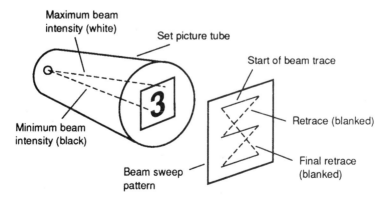

FIGURE 2.2   Electron beams in camera and receiver picture tubes.

At the TV set, the picture-tube beam follows the camera beam in both position and intensity. In the example of Fig. 2.2, both electron beams increase during the sweep across any of the white background. Likewise, both beams decrease during the sweep across any portion of the numeral 3. The TV set picture tube reproduces the numeral 3 in black and white (black, or minimum beam intensity, on the numeral portion of the sweep; white, or maximum intensity, on the card background portion).

2.2 BASIC BLACK AND WHITE TV SET CIRCUITS

Figure 1.1 shows the block diagram of a black and white TV set. The diagram shown is a *composite* of several types of sets and is presented as a point of reference for troubleshooting. Note that many of the discrete-component circuit groups of Fig. 1.1 are found as ICs in present-day sets. However, the circuit groups generally have essentially the same functions, as well as inputs and outputs, so Fig. 1.1 can be used for basic troubleshooting.

As an example, the amplifier, mixer, and oscillator shown as part of the tuner group in Fig. 1.1 are often part of a replaceable package in present-day (IC) sets. Individual tuner components or circuits are not available. However, the tuner still has an input (from the antenna), an output (sound and video in IF form), and an AGC input. These input-output points must be checked during troubleshooting, whether the set has discrete components, ICs, or plug-in modules.

The following paragraphs of this section describe the principles of all black and white TV sets and monitors. The detailed circuit descriptions (and troubleshooting procedures) for a similar discrete-component set are given at the end of this chapter. Digital-IC TV set descriptions are given in Chap. 4.

2.2.1 Low-Voltage Power Supply

The basic function of the low-voltage power supply (not shown) is to provide direct current to all circuits of the set, except the high voltages required by the picture tube. The high voltages are supplied by the *flyback circuit* (more properly known as the high-voltage power supply and horizontal-output circuit), as described in Sec. 2.2.2. The low-voltage power supply consists essentially of a rectifier (typically a full-wave bridge), followed by a regulator (typically zener and transistor regulator).

2.2.2 High-Voltage Power Supply and Horizontal Output

Although high-voltage power supply and horizontal output circuits have more than one function in all TV sets, the circuits do not have the same functions in all sets. The main functions are (1) to provide a high voltage for the picture tube and (2) to provide a horizontal deflection voltage (horizontal sweep) to the picture-tube deflection yoke.

In many sets, the circuits also supply a *boost voltage* for the picture-tube focus and accelerating grids. In some sets, the circuits also supply a voltage for the video output transistor (Sec. 2.2.8) since the output transistor often requires about 40 to 70 V (instead of the 12 V available from the low-voltage power supply). Often, the circuits also supply an automatic phase control and automatic frequency control (APC/AFC) signal to lock the horizontal oscillator frequency (Sec. 2.2.3) and an AGC signal (Sec. 2.2.9).

The circuits generally have only one input, no matter what combination of outputs are provided. The circuits receive pulses from the horizontal driver (Sec. 2.2.3). These pulses are at 15,750 Hz and are synchronized with the TV broadcast.

It is important that you note and remember inputs and outputs in any video device being serviced. As discussed in Chap. 1, the troubleshooting technique in this book is based on a comparison of inputs and outputs at each circuit group or IC. Let us see how this applies to the HV-supply and horizontal output circuits.

Assume that the input to the circuit is normal (15,750-Hz sync pulses of the correct amplitude) but that one or more of the outputs (high voltage, horizontal sweep, boost voltage, AGC, or APC/AFC, etc.) is absent or abnormal. This pinpoints trouble to some part of the HV-supply and horizontal output circuits (or to the IC that replaces these circuits).

As discussed in Chap. 1, it is important that you note and remember adjustment controls for all circuits of the video device, since adjustment of controls is a part of any troubleshooting technique. The only adjustment these circuits typically have is the *horizontal driver* that sets amplitude of the horizontal sweep. In some sets, the horizontal driver also sets the amount of high voltage available to the picture tube. The amount of high voltage is determined by the size and type of picture tube. Typically, a 25-in picture tube requires 25 to 30 kV.

2.2.3   Horizontal Oscillator and Driver

The horizontal oscillator and driver circuits provide drive signals to the HV-supply and horizontal output (Sec. 2.2.2). These 15,750-Hz signals are synchronized with the picture transmission by sync pulses from the sync separator (Sec. 2.2.5).

In addition to the one basic input and output, most horizontal oscillator and driver circuits have a feedback signal that is taken from the output and fed back to the input. The feedback signal provides an APC/AFC system for the horizontal sweep circuits. The APC/AFC system ensures that the horizontal sweep signals are synchronized (both for frequency and phase) with picture transmission, despite changes in line voltage and temperature or minor variations in circuit values.

The sync pulses are compared with the horizontal sweep signals (for frequency and phase) in the phase-detector portion of the circuits. Deviations of the horizontal sweep signals from the sync pulses cause the horizontal oscillator to shift in frequency or phase as necessary to offset the initial (undesired) deviation.

Although there are many types of APC/AFC circuits used for control of the horizontal oscillator (typically a blocking oscillator in discrete-component circuits), the oscillator is not triggered directly by the sync pulses. Instead, the sync and comparison pulses (fed back from the horizontal sweep) produce a variable dc control voltage that is applied to the horizontal oscillator. Any change in control voltage shifts the horizontal oscillator frequency and phase as necessary to maintain a phase and frequency lock.

In addition to automatic changes in the control voltage, the voltage can be manually set by the horizontal adjustment controls. These controls are not standard on all TV sets. Let us consider some typical examples.

As a minimum, there is a *horizontal hold* control, which permits manual adjustment of the control voltage applied to the horizontal oscillator. Some sets also include an *AFC control*, which also sets the oscillator control voltage. When both controls are used, the AFC is generally a back-panel adjustment (not readily accessible to the user), whereas the horizontal hold is a readily accessible user control.

There also may be a *horizontal stabilizer* (sometimes called the *horizontal frequen-*

cy control), which is a parallel-tuned circuit between the phase detector and horizontal oscillator. This circuit sets phasing of the oscillator. Another adjustment found on some sets is the *horizontal phase* (or phasing) control, which is a series-tuned circuit in the feedback line between the output and the phase detector. This phasing circuit is series-tuned to adjust the phase of the comparison pulse (from the output) to the phase detector. Note that the horizontal stabilizer and phase controls do not normally require adjustment, even during troubleshooting, unless parts have been replaced.

2.2.4 Vertical Sweep

The vertical sweep circuits provide a vertical deflection voltage (vertical sweep) to the picture-tube deflection yoke. The vertical circuits also supply a blanking pulse to the picture tube (usually through the video amplifier, Sec. 2.2.8). This blanks the picture tube during retrace of the vertical sweep. The vertical signals are at a frequency of 60 Hz and are synchronized with the picture transmission by vertical sync pulses from the sync separator (Sec. 2.2.5). The vertical sweep circuits have one input and two outputs.

Figure 1.1 shows that the vertical circuits include the vertical oscillator, driver, and output. (In most sets, the vertical deflection yoke is also considered part of the vertical circuits, just as the horizontal yoke is usually part of the horizontal output circuits.) In some sets, the functions of the vertical driver and vertical output are combined in one IC.

The number and type of controls are not standard for the vertical sweep circuits of all sets. However, as a minimum, there is a *vertical hold,* a *vertical size,* and a *vertical linearity* control. On some sets (particularly discrete-component sets), there is also a *vertical bias* control.

The vertical hold control sets the frequency of the vertical oscillator (60 Hz) and thus synchronizes the oscillator (generally a blocking oscillator) with the input sync pulses. The vertical size control is essentially a gain control used to set the amplitude of the vertical oscillator signal (a sawtooth sweep) applied to the driver and output. The vertical linearity control determines the linearity of the sawtooth sweep (or determines the portion of the sweep used) applied to the output and vertical deflection yoke. If the sawtooth sweep is not linear (or a nonlinear portion of the sweep is used), the picture is not linear.

The vertical bias control (when used) sets bias on the vertical output transistor. Adjustment of the bias control can affect both vertical size and linearity to some extent. However, the vertical bias control does not normally require adjustment, even during troubleshooting, unless the vertical output transistor (or related parts) has been replaced. The primary function of the vertical bias control is to compensate for variations in replacement vertical output transistors.

2.2.5 Sync Separator

The sync-separator circuits function to remove the vertical and horizontal sync pulses from the video circuits (Sec. 2.2.8) and apply the pulses to the vertical (Sec. 2.2.4) and horizontal (Sec. 2.2.3) sweep circuits, respectively. The sync-separator circuits also function as clippers and/or limiters to remove the video (picture) signal and any noise so that the sweep circuits receive only sync pulse.

Sync-separator circuits are not standardized, particularly in discrete-component TV. In some IC sets, the sync circuits are in the same IC as the video circuits. No matter what system is used, the vertical sync pulses are applied through a capacitor-resistor low-pass filter (often called the *vertical integrator*) to the input of the vertical sweep circuit. The horizontal sync pulses are applied to the AFC (phase detector) of the horizontal sweep circuits where the pulses are compared with the horizontal sweep (as to frequency and phase, see Sec. 2.2.3).

Typically, there are no adjustment controls in the sync separator. A possible exception is on some discrete-component sets where a bias control sets the level of the clipping and limiting action.

2.2.6 RF Tuner

The functions of an RF tuner, or tuner package, in a TV set are essentially the same as the RF sections in other AM/FM receivers. The basic TV tuner consists of an RF amplifier, mixer, and oscillator. The TV channels consist of two carriers, the FM-sound carrier and the AM-picture (or video) carrier. Both carriers are amplified by the RF amplifier. Both carriers are mixed with signals from the oscillator to produce two IF signals in the mixer. Thus, the RF tuner has one input and one output, even though there are two carriers.

The present trend in TV sets is to use some form of *frequency synthesis* (FS) in the tuner. This FS tuning is also known as *digital tuning, quartz tuning,* or *synthesized tuning* and is discussed in Chap. 4. In this section, we are concerned with basic tuner functions, whether the tuner is a replaceable package or composed of discrete components in a shielded module.

The non-FS tuner (such as shown in Fig. 1.1) receives automatic gain control (AGC) signals from the AGC circuits (Sec. 2.2.9). The AGC signals are generally applied only to the RF amplifier (not the mixer and oscillator) and function to control gain of the RF amplifier in the presence of carrier-signal variations. For example, if the carrier-signal strength increases, the AGC signals decrease the RF amplifier gain and thus offset the initial increase in carrier strength.

Some non-FS tuners also receive automatic fine-tune (AFT) signals (generally from the video detector). The AFT signals are applied to the oscillator and shift the oscillator frequency as necessary to keep the RF circuits tuned to the exact channel frequencies. For example, if the tuner drifts below the selected channel frequency, the AFT signals increase the tuner frequency and thus offset the initial frequency drift.

In a typical non-FS tuner, the tuning is accomplished by slug-tuned coils, with one set of coils for each channel. Generally there are three coils for each channel (one each for the RF amplifier, mixer, and oscillator). However, the coil arrangement varies with the type of RF tuner circuit.

In addition to (or instead of) the AFT circuits, many RF tuners have manual fine-tune circuits. The fine tune is a front-panel (user) adjustment control. Typically, the fine-tune control is a slug-tuned coil in parallel with the oscillator coil.

Some discrete-component tuners (as well as tuner packages) have other adjustments, such as traps and filters at the input and output. However, there is no standardization of such adjustments.

In some older sets, the UHF tuner is physically separate from the VHF tuner (in pre-1962 sets, you may not find any UHF tuner, if you can find a pre-1962 set). The functions of UHF tuners are essentially the same as for the VHF tuners, although there are circuit differences. For example, many UHF tuners do not have RF amplifiers, and the mixer uses a diode (rather than a transistor). Some UHF circuits are tuned by resonant cavities rather than tuning coils and capacitors, and some UHF tuners do not have AGC or AFT. Further, some older UHF tuners have continuous tuning, similar to an AM/FM broadcast radio, rather than fixed channel-by-channel tuning (in steps).

Many non-FS tuners are of the turret type, where a separate set of drum-mounted coils is used for each channel, and the entire drum rotates when the channel is selected. Other tuners are of the switch type, which have series-connected coils mounted on wafer switches.

Note that the RF tuners of many older TV sets will not accommodate all TV channels, particularly the cable-TV or CATV channels. This is not true of the FS tuners described in Chap. 4.

2.2.7 IF and Video Detector (VIF)

The IF and video-detector circuits amplify both picture and sound signals from the RF tuner, demodulate both signals for application to the video (Sec. 2.2.8) and sound IF amplifiers (not shown), provide signals to the AGC circuits (Sec. 2.2.9), provide AFT signals to the RF tuner (Sec. 2.2.6), and trap (or reject) signals from adjacent channels.

The present trend in TV sets is to combine most of the IF and video-detector functions into one or two ICs (often called the VIF section or module). The only parts external to the ICs are adjustment controls (typically slug-tuned coils for transformers and traps, since it is not practical to fabricate coils within an IC).

In many sets, both picture and sound outputs from the video detector are applied to the video amplifier circuits where both signals are amplified by at least one stage of the video amplifier. At this point, the sound signals are applied to the input of the sound-IF (SIF) amplifier, and the picture signals are applied to the remaining stages of the video amplifier (and to the sync separator). In other sets, the sound signals from the video detector are applied directly to the SIF amplifiers.

In addition to the signal input and output, most stages of the IF amplifier receive an AGC signal from the AGC circuits. These signals, which are in the form of a varying dc bias voltage, are the same as those applied to the RF tuner. The AGC circuits also receive video and sync signals from the IF amplifiers (taken from the last IF amplifier, before video detection).

In those sets with AFT, a portion of the video-detector output (in the form of an AFT voltage) is applied to the oscillator within the RF tuner. As discussed in Sec. 2.2.6, these AFT signals lock the oscillator onto the frequency of the selected channel and thus provide automatic fine tuning.

The VIF circuits do not usually have any operating or adjustment controls, as such. However, the circuits may contain a number of IF transformers which must be aligned to provide proper IF bandwidth. Generally, there is one transformer at the input and output of the IF amplifier, as well as one transformer between each IF stage. Thus, there are four transformers in a three-stage discrete-component IF amplifier circuit. The transformers are usually slug-tuned, although some older sets use variable capacitors.

As with most FM receivers, the IF stages are stagger-tuned. That is, each IF transformer is tuned to a different frequency (one tuned to the high end, one to the low end, and the remainder to various points in between). This gives the IF amplifiers sufficient bandwidth to pass both the picture and sound signals.

In addition to the IF transformers, there are a number of traps in most VIFs. These traps are tuned circuits (again with a fixed capacitor and slug-tuned coil) used to reject and remove signals from adjacent TV channels. Traps are necessary since TV signals may be broadcast simultaneously on adjacent channels, and the combination of wideband broadcast channels and wideband IF circuits may result in an overlap of signals.

2.2.8 Video Amplifier and Picture Tube

The video-amplifier circuits have several functions, and there are many circuit configurations in use. In a typical set, the video-amplifier circuit has three inputs and three outputs. The primary input is the demodulated picture and sound signal from the video-detector output (Sec. 2.2.7). Secondary inputs are blanking pulses from the vertical and horizontal sweep circuits. The primary output is the video signal (picture and sync) applied to the picture tube. This same video output is applied to the sync-separator and SIF circuits.

Note that some sets do not have horizontal blanking pulses. Instead, the brilliance of the picture tube is kept low enough so that the electron-beam horizontal retrace cannot be seen.

In the composite circuit of Fig. 1.1, the three signals (picture information and vertical and horizontal blanking pulses) are applied through the driver and output stages to the picture tube (usually at the cathode). These three signals control the picture-tube electron-beam intensity.

The pulses blank the picture tube (cut off the electron beam) during the retrace period of the horizontal and vertical sweeps. During the trace periods of both sweeps, the picture signal varies the intensity of the electron beam so as to paint a picture on the tube screen. As discussed, the picture-tube beam "follows" the broadcast camera tube beam, in both intensity and position, so as to reproduce the picture focused in the camera lens.

The sound signal is prevented from being applied to the picture tube by a trap between the video output stage and the picture tube input. (In most sets, the input is at the cathode, but it may be at a grid on some picture tubes.) The trap (often called the sound, SIF, or 4.5-MHz trap) is usually a parallel fixed capacitor and slug-tuned coil, tuned to 4.5 MHz.

The sound signal is applied to the input of the SIF circuits after amplification by the first stage of the video amplifier (often called the driver or video amplifier). The picture and sync-pulse information does not pass to the audio section since the SIF circuits are tuned to 4.5 MHz (in a properly functioning set).

The picture signal and sync pulses are applied to the input of the sync-separator circuits (Sec. 2.2.5) after amplification by the video driver. The sound signal is also present at this point. However, since the sound information is FM and is at a frequency far removed from either the horizontal or vertical sweep frequencies (4.5 MHz compared to 60 and 15,750 Hz), the sound information does not affect the sync pulses. As discussed in Sec. 2.2.5, the picture signal is removed in the sync separator by clipping and limiting action. Only the sync pulses are applied to the sweep circuits.

In most black and white TV sets, the video-amplifier and picture-tube circuits have at least two controls: *contrast* and *brilliance* (or *brightness*). There are many configurations for these two controls.

The brilliance or brightness control determines the amount of fixed bias applied to the picture-tube cathode. This sets the intensity of the electron beam (with or without a signal present). The contrast control determines the amount of drive signal applied to the picture tube. An increase in drive signal produces an increase in contrast. In the circuit of Fig. 1.1, the contrast control is, in effect, a volume or gain control between the video driver and video output stages.

2.2.9 AGC

There are many types of AGC circuits. Most discrete-component sets use keyed saturation-type AGC. The RF-tuner and IF-stage transistors connected to the AGC line are forward-biased at all times. On strong signals, the AGC circuits increase the forward bias, driving the transistors into saturation, thus reducing gain. Under no-signal conditions, the forward bias remains fixed.

Keyed-AGC circuits have two inputs. One is from the IF circuit (Sec. 2.2.7) output and is at the IF center frequency. The other input is from the horizontal output circuit (Sec. 2.2.2) at a frequency of 15,750 Hz. The AGC output is a varying dc bias voltage (controlled by bursts of IF signals).

A portion of the IF signal is taken from the IF amplifiers and is pulsed, or "keyed," at the horizontal-sweep frequency. The resultant keyed bursts of signal control the amount of dc voltage produced on the AGC line. The AGC output is applied to both the IF amplifiers and the RF tuner. A few AGC circuits are provided with an AGC control that sets the level of AGC action.

2.3 SAFETY PRECAUTIONS FOR VIDEO TEST EQUIPMENT

In addition to a routine operating procedure, certain precautions must be observed during the operation of any test equipment used in video service. Many of these precautions are the same for all types of test equipment; others are unique to special test instruments such as meters, scopes, and signal generators. Some of the precautions are designed to prevent damage to the test equipment or to the video circuit. Other precautions are to prevent injury to you. Where applicable, special safety precautions are included throughout the various chapters of this book.

The following general safety precautions should be studied thoroughly and then compared to any specific precautions called for in the test equipment service literature and in the related chapters of this book. Refer to Sec. 1.6.13 for general video service precautions.

1. Many service instruments are housed in metal cases. These cases are connected to the ground of the internal circuit. For proper operation, the ground terminal of the instrument should always be connected to the ground of the video device being serviced. Make certain that the chassis of the video device is not connected to either side of the ac line or to any potential above ground. If there is any doubt, connect the video device to the power line through an *isolation transformer.*

2. Remember that there is always danger when servicing video equipment that operates at hazardous voltages (such as the high voltages used by picture tubes). Keep this firmly in mind as you pull off a TV set back panel and apply power through a "cheater" cord. Always make some effort to familiarize yourself with the set before servicing it, bearing in mind that high voltages may appear at unexpected points in a defective receiver.

3. It is good practice to remove power before connecting test leads to high-voltage points. It is preferable to make all service connections with the power removed. If this is impractical (as is usually the case), be especially careful to avoid accidental contact with circuits that are grounded. Working with one hand away from the equipment and standing on a properly insulated floor lessens the danger of electrical shock.

4. Capacitors may store a charge large enough to be hazardous. Discharge filter capacitors before attaching test leads (but after you have removed the power).

5. Leads with broken insulation offer the additional hazard of high voltages appear-

ing at exposed points along the leads. Check test leads for frayed or broken insulation before using the leads.

6. To lessen the danger of accidental shock, disconnect test leads immediately after the test is completed.

7. Remember that the risk of severe shock is only one of the possible hazards. Even a minor shock can place you in danger of more serious risks, such as a bad fall or contact with a higher voltage.

8. The experienced service technician guards continuously against injury and does not work on hazardous circuits unless another person is available to assist in case of accident.

9. Even if you have considerable experience with test equipment used in service, always study the service literature of any instrument with which you are not thoroughly familiar.

10. Use only shielded leads and probes. Never allow your fingers to slip down to the metal probe tip when the probe is in contact with a "hot" circuit.

11. Avoid vibration and mechanical shock. Most electronic test equipment is delicate.

12. Study the video circuit being serviced before making any test connections. Try to match the capabilities of the instrument to the circuit being serviced.

2.4 SIGNAL GENERATORS IN BASIC VIDEO SERVICE

The signal generator is an indispensable tool for video service. Without a generator, you depend entirely on broadcast signals and are limited to signal tracing only. This means that you have no control over frequency, amplitude, or modulation of such signals and have no means for signal injection.

With a signal generator of the appropriate type, you can duplicate transmitted signals or produce special signals required for alignment and test of all circuits within the video equipment. Also, the frequency, amplitude, and modulation characteristics of the signals can be controlled so that you can check operation of a video device under various signal conditions (weak, strong, normal, abnormal, etc.).

In addition to conventional RF, pulse, and audio generators, there are generators designed specifically for use in video service. For example, in consumer video, there are sweep, marker, analyst, pattern, and color generators. Often the functions of these generators are combined. For example, several manufacturers produce a sweep and marker generator. Similarly, the analyst and color generator functions are often combined in a single instrument. (NTSC color generators are described in Chap. 3.) The purpose, operating principles, and typical characteristics of the remaining generators are described in the following sections.

2.4.1 Sweep and Marker Generator

The main purpose of the sweep and marker generator in video service is *sweep-frequency alignment*. A sweep and marker generator capable of producing signals of the appropriate frequency is used with a scope to display the *bandpass characteristics* of a video circuit (tuner, IF, video amplifier, etc.).

The *sweep portion* of the generator is essentially an FM generator. When the sweep generator is set to a given frequency, this is the *center frequency.* The output varies back and forth through this center frequency. The rate at which the frequency modulation takes place is typically 60 Hz. The sweep width, or the amount of variation from the center frequency, is determined by a control, as is the center frequency.

The *marker portion* is essentially an RF generator with highly accurate dial markings that can be calibrated precisely against internal or external signals. Usually, the internal signals are crystal controlled. Marker signals are necessary to pinpoint frequencies when making sweep-frequency alignments and are usually produced by a built-in *marker adder.* The basic sweep-frequency alignment procedure is described in Sec. 2.4.2.

In addition to the basic sweep and marker outputs, the generator may have other special features. For example, the generator may provide a *variable bias* to disable the AGC circuits and a *blanking circuit* that permits a zero reference line to be observed on the scope during the retrace period.

2.4.2 Basic Sweep-Frequency Alignment Procedure

The relationship between the sweep and marker generator and scope during sweep-frequency alignment is shown in Fig. 2.3. If the equipment is connected as shown in Fig. 2.3*a*, the scope sweep is triggered by a sawtooth output from the generator. The scope's internal sweep is switched off, and the scope sweep selector and sync selector are set to external.

Under these conditions, the scope sweep represents the total sweep spectrum as shown in Fig. 2.3*c*, with any point along the horizontal trace representing a corresponding frequency. For example, the midpoint on the trace represents 15 kHz. If you want a rough approximation of frequency, adjust the horizontal gain control until the trace occupies an exact number of scale divisions on the scope screen (such as 10 cm for the 10- to 20-kHz sweep signal). Each centimeter division then represents 1 kHz.

If the equipment is connected as shown in Fig. 2.3*b*, the scope sweep is triggered by the scope's internal circuits (both the sweep and sync selectors are set to internal). Certain conditions must be met to use the test connections shown in Fig. 2.3*b*. If the scope has a triggered sweep (Sec. 2.5), there must be sufficient delay in the vertical input, or part of the response curve may be lost. If the scope is not a triggered sweep, the generator must be swept at the same frequency as the scope (usually the line frequency of 60 Hz). Also, the scope or generator must have a *phasing control* so that the two sweeps can be synchronized. If the phase adjustment is not properly set, the sweep curve may be prematurely cut off, or the curve may appear as a double or mirror image, as shown in Fig. 2.3*d*.

As shown in Fig. 2.3*c*, the markers provide accurate frequency measurement. Although some older generators have variable markers, present-day generators have a number of markers at precise, crystal-controlled frequencies. Such fixed-frequency markers are illustrated in Fig. 2.3*e*, which shows the bandpass response curve of a typical video circuit (a VCR tuner and VIF package). The markers can be selected (one or several at a time) as needed.

The response curve (scope trace) depends on the video circuit under test. If the circuit has a wide bandpass characteristic, the generator is set so that the sweep is wider than that of the circuit. (The bandpass of the circuit shown in Fig. 2.3*e* is about 6 MHz.) Under these conditions, the trace starts low at the left, rises toward the middle, and then drops off at the right.

The sweep and marker-scope method of alignment tells at a glance the overall bandpass characteristics of the circuit (sharp response, irregular response at certain

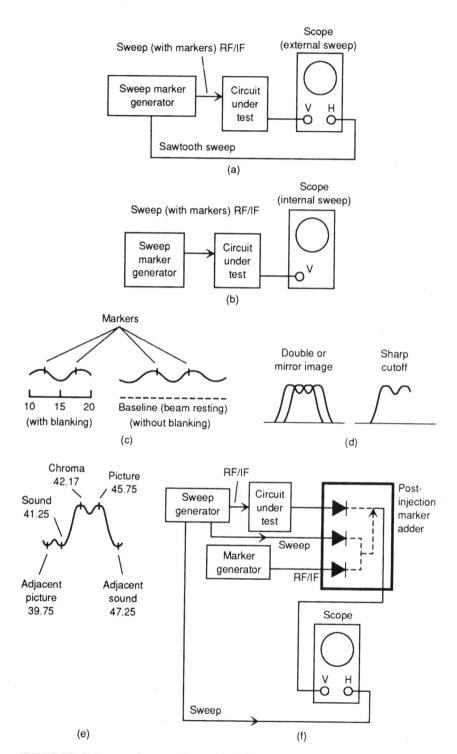

FIGURE 2.3 Basic sweep-frequency alignment procedures.

frequencies, and so on). The exact frequency limits of the bandpass can be measured with the markers (often called *pips*).

2.4.3 Direct-Injection versus Postinjection

There are two basic methods for injection of marker signals. With *direct-injection,* the sweep-generator and marker-generator signals are mixed before the signals are applied to the circuit. This method is sometimes called *preinjection* and has generally been replaced by postinjection.

With postinjection, as shown in Fig. 2.3*f,* the sweep-generator output is applied to the circuit. A portion of the sweep-generator output is also mixed with the marker-generator output in a mixer-detector circuit known as a *postinjection marker adder.* The mixed and detected output from both generators is then mixed with the detected output from the circuit. The scope vertical input represents the detected values of all three signals (sweep, marker, and circuit output).

Most present-day sweep and marker generators have some form of built-in postinjection marker-adder circuits. The postinjection (sometimes known as *bypass injection*) method for adding markers is usually preferred for consumer-electronics video service because it minimizes the chance of overloading the circuits and permits use of a narrowband scope. At one time, postinjection marker-adders were available as separate units, and they are still in use today.

2.4.4 Typical Sweep and Marker Generators

The following are the features found on a typical sweep and marker generator.

There are RF outputs, with equivalents of all IF and chroma (color) markers (Fig. 2.3*e*) available for one or more VHF TV channels (typically 3 and 4). This makes it possible to connect the RF output to the antenna terminals of a TV set or VCR and, without further input reconnections, evaluate alignment conditions of all tuned signal-processing video circuits.

There are several (typically 10) *crystal-controlled markers,* and postinjection markers can be added. All markers can be used simultaneously or individually. The markers shown in Fig. 2.3*e* are typical.

A *video sweep output* permits direct sweep alignment of the video circuits where specified by the manufacturer. With some generators it is necessary to use the IF sweep for signal injection and then monitor the video circuits for response.

The generator may include *pattern polarity reversal* and *sweep reversal* features which permit you to match scope displays shown in service literature (positive or negative, left- or right-hand sweep, etc.).

In some generators, the *markers can be tilted* to horizontal or vertical positions, permitting easy identification. For example, if the sides of a bandpass display are steep (vertical), a horizontal marker is easier to identify. A vertical marker shows up better on the flat top (horizontal) of the same pattern.

The generator may also include a number of features that are primarily for TV service: built-in amplifiers and filters, marker outputs for spot alignment of traps and bandpass circuits, and visual reproductions of idealized alignment curves on the front panel to indicate desired marker positions.

2.4.5 Analyst and Pattern Generators

An analyst generator is used for *signal substitution* (a form of signal injection). A bench-type analyst generator provides outputs that duplicate all essential signals in a TV set. Such generators have RF, IF, composite video (including sync), sound IF, audio, separate sync, flyback and yoke test signals, etc.

The RF, IF, and video signals are usually in the form of a pattern (or patterns), typically lines or bars (horizontal and vertical), dots, crosshatch lines, square pulses, blank rasters, or color bars (Chap. 3). Some typical patterns are shown in Fig. 2.4. Such patterns can be used in video troubleshooting. For example, with a typical analyst generator, you can inject a black and white crosshatch pattern at some particular channel into the antenna of a VCR and record the pattern. Then you can repeat the process using the color pattern. You then play back the recorded patterns on a known-good TV monitor.

If the display is a clear, sharp crosshatch pattern, followed by a good color pattern, you know that the VCR is capable of recording and playing back a good picture. If only black and white is good, you know there are problems in the color circuits. If there is no playback, you can play the recorded tape on a known-good VCR. If there still is no playback, the problem is in the record circuits of the suspected VCR. If there is a good playback, the problem is in the playback circuits of the suspected VCR.

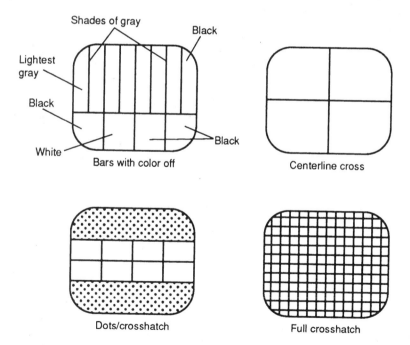

FIGURE 2.4 Typical analyzer and NTSC generator patterns.

Thus, any type of analyst and pattern generator can serve to pinpoint defective circuit areas during VCR troubleshooting.

At one time, the *test-pattern generator* was popular for black and white TV service. Such generators differ from present-day analyst and pattern generators in that the test-pattern generator reproduces positive transparencies of various pictures that may be inserted into the generator. In effect, the test-pattern generator is a miniature TV station capable of reproducing still pictures. Such generators have generally been replaced by the analyst generators described here or by combination pattern and color generators described in Chap. 3.

2.4.6 Typical Analyst and Pattern Generators

The typical analyst and pattern generator produces several patterns, including color, at RF, IF, and video frequencies. In effect, the instrument generates many signals normally transmitted by a TV station, and those produced within a TV set, for point-to-point troubleshooting. For example, a typical instrument generates:

VHF and UHF signals on various channels for testing the RF tuner (and overall performance)

IF signals from 20 to 48 MHz for testing the IF portion of TV sets and VCRs

A 4.5-MHz sound channel test signal that is frequency modulated by a 1-kHz tone

A separate 1-kHz audio tone

Positive and negative composite video signals (including the sync pulses) for injection into video stages

NTSC color bars such as described in Chap. 3

2.5 OSCILLOSCOPES IN BASIC VIDEO SERVICE

Ideally, a scope used for consumer video service should have a bandwidth of 25 MHz or more, although you can probably get by with a 10-MHz scope (or even a 6-MHz scope). Minimum sweep time should be 0.1-μs per division, although a 0.2-μs sweep will probably do the job.

The scope should have a *triggered sweep* in addition to conventional internal and external sweep synchronization. With a triggered sweep, the scope is synchronized by the signals applied to the vertical input. The sweep remains at rest until triggered by the signal. This assures that the signals are always synchronized, even when the waveform is of varying frequency. The triggered sweep threshold should be fully adjustable so that the desired portion of the waveform can be used for triggering.

A *dual-trace scope* facilitates some measurements (such as timing measurements required in VCR and camcorder servo adjustment) but is not absolutely required. Such refinements as *calibrated voltage scales, calibrated sweep rates, Z-axis input* (for intensity modulation), and *sweep magnification and illuminated scales* are, of course, always helpful in video work (as they are in any type of electronic service).

Many scopes have special provisions for TV and VCR service. Such features include the *TV sync or sweep* (usually identified as TV horizontal and TV vertical, or TVH and TVV, or some similar term) and the *vectorscope* provision.

2.5.1 TVH and TVV Sync Sweeps

The TVH and TVV functions are usually selected by means of front-panel switches and permit *pairs* of vertical or horizontal sync pulses to be displayed on the scope screen. As shown in Fig. 2.5, the signal applied to the vertical input (containing 60- and 15,750-Hz sync pulses) is also applied to a sync separator. This separator operates similarly to a TV sync separator (Sec. 2.2.5) and delivers two separate outputs (if both sync inputs are present). If only one sync input is present, only one output is available. Either way, the selected output (60 or 15,750 Hz) is used to synchronize the horizontal sweep trigger.

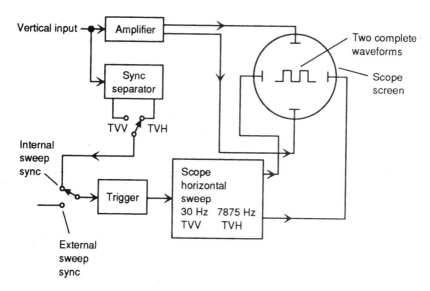

FIGURE 2.5 TVH and TVV sync sweep connections.

When TVV is selected, the scope horizontal sweep is set to a rate of 30 Hz, and the horizontal trigger is taken from the TVV output of the sync separator. With the horizontal sweep at 30 Hz and the trigger at 60 Hz, there are two pulses for each sweep and two vertical sync pulses (or two complete vertical displays) appear on the screen.

When TVH is selected, the scope horizontal sweep is set to a rate of 7875 Hz, and the trigger (15,750 Hz) is taken from the TVH output of the sync separator. Again, two pulses or displays are presented on the screen since the sweep is one-half the trigger rate.

The TVH and TVV features not only simplify observation of waveforms (you always get exactly two complete displays), but the features can also be helpful in trouble localization. For example, assume that you monitor waveforms at some point where *both* vertical and horizontal sync pulses are *supposed to be* in the circuit. If you find two steady displays on the TVV position but not in the TVH position, you know that there is a problem in the horizontal circuits or that horizontal sync pulses are not getting to the point being monitored.

2.5.2 Vectorscopes

A vectorscope is used in service of color video circuits. For that reason, we describe vectorscope operation in Chap. 3. The vectorscope is sometimes used in conjunction with a color generator to check the response of a color TV set to color signals. The vectorscope is used more extensively in color TV-broadcast work.

A vectorscope permits the phase relationship (and amplitude) of color signals to be displayed as a *single pattern* on the scope screen. When used for TV service, the vectorscope monitors the color signals directly at the color-gun inputs of the color picture tube. This makes the vectorscope adaptable to any type of color circuit (any type of color demodulation). By comparing the vectorscope display against that of an ideal display (for phase relationship, amplitude, general appearance, and so on), the condition of the TV color circuits can be analyzed. A conventional scope can be used as a vectorscope. This requires special connections to the horizontal and vertical deflection system, as described in Chap. 3.

2.6 MISCELLANEOUS TEST EQUIPMENT

In addition to a good scope, a sweep and marker generator, and a color generator (preferably with pattern and analyst features), most video service can be performed with conventional test equipment found in other electronic service fields. These include meters, transistor and diode testers, RC-substitution boxes, and assorted adapters, clips, and probes.

Specialized video equipment such as VCRs and camcorders require some equally specialized test equipment (and tools). These instruments are discussed in the related chapters. In the following sections, we concentrate on those items of test equipment that apply to a wide variety of video equipment.

2.6.1 Probes (for Meters and Scopes)

Most meters and scopes used in video service operate with some type of probe. In addition to providing for electrical contact to the video circuit under test, probes modify the voltage being measured to a condition suitable for display on a scope or meter. For example, assume that the picture-tube high voltage must be measured and that this voltage is beyond the maximum input limits of the meter or scope. A *voltage-divider probe* can be used to reduce the voltage to a safe level for measurement. The voltage is reduced by a fixed amount (known as the *attenuation factor,* typically by 10:1, 100:1, or 1000:1).

Basic Probe. In the simplest form, the basic probe is a *test prod* (possibly with a removable alligator clip). Basic probes work well on video circuits carrying direct current or audio. If, however, the circuit contains high-frequency signals, or if the gain of the scope or meter is high, it may be necessary to use a special low-capacitance probe.

Hand capacitance in a simple probe can cause hum pickup. This condition is offset by shielding in low-capacitance probes. Of more importance, the input impedance of a meter or scope is connected directly to the circuit under test by a simple probe. Such input impedance can disturb circuit conditions.

Low-Capacitance Probes. The low-capacitance probe contains a series of capacitors and resistors that increase the meter or scope impedance. In most low-capacitance probes, the resistors form a divider (typically 10:1) between the circuit under test and the meter or scope input. Thus, low-capacitance probes serve the dual purpose of capacitance and voltage reduction. When low-capacitance probes are connected at the inputs of meters or scopes, remember that the voltage indications are one-tenth (or whatever value of attenuation is used) of the actual value.

High-Voltage Probes.   Most high-voltage probes are *resistance-type voltage-divider* probes. Such probes are similar to the low-capacitance probe except that the frequency-compensating capacitor is omitted. Usually, the conventional resistance-type probe is used when a voltage reduction of 100:1, or greater, is required and when a flat frequency response is of no particular concern.

High-voltage probes for video service must be capable of measuring potential at or near 30 kV (typical), usually with a 1000:1 voltage reduction. In certain isolated cases, the resistance-type probe is not suitable because of stray conduction paths set up by the resistors. A *capacitance-type* probe can be used in those cases. Such probes contain two (or more) capacitors, with values selected to provide the desired voltage reduction and to match the input capacitance of the meter or scope.

Special High-Voltage Probe.   There are high-voltage probes that need not be connected to a scope or meter. Such probes have a built-in meter that permits picture-tube voltages to be measured directly.

Radio-Frequency Probes.   An RF probe is required when the signals to be measured are beyond the frequency capabilities of the scope or meter. RF probes convert (rectify) the RF signals into a dc voltage that is equal to the peak RF voltage (or possibly equal to the RMS of the RF voltage). The dc output of the probe is applied to the meter or scope input and is displayed as a voltage readout in the normal manner.

Demodulator Probes.   The circuit of a demodulator probe is essentially like that of the RF probe. However, the demodulator produces both ac and dc output. The RF carrier frequency is converted to a dc output voltage equal to the RF carrier. If the carrier is modulated, the modulating voltage appears as ac (or pulsating dc) voltage at the probe output.

The meter or scope is set to measure direct current, and the RF carrier is measured. The meter or scope is then set to measure alternating current, and the modulating voltage (if any) is measured. Generally, demodulator probes are used for signal tracing, and the output is not calibrated to any particular value. However, some demodulator probes produce calibrated outputs equal to both carrier and modulation (and may be described as RF and demodulator probes).

2.6.2   Frequency Counters

The two most common uses for a counter in video work are (1) to check or adjust the various 3.58-MHz oscillators in the chroma or color circuits of TV sets, VCRs, and camcorders, etc., and (2) to measure timing of VCR or camcorder servo systems. Counters can also be useful when checking the clock oscillators of microprocessors (such as found in digital video circuits, Chap. 4).

In reality, the so-called 3.58-MHz oscillator is locked (in frequency) to a color TV broadcast at a frequency of 3.579545 MHz. The oscillator remains locked at this frequency, even though the phase and color hue may shift. A seven-digit counter is required to get the full-frequency resolution.

In addition to seven-digit resolution, the counter must also be capable of reading low frequencies. For example, many of the servo-sync signals found in VCRs and camcorders are in the 30-Hz range (or even the 15-Hz range).

2.6.3 Picture Tube (CRT) Testers and Rejuvenators

These instruments provide for test and possible rejuvenation of picture tubes. The instruments are also known as tube or CRT renewers. With these units it is possible to check each element of the picture tube for such factors as shorts, opens, leakage, and proper emission. It is also possible to rejuvenate some picture tubes by application of high voltages, heavy currents, etc., to the proper elements.

Because of the highly specialized nature of testers and rejuvenators, and because you receive a detailed set of instructions with the instrument, we do not describe the units here. However, you should be aware of their use, particularly in service shops specializing in color TV and terminals with color graphics.

2.6.4 Video Receiver-Monitor and Monitor-Type TV Sets

If you are planning to be in VCR, camcorder, and video-disc player service on a full-scale basis, you should consider a receiver-monitor such as is used in studio or industrial video work. These receiver-monitors are essentially TV receivers with video and audio *inputs and outputs* brought out to some accessible point (usually on the front panel). There are also *monitor-type* TV sets designed specifically for VCRs, video-disc players, and video games.

The *output connections* make it possible to monitor broadcast video and audio signals as the signals appear at the output of a TV IF section (the so-called *baseband signals,* generally in the range of 0 to 4.5 MHz, at 1 V peak-to-peak for video and 0 dB, or 0.775 V, for audio). These output signals from the receiver-monitor can be injected into the VCR at some point in the signal flow past the tuner IF. (Note that monitor-type TV sets do not generally provide baseband outputs.)

The *input connections* on either a receiver-monitor or a monitor-type TV make it possible to inject video and audio signals from the VCR (before the signals are applied to the RF output unit) and monitor the display. Thus, the baseband output of the VCR can be checked independently from the RF unit.

If you do not want to go to the expense of buying an industrial receiver-monitor or a monitor-type TV, you can use a standard TV to monitor the VCR. Of course, with a TV set, the VCR video signals are used to modulate the VCR RF unit. The output of the RF unit is then fed to the TV antenna input. Under these conditions it is difficult to tell if faults are present in the VCR video or in the VCR RF unit. Similarly, if you use an NTSC generator for a video source, the generator output is at an RF or IF frequency, not at the baseband video frequencies (on most NTSC generators).

2.7 BASIC TV ALIGNMENT PROCEDURES

The general procedure for alignment of split-sound and intercarrier types of TV sets is the same, the major differences being in the number of intermediate frequencies used and the specific frequencies involved. Most modern TV sets are of the intercarrier type (one set of IF amplifiers for both picture and sound). Some very old TV sets are of the split-sound type, where separate IF circuits are used for picture and sound.

No matter what IF systems and frequencies are used, there are four separate steps for overall alignment of black and white sets (and the black and white circuits of color sets): (1) tuner or RF alignment, (2) picture (video) IF alignment, (3) trap alignment, and (4) sound IF and FM detector alignment. They are described in the following sections. Keep in mind that these descriptions are general in nature and that the steps apply primarily to sets with discrete-component circuits. However, present-day sets with replaceable circuit boards still require alignment. The following basic procedures can be used as a guide in performing alignment for any set. However, always follow the specific procedures recommended in the service literature.

2.7.1 RF Tuner Alignment

Figure 2.6 shows the basic test connections and typical waveshape for postinjection alignment of an RF tuner. Figure 2.7 shows the direct-injection alignment connections. The primary purpose of alignment is to obtain a response curve of proper shape, frequency coverage, and gain. Most RF tuners merely require "touch-up" alignment in which relatively few of the adjustments are used. The digital and frequency-synthesis tuners described in later chapters often do not require alignment (as such).

Complete front-end alignment includes alignment of the antenna-input circuits and adjustment of the amplifier and RF-oscillator circuits. The antenna input circuits are usually aligned to give a response curve which has a sharp drop-off slightly below Channel 2 and which is flat up through the highest channel involved. Alignment of the amplifier and oscillator stages includes setting the oscillator frequencies for all channels, setting one or more traps to the correct frequency, and adjustment of tracking between RF amplifier and oscillator. Alignment is done by setting adjustments so that the waveshape on the scope resembles the waveshape shown in service literature (such as shown in Fig. 2.6c).

The marker-generator signals are used to provide frequency-reference points as aids in shaping the curve. For example, with the sweep generator set to deliver an output on Channel 3, markers are injected at 61.25 and 65.75 MHz (picture and sound carrier frequencies for Channel 3) as shown in Fig. 2.6c. The markers on the curve show the separation between the picture and sound carriers of 4.5 MHz. Since the RF tuner must pass both sound and picture channels, a tuner bandpass of about 6 MHz is required.

If you are working with a supposedly good tuner, check alignment by observing the response curves for each channel. Curves for individual channels should be examined and compared with those shown in the service literature. If a particular response curve indicates that alignment is required, follow the recommended procedure.

If you are working with a suspected defective tuner, leave the tuner set to only one channel position until the trouble has been located and cleared. Then other channel positions can be compared with the initial one for sensitivity, switching noise, and general performance.

RF Tuner Alignment Notes.   The following notes describe typical alignment and test connections for an RF tuner.

1. Do not start alignment until the set, *and all test equipment,* has warmed up. This applies even to solid-state sets and test equipment.

Leave the tuner oscillator in operation during alignment. It is possible to make some tuner adjustments with the oscillator disabled. However, the lack of oscillator injection voltage at the mixer alters the mixer bias, resulting in an increase in amplitude of the response curve and distortion of the waveshape.

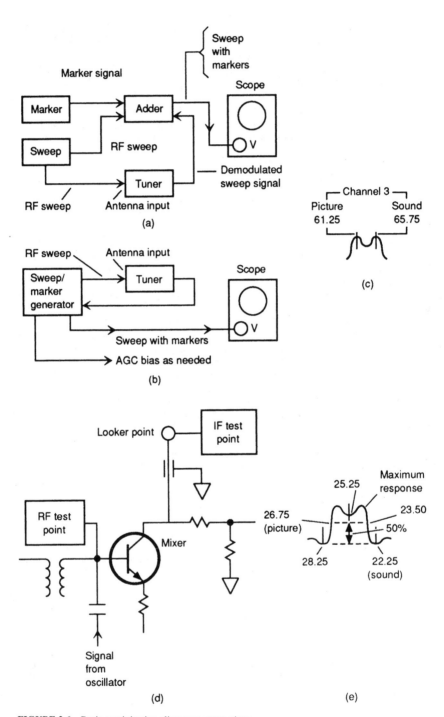

FIGURE 2.6   Basic postinjection alignment connections.

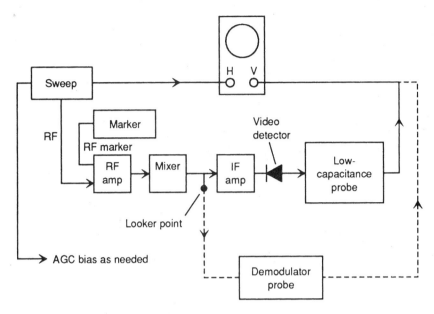

FIGURE 2.7   Basic direct-injection alignment connections.

If there has been extensive work on the RF tuner (particularly by untrained "techni-cians"), it may be necessary to check the oscillator frequency on one or more channels with a counter. However, this is generally not required.

If you find serious misalignment of the tuner when making the test setup or if you find considerable difficulty or failure in alignment, this usually indicates a defect in the tuner. Likewise, if alignment fails to produce correct tuner curves, you should sub-mit the tuner to the basic troubleshooting procedures described in Chap. 1 (or, more realistically, replace the tuner at outrageous expense to the customer).

2. Connect the equipment as shown in Fig. 2.6 for postinjection alignment. The connections in Fig. 2.6*a* assume that the sweep generator, marker generator, and mark-er adder are three separate units, while Fig. 2.6*b* assumes that a single sweep and marker generator (with adder) is used (typical for present-day test equipment). With either arrangement, it is necessary to connect the output to the scope vertical input.

The output test point may be across the load resistor of the video detector (IF out-put) or at the RF tuner output (at the looker point). If you connect at the video detec-tor, you get an "overall" response curve (RF, IF, detector). By connecting directly to the looker point, you can check the shape of the tuner curve independently of other circuits (which is usually the better method if you suspect problems in the tuner).

Figure 2.6*d* shows two typical output test points for an RF tuner. Note that signals at the base test point are at the RF frequency, whereas signals at the collector test point are IF. With either test point, the signals are demodulated (in the marker-adder or generator) for presentation on the scope vertical input.

3. Connect the equipment as shown in Fig. 2.7 for direct-injection alignment. With this arrangement, both the sweep and marker generator outputs are connected directly to the RF tuner input (at the antenna terminals). The output to the scope can be taken from the video detector (for an overall response curve) or at the tuner looker point.

Note that if the video-detector output is chosen, a low-capacitance probe can be used with the scope (since the video detector demodulates both the sweep and marker signals). However, if you monitor the output at the tuner looker point (to get a separate tuner response curve), you must use a demodulator probe because the tuner output is usually at a frequency beyond the bandpass of the scope.

4. With either alignment method, it is necessary to disable the AGC line. Figures 2.6 and 2.7 assume that bias voltages to disable the AGC circuits are available from the sweep and marker generator (which is usually the case with present-day generators). Always use the recommended bias voltage (amplitude, polarity, etc.) and apply the bias at the recommended test point. In some sets, there is a common AGC line for both tuner and IF amplifiers. In other sets, the tuner bias is taken from a separate line.

5. With the test connections made as shown in Figs. 2.6 and 2.7, adjust the sweep and marker generator to the appropriate channel. Set the sweep width to maximum or as recommended in service literature. Typically, sweep width is 10 to 15 MHz. Adjust the scope controls for a trace similar to that shown in Fig. 2.6 (or as shown in literature).

6. The test setup is now complete and ready for test and/or alignment as described in the service literature. Typical alignment procedures are described in Chap. 3.

2.7.2   IF Amplifier and Trap Alignment

As in the case of the tuner, the primary purpose of IF-amplifier alignment is to get a response curve of proper shape, frequency coverage, and gain. The purpose of trap alignment is to remove undesired signals from the IF circuits. Again, most IF amplifiers require only touch-up alignment. Of course this may not be true if there has been extensive work in the IF circuits.

If a TV set is to give wideband amplification to the television signal, the picture (video) IF system of the set must pass a frequency band of about 3.5 to 4 MHz. This is necessary to ensure that all video information is fed through to the picture tube and that the resultant picture has full definition. (The bandpass of color sets must be essentially flat to beyond 4 MHz to ensure that the color information contained in the color sidebands is not lost. Likewise, the bandwidth of S-VHS TV sets must be even greater, as discussed in Chap. 5.)

The desired bandpass is obtained by proper alignment of the IF adjustment controls (typically coil tuning slugs but possibly trimmer capacitors), with sweep and marker signals fed to the IF amplifiers so that a response curve (with markers) is produced on the scope screen. The basic sweep and marker generator techniques described in Sec. 2.7.1 are used. However, input to the IF amplifiers is at the RF-tuner looker point or at a separate IF input cable (in some sets). Alignment is done by setting adjustments so that the waveshape on the scope resembles the waveshape shown in service literature. Typical alignment procedures are described in Chap. 3.

IF Alignment Notes.   The following notes describe typical alignment problems and techniques. These should be compared with any notes found in literature and with example procedures described in Chap. 3.

Figure 2.6*e* is a typical IF amplifier waveshape obtained with a sweep and marker generator setup. Note that the frequency relationship of the sound carrier to the picture carrier is sometimes reversed in the IF amplifiers because the tuner oscillator usually operates at a frequency higher than that of the transmitted carrier.

For example, assume that (for Channel 9) a picture carrier is transmitted at 211.25

MHz and that the corresponding sound carrier is transmitted at 215.75 MHz. Further assume that the IF frequency is 26.75 MHz (typical for older black and white sets). If the tuner oscillator is at 238 MHz, this signal combines with the picture carrier of 211.25 MHz to produce a difference-IF signal of 26.75 MHz. However, the sound carrier of 215.75 MHz also combines with the 238-MHz oscillator signal to produce a difference-IF signal of 22.25 MHz, which is 4.5 MHz *below* the picture IF.

Some sweep and marker generators compensate for this condition by sweeping from high to low frequencies. However, always consult the service literature for proper response curve-shape marker frequencies and check the generator literature for the method of producing the display.

No matter how the display is obtained, the following two characteristics of the IF response curve (Fig. 2.6e) should be noted: (1) the amplitude of the picture carrier is set at about 50 percent of the maximum response, and (2) the sound-carrier amplitude must be at 1 percent (or less) of the maximum response.

The *sound carrier* is kept at this low level to prevent interference with the video signal. The *skirt selectivity* (or selectivity at the high and low extremes of the approximate 6-MHz bandpass) of the IF response is made sharp enough to reject the sound part of the composite signal. In some sets, an absorption circuit (consisting of a trap tuned to the sound IF) is used to get the necessary selectivity. In other sets, additional traps are tuned to frequencies of *adjacent sound and picture channels*. All of these traps have a pronounced effect on the shape of the response curve.

The *picture carrier* is placed at about 50 percent of maximum response because of the nature of the TV broadcast transmission. If the IF circuits are adjusted to put the carrier too high on the response curve, the effect is a general decrease in picture quality caused by the resulting low-frequency attenuation. If the picture carrier is placed too low on the curve, there is a loss of low-frequency video response (that usually shows up as *poor definition* in the picture). *Loss of blanking and poor sync* can also occur when the picture carrier is too low on the response curve.

Note that in addition to the picture and sound frequencies, two additional marker frequencies are shown on the curve of Fig. 2.6e (23.5 and 28.25 MHz). These marker frequencies are used to align the IF traps. Most IF amplifiers have one or more traps, depending on the type of set. The setup for alignment of the IF traps is essentially the same as for alignment of the IF stages. However, certain problems may occur.

Because the IF amplifier response is very low at the trap frequencies (when properly adjusted), the marker may be difficult to see on the response curve. In some cases, traps are set with a meter rather than a scope. The meter is connected to measure the signal at the video-detector load resistor (IF output), the marker generator is set for the trap frequency, and the trap is tuned for a minimum voltage reading on the meter.

Typically, the traps are set first, then the IF amplifier circuits are aligned. Since any adjustment of the circuits usually detunes the traps (slightly), the traps may require touch-up. Always follow the exact procedures given in service literature.

2.7.3 Alignment of 4.5-MHz Sound Traps

The video-amplifier circuits of all TV sets (black and white as well as color) have a 4.5-MHz sound trap. This trap is tuned to reject any sound signals (at 4.5 MHz) that may pass through the video amplifier (so as to prevent the signals from appearing on the picture tube). Adjustment of the sound trap is usually not critical for black and white sets. However, the problem of *sound in picture* for a color set may be quite critical. For this reason, the problem of sound-trap adjustment is discussed fully in Chap. 3.

For a typical black and white set, the sound trap can be adjusted simply, without test equipment, once the RF and IF stages are aligned. To adjust the trap, tune in a strong TV broadcast signal, and set the contrast control at maximum. Adjust the fine-tuning control until a beat pattern (herringbone) is visible on the picture-tube screen. Then adjust the sound trap (usually a coil tuning slug) for *minimum beat interference.* Reset the fine tuning, making sure that the beat interference is completely gone when the fine tuning is set back to normal (for best picture and sound, usually at midrange). Do not try this on a color set or on any set with frequency-synthesis tuning.

2.7.4 Checking Response of Individual Stages

Although it is not usually necessary to check individual IF stages in present-day IC video equipment, the need may arise in certain circuits. The response of individual stages, or of two or more stages together, may be checked using a sweep-marker generator combination. Basically, the sweep signal is fed into the stage *immediately preceding* the stage being checked. The response curve is then checked on a scope connected across the video-detector load resistor. As in the case of other IF circuit tests, the AGC line must be disabled for accurate results.

1. Connect the equipment as shown in Fig. 2.8. Disable the AGC line using the bias recommended in service literature.
2. Connect the marker generator (or the marker output of a combined sweep-marker generator) to the RF tuner looker point.
3. Connect the sweep generator to the input of the IF amplifier stage ahead of the stage being checked. This isolates the test equipment from the stage being checked. Note that the sweep-generator output should not be connected to the input of the stage being checked because even a slight loading of the input circuit may cause a change in circuit impedance and result in distortion of the normal response characteristics. The test connections in Fig. 2.8 (sweep generator at test point B) are to check response of the third IF amplifier.

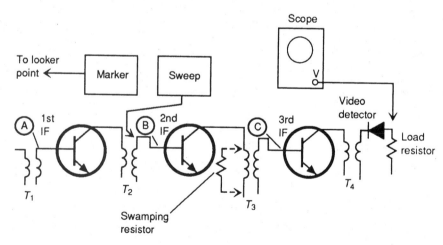

FIGURE 2.8 Checking response of individual stages.

4. Connect a resistor (typically 500 to 1000 Ω) across the primary of the transformer ahead of the stage being checked (primary of T_3). The resistor acts to "swamp" the primary winding and prevents inductive reactance of the winding from affecting the bandpass characteristics of the stage being checked (third IF). With the connections in Fig. 2.8, the response curve is determined primarily by the third-IF bandpass characteristics but is also affected by the video-detector response.

5. To check the bandpass characteristics of the video detector only, move the generator from point B to point C, and place the swamping resistor across the primary of T_4.

6. To check response of the second IF, third IF, and video-detector stages together, move the generator to point A, and connect the swamping resistor across the primary of T_2.

2.7.5 Checking Overall Video Sweep Response

It is possible to check the overall response of a TV set (and a VCR) from the RF tuner to the picture tube using sweep techniques. Such techniques permit observation of the true overall frequency response, including the effect of the video-detector load circuit. When RF, IF, and video circuits are checked separately, the effect of the video-detector load may change, and this change cannot be observed (except by an overall check). Overall response is especially important in the alignment of color sets.

The use of sweep techniques for overall response checks has generally been replaced by the use of analyst or NTSC generators. As described in Chap. 3, a typical NTSC generator produces RF carrier outputs on TV channels (typically Channels 3 and 4), as well as IF outputs (typically at 45.75 MHz). The carriers can be modulated by a number of patterns (color bars, staircase, rasters, etc.) that appear on the picture-tube screen. Thus, the responses of all circuits from the antenna to the picture tube are checked simultaneously. NTSC generators are described throughout Chap. 3. The use of such generators for overall checks is described in Sec. 2.9.

2.8 TROUBLESHOOTING BASIC VIDEO CIRCUITS WITH SIGNAL INJECTION VERSUS SIGNAL TRACING

As discussed in Chap. 1, there are two basic methods for troubleshooting video circuits: signal tracing and signal injection. For example, in troubleshooting TV and VCR video circuits with true signal tracing, you must rely on the signals broadcast by the TV station and on the signals developed by the TV set in response to broadcast signals.

In simple terms, you trace through all circuits of the set, checking that the waveforms, pulses, signals, etc., are as they appear in service literature. You compare the scope patterns obtained from the set with the standard patterns (found in literature) for amplitude, shape, frequency, etc. An absent or abnormal waveform indicates problems in the circuits being measured.

With true signal injection, you must have an analyst or NTSC generator to duplicate the broadcast signals, as well as the signals produced by the set. All analyst and NTSC generators are not the same, so it is difficult to generalize about the basic signal-injection technique. However, in the simplest of terms, you inject precise signals into all sections of the TV set and note the pattern produced on the picture-tube screen.

We do not attempt to promote either signal tracing or signal injection as the better method. This controversy has gone on for many years and eventually boils down to a matter of choice. However, it is generally conceded by most TV technicians that signal injection is the quickest (and thus the most profitable). For that reason, we devote most of this chapter to signal injection using analyst and NTSC generators. In Sec. 2.9, we describe the use of analyst and NTSC generators without separate sync signals. In Sec. 2.10, we cover troubleshooting with more advanced generators.

2.9 BASIC TROUBLESHOOTING WITH AN ANALYST OR NTSC GENERATOR

This section is devoted to basic troubleshooting with an analyst or NTSC generator. With such a generator, you can inject test patterns (lines, dots, crosshatch, pulses, and color signals) at RF, IF, and video frequencies in stage-by-stage troubleshooting techniques. You apply the signals at appropriate test points throughout the circuits and observe the results on the picture-tube screen.

As a supplemental troubleshooting procedure, you can use a scope to view the test-pattern waveform as the generator signal is processed through the set. With an analyst or NTSC generator, you have the advantage of being able to inject the signal directly to the stage being checked.

Any of the test patterns can be used in this procedure. However, *pulse and/or staircase waveforms* are especially useful when checking IF and video stages. The pulse is a simple, sharply defined trace. Thus, distortion or loss of gain in the waveform is readily apparent. The staircase pattern is valuable for checking linearity in any video amplifier. (Nonequal steps monitored at the output of a video circuit represent nonlinear distortion since all of the staircase steps are equal at the input.) Staircase patterns are discussed further in Chap. 3.

The method described in this section starts by connecting an RF signal from the generator to the antenna terminals to evaluate the overall performance of the set and to note the particular picture problem (no picture, smeary picture, etc.). Then you inject signals back to the video and chroma circuits (in color sets), working forward through the IF stages until the picture problem appears. In this way, you isolate the defective circuit.

Of course, the defective stage can be found just as easily by starting at the mixer input and working back through the set until a normal picture is found. The technique you use is a matter of individual preference (and the capabilities of the generator).

2.9.1 Troubleshooting Modular TVs and Monitors

As discussed in other chapters, present-day sets with various stages (or parts of stages) on plug-in modules are serviced in a manner different from discrete-component sets. Many times the trouble can be corrected quickly and easily by replacing the defective module. An analyst or NTSC generator can be of great help to you in determining what stage or module is defective. The generator is especially useful in isolating a problem that is not on a plug-in module but is in the master board that the modules plug into.

The troubleshooting procedures with signal injection and signal tracing on modular sets are similar to those used with any other set. The basic difference is that to isolate a defective module, the test points specified on the master board (or on the module socket), representing the input and output points of the modules, are particularly important. On some modules, the components are accessible and you can use standard troubleshooting techniques to locate defective components. Other modules are sealed, however, and cannot be replaced.

2.9.2 Checking Overall TV and Monitor Performance

1. Connect the analyst or NTSC generator to the antenna terminals as shown in Fig. 2.9*a* to check overall performance of the TV or monitor. Set the tuner and generator to the same channel (3 or 4). Apply RF with a pattern to the set.

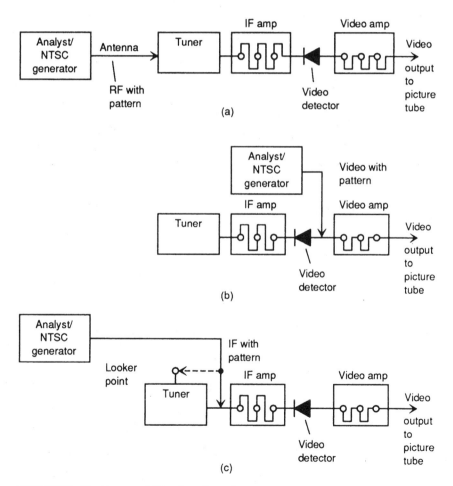

FIGURE 2.9   Checking overall TV and monitor performance.

2. Adjust the set fine tuning to get the best pattern. Adjust the set brightness, contrast, and other controls as necessary.

3. If the set is operating normally and has had proper setup adjustments (AGC, horizontal and vertical oscillators, focus, color killer, etc.), the pattern displayed on the picture tube should be good. If not, proceed to Sec. 2.9.5 (video amplifier check).

2.9.3 Checking AGC

If the generator has a variable RF output, the overall AGC function can be checked using the same connections shown in Fig. 2.9a. Connect a meter to the AGC test point at the tuner (or at the IF stages), and vary the RF output of the generator. If the AGC system is operating, the voltage at the test point will vary as the generator RF-level control is adjusted.

If the generator does not have a variable RF output, the AGC circuits can be checked by monitoring the AGC line voltage while applying and removing the RF signal. There should be an abrupt change in AGC voltage when the RF is applied and removed.

2.9.4 Checking Sync Lock-In Range

If the generator has a variable sync, you can determine how well the set holds sync (horizontal and vertical) over a wide range of sync levels using the connections shown in Fig. 2.9a. In this way, you can determine if the set sync is "touchy" (overly sensitive) to sync-level variations.

2.9.5 Checking Video Amplifier

Connect the generator to the video amplifier input as shown in Fig. 2.9b. Do not reconnect the tuner to an antenna. Apply video with a pattern to the set.

1. If the generator has a variable video output or a video output where the polarity can be reversed, adjust the video control as necessary. For example, if the video control is set to the wrong polarity, the picture will be dark and there is no sync. Turn the control in the proper direction until the picture begins to "tear," and then back off the control just to the point where the best picture is obtained.

2. If the picture is about normal, the video amplifier (from video detector to picture tube) is good. Proceed to a check of the IF and tuner circuits as described in Sec. 2.9.7. However, if the picture has the original problem discovered during the overall response check (Sec. 2.9.2), you have isolated the trouble as being in the video amplifier.

3. You can further isolate the defective circuit within the video amplifier by additional signal injection into test points at each video stage (input and output of each video amplifier in a discrete-component set). As the signal is applied to various test points in the video amplifier, it may be necessary to readjust the video-level control (one of the advantages of variable video). In most sets, the video-sync polarity is inverted (reversed) at each stage of the amplifier. (You can also inject the signal at the video detector and signal trace through the video-amplifier circuits using a scope, as described in Sec. 2.9.8.)

2.9.6 Checking Chroma Amplifier

A rough check of the chroma amplifier in many color sets can be made by injecting a color-bar pattern at the video frequency into the chroma-amplifier input (video-amplifier output). The sync may not be too stable when the signal is injected at that point because you are beyond the sync takeoff point. However, the pattern color should be fairly good.

If you find trouble in the chroma amplifier, you can trace the color-bar signal using a scope. Refer to Chap. 3 for detailed descriptions of color troubleshooting procedures.

2.9.7 Checking IF Amplifier and Tuner

1. Connect the generator to the RF-tuner looker point or to a test point at the input of the IF amplifier as shown in Fig. 2.9c. On some sets, you can unplug the cable connecting the tuner to the IF amplifier and apply the generator signal to the cable. Apply IF (typically 45.75 MHz) with a pattern to the set.

2. Adjust the generator level control (if any) for the best picture. Observe various test patterns, including color bars. If you get normal patterns, and the problem noted in Sec. 2.9.2 (overall response) is not apparent, the IF-amplifier stages are operating properly, and you have isolated the problem to the tuner. However, if the picture still has the original problem, you have isolated the trouble to the IF amplifier (or possibly the video-detector diode).

3. You can further isolate the defective area by additional signal injection into test points at each IF stage (input and output of each IF amplifier in a discrete-component set). (You can also inject the signal at the IF input and signal trace through the IF-amplifier circuits using a scope, as described in Sec. 2.9.8.)

2.9.8 Signal Tracing with an Analyst or NTSC Generator

When an analyst or NTSC generator is used in signal tracing, the generator serves as the signal source (in place of TV broadcast signals). The basic procedure consists of applying a test signal at the antenna or looker point and then using a scope to view the pattern as the signal is processed through the IF, video, and chroma stages. When you reach a test point where the waveform has disappeared, is badly distorted, or does not have sufficient gain, you have localized the trouble.

The scope must have a low-capacitance probe (for the signal tracing of circuits after the video detector) and a demodulator probe (for signal tracing ahead of the video detector). Before you start, connect the scope probes directly to the generator output. In this way, you know what pattern is produced by each of the probes before the signals are processed.

It is usually recommended that the generator be set to produce a pulse or staircase pattern when tracing through the IF and video circuits (or any amplifier). If you are signal tracing in the chroma-amplifier circuit, a color-bar pattern must be used (to be of any real value).

When signal tracing a set with a bad picture, inject the signal at the looker point or the input to the IF. You could connect the signal to the antenna terminals. However, if the picture is really poor, it is difficult to get correct fine tuning, and this can result in misleading waveforms. Although you can inject the signal at other test points within the IF amplifier or video stages (as described for signal injection), the scope waveform may have some distortion because of circuit loading (or "hash" because of pickup through the test lead).

If the generator has a variable sync control, such a provision can be extremely useful in signal tracing a set with poor sync or a complete loss of sync. You can evaluate performance of the circuit by observing the sync pulse on the waveform at various test points and noting how the sync pulse appears at different sync levels.

Some service literature recommends that the horizontal circuits be disabled during testing and troubleshooting. This is done so that the horizontal signals do not interfere with test signals. When this recommendation is followed, it is necessary to supply a bias voltage to the AGC line (since most sets use some form of keyed AGC that requires signals from the horizontal circuits). If the AGC is not operating, the waveforms obtained during signal tracing may be distorted. Always follow the service literature recommendations concerning AGC bias.

2.10 ADVANCED TROUBLESHOOTING (SIGNAL SUBSTITUTION) WITH AN ANALYST OR NTSC GENERATOR

With an analyst or NTSC generator, you can inject black and white test patterns and/or color bars at RF, IF, and video frequencies for stage-by-stage (or IC-by-IC) troubleshooting. Additional outputs are available for injection into the sound-IF (SIF) and audio circuits of the set. With a true analyst generator, you can also inject separate sync signals as well as signals for test of the vertical and horizontal sweep circuits and yoke deflection coils.

As a supplemental troubleshooting procedure, you can use a scope to view the pattern waveforms as the signals are processed through the set. However, the primary advantage of a full analyst generator is signal injection, where the picture tube (and loudspeaker) are used to troubleshoot the circuits.

The signal-injection technique is also known as signal substitution. In effect, the generator signals are substituted for the missing (or abnormal) signal in the set and are used to restore operation. The signal is usually injected nearest the picture tube first and then moved (one stage or one IC at a time) toward the antenna until signal substitution does not restore operation. Note that this procedure for locating the defective stage is opposite to that described in Sec. 2.9, where the signal is initially injected at the front end (tuner or antenna).

Figure 2.10 shows a block diagram of a typical black and white TV set, illustrating 23 typical signal-injection points that may be used for locating a defective stage. Note that Fig. 2.10 shows all of the stages as individual circuits. This is done to illustrate the functional relationships of the stages along the signal path.

In present-day sets, many of the individual stage functions are combined into ICs or other modules. For example, the RF amplifiers, mixers, and oscillators are combined into a tuner package, while the IF stages and video detector are combined into a single IC. However, the basic signal-substitution technique remains the same (but with fewer test points). The service literature for most TV sets shows the input-output test points for all packages, ICs, and modules in the signal paths.

No matter what circuit configuration is used, to successfully use the generator, you need only know which signal to inject at each stage and the approximate amplitude of the signal. The steps in the remaining paragraphs of this section give you the specific

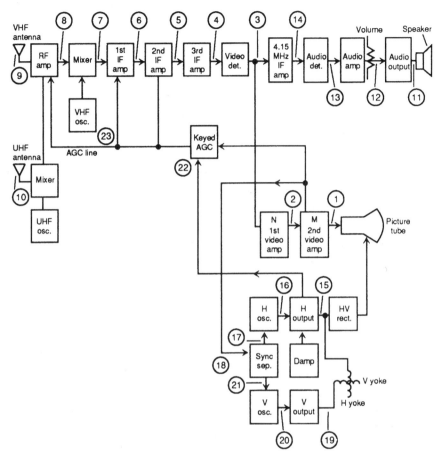

FIGURE 2.10 Signal-injection test points in basic TV set.

information for such signal injection or substitution. During most signal-substitution procedures, it is preferable to use as many outputs from the generator (simultaneously) as necessary to restore operation.

2.10.1 Basic Signal-Substitution Techniques

The basic procedure for troubleshooting by signal substitution may be summarized as follows:

1. First you analyze the symptoms to determine the group of stages (or the IC or module) that should be checked. The relationship between trouble symptoms and areas of the set where the trouble is most likely to occur is essentially the same as described in Chap. 1 (and at the end of this chapter). However, the sequence used to isolate trouble with signal substitution differs from that when signal tracing is

used. Thus, the symptom analysis described in this section is not exactly the same as that covered in Chap. 1 and later sections in this chapter.

2. Next, you inject a signal of the proper type into the suspected stage or IC or module *farthest* from the antenna.

3. If proper operation is restored, you inject a signal into the next stage nearer the antenna. You must inject the proper type and level of signal to simulate normal operating conditions in each stage. You continue this stage-by-stage (or IC-by-IC) injection until you find a point where signal injection does not restore proper operation. You have now located the defective stage or IC.

4. Next, you inject a signal at each part *in series with the signal path* (such as coupling capacitors and transformers) until you have isolated the trouble to as small an area as possible.

5. Finally, you use voltage and resistance checks to locate the specific trouble in the area isolated by the signal-injection process.

6. When you become more familiar with your particular generator, you may wish to skip stages and check a complete section at a time. On sets with replaceable modules, it may be advantageous to check from module to module (output of one module to input of the next module).

Example of Signal Substitution with an Analyst.   Assume that you are servicing a set with a block diagram similar to Fig. 2.10.

1. The symptoms are no video, audio normal, and raster normal. These symptoms tell you that the horizontal and vertical sections are operating because the raster is present. Since sound is normal, you can assume that all stages from the antenna to the video detector are operating. The stages that require checking are the picture tube, second video amplifier, and first video amplifier.

2. You inject a *high-amplitude sync signal* directly into the picture tube (point 1) and find that video bars are displayed on the picture tube screen. This proves that the picture tube is good (or at least operating).

3. You inject a *maximum video signal* at the input to the second video amplifier (point 2) and find that a test pattern is displayed on the screen. This proves that the second video amplifier is good.

4. You then reduce the video-signal amplitude until the picture is barely visible. (The next step injects the signal at the first video amplifier. If the first video stage amplifies normally, it should provide a much stronger picture when the signal is applied.)

5. You inject a low-level video signal at the input of the first video amplifier (point 3) and find that the test pattern disappears completely. You have found the defective stage. The trouble lies between the input of the first video amplifier and the input of the second video amplifier.

6. Next, you refer to a schematic of the defective stage (Fig. 2.11). You inject a *maximum video signal* at point 2A and find that the test pattern is again displayed. This proves that the coupling capacitor C_5 is good.

7. You inject a *maximum video signal* at the output of the first video amplifier (point 2B) and find that the test pattern is still displayed. This means that Q_1 is not providing an output signal.

8. Finally, you make voltage measurements and find that the voltages at Q_1 are not

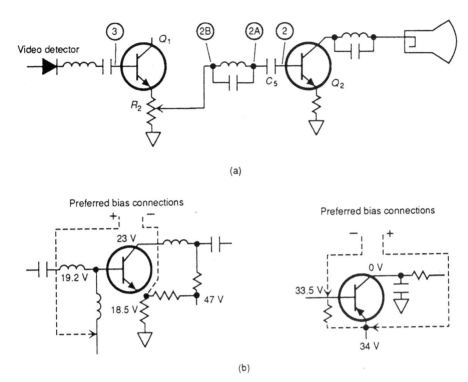

(a)

(b)

FIGURE 2.11 Example of signal substitution with an analyst or NTSC generator.

correct. For example, assume that R_2 is not properly grounded because of a bad PC board connection (not properly soldered).

2.10.2 Troubleshooting Video Transistor Circuits with Signal Substitution

Here are some notes for troubleshooting transistor circuits in video equipment when signal substitution is used:

Bias: When applying a bias voltage (as is sometimes required during signal injection), always apply bias to provide forward bias across the base-emitter junction (base positive in respect to emitter in NPN and negative in PNP).

Bias reference point: Be very careful of the bias reference point used in signal injection troubleshooting. Usually, it is preferable to use the emitter of the transistor as reference, rather than PC ground. In both examples in Fig. 2.11*b*, PC ground cannot be used as reference. If used, even a low or zero bias places a substantial difference of potential (18.5 V in one case and 34 V in the other) across the base-emitter junction.

Excessive injection voltages: Too much voltage can burn out a transistor, especially when applied across the base-emitter junction. If variable, keep injection voltages low unless you have checked the service literature and are absolutely sure that higher levels are normally used in the circuit.

Sync signals: When injecting sync signals, keep the sync control at zero when not in use. Then, if you happen to make a circuit connection before you adjust the level, the connection will not burn out a transistor. After injecting the sync signal, make sure the amplitude is reduced to a safe level before changing the point of injection. A safe level in one circuit may not be safe in another circuit. When possible, inject signals at a point that routes the sync through a series resistor into the transistor base. The resistor helps absorb the sync signal and protects the transistor.

2.10.3 Troubleshooting Tuned Amplifiers with Signal Substitution

Tuned amplifiers are stages which amplify signals within a specific frequency band and block signals of other frequencies. The stages are usually coupled. For example, tuned stages within a TV set are the RF, IF, 4.5-MHz sound IF, and color IF amplifiers.

The following steps describe the general approach for troubleshooting tuned amplifiers with signal substitution:

1. Inject a signal of the correct frequency (RF, IF, color, etc.). Typically, analyst or NTSC generators provide a variety of signals at frequencies commonly found in TV sets. Usually, some of the signals are variable, while others are fixed (in amplitude).

2. Inject the signal at the *input of the suspected stage that is farthest from the antenna.* The input is preferred because the low-impedance signal source of the generator is less likely to load the tuned circuit (which may affect frequency response). If no display appears on the picture-tube screen (or loudspeaker in the case of a sound signal), you have probably located the defective stage. If you get a good display (test pattern, color, or sound), reduce the amplitude of the injected signal until the display is barely visible.

3. Move the injection point to the input of the next stage nearer the antenna. If that stage is operating properly, the display will be much brighter (because of stage amplification). This method not only locates inoperative stages but weak stages as well.

4. Reduce the signal amplification again to the point where the display is barely visible. Note the *difference in setting of the amplitude adjustment.* This difference is a relative indication of stage gain. After checking a few TV sets, you will soon learn the normal gain to be expected at various stages (at least if you service similar sets). This helps you to spot weak stages.

5. Continue to move the signal-injection point toward the antenna, one stage at a time, until no display or a poor display is produced. This should occur when you locate the defective stage. Now you must locate the defective part within the stage.

6. Start by checking the transformers. Inject the signal at the *secondary and primary* of the stage transformers. If the display is normal with a signal injected at the secondary but not at the primary, the transformer is probably defective. If the transformer is good, proceed with voltage and resistance checks.

2.10.4 Troubleshooting Untuned Amplifiers with Signal Injection

The following steps describe the general approach for troubleshooting untuned amplifiers with signal substitution.

Untuned amplifiers differ from tuned amplifiers in that untuned amplifiers amplify a *wide band of frequencies.* Resistive-capacitive coupling is normally used between untuned amplifier stages, which in a TV set include audio and video amplifiers.

Troubleshooting of untuned amplifiers is performed by injecting a signal at the *output of the stage farthest from the antenna* to produce normal operation. For video signals, the signal amplitude is decreased until the display is barely visible; then the injection point is *changed to the input of the stage.* If the stage is operative, the amplification should produce a brighter display.

Continue moving the injection point toward the antenna until no output is obtained. This is the defective stage. Proceed with voltage and resistance checks to locate the defective part.

2.10.5   Troubleshooting Pulse Stages with Signal Substitution

The following steps describe the general approach for troubleshooting pulse amplifiers with signal substitution.

Pulse circuits have a short duration of operation compared to "off time" and are thus often difficult to troubleshoot. Pulse circuits in a TV set include sync separators and amplifiers, keyed AGC circuits (burst amplifier in color sets), and the horizontal and vertical sweep circuits.

In any pulse circuit, waveshape is important and often critical. Not only must the pulses be present at a given point in the circuit but the pulses must be of the correct amplitude, duration, and frequency (as well as shape). An analyst or NTSC generator can provide an advantage over signal tracing with a scope when troubleshooting pulse circuits. Of course, the generator must produce pulses of the correct type.

For example, with the proper pulses, you can substitute the type of signal actually used in the stage during normal operation and then check the display. (If normal operation is restored when you apply a pulse of the correct type to some point in the circuit, you know that the stages between that point and the picture tube are good.) As in the case of tuned and untuned amplifiers, you inject signals at every point in the series path of the pulse signal until the point is located at which the display is abnormal.

One special precaution when injecting pulse signals is that the *correct polarity* should be observed. Some stages require a positive pulse, while others require a negative one. Also, *amplifiers usually invert the signal*; thus the polarity must be reversed as the injection point is changed from one stage to the next or from the output to the input of a stage. Analyst or NTSC generators often provide video and sync-pulse signals with reversible polarities.

2.10.6   Analyzing Trouble Symptoms with Signal Substitution

As discussed in Sec. 2.10.1, symptom analysis described in this section is not identical to that covered in Chap. 1, and later in this chapter since the sequence used to isolate trouble with signal substitution differs from that when signal tracing is used. Observe the following steps before using the notes in this section to isolate trouble with signal substitution:

1. Always start troubleshooting by injecting a VHF test signal at the antenna terminals of the TV set and checking all symptoms.

2. Use the test pattern from the generator, not a picture from the TV station. The picture broadcast by a TV transmitter changes continuously (unless the picture is a color bar or other pattern), but the pattern from a generator should be reliable. If the display does not match the generator pattern, this condition is easily detected. On the other hand, small irregularities are not seen easily with a transmitted picture.

3. Analyze the symptoms thoroughly. A careful check of symptoms directs you to specific stages that could cause the trouble. Do not rely on another person's descriptions of the symptoms. You may gain additional information that saves you time by making your own observations. Check both video and sound, and if it is a color set, check color reception (of a color-bar pattern).

4. Learn to analyze the generator patterns. Chapter 3 includes discussions of how generator patterns can be used to evaluate trouble symptoms. Remember that a test pattern shows much more than merely whether the TV set is operating or not. When properly interpreted, the pattern can provide a complete check of overall performance and show whether adjustments are necessary.

No Raster, No Sound. The most probable cause of this symptom is the absence of one or more (or all) low voltages (assuming that the set is plugged in, turned on, and a fuse is not blown). Refer to Sec. 2.17.

No Raster, Sound Normal. This symptom appears whenever the picture tube does not have the proper electron beam to illuminate the screen. This could be caused by a bad picture tube or absence of voltage on the cathode or grids of the picture tube, but the symptom is more often caused by the absence of high voltage on the picture-tube anode. In turn, high voltage depends on the horizontal sweep signal. There are quite a number of stages in which the trouble could be located. Refer to Sec. 2.19, Troubleshooting Vertical and Horizontal Sweep Circuits.

No Vertical Deflection. This symptom is usually caused by a defective vertical oscillator, vertical output stage, vertical output transformer and associated parts, or vertical yoke. Refer to Sec. 2.19.

No Video, Sound Normal. This symptom is normally caused by a defective video amplifier or a bad picture tube. Because the sound and video separation point varies from one set to another, it is also possible that the video detector stage is at fault. In other sets, only the final video amplifier can produce the symptoms. The following signal-injection procedures should locate the defect no matter where the sound and video separation occurs:

1. Inject a maximum video test signal into the input of the final video amplifier (2, Fig. 2.10). If no video is produced on the picture tube, the trouble is between the input of the final video amplifier and the picture tube. Go to step 2. If video is produced (the test pattern appears on the picture tube, probably out of sync), the trouble is before the final video amplifier. Go to step 4.

2. Inject a maximum sync signal directly into the picture tube (1, Fig. 2.10). If no diagonal bars are seen, the picture tube is defective or the circuit is shorted. Check for shorts. Note that the sync signal from most generators produces diagonal bars, but not a test pattern. The video signal from the generator produces a test pattern, as do the RF and IF signals. If diagonal bars are produced on the screen, the picture tube is capable of displaying video. Go to step 3.

3. Inject maximum sync signals at each series part from the picture tube to the output of the final video amplifier. If diagonal bars are not displayed for all injection

points, check at which part the display is lost. If diagonal bars are still displayed with a sync signal injected at the final video-stage output, but not at the input, the stage is suspect. Check all related voltages and resistance.

4. Inject a maximum video signal at the output of the preceding video-amplifier stages. If no video (test pattern) is displayed on the picture tube, check the coupling parts between the first and final video-amplifier stages (Fig. 2.11*a*). If video is displayed, go to step 5.

5. Inject a medium-amplitude video signal at the input of the first video-amplifier stage. If no video is displayed on the picture tube, the first video amplifier is suspect. Check all related voltages and resistances. If video is displayed, sound and video separation occurs ahead of the video detector. Follow the procedure outlined in the no-video, no-sound trouble symptoms (next).

No Video, No Sound. This symptom is caused by a malfunction between the antenna terminals and the point where the video and sound are separated. This usually includes the tuner, IF amplifiers, and video detector. The following signal-injection procedures should locate the defect no matter where the sound and video separation occurs:

1. Inject an IF test signal at the input of the third IF amplifier or the last IF amplifier (5, Fig. 2.10). Of course, if the IF-amplifier circuits are contained within an IC, inject the signal at the IF input of the IC. If no test pattern is displayed or the pattern is unusually weak, the IF stage (or IC) and detector are suspect. Check all related voltages and resistances.

If the test pattern is displayed properly, the trouble is toward the antenna. Go to step 2.

2. Move the IF test signal to the input of the second IF amplifier or to the stage ahead of the final IF amplifier (6, Fig. 2.10). Measure the relative gain of the amplifier (Sec. 2.10.3, step 4). If no test pattern is displayed or gain is unusually low, the second IF amplifier is suspect. Inject the IF signal at each part in series with the signal path to locate the point where signal is lost. Check all related voltages and resistances.

If the test pattern is displayed properly, the stage is good. Go to step 3.

3. Repeat the procedures of step 2, but this time inject the signal at the input of the first IF amplifier (7, Fig. 2.10). If the test pattern is displayed properly, the stage is good. Go to step 4.

4. If the test pattern is still displayed with the IF signal injected at the first IF amplifier or at the input to an IC IF circuit, the RF tuner is suspect. If the RF tuner is not a sealed package and there are replaceable parts, the mixer, oscillator, and RF amplifier can be checked individually. (This is not usually the case in present-day sets.) The following steps describe the procedures for checking individual stages in the RF tuner.

To test the mixer tuner, inject an IF signal into the mixer input (8, Fig. 2.10). If the mixer is operating, the signal is passed, and the test pattern is displayed. If the mixer is defective, no test pattern is seen. However, in some circuits, the mixer may pass an IF signal even if the mixer is defective.

5. To test the oscillator, inject an RF test signal at the mixer input (8, Fig. 2.10). If the oscillator is operating and on the proper frequency, the injected signal and the oscillator signal mix to produce the IF signal, and the test pattern is displayed. If the oscillator or mixer are defective, no test pattern is seen.

6. If the mixer and oscillator are good, the trouble is in the RF amplifier. Inject an RF test signal at various points in the circuit to localize the malfunction. Always inject an RF signal on the same channel to which the set is tuned. Try several channels if the generator is capable of producing RF test pattern signals on more than one channel. Typically, an analyst or NTSC generator produces RF signals on at least two channels.

7. If there is no video and no sound on UHF channels only, the trouble is in the UHF tuner or the UHF portion of the tuner package.

No Vertical Sync.   With a loss of vertical sync, the picture rolls no matter where the vertical hold (if any) is set. This symptom is caused by the absence of vertical sync pulses at the vertical oscillator. Since the picture has horizontal sync, the trouble should be only in a section of the sync-separator stage or in a vertical sync amplifier stage (if used).

Inject sync signals at points 18 through 21 of Fig. 2.10 or as described in Sec. 2.19. The required signal may be positive or negative polarity; try both. Remember to reverse polarity when changing the injection point from the output to the input of a stage.

Inject signals of sufficient amplitude to simulate normal operating values (refer to the schematic for normal peak-to-peak signal values). If too much signal is required to restore proper operation, check for a part that has changed value drastically. Also check the gain of the sync-separator stage or stages. The separator may provide sufficient signal to produce horizontal but not vertical sync. Check relative gain by injecting the minimum sync signal that provides vertical sync at the output and input of a stage and note the difference in generator sync-control setting.

No Horizontal Sync.   With this trouble, diagonal lines appear on the screen, but a picture cannot be locked in with the horizontal hold control (if any). Since no picture is visible, the presence or absence of vertical sync cannot be easily detected. However, this is of little consequence, since the following procedure locates the trouble source in either case.

The symptom could be caused by a defective sync-separator stage or the AFC circuit which precedes the horizontal oscillator.

1. Inject a sync signal at the sync input of the horizontal oscillator (17, Fig. 2.10) or as described in Sec. 2.19. The required signal may have positive or negative polarity; try both. Refer to the schematic and adjust the sync control of a typical level of sync-pulse signal. Horizontal sync should be restored. Readjust the horizontal hold control (if any) as required.

2. Inject a sync signal into the horizontal phase detector. If sync is not restored, the phase detector is defective. If normal operation is restored, go to step 3.

3. Inject a sync signal at the output and input of the sync-separator stage. If the stage is operational, check stage gain. If no sync is obtained, the sync separator is defective. If sync is restored, check coupling parts of the sync-separator stage.

Overloaded Video.   With this problem, the picture appears negative and out of sync at both normal and high signal levels (injected at the antenna terminals). When the antenna signal is decreased, the normal test pattern appears (but with considerable snow). This symptom is usually caused by an RF, IF, or video amplifier that is overdriven, which results in clipping of the signal at normal and high signal levels.

The AGC circuits are a good place to start checking for an overloaded-video symptom since a faulty AGC allows amplifiers to operate at maximum gain.

1. Inject an IF signal (test pattern) to the third (or final) IF amplifier. Use a maxi-

mum IF signal. In most discrete-component sets, the AGC circuits do not control gain of the third or final IF. Thus, if the overloaded-video symptom appears, the trouble is not in the AGC. Inject a video signal (test pattern) into each video amplifier stage. Make voltage and resistance checks in the defective stage to locate the defective part. Look for leaky capacitors or changed resistor values that might shift the stage bias point (Chap. 1).

If a normal test pattern is displayed with the IF-signal injected at the third (or final) IF amplifier, it is possible that the trouble is in the AGC circuits, although not conclusive. Go to step 2.

2. Inject an RF signal (test pattern) into the antenna terminals, and adjust the generator controls for an overload symptom.

3. Apply a bias to the AGC line. Increase the bias and observe the test-pattern display. In typical discrete-component circuits, the picture should be dim or cut off with the bias control set to zero. As the bias is increased, normal operation should be restored; then the overloaded-video symptom should return.

If varying the bias control affects the picture as described, the trouble is in the AGC circuits. The amplifiers have demonstrated that they can be controlled with proper AGC.

If the overloaded-video symptom continues to appear while the bias is adjusted, either the trouble is unrelated to AGC or the AGC line is completely shorted. Go to step 5.

4. Remove the bias and inject a keying pulse (sync output of the generator) to the keyed AGC stage. If this step produces normal operation, check for trouble in the AGC winding of the flyback transformer or in the coupling parts between the flyback transformer and the keyed-AGC stage.

5. In a discrete-component set, disconnect the base of the second IF from the AGC circuit, and connect a bias in place of the AGC. In an IC circuit, disconnect the AGC line from the IC and replace the line with a bias. If there is a short in the AGC line, the short is now disconnected from the stage or IC.

Vary the bias. If the variable bias affects the picture as described in step 3, repeat the procedure for the remaining IF stages and the tuner. If the overloaded video symptom does not appear continuously at any of the stages, the trouble is in AGC.

If the overloaded-video symptom reappears and the bias has no effect on the picture in any of the stages (IF or RF), check for a defective part in that stage. Be especially alert for a leaky capacitor that may shift the amplifier bias (Sec. 1.16).

2.11 ANALYZING THE COMPOSITE VIDEO WAVEFORM

Probably the most important waveform in TV service is the composite waveform consisting of the video signal, the blanking pedestals, and the sync pulses. Figure 2.12 shows typical scope traces when you are observing composite video signals synchronized with horizontal-sync and vertical-blanking pulses. The trace shown is one horizontal line of an NTSC color-bar signal. NTSC generators and color signals are discussed fully in Chap. 3. Here, we are concerned with the use of such signals in basic black and white TV troubleshooting.

You can observe the composite video signals at various stages in the TV set to determine whether the circuits are performing normally. A knowledge of the waveform

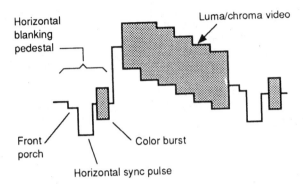

FIGURE 2.12   Analyzing the composite video waveform.

makeup, the appearance of a normal waveform, and the causes of various abnormal waveforms help you locate and correct many problems. You should study such waveforms in a set known to be in good operating condition, noting the waveform at various points in the video amplifiers.

Note that the pattern of Fig. 2.12 may appear inverted. The output of some video detectors is positive, whereas other video detectors produce negative outputs. Also, it is possible to invert the pattern for a positive or negative display (whichever is convenient) by adjusting the scope controls (on most scopes).

The pattern of Fig. 2.12 assumes that a color signal from an NTSC generator is applied to the antenna terminals. The video portion of the pattern contains black and white luminance levels as well as the chroma and luma levels for the standard NTSC colors. As discussed in Chap. 3, many NTSC generators also produce a signal similar to that of Fig. 2.12 but without the color burst and chroma levels.

Figure 2.12 also assumes that the composite video waveform is being monitored at the video-detector output. This is the usual starting point. However, the composite video may also be checked at other points in the video circuits by moving the scope probe to those points. (Note that the waveforms shown in Fig. 2.12 are theoretical and will almost never appear exactly as shown in any circuit, even at the video detector output.)

2.12 ANALYZING THE HORIZONTAL SYNC PULSE

The IF-amplifier response of a TV set can be evaluated to some extent by careful observation of the horizontal sync pulse waveform. The appearance of the sync waveform is affected by the IF amplifier bandpass characteristics. Some typical waveform symptoms and the relation to IF-amplifier response are shown in Fig. 2.13a. Sync waveform distortions produced by positive or negative limiting in IF-overload conditions are shown in Fig. 2.13b. An expanded horizontal sync pulse and blanking pedestal of a color signal are shown in Fig. 2.13c.

Note the waveforms shown in Fig. 2.13c. For example, the tips of the sync pulse and blanking pedestal are rounded off when there is a loss of high-frequency response in the IF stages (or in the overall response of the set). When there is excessive high-

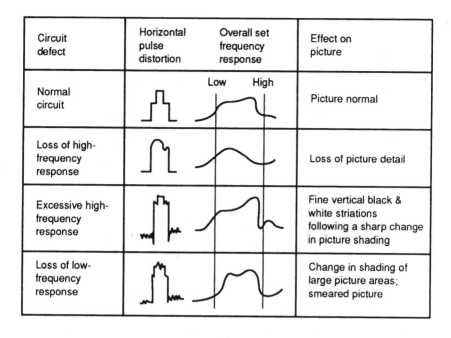

(a)

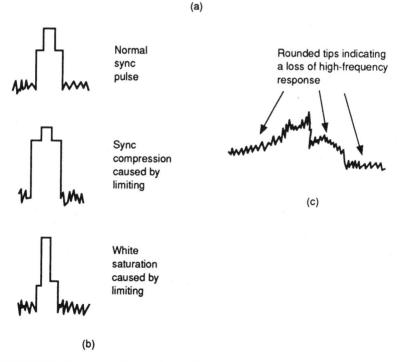

(b)

(c)

FIGURE 2.13 Analyzing the horizontal sync pulse.

frequency response or a general loss of low-frequency response, the pulse tips become sharp.

2.13 VERTICAL INTERVAL TEST SIGNAL (VITS)

Many broadcast TV signals contain a built-in test signal (the VITS) that can be a very valuable tool in troubleshooting TV sets. The VITS can localize trouble to the antenna, tuner, IF, or video sections and shows when realignment may be required.

The basic purpose of the VITS is for use by broadcast engineers who monitor the signals to check that transmissions meet established standards. On some older sets (such as Sylvania) a portion of the VITS is used to control sync of the 3.58-MHz oscillator found in color sets (Chap. 3). The circuits involved are called vertical interval reference (VIR) circuits and can be selected by the user. (User control is provided since not all broadcast signals contain the VITS signal.)

The following procedures show how you can analyze and interpret scope displays of the VITS to troubleshoot TV sets. The VITS is transmitted during the vertical blanking interval. On the TV set, the VITS can be seen as a bright white line (and/or series of dots) above the top of the picture. However, the vertical hold, linearity, and height controls may require adjustment to view the vertical blanking interval.

On some sets with an interval vertical-retrace blanking circuit, the vertical blanking pulse must be disabled to see the VITS. The procedures are covered in the following sections. On other sets, without interval vertical blanking, it is only necessary to advance the brightness control. For troubleshooting, it is not necessary to see the VITS on the picture-tube screen. Instead, you monitor the VITS signal on a scope after the signals pass through the set.

2.13.1 The VITS Pattern

The transmitted VITS is a precision sequence of specific frequency, amplitude, and waveshape, as shown in Fig. 2.14. The first frame of the VITS (line 17) begins with a "flag" of white video, followed by sine-wave frequencies of 0.5, 1.5, 2.0, 3.0, 3.6 (3.58), and 4.2 MHz. This sequence of frequencies is called the *multiburst*. The first frame of field 2 (line 279) also contains an identical multiburst. This multiburst portion of the VITS is of the most value in troubleshooting.

The second frame of the VITS (lines 18 and 280), which contains the sine-squared pulse, window pulse, and the staircase of 3.58-MHz bursts at progressively lighter shading, is valuable to broadcast engineers but has less value to the service technician. (If you are not familiar with TV fields and frames, refer to Sec. 3.2.)

As seen on the TV screen, field 1 is interlaced with field 2 so that line 17 is followed by line 279, and line 18 is followed by line 280. The entire VITS appears at the bottom of the vertical blanking pulse and just before the first line of video.

Each of the multiburst frequencies is transmitted at *equal strength*. By observing the *comparative strengths* of these frequencies after the signal is processed through the set, you can check the frequency response of the set at the normally used video frequencies.

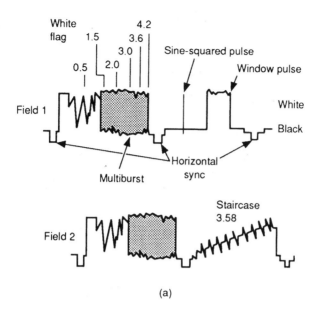

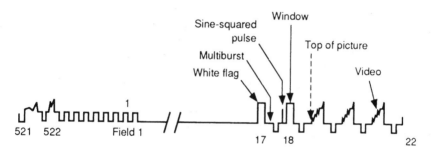

(a)

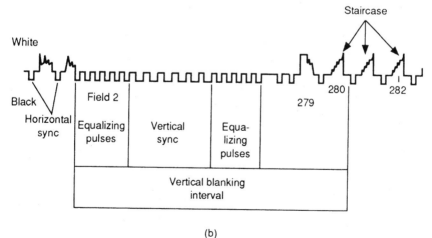

(b)

FIGURE 2.14 The VITS patterns.

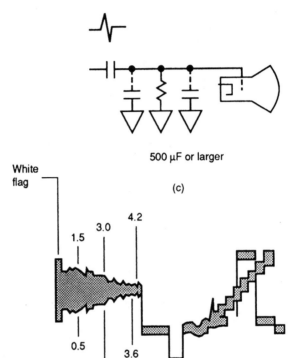

FIGURE 2.14 *(Continued)* The VITS patterns.

2.13.2 Typical VITS Test Connections and Measurement Procedures

The procedures for monitoring VITS as the signals are processed by the set vary, depending on the circuits. However, the following steps can be modified as necessary to accommodate all types of TV sets:

1. Connect the scope probe to the output of the video detector or other desired point in the video section. Start with the video-detector output.

2. Set the scope sweep controls for a triggered sweep.

3. If the set has a vertical retrace-blanking circuit, disable the circuit, as shown in Fig. 2.14c. The retrace-blanking circuit is bypassed to ground by the large-value capacitor. On sets without vertical-retrace blanking, simply advance the brightness control.

4. Operate the scope controls as necessary to get a VITS pattern similar to that shown in Fig. 2.14d. Note that field 1 is superimposed on field 2.

5. Typically, you must advance the scope intensity control to full on. Then adjust the sweep-sync control slowly to the point where triggering barely starts and a horizontal pattern is formed. Make very slight adjustments of the sweep-sync control until the VITS pattern is obtained. Because of the constantly varying video signal, sync lock-in of the VITS can be quite critical. (On some scopes, additional fine adjustment of the sync level can be obtained by slight readjustment of the vertical gain control.)

It is normal for strong video signals to override the VITS pattern occasionally. This effect varies with different televised scenes and may be more noticeable on certain channels. However, with careful adjustment of controls, a good VITS pattern can usually be maintained. On some sets, you can get more solid lock-in of the VITS pattern using external sync as described in step 8.

6. Using the scope sweep and horizontal gain controls, you can select the portion of the VITS pattern you want to view and then expand the pattern as desired. Generally, you are interested only in the multiburst. Figure 2.15a is an expanded trace

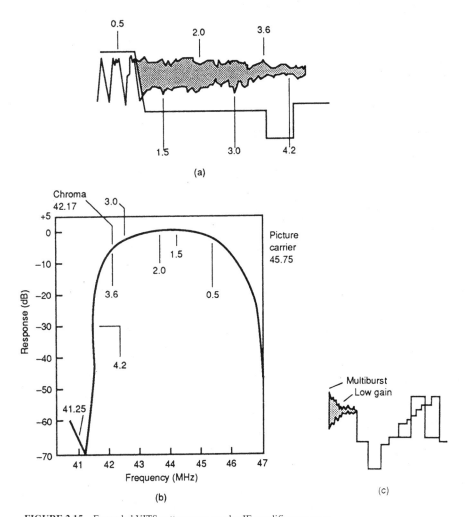

FIGURE 2.15   Expanded VITS pattern versus color IF-amplifier response.

showing just the multiburst portion of the VITS pattern. Note that the multiburst portions of the VITS are contained on only 2 horizontal lines of the 525-line raster and that the staircase and other portions are on only 1 line. The VITS appears substantially dimmer than the rest of the composite video pattern. The bright vertical line pattern (shown to the right in Fig. 2.15a) may be distracting. You can minimize this by adjusting the horizontal position control to place the bright area just beyond the right side of the screen edge.

7. Alternatively, you can display the VITS along with vertical blanking pulses. Set the scope sweep controls as you would to monitor the vertical composite-video waveform (Sec. 2.11). Usually this involves setting the scope sweep control to vertical or TVV or to a similar position. Then you operate the scope sweep-expansion controls to get a display.

8. As another alternative, you can display the VITS alone (as with a triggered sweep, steps 1 through 6) but with an *external sync* to lock in the VITS. Locate a point in the set where a negative-going (or positive if the set has positive sync pulses) vertical sync pulse is available. A convenient point on most sets is the appropriate wire in the convergence board that leads to the vertical winding of the deflection yoke. Be sure the set's vertical and horizontal controls are set to provide a stable, in-sync picture. Connect the vertical pulse from the set to the external-sync input of the scope. Operate the scope controls to lock in the VITS pattern and proceed as described in steps 1 through 6.

2.13.3 Analyzing the VITS Waveform

All frequencies of the multiburst are transmitted at the same level but are not equally passed through the set circuits because of the response curve of those circuits (RF, IF, and video). Figure 2.15b shows the desired response for an "ideal" color set, identifying each frequency of the multiburst and showing the allowable amount of attenuation for each. Remember that −6 dB equals half the reference voltage (the 2.0-MHz modulation should be used for reference).

Figure 2.15b represents the IF-amplifier response curve of a color set, with the curve frequencies given in terms of typical IF range (41 to 47 MHz), and the VITS modulation frequencies superimposed at corresponding points. For example, the 3.6-MHz VITS modulation occurs at the chroma or color frequency of 42.17 MHz.

In an ideal color set (and a good black and white set) the 2.0- and 1.5-MHz modulation signals should be at about the same level, the 0.5- and 3.0-MHz signals should be about 3 dB down from the 2.0-MHz reference level, the 3.6-MHz (chroma) signal should be about 6 dB down from (or about one-half) the 2.0-MHz reference level, and the 4.2-MHz modulation signals should be about 30 dB down from the 2.0-MHz reference. Compare this with Fig. 2.15a, which is a normal VITS pattern in a good set. Then compare both patterns with that of Fig. 2.15c, which is a VITS pattern (measured at the video detector) from a set with a defective tuner.

Note the loss of gain on the higher-frequency (right-hand side) multiburst signals shown in Fig. 2.15c. In this example, the set has no color, but a good black and white picture. Since the VITS shows low gain in the chroma frequencies (3.6 MHz) at the video detector, you know the problem is not in the color bandpass amplifier but must be ahead of the video detector (in the tuner or IF amplifiers).

2.13.4 Localizing Trouble with the VITS Waveform

To localize trouble, start by observing the VITS at the video detector. This localizes trouble to a point either before or after the detector. If the multiburst is normal at the video detector, check the VITS on other channels.

If some channels are good but others are not, you probably have a defective tuner or antenna-system troubles. Do not overlook the chance of the antenna system causing "holes" or poor response on some channels.

If the VITS is abnormal at the video detector on all channels, the trouble is in the IF amplifier or tuner. If the VITS is normal at the video detector for all channels, the trouble is in the video amplifier. Look for open peaking coils, off-value resistors, bad solder connections, partial shorts, etc.

As an example of localizing trouble with the VITS, assume that you have a set with a poor picture (weak, poorly defined) but that the set is fully operative and that the antenna and/or cable system is known to be good. Further assume that the VITS pattern at the video detector is about normal, except that the 2.0-MHz burst is low compared to the bursts on either side. This suggests that an IF trap is detuned, chopping out frequencies about 2 MHz below the picture-carrier frequency.

Switch to another channel carrying VITS. If the same condition is seen, your reasoning is good, and the IF amplifier requires adjustment. If the poor response at 2.0 MHz is not seen on other channels, possibly an FM trap in the tuner input is misadjusted, causing a drop in signal on only one channel. Other traps at the input (RF tuner) could similarly be misadjusted or faulty.

One final note concerning VITS signals. The VITS patterns broadcast by all stations are not identical, although the patterns all resemble those shown here. This may cause a problem when switching from one channel to another during the trouble-localization process. Be careful that a different VITS pattern does not lead you to believe a defect exists on one or more channels. Compare the VITS pattern with the picture. For example, if you get an odd VITS pattern on one channel, and the picture is poor on that channel, you have definitely localized trouble to the tuner.

2.14 TESTING VIDEO TRANSFORMERS, YOKES, AND COILS

Transformers, yokes, and coils are often difficult components to test. Of course, if one of these components has a winding that is completely open or shorted or that has a very high resistance, this shows up as an abnormal voltage or resistance during troubleshooting. However, a winding with a few turns shorted, or leakage to another winding, may produce voltage or resistance indications that appear close to normal and pass undetected.

A scope can be used to perform a *ringing test* on such components. In this procedure, a pulse is applied to the component under test. The condition of the component can then be evaluated by the amount of *damping* observed in the waveform.

The best way to get reliable results from a ringing test is to compare the ringing waveform of the part being tested with the waveform of a known-good duplicate part. However, since duplicate parts are rarely available for comparison, you must judge the part being tested by a study of the waveform. To gain experience in evaluating the ringing waveform, it is helpful to try the procedure several times, both with good parts and with parts that you have purposely shorted with various resistances.

2.14.1 Basic Ringing-Test Procedure

Figure 2.16 shows the various ringing-test connections as well as typical ringing-test waveforms. The basic ringing-test procedure is as follows:

1. Remove power from the part or circuit to be tested. Do not apply power to the circuit at any time during the test procedure.

2. Disconnect the part to be tested from the related circuit. Although many parts can be tested in circuit, it is usually necessary to remove circuit connections, especially where diodes and transistors have a loading effect on the part.

3. Connect the scope to the part terminals as shown. Use a low-capacitance probe.

4. Connect a pulse source to the part through a capacitor as shown in Fig. 2.16a or through a few turns of wire as shown in Fig. 2.16b. All scopes have circuits that produce pulses suitable for use in ringing tests (if you do not have a pulse generator). These pulses are the same ones used to trigger the scope horizontal sweep. Unfortunately, the pulses are not always accessible without modification of the scope. However, many scopes designed specifically for TV service have an external terminal (either front or rear panel) that provides for connection to the pulse source.

5. Operate the scope controls to produce a ringing waveform as shown in Fig. 2.16c or 2.16d. The waveform of Fig. 2.16c is a typical ringing pattern for a normal coil or transformer winding. The waveform of Fig. 2.16d is representative of patterns obtained from defective coils and transformers (with shorted turns or leakage in the windings). Remember that these waveforms are typical. Experience is necessary to judge the waveforms when no duplicate parts are available for comparison.

6. To help you make the judgment, connect a resistance across the part terminals (or make a loop consisting of a few turns of solder and pass the loop around the part). Note any change in the ringing pattern. There should be a drastic change in the pattern when the resistor (or solder coil) is added. If not, the part is probably defective.

2.14.2 Ringing Test for High-Voltage (Flyback) Transformer

To test the high-voltage transformer in the horizontal system of a TV set, disconnect the transformer leads. Using the procedure described in Sec. 2.14.1, get a ringing waveform across the transformer primary. Typical test connections are shown in Fig. 2.16e.

With the ringing pattern established, connect a short across each of the transformer secondary windings in turn, and check the ringing pattern. If a significant change in the waveform is noted for all windings, the transformer is probably good. Little or no waveform change indicates that the transformer is defective.

2.14.3 Ringing Test for Picture-Tube Yoke

To test the horizontal and vertical windings of the deflection yoke, disconnect the yoke from the circuit. Using the procedure described in Sec. 2.14.1, get a ringing waveform across each winding in turn. Typical test connections are shown in Fig. 2.16f.

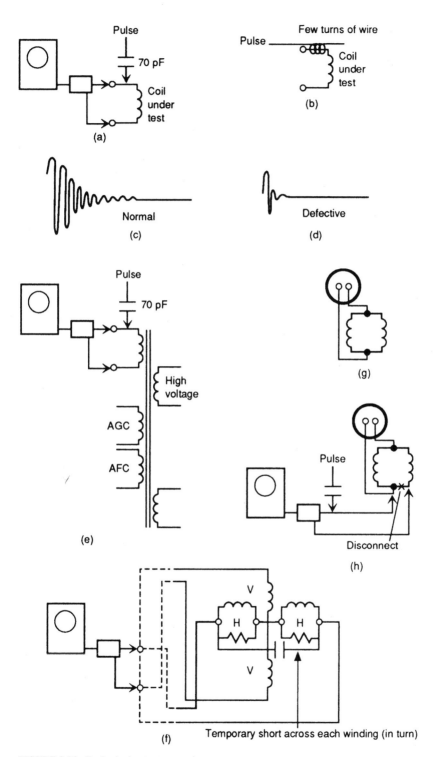

FIGURE 2.16 Basic ringing-test connections.

Note that the horizontal winding of the yoke in Fig. 2.16*f* consists of two sections. Alternately short each section. If the winding is good, the effect on the waveform should be the same as each section is shorted.

The vertical winding of the yoke also consists of two sections, with a damping resistor connected across each section. Disconnect these damping resistors, as well as the capacitor that is in parallel across both vertical sections. Then alternately short each section. If the winding is good, the effect on the waveform should be the same as each section is shorted.

On some yokes, the two sections of each winding are connected in parallel as shown in Fig. 2.16*g* and *h*. In such cases, disconnect the sections at one end. Connect the scope across both sections, as shown in Fig. 2.16*h*, and then short each section in turn. Again, the effect on the waveform should be the same as each section is shorted.

2.14.4 Ringing Test for Picture-Tube Coils

Use the procedure described in Sec. 2.14.1 to test the width, linearity, focus, etc., coils of a picture tube if you suspect shorted turns or leakage. The solder loop usually produces the best results. Again, if the solder loop causes a significant change in the waveform, the coil under test is probably good. However, *experience* is the best judge when making any ringing test.

2.15 INTERPRETING TEST PATTERNS

Most TV stations broadcast test patterns at some time (usually at the beginning and end of the broadcast day). At one time, the test patterns were black and white and contained elaborate displays of circles, horizontal and vertical grid lines, diagonal lines, and so on. Today, the test patterns are in color and resemble those produced by NTSC generators (color bars). The interpretation of such patterns is described fully in Chap. 3.

2.16 ADJUSTING TV AND MONITOR PICTURE WITH TEST PATTERNS

It is difficult to adjust centering, size, and linearity of a black and white TV picture without the aid of a test pattern. Using the picture that is normally transmitted by a station, no image remains fixed long enough to permit complete adjustment. The most common problem with adjustment using a broadcast picture is nonlinearity (even if you get good centering and size). The NTSC generators used for color TV also produce a number of patterns suitable for black and white sets. The procedures for using these patterns are described in Chap. 3.

2.17 TROUBLESHOOTING LOW-VOLTAGE POWER-SUPPLY CIRCUITS

Figure 2.17 is the schematic of a typical low-voltage power supply for a portable black and white TV. The circuit consists essentially of a full-wave bridge rectifier

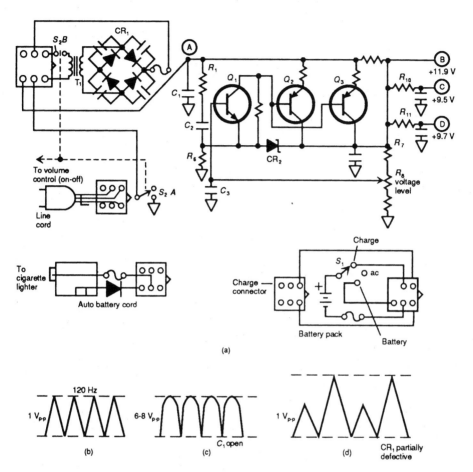

FIGURE 2.17   Low-voltage power supply for portable black and white TV.

CR_1, followed by a solid-state (zener CR_2 and transistors Q_1, Q_2, and Q_3) regulator. As with many portables, the set can be operated with self-contained rechargeable batteries. The set can also be operated from an automobile battery (12 V) using an adapter plug connected to the cigarette lighter.

2.17.1   Basic Circuit Theory

The circuit provides direct current at about 10 to 12 V. This voltage is sufficient for all circuits in the set except for the high voltage that is required by the picture tube. Typically, high voltages are supplied by the flyback circuit (Sec. 2.18).

The line voltage is dropped to about 12 V by T_1 and is rectified by CR_1. The output from CR_1 is regulated by Q_1, Q_2, and Q_3, as well as CR_2, and is distributed to three separate circuit branches, each at a slightly different voltage level. Operation of the regulator is standard. The emitter of Q_1 is held constant by CR_2, whereas the base of

Q_1 depends on the output voltage. Any variations in output voltage (resulting from changes in input voltage or variations in load) change the base voltage in relation to the emitter voltage. These changes appear as a variation in Q_1 collector voltage and, consequently, in Q_2 and Q_3 base voltage.

Transistors Q_2 and Q_3 are connected in parallel with the output from CR_1 and the load. Thus, Q_2 and Q_3 act as variable resistors to offset any changes in output voltage. The level of output voltage is set by R_8.

When the set is to be operated on battery power, switch S_1 is set to battery, the ac power plug is removed, and the battery-pack cord is plugged in. When the battery is to be recharged, S_1 is set to charge, and the power cord is plugged into a special connector on the battery pack. Although there is no standardization, rechargeable batteries provide about 4 to 6 hours of operation and require about 8 to 12 hours of recharging. When the set is to be operated from an automobile cigarette lighter, the ac power plug is removed and the automobile battery cord is plugged in.

2.17.2 Recommended Troubleshooting Approach

If the symptoms indicate a possible defect in the low-voltage supply, the obvious approach is to measure the dc voltage. If there are many branches (three branches in Fig. 2.17), measure the voltage on each branch. If any of the branches are open (say an open R_{10} or R_{11}), the voltage on the other branches may or may not be affected. However, if any of the branches are shorted, the remaining branches are probably affected (the output voltage is lowered).

Current Measurements. It is not practical to measure currents in present-day TV circuits where virtually all parts are mounted on PC boards. Likewise, it is not practical to disconnect each output branch, in turn, until a short or other defect is found. Resistance measurements are far more practical.

Resistance Measurements. If one or more output voltages appear to be abnormal, and it is not practical to measure the corresponding current, remove the power and measure resistance in each branch. Compare actual resistances against those in the service literature, if available. If you have no idea of the correct resistance, look for obvious low resistance (a complete short or a resistance of only a few ohms).

Isolation Transformer and Substitute Supply. Always use an isolation transformer when checking any solid-state circuit. As a convenience, have a 12-V dc supply to substitute for the low-voltage supply circuit in the set. A 12-V battery eliminator makes a good source if a conventional supply is not available. However, ideally, the substitute supply should be adjustable and have a voltmeter or ammeter to monitor power applied to the set.

Battery Operation. If the set can be operated on batteries, switch to battery operation and see if the trouble is cleared. If so, the problem is definitely localized to the low-voltage supply. Likewise, if the set can be operated from an automobile battery or similar arrangement, switch to that mode and check operation.

Zener Replacement. If it becomes necessary to replace a zener in the regulator circuit, always use an exact replacement. Some technicians replace zeners with a slightly different voltage value and then attempt to compensate by adjusting the regulator circuit. This may or may not work.

Scopes.   Do not overlook the use of a scope in troubleshooting power supplies. Even though you are dealing with dc voltages, there is always some ripple present. The ripple frequency and the waveform produced by the ripple can help in localizing possible troubles, as discussed in Sec. 2.17.3.

2.17.3 Typical Troubles

The following sections discuss symptoms that could be caused by defects in the low-voltage supply.

No Sound and No Picture Raster.   When there is no raster on the picture-tube screen and no sound whatever, it is likely that the low-voltage supply is totally inoperative or is producing a very low voltage. If the circuits are producing a voltage about 50 percent of normal (say 6 V in a 12-V system), there is some sound, even though the raster may be absent. The most likely defects are filter capacitor leakage, a defect in the regulator (completely off), or a short in the output line.

Start the troubleshooting process by making voltage measurements at test points A, B, C, and D (or their equivalents in the set you are servicing). If the voltage is absent at all test points, check all fuses and switches as well as CR_1 and T_1 (right after you make sure the set is plugged in). If the voltage at B, C, and D is absent or very low but the voltage at A is high, suspect the regulator. If all of the voltages are low, look for a short in one or more of the output lines.

To check the regulator, first make sure that all three transistors are forward-biased. Typically, the base of Q_1 is about 2 V, with an emitter voltage of 1.5 to 1.8 V. The bases of Q_2 and Q_3 are about 11 V, with the emitters at 11.5 V. In any event, all three transistors must be forward-biased for the regulator to operate. Check the transistors as described in Chap. 1.

Note that it is possible for Q_1 to cut off while Q_2 and Q_3 are forward-biased. However, such a condition quickly points to a fault in Q_1 or the associated parts.

A common fault in discrete-component regulators is a base-emitter short in the current-carrying transistor Q_2 or Q_3. Since these transistors are in parallel, it may be difficult to tell which transistor is at fault. If necessary, disconnect each transistor and check it separately. If either Q_2 or Q_3 has a base-emitter short, the regulator is cut off, and both the base and emitter are high.

No Sound, No Picture Raster, and Transformer Buzzing.   When there is no sound or picture, with transformer buzzing, there is probably an excessive current being drawn in the supply. The most likely defects are a short circuit ahead of the regulator (test point A, for example), shorted rectifiers, or shorted turns in the transformer.

Again, start by making voltage measurements at points A, B, C, and D. Note that prolonged excessive current causes one or more fuses to blow. Thus, with these symptoms, the assumption must be an excessive current that is still below the rating of the fuses. This normally results in a very low voltage (but not an absent voltage).

Unlike the previous symptoms, the regulator is probably operating, so concentrate on shorts (particularly the rectifiers) and possibly shorted transformer turns. Both T_1 and CR_1 can be checked by substitution or resistance test, whichever is convenient. If this does not localize the problem, check for short circuits in each branch of the output. Also look for overheated parts, such as the transformer, or burned PC wiring.

Distorted Sound and No Raster.   These symptoms are similar to those previously described, except that there is some sound (often with intercarrier buzz). Start by making voltage measurements at test points A, B, C, and D. Then measure the amplitude and frequency of the waveforms at the same test points, particularly at the bridge output (test point A) and output (test point B). An analysis of the waveforms can often pinpoint trouble immediately.

The normal waveform at the bridge output is a sawtooth (almost) at twice the line frequency (usually 120 Hz), as shown in Fig. 2.17b. If the filter and regulator are operating properly, the 120-Hz signal is suppressed at the final output. However, there may be some line-frequency (60-Hz) ripple at the regulator output (all three branches).

If there is a strong 120-Hz waveform at the regulator output, this indicates that the filter and/or regulator are not suppressing the 120-Hz bridge output. The most likely cause is an open C_1 or a regulator defect. Of course, C_1 could be leaking, but excessive leakage or a short will blow the fuse.

The waveform at point A also indicates the condition of C_1. If the waveform is a 120-Hz half sine wave similar to Fig. 2.17c, rather than the sawtooth of Fig. 2.17b, C_1 is probably open. This is confirmed further if the waveform amplitude at test point A increases from a typical 1 V (or less) to several volts, as shown in Fig. 2.17c.

The condition of CR_1 is also indicated by the waveform at A. If the waveform is 60 Hz (line frequency), one-half of CR_1 is defective (most likely shorted or open). If the waveform is not symmetrical (Fig. 2.17d), one diode of CR_1 is probably defective (most likely leaking).

Note that if any of the diodes in CR_1 are shorted, T_1 will probably run very hot. Thus, a hot T_1 does not necessarily indicate a bad T_1. However, if the waveform at test point A is good (indicating a good CR_1 and C_1) but T_1 is hot, the most likely problem is a defective T_1 (possibly shorted turns).

Note that all four diodes in CR_1 are paralleled with capacitors to protect the diodes in case of sudden voltage changes that might exceed the breakdown voltage. If one of these capacitors is shorted, this can give the appearance of a shorted diode. If CR_1 is a sealed package with self-contained capacitors, the entire package must be replaced.

If the capacitors can be replaced separately from the diodes, make sure that both the capacitor and diode are good before replacing either. For example, if the capacitor is open, the corresponding diode may be damaged. If the diode is replaced, the trouble is repeated unless the capacitor is also replaced.

Remember that an increase in load (say, because of a short or partial short) causes an increase in ripple amplitude, even with a good filter and regulator, because a larger load (lower load resistance) causes faster discharge of the filter capacitor between peaks of the sine-wave pulses from the rectifier.

Picture Pulling and Excessive Vertical Height.   Thus far, we have discussed symptoms and troubles that result from a low power-supply output because of defective parts or shorts. It is possible that the power supply can produce a high output voltage, resulting in picture pulling (the raster is stretched vertically and is bent). Hum bars across the picture screen are usually present with this condition.

Assuming that the trouble is definitely in the low-voltage supply, the most likely cause is in the regulator. A defect in the rectifier, filter, or transformer usually results in a low supply output. It is possible for the regulator to be cut off, causing the output to increase in voltage. The logical approach is to start the troubleshooting process by making voltage and waveform measurements at all test points.

If the voltage at test point B is nearly the same as at test point A (within about 0.5 V), the regulator is probably cut off. The voltage at test points C and D will also be high. The condition can be confirmed further if the waveform at test points B, C, and D is at 120 Hz (indicating that the regulator is not suppressing the 120-Hz bridge output). With the trouble definitely pinned down to the regulator, test all of the transistors (Q_1 through Q_3) and CR_2.

Low Brightness and Insufficient Height.   When the low-voltage supply drops below normal but is not really low, one of the first symptoms is low picture brightness. Picture height and width are also reduced, but height is generally reduced more. There are several possible causes for a small reduction in supply voltage. One of the bridge-rectifier diodes (or the related protective capacitor) can be shorted. Or the diode can be open, but this is less likely. Capacitor C_1 can be leaking. There may be a partial short in one of the output lines, thus dropping the voltage at all outputs by a small amount.

Start by checking the voltages at test points B, C, and D. Try to correct a low-voltage condition at all outputs by adjustment of R_8. It is possible that the low output is caused by component aging. Look for trouble in CR_1 or C_1 if R_8 must be set to an extreme for correct voltage output.

Use the procedures previously discussed to check CR_1 and C_1. For example, a waveform similar to Fig. 2.17c at test point A indicates an open C_1. The waveform in Fig. 2.17d indicates a partial defect in CR_1. Remember that a poor solder joint or a break in PC wiring can simulate a defective (open) C_1.

Insufficient Height and Width, Weak Sound, and Snowy Picture.   These are also symptoms of below-normal voltage from the low-voltage supply. Again, start by checking voltages at test points B, C, and D, and try adjusting R_8.

Also measure the waveform and voltage at test point A. If the indications are correct at A but the voltages at B, C, and D are low (and cannot be adjusted), the regulator is at fault. This means that you must dig into the circuit, looking for such problems as short circuits (from solder splashes) and breaks in PC wiring. Let us consider a few examples.

When troubleshooting any circuit such as the regulator in Fig. 2.17, voltage and resistance checks are generally more useful than scope checks. Likewise, capacitor failures are more likely than resistor failures (which is usually true in most video equipment).

Assume that R_7 is open (broken PC wiring, cracked resistor, etc.). This shows up in several ways. First, R_8 has little or no control of the regulator circuit (little effect on output voltages). The base voltage of Q_1 is abnormal, as is the resistance at test point B. However, the waveforms throughout the regulator circuit can remain normal (or so close to normal as to be unnoticed).

Now assume that C_3 is shorted. This cuts off Q_1 and provides a partial short across the output lines. Such a defect is generally easy to locate since the voltage at the base of Q_1 (normally 2 V) is zero. Likewise the resistance from the base of Q_1 to ground is zero.

When tracking troubles by means of voltage and resistance measurements, it is often necessary to check several components for one abnormal reading. For example, assume that the emitter of Q_1 shows a high reading (say 3 V instead of 1.5 V). There are three logical suspects. First, C_2 can be leaking. This applies a large voltage to the emitter of Q_1 through R_1. Next on the list is a defective R_6 where the resistance has increased (say because of a partial break in the composition resistance material). Of course, CR_2 is supposed to overcome minor changes in R_6 resistance. However, a large change in R_6 can prevent CR_2 from maintaining the desired 1.5 V at the emitter of Q_1. Finally, CR_2 can be at fault, even though zeners are generally rugged.

2.17.4   Typical Low-Voltage Circuits

Figure 2.18 shows the low-voltage supply circuits for a Sony 13-in TV set, as well as the FBT high-voltage circuits, in basic block form. Figures 2.19 through 2.21 later in

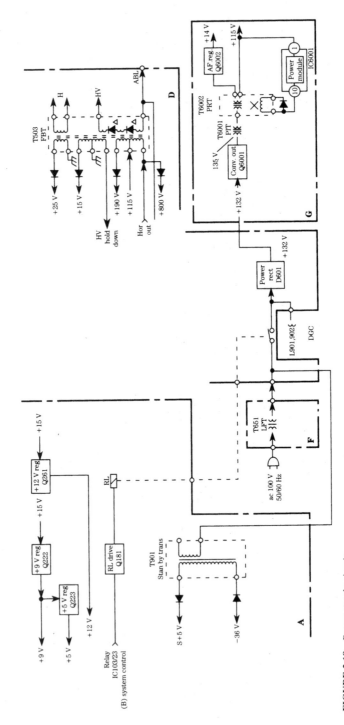

FIGURE 2.18 Power supply circuits.

the chapter show the circuits in greater detail. The high-voltage circuits are discussed in Sec. 2.19.

The relay-control signal from system control (Sec. 2.20) is applied to the power relay RY_{131} through relay drivers. In turn, RY_{181} controls line ac to the main supply, rectifier D_{601}, and to the degaussing coils. The ac power is applied to a standby transformer T_{901} (to generate +5 and −36 V) at all times.

The rectified +135-V output from D_{601} is applied to a switching supply that generates +14- and +115-V outputs. The switching supply also produces the overvoltage protection (OVP) signal that cuts off the supplies in the event of an overvoltage condition.

Line Input Circuits. As shown in Fig. 2.19, ac line power is applied to the standby supply (Fig. 2.20) at all times when the line cord is plugged in. The ac power is applied to the degaussing coils L_{901} and L_{902} through thermal switch THP_{601} and to main power rectifier D_{601} when RY_{181} is actuated by a power turn-on command from system control. The thermal characteristics of THP_{601} provide momentary degaussing of the CRT each time the set is turned on.

Auxiliary and Standby Power Circuits. As shown in Fig. 2.20, ac line power is applied to standby transformer T_{901}. The output from one secondary of T_{901} is rectified by D_{182} and produces the standby −36 V for the tuning memory (Chap. 4). The output from the other secondary of T_{901} is rectified by D_{183} and D_{184} to produce the standby 5-V (S_5-V) supply. The 8 V is also used to power relay RY_{181}.

Transistors Q_{222} and Q_{223} receive 15 V from the FBT and horizontal stages (Sec. 2.19) and regulate the 15 V down to 5 and 9 V. Transistor Q_{221} also receives 15 V from the FBT and regulates the 15 and 12 V.

The power-on command from system control is applied to relay RY_{181} through Q_{183} and Q_{181}. When actuated, the normally open contacts of RY_{181} close, applying line power to the degaussing coils and main power supply (Fig. 2.19), as described. The pulse produced by the power-on command is applied to Q_{251} through C_{259}. This mutes the audio stages (Sec. 2.18.4) for ½ to 1 s during power-on.

OVP is provided by the switching supply (Fig. 2.21) in the form of the OVP input to Q_{181}. Should an overvoltage condition occur, the OVP signal goes low, turning Q_{181} off. As a result, relay RY_{181} is deenergized, and the contacts open to remove power from all but the standby circuits.

Switching Supply. As shown in Fig. 2.21, 135 V from the main supply is applied to power-input transformer T_{6001} and transistor Q_{6001}. One secondary winding of T_{6001} and Q_{6001} forms a 71-kHz oscillator. The oscillator signal is induced across T_{6001} windings and applied to power-regulator transformer T_{6002}. The oscillator signal at R_{6007} is rectified by D_{6003}, producing 17 V at series regulator Q_{6002}. In turn, Q_{6002} regulates the 17 V down to 14 V. The output from the other winding of T_{6002} is rectified by D_{6002} and D_{6007}, producing 115 V, which is the primary B⁺ for the horizontal-output stages (Sec. 2.19).

Regulation of the 14- and 115-V lines is performed by the IC_{6001} module. Reference voltages are set across R_{6006}, R_{6005}, and R_{6004} at pins 8, 6, and 1, respectively. These reference voltages control the condition of two transistors within IC_{6001}. In turn, this controls the current through the control winding of T_{6002}. If the output current of the 14- or 115-V lines tends to increase (lowering the line voltage), the internal transistors of IC_{6001} conduct more. The current through pin 10 of IC_{6001} and the control winding of T_{6002} increases, lowering the winding inductance and increasing the level of the line voltages to offset the initial drop in line voltage. The opposite occurs if the line voltages tend to increase.

The 115 V at pin 1 is applied across a zener within IC_{6001}, setting the reference

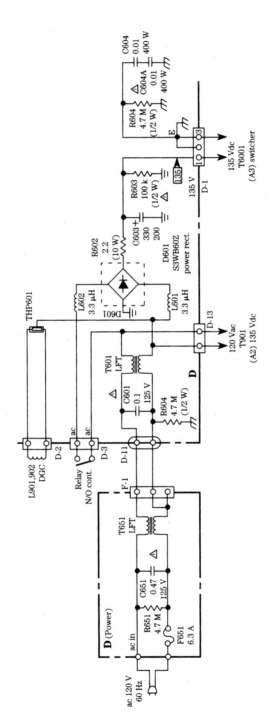

FIGURE 2.19 Line input circuits.

2.60

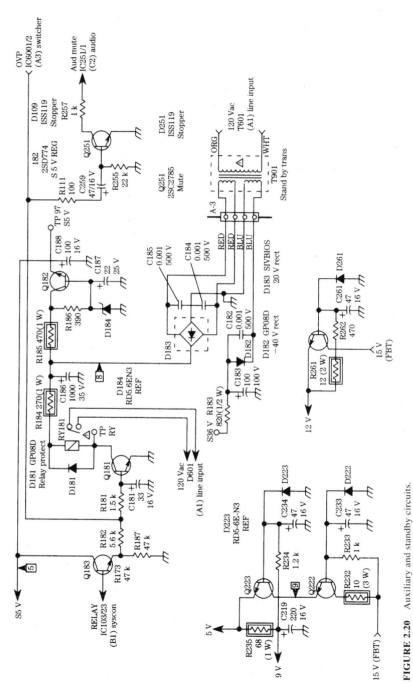

FIGURE 2.20 Auxiliary and standby circuits.

2.61

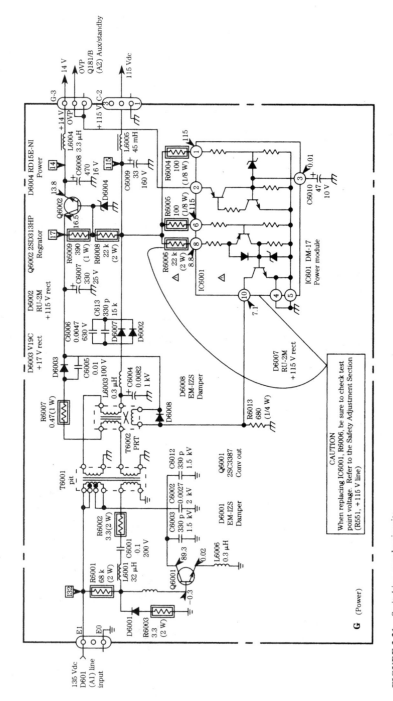

FIGURE 2.21 Switching supply circuits.

2.62

level of the remaining two internal transistors. Should an OVC occur, such that the breakdown voltage of the zener is exceeded, the internal transistors turn on, making pin 2 of IC_{6001} low. This low output is the OVP signal applied to Q_{181} to turn off the main power by opening RY_{181}.

2.18   TROUBLESHOOTING VIF AND SIF CIRCUITS

This section is devoted to TV-set circuits between the front-end tuner and audio and video processing stages. Such circuits are generally known as the VIF (video IF) and SIF (sound or audio IF) stages and usually include both video and a sound detector (although there are many different configurations in present-day sets). In some literature, the VIF is called the PIF, or picture IF. No matter what configuration is used, or what it is called, the basic function of the IF and video and sound-detector circuits is to amplify both picture and sound signals from the tuner, demodulate both signals for application to the video and sound processing circuits, and trap (or reject) signals from adjacent channels. The IF circuits also provide feedback signals to control the channel-changing functions of the front-end circuits.

2.18.1   VIF and SIF Basics

Figure 2.22 shows the signal path through the IF and video-detector stages of a typical discrete-component TV set. (The IF and video-detector functions are usually com-

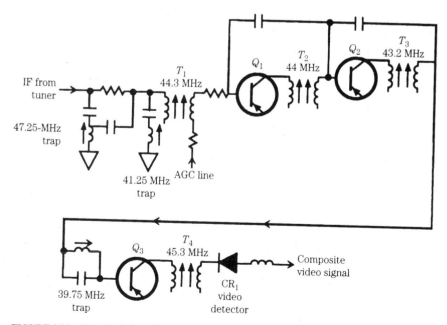

FIGURE 2.22   Signal path through IF and video detector in discrete-component sets.

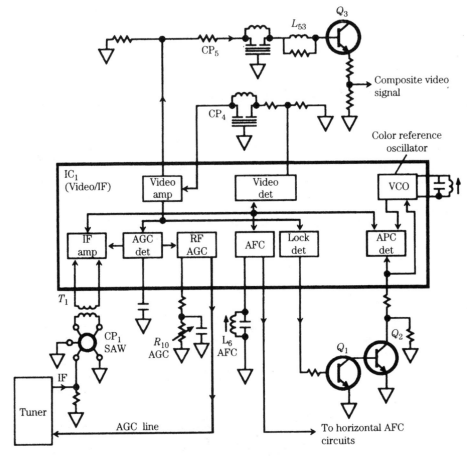

FIGURE 2.23   Signal path through IF and video circuits of IC set.

bined into one IC in present-day sets.) In most cases, the IF, video-detector, and sound-detector functions are combined into an IC such as shown in Fig. 2.23, along with AFC, AGC, and APC functions.

In the circuit of Fig. 2.22, all stages are forward-biased and the first two stages Q_1 and Q_2 receive forward bias from the AGC circuit. The same AGC line is connected to the tuner. On strong signals, the forward bias is increased, driving Q_1 and Q_2 into saturation and reducing gain.

Each stage is tuned at the input and output by corresponding transformers T_1 through T_4. The stages are stagger-tuned (each transformer is tuned to a different peak frequency). This provides the overall IF circuit with a bandwidth of about 3.25 MHz. Three traps are used. The 41.25- and 47.25-MHz traps are series-resonant, whereas the 39.75-MHz trap is parallel-resonant. The traps are adjusted for a minimum output signal at the video detector when a signal of the corresponding frequency is injected at the IF input. Output of the IF stages can be measured at CR_1. The video output is typically about 1 V, with the picture signals about 0.25 to 0.5 V.

In the circuit of Fig. 2.23, IF output from the tuner is applied to an IF amplifier within IC_1 through SAW filter CP_1 and transformer T_1. The IF output is applied to a video detector in IC_1. The composite output from the detector is applied through filter CP_4 to a video amplifier within IC_1. (The detector output is also applied to a sound-IF amplifier in IC_1, not shown.)

The video amplifier output is applied to the picture-tube and video and chroma circuits through CP_5, L_{53}, and Q_3. The video-amplifier output is also applied to an AGC detector within IC_1. This AGC circuit controls both the IF amplifier within IC_1 and the tuner (through the RF amplifier in IC_1). The AGC circuit is adjusted by R_{10}.

The output of the IC video detector is applied to an AFC circuit in IC_1. This circuit is adjusted by L_6 and provides pulses to the horizontal AFC circuits (Sec. 2.19). The IC video-detector output is also applied to an APC circuit that controls operation of the synchronous detector in IC_1 (in conjunction with an IC_1 lock detector that receives signals from the IC_1 video amplifier). Notice that the synchronous detector is not always found in the IF IC but can be located in a separate IC.

2.18.2   Typical VIF and SIF Circuits

Figure 2.24 shows the circuits involved in processing the IF signal (of a 13-in Sony) from the tuner to develop the composite video output, as well as the sound output. The IF signal developed by the tuner is applied to the VIF stages within IC_{201}. The video is processed, and composite video is separated from the carrier and the sound signal. In this particular set, the composite video is applied to the video and chroma processing stages (Sec. 3.11) through switches in the RGB-interface stages (Sec. 3.12).

The IF stages within IC_{201} return the RF AGC control voltage to the tuner and thus control RF and IF gain. The VIF stages within IC_{201} provide AFT up-down correction signals to the tuning-control microprocessor IC_{103} (Chap. 4). IC_{301} acts on these AFT correction signals to correct the PLL data and provide compensation for local-oscillator drift in the tuner.

The SIF signal is separated from the composite video output of IC_{201} and is returned to SIF circuits within IC_{201}. These SIF circuits include an FM discriminator to remove FM audio from the 4.5-MHz sound carrier. In this particular set, the audio is then routed to the speaker and earphone jack through switches in the RGB interface stages (Sec. 3.12), to an audio preamplifier, and to an audio power amplifier.

2.18.3   VIF Circuits

As shown in Fig. 2.25, the VIF signal developed by the tuning system (Sec. 4.1) is applied to VIF preamplifier Q_{221}. The amplified IF signal is then applied to SWF_{201} through C_{228}. SWF_{201} is a 45.75-MHz SAW filter. The VIF signal from SWF_{201} is applied to the VIF amplifier in IC_{201} through pins 1 and 2. The VIF signal is further amplified and sent to the synchronous-detector stage within IC_{201}. The detector removes the 45.75-MHz carrier, leaving the composite video and the SIF signal. Both of these signals are processed by an automatic noise canceling/white noise limiting (ANC/WNL) circuit and are output from IC_{201} at pin 20. The VIF detector is tuned by T_{202}.

The processed video-audio signal at pin 20 of IC_{201} is applied through CR_{201} to video buffer Q_{201}. CR_{201} is a ceramic filter connected as a trap to remove the 4.5-MHz SIF signal from the composite video being buffered by Q_{201}. The buffered output from Q_{201} is applied to the video and chroma stages (Sec. 3.11) through switches in the RGB interface stages (Sec. 3.12).

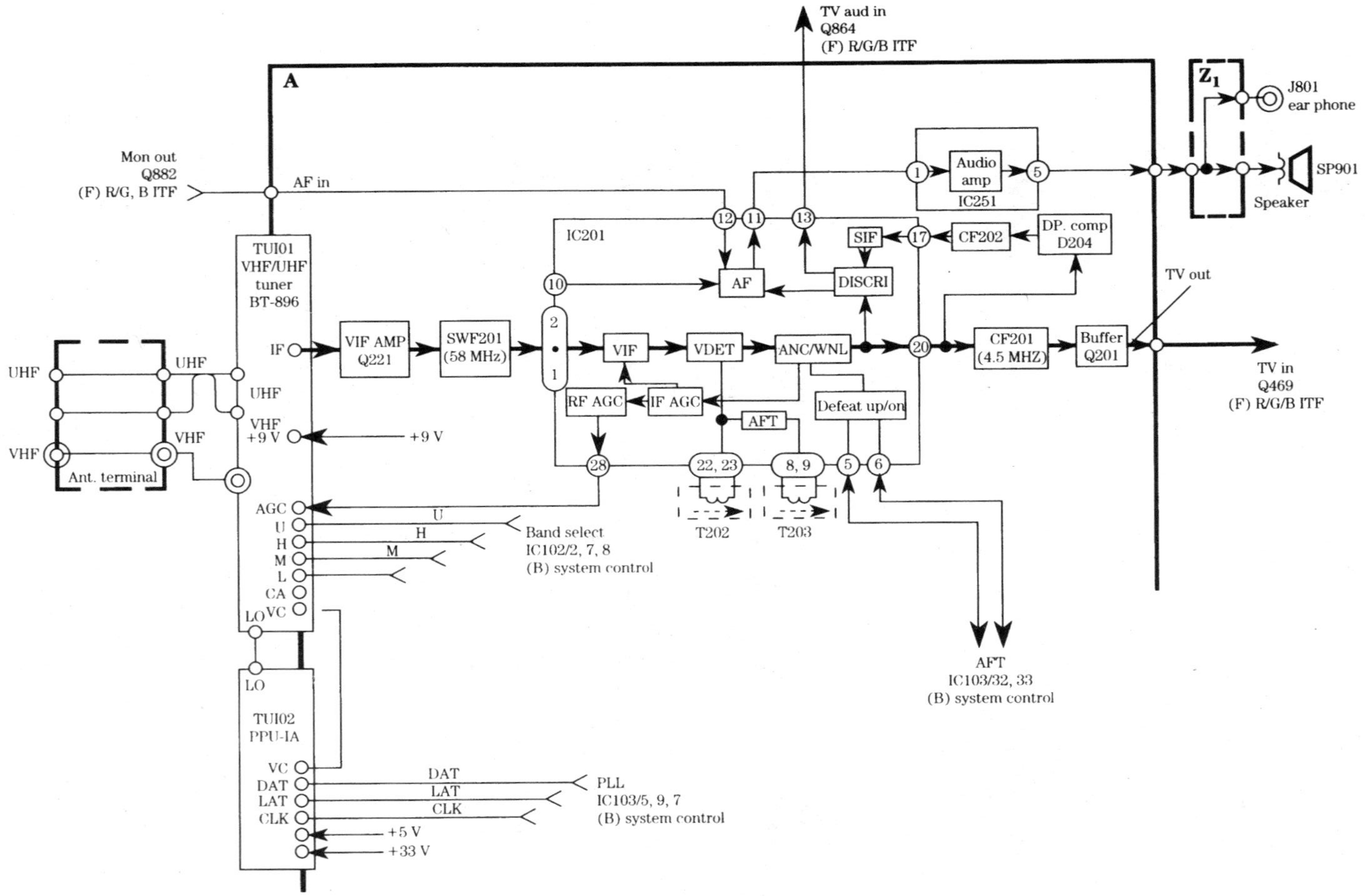

FIGURE 2.24 VIF and audio circuits.

2.66

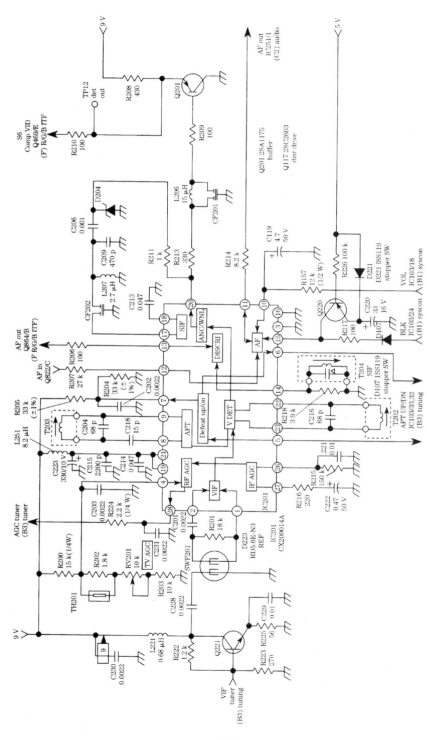

FIGURE 2.25 VIF circuits.

2.67

The output at pin 20 of IC_{201} is also applied through CF_{202} back to SIF circuits within IC_{201}. CF_{202} is a ceramic filter connected in a bandpass configuration to pass only the 4.5-MHz SIF signal to pin 17 of IC_{201}. Diode D_{204} serves to limit this signal.

The 4.5-MHz SIF signal is amplified and applied to an FM discriminator within IC_{201}. The discriminator removes the 4.5-MHz carrier, leaving only the audio. This audio is applied to a preamplifier within IC_{201} through pin 13, switches in the RGB interface stages, and pin 12. The preamplified audio is then applied to audio power amplifier IC_{251} through pin 11 of IC_{201} and muting transistor Q_{251} (Sec. 2.18.4). The discriminator is tuned by T_{204}.

The audio preamplifier within IC_{201} has two control inputs. The input at pin 10 is the volume-control voltage developed by microprocessor IC_{301}. As discussed in Sec. 2.20, the control voltage at pin 10 varies from about 0 to 5 V and sets the level of the audio at both the speaker and earphone jack.

The input at pin 15 is a muting signal. A high at pin 15 completely turns off the audio amplifier in IC_{201}, removing the output signal at pin 11, to effectively mute the entire audio system. As discussed in Sec. 2.20, a blanking signal is developed at pin 24 of IC_{103} during channel changes. This positive signal is applied to pin 15 through D_{107} and mutes the audio.

Transistor Q_{220} provides audio muting when power is off. When power is on, the +5-V line is high, Q_{320} is off, and C_{221} charges through D_{221}. When power is turned off, the +5-V line goes low and Q_{220} is forward-biased and turns on. This dumps the charge on C_{221} into pin 15, providing a momentary muting pulse to prevent speaker "pop" at power-off.

The detected video is applied to the IF AGC stages within IC_{201}. This is a peak AGC cycle that samples the level of the horizontal-sync pulses to determine signal strength and produces a corresponding dc-correction voltage to the internal video IF amplifiers, as well as to the RF AGC stage within IC_{201}.

The RF AGC is applied to the tuner AGC terminal through pin 28 of IC_{201}. This voltage controls gain of the RF amplifiers within the tuner and varies between 2 and 8 V (at pin 28 of IC_{201}). This is a reverse AGC system where an increase in the IF signal increases the AGC voltage, driving the tuner RF amplifiers into saturation and reducing gain.

The filter networks at pins 26 and 27 of IC_{201} control response time of the IF AGC. In turn, this also controls the response of the RF AGC stages to prevent overreaction to sudden signal changes. The RF AGC level is set by RV_{201}.

The center-frequency (45-MHz) VIF signal within IC_{201} is sampled, and AFT correction pulses are produced at pins 5 and 6. If the IF signal varies in frequency above or below the 45.75-MHz center frequency, the AFT up-down correction pulses are directed to microprocessor IC_{103} at pins 32 and 33, as described in Sec. 4.1. (IC_{103} performs AFT correction to the PLL as necessary to return the VIF signal to 45.75-MHz.) The AFT circuit is tuned by T_{203}.

2.18.4 Audio Circuits

As shown in Fig. 2.26, the audio recovered from the VIF signal in IC_{201} is applied to amplifier IC_{251} through C_{258}. IC_{251} amplifies the audio to a level suitable for driving the speaker and earphone jack. The external RC networks at pins 1 through 4 and 8 of IC_{251} provide negative feedback to control gain and shape frequency response to the desired characteristics. Notice that the 14 V at pin 6 of IC_{251} is taken from a switching power supply (Sec. 2.17.4) and is the only place where the 14 V is used. Thus, if there is a failure of audio only, check the 14-V source first. (When troubleshooting, always look for similar situations where failure can be isolated to a single source.)

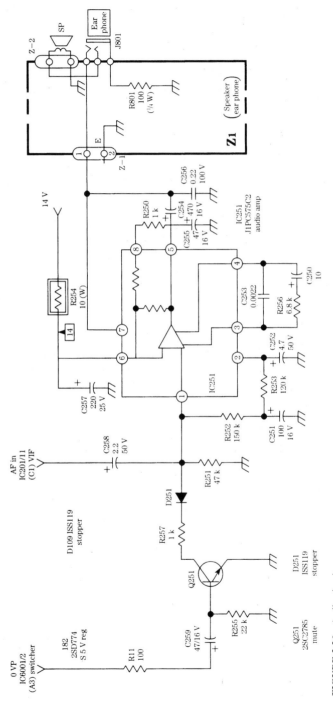

FIGURE 2.26 Audio circuits.

Transistors Q_{251} and C_{259} provide for muting on power-up. When power is first applied, the relay switching voltage (Sec. 2.17.4) is also applied through R_{111} to C_{259}. The rising voltage is differentiated by C_{259}, producing a momentary positive pulse to Q_{251}. This turns Q_{251} on, pulling the audio signal down to ground through D_{251} and effectively mutes the audio. This is a momentary effect that occurs for about $\frac{1}{2}$ s on power-up.

2.18.5 Troubleshooting VIF and SIF Circuits

The recommended approach for troubleshooting IF stages depends largely on the available test equipment and type of IF. Ideally, an analyzer or NTSC generator is used because such generators duplicate the signals normally found at the tuner output (IF input) and video-detector output (as well as several other signals).

If the picture display is good with a signal injected at the video-detector output (pin 20 of IC_{201}, Fig. 2.24), but not good with a signal injected at the IF input (pins 1 and 2), the problem is in the IF. A possible exception is when the AGC circuits are defective. This problem can be eliminated as a cause of trouble by clamping the AGC line with the appropriate voltage (say, with a fixed dc voltage between pin 28 of IC_{201} and ground). If the problem remains with the correct AGC bias applied, the problem is in the IF.

If an analyst or NTSC generator is not available, the next recommended test setup is a sweep generator (with markers) and a scope (Sec. 2.7). The sweep-generator signal is injected at the IF input (pins 1 and 2 of IC_{201}), and the signal is monitored at various points throughout the IF with a scope. The sweep generator is most effective when troubleshooting IF stages similar to those in Fig. 2.22 because such stages require alignment and trap adjustment. When the IF stages are combined with other functions in an IC, such as shown in Fig. 2.24, the analyst or NTSC generator is far more practical.

2.18.6 Typical VIF and SIF Troubles

The following sections discuss symptoms that could be caused by defects in the IF and video-detector circuits.

No Picture or Sound.   If a raster is present but there is no picture or sound, or the sound is very weak and noisy, inject an IF signal at the IF input. Also inject an IF signal at the IF output of the tuner (TU_{101} in Fig. 2.24). If operation is normal with the signal injected directly at IC_{201}, but not when injected at the TU_{101} output, suspect Q_{221} and/or SWF_{201}. If operation is normal with the signal injected at the TU_{101} IF output, the problem is in the tuner (Sec. 4.1) rather than in the IF.

If the fault appears to be isolated to the IF stages (or the IF IC), clamp the AGC line and repeat the test. In the circuits of Figs. 2.24 and 2.25, the AGC voltage (at pin 28 of IC_{201}) is in the range from 4 to 8 V and can be adjusted by RV_{201}. First try adjustment of RV_{201}. This might cure the problem. If not, clamp the AGC line with a fixed dc voltage. If this cures the problem, suspect trouble in the AGC circuits within IC_{201}.

Notice that the IF ICs in most sets provide AFT signals to the tuning system (such as the AFT up-down signals from pins 5 and 6 of IC_{201} to the tuner circuits of Fig.

4.4). If these signals are absent or abnormal, the tuner will not fine-tune at the channel, even though the channel is locked in by the H-sync signal as described in Sec.
4.1.12. This makes the tuner and/or tuning-system microprocessor appear to be bad
when the problem is actually in the IF IC. Obviously, if any circuit within the IF IC is
defective, the entire IC must be replaced (as is the case with any IC).

Poor Picture or Sound.   The basic procedure for troubleshooting a poor picture and
sound symptom (weak sound, poor contrast, etc.) is the same as for no sound or picture. However, it is sometimes possible to cure poor picture and sound with adjustment (such as the adjustment points shown earlier in Figs. 2.24 and 2.25).

Hum Bars and Hum Distortion.   Hum is generally the fault of the power-supply filters. One possible exception is leaking of vertical-blanking pulses into the IF stages or
IC (say, through an open decoupling or bypass capacitor).

Picture Smearing, Pulling, or Overloading.   Generally, these symptoms are associated with the video-processing circuits (Sec. 3.11) rather than the IF, particularly when
sound is good. However, if the IF stages are improperly aligned or adjusted, or if there
is a defective part making proper IF alignment impossible, the same symptoms can
occur even though the sound might be good.

To eliminate doubt, inject a composite video signal with an analyst or NTSC generator at the video-detector output (pin 20 of IC_{201}). If the picture problems are eliminated, suspect the IF IC. Next, clamp the AGC (and/or adjust the AGC) to see if the problem is cleared. If the problem remains with good AGC, replace the IF IC (or try IF
alignment and adjustment).

Intermittent Problems.   When *both* the picture and sound show intermittent problems, suspect both the tuner and IF. However, the AGC circuit or power supply could
be at fault. Monitor both the tuner output (IF output of TU_{101}) and the IF output (pin
20 of IC_{201}). Watch for changes when the intermittent condition occurs. As a minimum, monitor the IF IC video output, AGC line, all dc voltages, and any tuning control signals (from tuner microprocessor, memory, bandswitch IC, etc.). As in the case
of any intermittent condition, look for bad solder joints, breaks in PC wiring, intermittent capacitors, and even intermittent transistors. Also try tapping (not pounding) the
IF and tuner modules. This can sometimes quickly pinpoint an intermittent module.

2.19 TROUBLESHOOTING VERTICAL AND HORIZONTAL SWEEP CIRCUITS

This section is devoted to TV-set circuits that provide the vertical and horizontal
sweep for the CRT. Because such circuits are closely associated with the high-voltage
and flyback-transformer (FBT) stages, as well as the sync separator, these circuits are
also included in this section. The low-voltage power-supply circuits are covered in
Sec. 2.17.

Vertical and horizontal circuits have several functions, and there are many circuit
configurations. Thus, it is very difficult to present a typical configuration. However,
most vertical and horizontal circuits have certain inputs and outputs that can be monitored. If the inputs are normal, but one or more of the outputs are abnormal, the problem can be localized to the vertical and horizontal circuits.

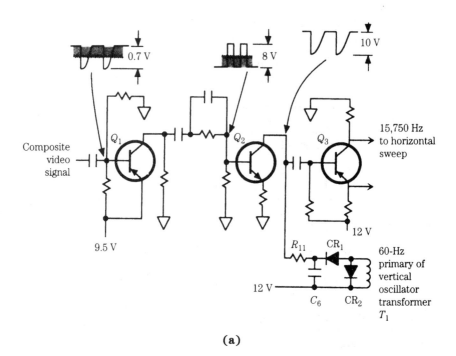

(a)

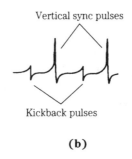

(b)

FIGURE 2.27 Signal path through sync-separator circuits.

2.19.1 Sync-Separator Basics

Figure 2.27 shows the signal path through the sync-separator circuits of a typical discrete-component TV set. In most present-day IC sets, the sync separators are within an IC, although they might also include external circuits.

The function of the sync separator is to remove the vertical and horizontal sync pulses from the video circuits and apply the pulses to the vertical and horizontal sweep circuits. The sync-separator circuits also function as clippers and/or limiters to remove the video signal (picture) and any noise. Thus, the sweep circuits receive sync pulses only and are free of noise and/or video signal (in a good set).

In the circuit of Fig. 2.27, the video input is negative-going and the input transistor

Q_1 is PNP. Thus, both the sync pulses and the signal or noise turn Q_1 on. However, with Q_1 biased near zero, the large sync pulses drive Q_1 into saturation at a level never reached by the signal and/or noise. The second sync separator Q_2 is reverse-biased so that the first portion (about 0.3 V typically) of Q_1 output is clipped. This further removes any signal and/or noise.

The output of Q_2 is applied to the low-pass filter (called the vertical integrator) of C_6 and R_{11}. The 60-Hz vertical signals are applied to the vertical-sweep circuit input through CR_1 and CR_2. These diodes pass the sync pulses to the vertical-sweep input but prevent the vertical oscillator pulses from passing into the sync circuits. The Q_2 output is also applied to Q_3, which acts as a phase splitter or phase inverter. The output from Q_3 is applied to the horizontal circuits.

2.19.2 Vertical-Sweep Basics

Figure 2.28 shows the signal path through the vertical-sweep circuits of a typical discrete-component TV set. In most IC sets, the vertical-sweep circuits (vertical oscillator and vertical drive or output) are within one or two ICs, although they might also include external circuits.

The vertical-sweep circuits provide a vertical-deflection voltage (vertical sweep) to the CRT deflection yoke. The vertical circuits also supply a blanking pulse to the CRT. This blanks the CRT during retrace of the sweep. The vertical-sweep signals are at a frequency of 60 Hz and are synchronized with the picture transmission by the sync signals taken from the sync separator. The vertical circuits of most IC sets also provide for distortion correction.

In the circuit of Fig. 2.28, the vertical oscillator is of the blocking type, operating at a frequency of 60 Hz and producing a pulse output of about 1 to 2 V. The oscillator pulse output is modified into a sawtooth sweep and is applied through the driver to the output stage. The final output is peaked sawtooth waveform. The sawtooth portion (about 4 to 5 V) is applied to the CRT deflection yoke, with the peaked portion (about 50 V) applied as a blanking pulse to the CRT.

The oscillator is locked in frequency by the sync pulses. Vertical hold R_2 is set so that the vertical sweep is locked with picture transmission. The Q_1 pulse output is converted to a sawtooth sweep by the C_1 and R_1 emitter network. The sweep is made linear through feedback and is adjusted with linearity control R_8. The amplitude of the sawtooth waveform to driver Q_2 (and thus the vertical size of the picture raster) is set by vertical-size control R_6. The emitter of Q_2 receives negative feedback from the yoke to provide a more linear output.

The deflection yoke requires heavy current. For this reason, output transistor Q_3 is of the power type, mounted on a heat sink (in discrete sets). Because there is always the danger of thermal runaway in power transistors, thermistor R_3 is included in the base circuit. Notice that if it is necessary to replace Q_3, it might be necessary to adjust vertical-bias control R_{15}. Most discrete sets have a vertical-bias control because the characteristics of a replacement transistor can be different from the original. In IC sets, the output transistor is replaced by an IC.

2.19.3 Horizontal-Sweep Basics

Figure 2.29 shows the signal path through the horizontal-sweep circuits of a typical discrete-component TV set. In most IC sets, the horizontal sweep circuits (horizontal oscillator, AFC, and horizontal drive) are within one or two ICs, with the horizontal

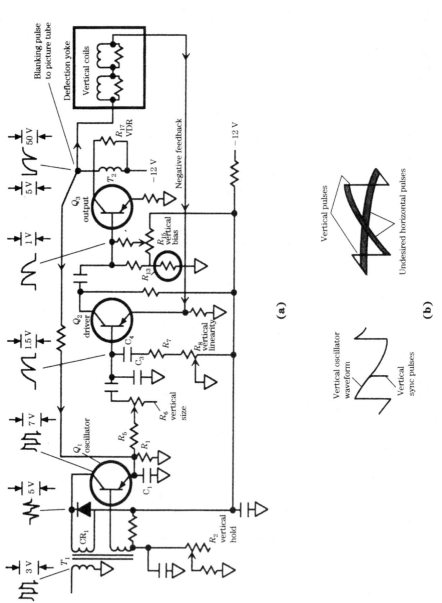

FIGURE 2.28 Signal path through vertical-sweep circuits.

2.74

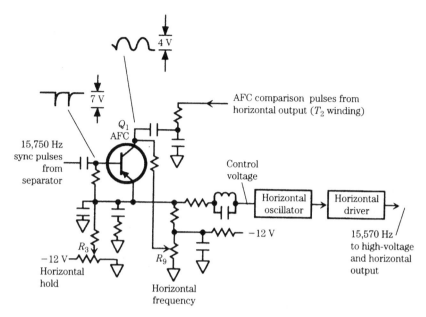

FIGURE 2.29 Signal path through horizontal-sweep circuits.

output in discrete form (possibly a power transistor with a built-in damper diode), as shown later in Figs. 2.31, 2.34, and 2.35.

The horizontal oscillator and driver circuits provide signals to the horizontal output and high-voltage circuits. The signals are at a frequency of 15,750 Hz and are synchronized with the picture transmission by sync signals from the sync-separator circuits.

Both discrete and IC horizontal-oscillator circuits include some form of AFC system to ensure that the horizontal-sweep signals are synchronized with picture transmission, despite changes in the line voltage and temperature (or minor variations in circuit values). The AFC action is produced by comparison of the sync signals with horizontal-sweep signals.

Deviations of the horizontal-sweep signals from the sync signals cause the horizontal oscillator to shift in frequency or phase as necessary to offset the initial (undesired) deviation. For example, if the horizontal sweep increases in frequency from the sync signals, the horizontal oscillator is shifted in frequency by a corresponding amount, but in the opposite direction (a decrease in frequency to offset the undesired change).

In the discrete circuit of Fig. 2.29, the Q_1 base is driven by the sync pulses, with the collector receiving comparison pulses from the horizontal output. The emitter waveform is a combination of both pulses. When there is a change in phase or frequency between the two sets of pulses, the emitter current changes. In turn, the control voltage applied to the horizontal oscillator changes in a direction that corrects the frequency and phase of oscillation.

In the IC circuit of Fig. 2.34 (shown later), the AFC circuit in IC_{501} receives sync pulses from pin 1 and AFC comparison pulses from the horizontal output. The dc output from the AFC circuit is applied to the horizontal oscillator (locked to the sync pulses). The horizontal-oscillator frequency can be adjusted (within narrow limits) by RV_{506}.

2.19.4   High-Voltage and Horizontal-Output Basics

Figure 2.30 shows the signal path through the high-voltage and horizontal-output circuits of a typical discrete-component TV set. Most present-day IC circuits include a few additional features, as discussed in Sec. 2.19.5. However, the basic elements are covered in Fig. 2.30. Figure 2.31 shows the relationship of the circuits to other stages in a typical IC set (the 13-in Sony).

The circuit of Fig. 2.30 provides a high voltage for the CRT and a sweep voltage for the horizontal yoke. In most cases, the circuit also supplies a boost voltage for the CRT focus and screen or accelerating grids and possibly a voltage for other components that require voltages higher than that which is available from the low-voltage supply. The horizontal-output circuits also supply an AFC signal to control the frequency of the horizontal oscillator. In most IC sets, an AGC winding is not required because the AGC signal is developed by the VIF and SIF circuits rather than by a keyed-AGC circuit found in discrete sets.

Although there is little standardization, most horizontal-output circuits have certain characteristics in common that must be considered during the troubleshooting process. The circuit receives pulses from the horizontal driver (at 15,750 Hz) synchronized with picture transmission. Transistor Q_1 is normally biased at or near zero so that one edge of the pulse drives the transistor into heavy conduction (near saturation). The opposite swing of the pulse cuts off Q_1. In effect, Q_1 is operated in the switching mode. Notice that Q_1 in Fig. 2.30 corresponds to Q_{502} in Fig. 2.35 (shown later). Also, T_1 corresponds to T_{501}, T_2 corresponds to T_{503}, and damper diode CR_2 is built into Q_{502}.

The Q_1 collector current is applied through a winding on the flyback transformer T_2, resulting in a pulse waveform at the other windings. The high voltage is rectified and applied to the CRT anode. The boost output is rectified and applied to the CRT focus and accelerator grid. On some sets, the horizontal-output transistor current is applied through the deflection yoke. On other sets, the transformer has a separate winding for the yoke. Either way, the horizontal-output pulses produce the horizontal sweep.

Because of the great variety in horizontal-output circuits, it is difficult to arrive at a typical theory of operation. However, the important point to consider in practical troubleshooting is that the damping diode CR_2 (built into Q_{502}, Fig. 2.35, shown later) starts to conduct when Q_1 is cut off. Diode CR_2 continues to conduct until Q_1 conducts.

During the horizontal forward scan (when the picture is displayed), CR_2 conducts and Q_1 is cut off from the start of the sweep to about midpoint. Then CR_2 is cut off, and Q_1 conducts for the remaining half of the sweep. This sequence is shown in Fig. 2.30b.

The scan sequence is important in troubleshooting some sets because any problems in the right-hand side of the picture are probably the result of defects in Q_1 (or related components), whereas trouble in the left-hand side is probably the result of a defective CR_2. Of course, where CR_2 is built into Q_1, both are replaced simultaneously if either is defective. Such is the case for a circuit as shown later in Fig. 2.35.

When Q_1 is cut off, the CRT is blanked, and current flows rapidly through the horizontal yoke in the opposite direction, pulling the electron beam back to the left side of the screen (known as the *horizontal retrace* or *flyback*). It is the pulse developed during the flyback interval that is used by the other windings on flyback transformer T_2.

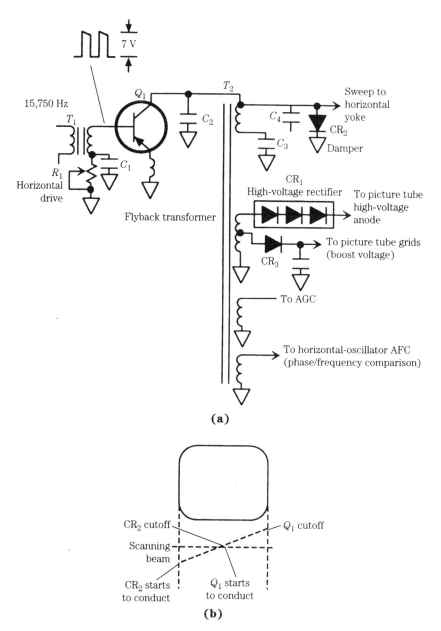

(a)

(b)

FIGURE 2.30   Classic high-voltage and horizontal output circuit.

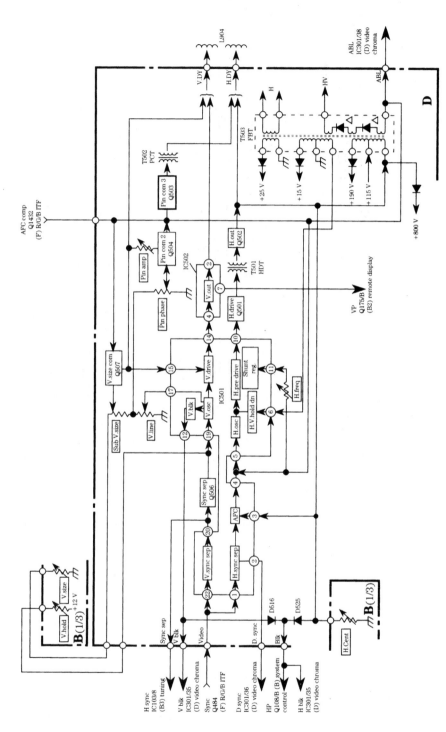

FIGURE 2.31 Horizontal and vertical sweep.

2.78

2.19.5   Basic IC Vertical and Horizontal Sweep

Figure 2.31 shows the circuits necessary to generate the vertical and horizontal drive pulses (synchronized to the transmitted sync signals) and produce a raster on the CRT screen. In this particular IC set (13-in Sony), many of the functions are performed in IC_{501}. Notice that the vertical-sweep signals applied to the vertical CRT yoke are output from IC_{502}. The horizontal yoke receives sweep signals from discrete components (drive Q_{501}, transformer T_{501}, and output Q_{502}).

2.19.6   Vertical Oscillator

Figure 2.32 shows the vertical-oscillator circuits for a typical IC set (13-in Sony). The main purpose of the circuit is to provide drive pulses for the vertical-output stages that produce sweep signals at the vertical yoke. In addition, the circuits separate the vertical and horizontal sync signals present in the video-detector output.

The composite video (via the RGB interface, Sec. 3.12) is applied to pin 22 of IC_{501} through C_{509}, R_{506}, and C_{508}, which act as a low-pass filter to remove video and chroma information (leaving only sync). The sync pulses at pin 22 are amplified and applied to pin 20. (Although the circuit between these two pins is called a separator, the function is more like that of an amplifier because both vertical and horizontal sync appears at pin 20.)

The sync information is applied to the tuning circuits (Fig. 4.4), and to sync separator Q_{506}. The output from Q_{506} is used to lock the vertical oscillator in IC_{501} with the transmitted sync. The vertical-oscillator frequency can be adjusted by RV_{505}. The amplitude of the vertical-drive signal to the vertical-output circuits (and thus the vertical height of the CRT display) is adjusted by RV_{504} and RV_{425} on the RGB interface board (Sec. 3.12).

A correction signal from the vertical yoke (Fig. 2.33) is applied to the vertical-drive circuit at pin 15 of IC_{501}. This feedback signal (known as the *parabola* or *S* correction signal) serves to maintain linearity of the drive signal at pin 14. The same feedback signal is applied to the pincushion-correction circuit as described next.

2.19.7   Vertical Output

Figure 2.33 shows the vertical-output circuits for a typical IC set (13-in Sony). The main purpose of the circuit is to provide sweep signals for the vertical yoke. In addition, the circuits provide signals necessary for pincushion correction in the horizontal sweep.

The vertical-drive signals from the vertical oscillator (Fig. 2.32) are applied to pin 4 of IC_{502}. The deflection-output circuits within IC_{502} are powered by the rectified output from flyback transformer T_{503}. The $+25$ V at D_{514} is doubled by D_{505} and C_{532} to produce a vertical-output supply voltage of about 40 to 45 V. It is important to note that the vertical-output stages of IC_{502} operate solely from the 25 V, taken from T_{503} (Fig. 2.35, shown later). This is done so that any variation in horizontal sweep will have a corresponding variation in the vertical sweep, thus maintaining a constant horizontal and vertical aspect ratio of 4:3.

The vertical-pulse output from pin 7 of IC_{502} is applied to the remote and display circuits. The vertical-sweep output from pin 2 of IC_{502} is centered by switch S_{501}, which has three positions. Each position applies a different dc level to the vertical yoke and thus moves the vertical display up or down.

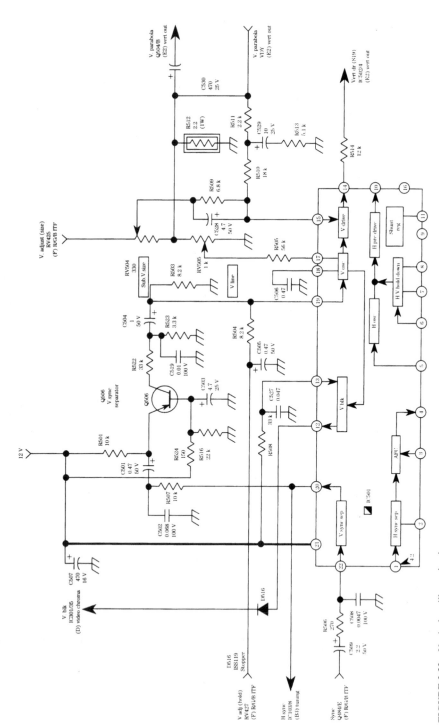

FIGURE 2.32 Vertical oscillator circuits.

2.80

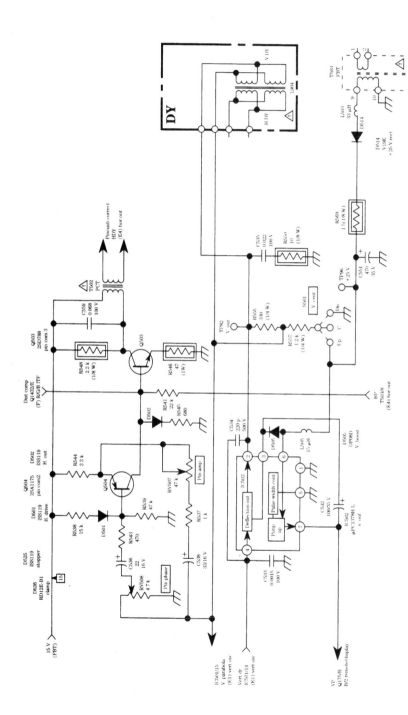

FIGURE 2.33 Vertical output circuits.

A sample of the vertical-output signal (the vertical parabola) is returned to the vertical-drive circuits to maintain linearity. The same signal is also applied to the horizontal yoke (Fig. 2.35, shown later) through Q_{504} and Q_{503}, and T_{502}, along with a distortion-compensation signal from the RGB interface (Sec. 3.12). The combined signals minimize the pincushion effect (where the CRT display appears to bend outward). RV_{507} and RV_{508} provide for pincushion adjustments.

2.19.8 Horizontal Oscillator

Figure 2.34 shows the horizontal-oscillator circuits for a typical IC set (13-in Sony). The main purpose of the circuits is to provide drive pulses for the horizontal-output stages that produce sweep signals at the horizontal yoke. In addition, the circuits provide for shutoff of the horizontal output and high-voltage stages in the event of excessive CRT voltages.

The composite video (via the RGB interface, Sec. 3.12) is applied to pin 1 of IC_{501} through C_{510}, R_{521}, and C_{511}, which act as a low-pass filter to remove video and chroma information (leaving only sync). The sync pulses at pin 1 are amplified and applied to the AFC circuits within IC_{501}. (Again, the circuit at pin 1 is essentially an amplifier, although it is called a separator.) A delayed sync (D-sync) output at pin 2 is applied to the video and chroma circuits (Fig. 3.26) for burst-gating (as described in Sec. 3.11).

The sync information from the separator is applied to the AFC and compared with a reference horizontal signal at pin 3. The AFC reference signal is taken from the horizontal-output transistor Q_{502} (Fig. 2.35). The AFC compares both signals and produces a correction signal at pin 4. The correction signal locks the horizontal oscillator in IC_{501} to the transmitted sync. The horizontal-oscillator frequency can be adjusted by RV_{506}.

2.19.9 Horizontal Output and High Voltage

Figure 2.35 shows the horizontal-output and high-voltage circuits for a typical IC set (13-in Sony). The main purpose of the circuit is to provide signals for the horizontal yoke. In addition, the circuits provide a number of sync and control signals to other stages.

The horizontal-drive signals from the horizontal oscillator/predriver (Fig. 2.34) are applied to the flyback transformer T_{503} and the horizontal yoke through horizontal drive Q_{501}, horizontal-pulse transformer T_{501}, and horizontal output Q_{502}. (Notice that Q_{502} has a built-in damper diode.) The output from Q_{502} applied to the yoke produces the horizontal sweep. This same output applied to one winding of T_{503} induces various voltages and signals at the other windings, as follows.

The ac output at pins 1 and 2 is used for the CRT heater. The dc output at pin 3 is the CRT high voltage. The ac output at pin 4 is rectified by D_{511} and used as the CRT screen-grid or accelerator-grid voltage. The ac output at pin 6 is rectified by D_{515} and used as the CRT focus voltage.

The ac output at pin 9 is rectified by D_{514} and used as the $+25$-V supply for the vertical output IC_{502} (Fig. 2.33). The ac winding at pin 7 is rectified by D_{510} and D_{513} and is used as the $+12$-V supply (Fig. 2.34).

The output at pin 8 is used by several circuits. First, the output is applied to pin 38 of IC_{301} (Fig. 3.26) and is used as the ABL, or automatic brightness limiter, discussed in Sec. 3.11. The same output is applied to the base of Q_{503} as part of the pincushion-correction function (Fig. 2.33). The output is also applied to the horizontal oscillator (Fig. 2.34) at pin 5 of IC_{501} as a feedback signal to hold the oscillator on frequency.

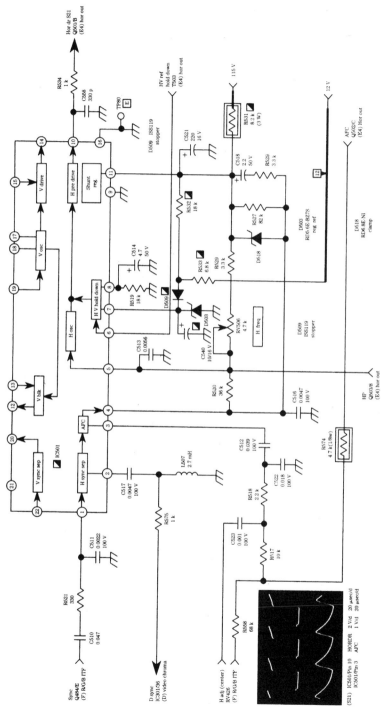

FIGURE 2.34 Horizontal-oscillator circuits.

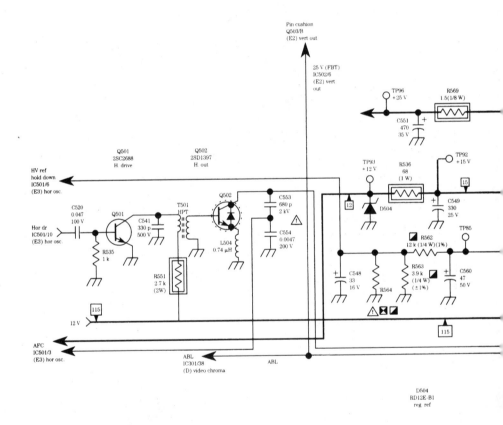

FIGURE 2.35 Horizontal output and high-voltage circuits.

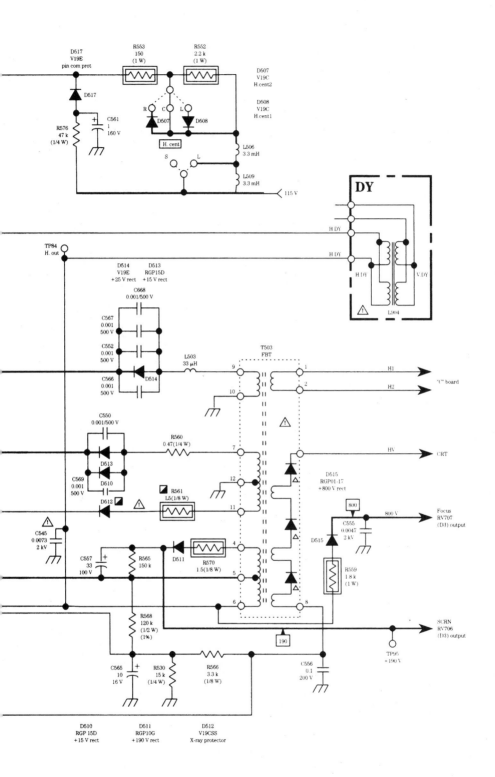

2.85

The output at pin 11 (rectified by D_{512}) is used as the high-voltage hold-down signal applied to pin 6 of IC_{501} (Fig. 2.34). The rectified output from D_{512} is compared to a fixed reference voltage at pin 7 of IC_{501}. If the voltage developed by T_{503} exceeds a safe level (where dangerous x-rays might be emitted), the voltage at pin 6 of IC_{501} varies sufficiently from the reference at pin 7, and the hold-down circuit in IC_{501} disables the horizontal oscillator. This removes all power from T_{503} and the CRT.

Once the high-voltage hold-down circuit is operated, it is necessary to remove power from the circuit and then restore power. If the malfunction is cleared, the circuits will operate normally when power is reapplied. If not, the hold-down circuit will again turn off the horizontal oscillator.

The horizontal-sweep output is centered by solder-bridge adjustments that select corresponding diodes D_{507} or D_{508}. The diodes apply different dc levels to the horizontal yoke and thus move the horizontal display left or right.

2.19.10   Troubleshooting Vertical and Horizontal Sweep Circuits

Because of the interdependence of these circuits, it is difficult to present a standard troubleshooting approach. Failure in one circuit can produce symptoms similar to those produced by failure in other circuits. For example, if the low-pass filter (vertical integrator) fails, there is a loss of vertical sync or poor vertical sync. This same symptom can be produced by failure of the vertical-sweep circuits as well. Also, operation of the sync-separator circuits depends on signals from other circuits. For example, if the video output is low, the sync-separator output is low (or possibly absent).

The first practical step in troubleshooting these circuits is to try adjusting the horizontal and/or vertical controls (Figs. 2.32 through 2.35). If this fails, proceed with the usual voltage and waveform measurements. Remember that if it is necessary to set a control to either extreme, or if the control must be reset repeatedly, look for marginal component failure (leaking capacitors, transistors, etc.).

2.19.11   Typical Sync-Separator Troubles

The following sections discuss symptoms that could be caused by defects in the sync-separator circuits.

No Sync.   If both horizontal and vertical sync are absent, the first step is to check for proper sync pulses at the input of the sync separator (from the video detector and possibly through a video amplifier, such as at pins 1 and 22 of IC_{501} in Fig. 2.31). If these pulses are abnormal (particularly if low in amplitude), both the horizontal and vertical sync outputs from the separator will be abnormal.

If the sync pulses to the separator are good, check the sync pulses at the end of the *last point common to both horizontal and vertical sync.* In the circuit of Fig. 2.32, check at pins 19 and 20 of IC_{501}. There should be both vertical and horizontal sync at pin 20 but only vertical at pin 19.

No Vertical Sync.   If vertical sync is absent or critical (requires frequent adjustment) but there is good horizontal sync, the problem is in the sync separator or vertical sweep. In the circuit of Fig. 2.32, if there is vertical sync at pin 20 of IC_{501}, but not at pin 19, suspect Q_{506}. On the other hand, if there is vertical sync at pin 19, but the vertical drive at pin 14 is not synchronized, suspect IC_{501}. Of course, before you pull IC_{501}, try correcting the problem by adjustment.

When measuring waveforms of vertical and horizontal sync pulses, it is convenient to set the scope sweep to 30 Hz (for vertical) and 7875 Hz (for horizontal). This displays two cycles of the corresponding pulses.

When measuring ahead of the vertical integrator or sync separator (the last point common to both horizontal and vertical sync), look for any noise or video signal that might have leaked through. For example, in the circuit of Fig. 2.32, there should be little noise or video at pin 22 of IC_{501} and no noise or video at pin 20. If there is excessive noise at pin 22, suspect the low-pass filter formed by R_{506} and C_{508} (this is not the vertical integrator). If there is no reduction in noise or video at pin 20 from that for pin 22, suspect IC_{501}.

When measuring the output of the vertical integrator or sync separator, look for kickback voltages or spikes from the vertical-sweep circuit. Such kickback produces a display similar to that of Fig. 2.27. Diodes CR_1 and CR_2 in Fig. 2.27 are included on some discrete circuits to prevent the kickback from reaching the separator. In the circuit of Fig. 2.32, where the vertical oscillator is part of IC_{401}, kickback is usually not a problem. However, if you have mysterious sync problems, look for vertical-oscillator signals that might have leaked from pin 19 back through Q_{506} to pin 20.

No Horizontal Sync.   If horizontal sync is absent or critical, but there is good vertical sync, the problem is in the horizontal portion of the sync separator or in the horizontal sweep. In the circuit of Fig. 2.34, check for horizontal sync at pins 1, 2, 4, 5, and 10. If there is good vertical sync, but the horizontal-sync pulses at pin 1 are absent or abnormal, suspect C_{510}, R_{521}, and C_{511}. (Again, R_{521} and C_{511} form a low-pass filter to remove noise and video from the sync line.) If there are good horizontal-sync pulses at pin 1, but not at pins 2, 4, or 5 or there is no drive output at pin 10, suspect IC_{501}. (Try adjustment before you pull IC_{501}.) Also look for problems in external components, such as R_{520} between pins 4 and 5.

Picture Pulling.   There are several forms of picture pulling. Often, the nature of the picture-pulling symptoms can pinpoint the trouble. For example, when picture pulling appears to be steady, and there is a bend in the image, this usually indicates that the horizontal AFC circuits are receiving distorted pulses. A possible cause is vertical kickback entering the sync separator and distortion of the horizontal output. Check the related pulses against those shown in the service literature.

If the picture pulling appears unsteady, and particularly if the pulling tends to follow the camera signal, this indicates poor sync separation. That is, the video output is not being clipped and limited sufficiently to remove all camera signals from the circuits. This problem is more likely in a circuit where the sync separator is discrete, such as shown in Fig. 2.32, than in a circuit where all the separator circuits are ICs.

Do not confuse picture pulling with distortion of the raster. If the raster is bent, the problem is not in the sync separator but in the horizontal or vertical sweep circuit (or possibly the yoke). If the raster edges are sharp, but the picture is pulling or bent, the problem can be in either the sweep or separator circuits. Check the raster alone by switching to an unused channel (or switch to the video function if the set is a TV monitor).

2.19.12   Troubleshooting Vertical-Sweep Circuits

Complete failure of the vertical sweep is an easy symptom to recognize and is usually easy to locate. With complete failure, there is no vertical sweep and the CRT display is a horizontal line. (With any TV set or monitor, do not operate the set when there is a complete failure of the vertical sweep. The bright horizontal line can burn out the CRT screen.)

In the case of complete vertical-sweep failure (all other functions normal), the first practical step is to check the waveforms at all points. If the oscillator waveform is bad, the problem is quickly localized to the oscillator stage. (However, in some vertical circuits, the oscillator does not go into oscillation unless there are sync pulses present, so always check for sync pulses first.) If all waveforms appear normal, but there is no vertical sweep, the deflection yoke is a logical suspect.

In IC sets, it is usually necessary to trace vertical-sweep waveforms through at least two ICs. For example, as shown in Fig. 2.31, the vertical oscillator and drive can be checked at pin 14 of IC_{501} (or at pin 4 of IC_{502}). The vertical-sweep output to the yoke can be measured at pin 2 of IC_{502}. (In this particular set, there is a vertical-output test point TP_{82}, as shown in Fig. 2.33.)

In the case of marginal failure (loss of sync, critical sync, distortion, vertical nonlinearity, line pairing or splitting, poor interlace, or lack of vertical height), the first step is to try correcting the problems with adjustment. If the problem is not eliminated or if it is necessary to set a control to an extreme, suspect marginal component failure. Look for leaking capacitors, transistors, or diodes; worn potentiometers; or open capacitors.

2.19.13 Typical Vertical-Sweep Troubles

The following sections discuss symptoms that could be caused by defects in the vertical-sweep circuits.

Insufficient Height.   The first step for this symptom is to adjust the vertical-height control (sometimes called the drive or size control). If the height control must be full on or nearly full on to get proper height, suspect the vertical drive. Notice that in some sets, there are two vertical-size adjustments. For example, as shown in Figs. 2.31 and 2.32, there is a vertical-size adjustment on the B-board and a subvertical-size adjustment RV_{504} on the main D-board. If it is impossible to get the proper vertical height by adjustment, look for leaking capacitors, worn controls, and leaking transistors. Again, check the vertical waveforms from oscillator to yoke.

Vertical-Sync Problems.   The first step for this symptom is to adjust the vertical hold (sometimes called the *lock* or *sync* control), such as the vertical-hold control shown connected to pin 19 of IC_{501} in Fig. 2.31. If the hold control must be fully on or fully off to get sync, or if it is necessary to readjust the control frequently, there is a problem in the vertical-sync circuits. Look for leaking capacitors, worn controls, and leaking transistors.

In the case of a very critical vertical sync (difficult to adjust or does not hold after adjustment), observe the amplitude of the vertical-sync pulse riding on the sawtooth portion of the vertical-oscillator or drive waveform (Fig. 2.28*b*). If the sync amplitude does not remain constant, look for problems in the sync separator (Fig. 2.32) rather than in the vertical-sweep circuit.

Vertical Distortion (Nonlinearity).   There are several forms of vertical distortion. Some are easy to recognize, such as *keystoning,* where one side of the picture is much larger than the opposite side. Keystoning is generally caused by a defect in the deflection yoke or related parts (Fig. 2.33), but it could be caused by a defect in the vertical output IC_{502}.

Other forms of vertical distortion are not so easy to recognize. In the extreme, there is compression of the picture at the top with the picture spreading at the bottom, or vice versa. Often the compression or spreading is slight. Use a crosshatch pattern from a TV generator to check linearity. For a quick check of linearity, adjust the vertical hold to produce slow rolling. Watch the blanking bar as the bar moves up or down on the screen. The blanking bar should remain constant (in vertical height) at the bottom, middle, and top of the screen.

Vertical distortion can be caused by improper adjustment, as well as by defective circuits. One problem here is the interaction of controls and components. For example, if the sawtooth sweep is low in amplitude (because of a defect not associated with nonlinearity), the height control can be adjusted to get proper height. These extremes in circuit resistance might make it impossible to get a linear picture no matter how the linearity control is adjusted.

If the problem cannot be cured by adjustment, the next step is to localize the fault with voltage and waveform measurements. Remember that the vertical sweep cannot be linear if the sawtooth waveform is not linear. Of course, if the sweep output from the final stage (such as at pin 2 of IC_{502} in Fig. 2.33) is good, but the vertical display is not linear, suspect the yoke and related parts.

Line Splitting. The problem of line splitting (line pairing or poor interlace) is often associated with vertical sweep. Although the trouble appears in the vertical sweep, the actual cause is often in related parts. For example, the trouble can be caused by an open capacitor in the sync separator or by horizontal pulses leaking back into the vertical sweep.

If you suspect horizontal pulses in the vertical sweep, check the sweep waveform with the scope retrace-blanking function disabled. Any horizontal pulses mixed with the sweep will appear as a pulse train on some portion of the vertical-sweep output waveform (Fig. 2.28*b*). If the horizontal pulses are present, the most likely causes are in the horizontal circuits, such as corona discharge from the high-voltage rectifier, leakage between leads (not too common in PC wiring, except where there are solder splashes or arc burns), and breakdown in the high-voltage section (resulting in spark discharge being picked up by the vertical circuits).

If there are no horizontal pulses in the vertical circuits, but there is definite line splitting, suspect the sync circuit, particularly the sync separator. When splitting is intermittent, suspect the high-voltage section (spark discharge, radiation from high-voltage lead, etc.).

2.19.14   Troubleshooting Horizontal-Oscillator Circuits

Failure of the horizontal-oscillator and related circuits (AFC and horizontal drive) can produce symptoms similar to those produced by failure in other circuits. For example, if the horizontal oscillator stops oscillating, there is no drive to the horizontal output. Thus, there is no high- or boost-voltage output, and the CRT screen is dark. The same symptom can be produced by failure of the horizontal output circuit, the low-voltage supply (Sec. 2.17), and the CRT. Also, in IC sets such as shown in Fig. 2.34, there are circuits (high-voltage hold-down) that shut the horizontal oscillator down when the CRT high voltages reach dangerous levels (and could produce x-ray radiation).

Operation of the horizontal oscillator depends on signals from other circuits. For example, the AFC circuit must have sync signals from the sync separator and comparison signals from the horizontal output. Because of these conditions, the only practical troubleshooting approach is to check waveforms of all outputs and inputs, followed by voltage measurements.

If the driver output is absent or abnormal with good sync and comparison-pulse inputs, the problem is in the horizontal-oscillator or driver sections. A waveform measurement at the input of the horizontal-driver or horizontal-oscillator output localizes the problem further. In the circuit of Fig. 2.34, IC_{501} contains both the horizontal oscillator and a predriver. The horizontal driver and output are in discrete form (Q_{501} and Q_{502}, Fig. 2.35). Generally, if the driver output is present but the symptoms point to a horizontal circuit failure (horizontal pulling, jitter, distortion, etc.), the problem is most likely in the AFC.

2.19.15 Typical Horizontal-Oscillator Troubles

The following sections discuss symptoms that could be caused by defects in the horizontal-oscillator or driver circuits.

Dark Screen.   If the CRT screen is dark (no raster), but sound is normal (indicating that the low-voltage supply is probably good), the first step is to measure the waveform at the horizontal-drive output (and at the predrive output, pin 10 of IC_{501}, Fig. 2.34). If the waveform is normal, the problem is in the horizontal output and high-voltage circuit or the CRT. If the waveform is not normal (weak, distorted, etc.), the problem is probably in the horizontal oscillator (or the high-voltage hold-down circuit has turned the oscillator off).

Before checking individual parts, there are some checks that help isolate the problem: waveforms at sync input (from sync separator), comparison-pulse input, horizontal-oscillator (or predriver) output, and voltage at all transistor elements and corresponding IC pins.

If the sync pulses are absent or abnormal, the problem is in the sync separator rather than in the horizontal section. If the comparison pulses are not normal, with a good output from the horizontal drive, the problem is in the horizontal output and high-voltage section.

If the sync and comparison pulses are both normal, but the horizontal output is absent or abnormal, the problem is probably in the AFC section or in the horizontal oscillator. Obviously, if any section of IC_{501} of Fig. 2.34 is defective, IC_{501} must be replaced (at great expense to the customer).

Narrow Picture.   The most logical cause of a narrow picture that cannot be corrected by adjustment is an insufficient horizontal drive. In circuits such as shown in Figs. 2.34 and 2.35, the first step is to isolate the problem to the IC (monitor at pin 10 of IC_{501}) or to the discrete components (monitor at Q_{501}, T_{501}, and Q_{502}).

Horizontal Pulling or Improper Phasing; Loss of Sync.   Horizontal pulling is present when the picture pulls and appears in diagonal form. If the picture pulls completely into diagonal lines, this indicates a complete loss of sync. The direction of the slant can provide a clue to the problem. If the lines slant to the right, the oscillator frequency is high, and vice versa. If the picture shifts to the right or left so as to be decentered, suspect incorrect phasing. That is, the horizontal oscillator is on frequency but not in exact phase with the sync signals.

It is possible that any of these problems can be the result of improper adjustment, so the first step is to adjust all horizontal controls. If this does not clear the problem, check all of the waveforms, transistor voltages, and related IC pins. Pay particular attention to the sync and comparison pulses applied to the AFC. If either of these is absent or abnormal, the AFC circuits cannot operate properly (even though the IC is good). For example, if the comparison pulses at pin 3 of the IC_{501} in Fig. 2.34 are absent, horizontal sync might not be completely lost. However, the horizontal-frequency adjustment RV_{506} will become very critical, and phase shift occurs (the picture is decentered).

Horizontal Distortion. There are many forms of horizontal distortion that can be caused by defects in the horizontal-oscillator and driver circuits. The so-called "pie-crust" distortion is a typical example. With pie crust, the picture image appears to be made up of wavy lines, even though there is no pulling, loss of sync, or jitter. Such distortion is almost always the result of marginal performance in a component rather than complete failure. The most likely defects are capacitors, particularly the capacitors that filter the control voltage from the AFC to the horizontal oscillator (such as C_{516} in Fig. 2.34).

2.19.16 Troubleshooting Horizontal Output Circuits

Many trouble symptoms caused by horizontal output circuits can also be caused by defects in other circuits. A dark screen (no raster) or insufficient width are two examples.

If the low-voltage supply is completely inoperative (almost zero), the screen can be dark. If the low-voltage supply is producing a low output, the picture width can be decreased. Of course, in either case, the sound is absent or abnormal. On the other hand, if the horizontal-drive circuits are defective (no drive signal to the horizontal output), there is no horizontal sweep or high voltage, even though the sound is probably normal. (In any service situation where the screen is dark, but there is normal sound, never overlook the possibility of a defective CRT.)

The most practical troubleshooting approach is to analyze the symptoms and then isolate the trouble to the horizontal output circuit by an input waveform check. That is, if the sound is normal (good low-voltage supply), measure the input waveform (from the horizontal drive) at the bases of Q_{501} and Q_{502} (Fig. 2.35). Check the waveform against the service literature. Figures 2.30 and 2.34 show some typical input waveforms from the horizontal oscillator.

If the input waveform is normal, the trouble is in the horizontal output circuit (unless the CRT is bad). Of course, if the input signal is not normal, the next step is to check the horizontal-drive circuits as described. If the input is good, make the following checks.

Dark Screen. In the case of a completely dark screen, the next obvious check is to measure the high voltage at the CRT anode. Then measure the accelerator grid and focus voltage from the boost circuit. If the voltages are present and normal, suspect the CRT. (Of course, if you have not already thought of it, check that the CRT filament or heater is on.)

High Voltage or Boost Voltage Absent. If either the high or boost voltage is absent or abnormal, this isolates the trouble to the corresponding circuit. If both voltages are absent (with the drive signal good), the problem is in Q_{501} and Q_{502} or related parts.

High Voltage Only Absent. If only the high voltage is absent, check for alternating current at the anode side of D_{515} (unless this is not recommended in the service literature). When measuring the high voltage (about 800 V in the circuit of Fig. 2.35), always use a meter with a high-voltage probe. Observe all of the usual precautions when measuring high voltages. Do not make an arc test with any solid-state or IC set.

Measuring Voltages. Measure the boost voltage with a meter and low-capacity probe. With the exception of CRT high voltage, most of the voltages in present-day sets can be measured using a low-capacity probe and meter (or scope). Observe any and all precautions given in the service literature. Typically, there are warning notes on the schematic such as "Do not measure high ac voltages."

Horizontal Output Transistor Measurements. It is often quite helpful if you can measure the emitter-collector current of the horizontal output transistor (Q_{502} in Fig. 2.35). Unfortunately, this is not always practical, so you must use dc voltage measurements to supplement the waveform checks. If the dc voltages and waveforms at Q_{502} are correct, it is reasonable to assume that the circuit is good up to FBT T_{503}.

Flyback and Yoke Checkers. The author has no recommendations regarding flyback and yoke checkers. If you use such checkers, make certain to follow the instructions and be sure that the checker is suitable for the set you are servicing. With a few exceptions, it is possible to service the horizontal circuits of most sets using waveform and voltage measurements.

Common Horizontal Output Troubles. The most common causes of trouble in horizontal output and high-voltage circuits (except for the set you are working on now) are capacitors, horizontal output transistor (because of the high currents), diodes, and transformers, in that order.

2.19.17 Typical Horizontal Output and High-Voltage Troubles

The following sections discuss symptoms that could be caused by defects in the horizontal output and high-voltage circuits.

Dark Screen. To review the troubleshooting sequence, if the screen is dark but sound is normal, suspect the horizontal circuits. Of course, the CRT can be defective, or the problem can be something simple (such as a lack of filament or heater voltage for the CRT).

Before checking individual parts, check the drive voltage waveform at the Q_{502} base and the sweep output at the Q_{502} collector; check for high voltage at the CRT anode (observing all precautions); check for boost voltage and any auxiliary voltages (focus, accelerator or screen grid, AFC, etc.).

The first check to be made depends on whatever is the most convenient. For example, it is logical to check the Q_{502} waveforms before checking the high voltage. In some sets, Q_{502} might be at a very inaccessible location, so the best bet is to check the high voltage first.

If the waveform at the base of Q_{502} is absent or abnormal, the problem is ahead of the horizontal output. If the base waveform is good, but the Q_{502} collector waveform is bad, Q_{502} is the first suspect. If the collector waveform is good, but one or more of the voltages from T_{503} are bad, check individual parts in the related circuit. Also check voltages at each of the Q_{502} elements, if practical.

Shorted capacitors (a common problem in the case of a dark screen) show up when dc voltages are measured. Open or leaking capacitors are not located so easily. Generally, an open or leaking capacitor produces an abnormal waveform.

Leaking transistors show up when waveforms are measured. The problem is confirmed further when the dc voltages are measured. Collector-base leakage in Q_{502} is a common problem. If leakage is bad enough to cause complete failure, which results in a dark screen, the dc voltages at the Q_{502} elements are incorrect.

Remember that Q_{502} operates at or near zero bias or possibly with reverse bias. If there is any substantial forward bias on Q_{502}, this is probably the result of leakage. Any collector-base leakage forward-biases a transistor. In the case of a normally cutoff Q_{502}, the undesired forward bias attenuates the collector waveform.

If the capacitors and Q_{502} appear to be in order, check the diodes. If the sweep output waveform (Q_{502} collector) is abnormal, check the damper diode (which is built into Q_{502} in this set). If the output voltages from T_{503} are abnormal (with a good sweep input from Q_{502}), check the corresponding diodes at the T_{503} terminals.

The high-voltage rectifier D_{515} is a series of diodes. If any of these diodes are shorted or develop excessive leakage, the remaining diodes can break down. Usually, the service literature recommends replacing all high-voltage diodes simultaneously (and often the diodes are all part of one package).

A shorted damper diode is usually easy to pinpoint because the Q_{502} collector waveform and dc voltage are abnormal. An open damper diode usually does not produce a dark screen. If any of the other diodes are defective, this shows up as absent or abnormal output voltages.

If the flyback transformer T_{503} has an open winding or if a winding is shorted, the problem is usually self-evident. However, if there is only a partial short, leakage between windings, or a high-voltage arc, it might be difficult to check. Substitution is the only sure check, unless you have a flyback-transformer tester suitable for the circuits. Unfortunately, replacement of the flyback transformer is not an easy job. So do not try substitution except as a last resort (when all other parts are known to be good).

Picture Overscan.   This symptom occurs when the picture becomes dim, even with the brightness control fully on, and there is an enlargement of the picture or raster. Usually, the enlargement is uniform, but there might be some defocusing.

In some circuits, the brightness control operates in reverse. That is, rotating the control for an increase in brightness produces a decrease, after reaching a critical point of control. If picture overscan is not accompanied by insufficient width, suspect the high-voltage circuit only.

The first obvious test is to measure the high voltage (observing all precautions). If the high voltage is normal (not likely), try a new CRT. Also, there might be a corona problem (leakage from the high-voltage lead), especially in large-screen TV sets. The only practical cure is to replace the lead. If the high voltage is low, check any high-voltage filter capacitor for leakage.

Narrow Picture.   The most logical cause of a narrow picture (that cannot be corrected by adjustment of width controls) is insufficient horizontal drive. This is usually accompanied by other symptoms, such as decreased brightness, picture distortion, and the like. Very often, a narrow picture is the result of a marginal breakdown rather than a complete breakdown. For example, if Q_{502} has some collector-base leakage, Q_{502} is forward-biased, and the sweep output is decreased.

The most likely defects depend on symptoms that accompany the narrow picture. For example, if there is distortion on the left-hand side of the picture, look for an open or leaking damper diode (Fig. 2.30). If there is right-hand distortion, check for a defective Q_{502}. (In the circuit of Fig. 2.35, the damper diode and Q_{502} are in the same package, but this is not true in all sets.)

The first step is to measure the Q_{502} collector waveform (the sweep waveform to the horizontal yoke). This waveform rarely, if ever, is normal when the picture is narrow.

Waveform measurements should be followed by voltage measurements at the Q_{502} elements.

Do not overlook insufficient horizontal drive, especially when waveform and voltage measurements appear about normal. Pay particular attention to drive pulses at the base of Q_{502}. Even a slight drop in amplitude can reduce sweep output to the horizontal yoke.

Foldback or foldover. Horizontal foldback usually occurs only on one side of the picture screen. A portion of the picture is folded back on one edge of the display. In some rare cases, there is a fold in the center. If there is a foldover (with good drive signal), all of the horizontal sweep components are suspect. Again, look for Q_{502} problems if foldback is on the right and damper-diode problems if the foldback is on the left.

The first test is to measure both the base (drive) and collector (output) of Q_{502}. Then measure the horizontal-yoke waveform (if different from the Q_{502} collector waveform, as is the case in some sets). Invariably, the yoke waveform is distorted. Remember that the horizontal sweep system is essentially a resonant circuit. The inductance of T_{503} combines with C_{545} to resonate at about 50 kHz (typical). If the resonant frequency is not correct (generally low), the waveform is distorted, resulting in foldback.

Nonlinear Horizontal Display. Any nonlinearity in the horizontal display is almost always found with at least one other problem (such as narrow picture, overscan, blooming, lack of brightness, foldback, etc.). Thus, the most likely causes of nonlinearity are the same as for other symptoms. Also, the troubleshooting and test sequence should be the same.

Horizontal nonlinearity should not be confused with the *keystone* effect (where the picture is wider at the top than at the bottom, or vice versa). The keystone effect is almost always caused by a problem in the horizontal yoke. (One set of horizontal-deflection coils has shorted turns and is unbalanced with the other set of coils.)

2.20 TROUBLESHOOTING SYSTEM-CONTROL CIRCUITS

This section is devoted to circuits that provide control of such functions as servicing the keyboard (front-panel controls), the remote-control input processing, and tuning operations (Sec. 4.1), including band selection, tuning memory, and phase and frequency lock. In most sets, such functions are under control of a microprocessor (also called a *controller, system controller, microcontroller, syscon,* etc.). In addition to the functions mentioned, the microprocessor usually handles the on-screen display, power on-off, audio volume control, and picture-level control.

System-control circuits have several functions, and there are many circuit configurations. Thus, it is very difficult to present a typical configuration. However, most system-control circuits have certain inputs and outputs that can be monitored. If the inputs are normal, but one or more of the outputs are abnormal, the problem can be localized to a specific part or IC. If you are not familiar with basic microprocessor functions, read *Lenk's Digital Handbook* (McGraw-Hill, 1993).

2.20.1 System-Control Basics

Figure 2.36 shows the basic system-control circuits for a typical TV set (13-in Sony). Figure 2.37 shows some additional details. The following sections describe the basic system-control functions and troubleshooting approach.

2.20.2 Reset Function

The IC_{103} reset function is under control of Q_{106} and Q_{107}. This circuit combination operates when power is initially applied and each time the power is turned on and off.

When the set is first powered up or plugged in, Q_{106} and Q_{107} are normally off. As the standby 5-V (S5-V) supply comes up, IC_{103} is reset, and (eventually) the breakover characteristics of zener D_{101} are exceeded. This causes Q_{106} and Q_{107} to turn on and applies S5-V to the reset and input at pin 3 of IC_{103}. The zener action of D_{101} ensures that the IC_{103} reset input remains low long enough to fully reset IC_{103}.

During normal operation, the power on-off function is controlled by Power switch S_{01} on the H-board. When S_{01} is pressed, a low is applied to the power input at pin 37, instructing IC_{103} to energize the power-on relay. Pin 23 of IC_{103} goes high. At the same time, the low at S_{01} grounds one end of C_{113}. This pulls reset pin 3 low for the charge time of C_{113}, resetting IC_{103}.

2.20.3 Master Clock

The master clock signal is generated within IC_{103}. Crystal X_{101} at pins 1 and 2 determines the operating frequency. For this set, the frequency is 2 MHz, with a sine wave of about 4 V_{p-p} at pins 1 and 2. The master clock signal should be available once power is applied to IC_{103}, and IC_{103} is reset. This master-clock or system-clock signal synchronizes or coordinates all of the IC_{103} microprocessor functions. It is essential that the frequency, amplitude, and waveform be correct for the set to operate properly.

2.20.4 Keyboard Service

The keyboard (or customer control) inputs shown in Figs. 2.36 and 2.37 are serviced using the scan 03 output lines at pins 13 through 16, and the key 01 input lines at pins 38 and 39. These input-output, or I/O, lines work with the keyboard matrix components located on the H-board.

When any key is pressed, an oscillator within IC_{103} is turned on, generating the appropriate scanner signals (positive-going pulses). The scanner outputs at pins 13 through 16 of IC_{103} are connected to one contact of the specific keyboard switches. The remaining contact of each switch is connected to a specific input at pin 38 or 39.

When a given key is pressed, a scanner signal is applied to the corresponding input through the closed switch contacts. A specific scanner signal is applied to a specific

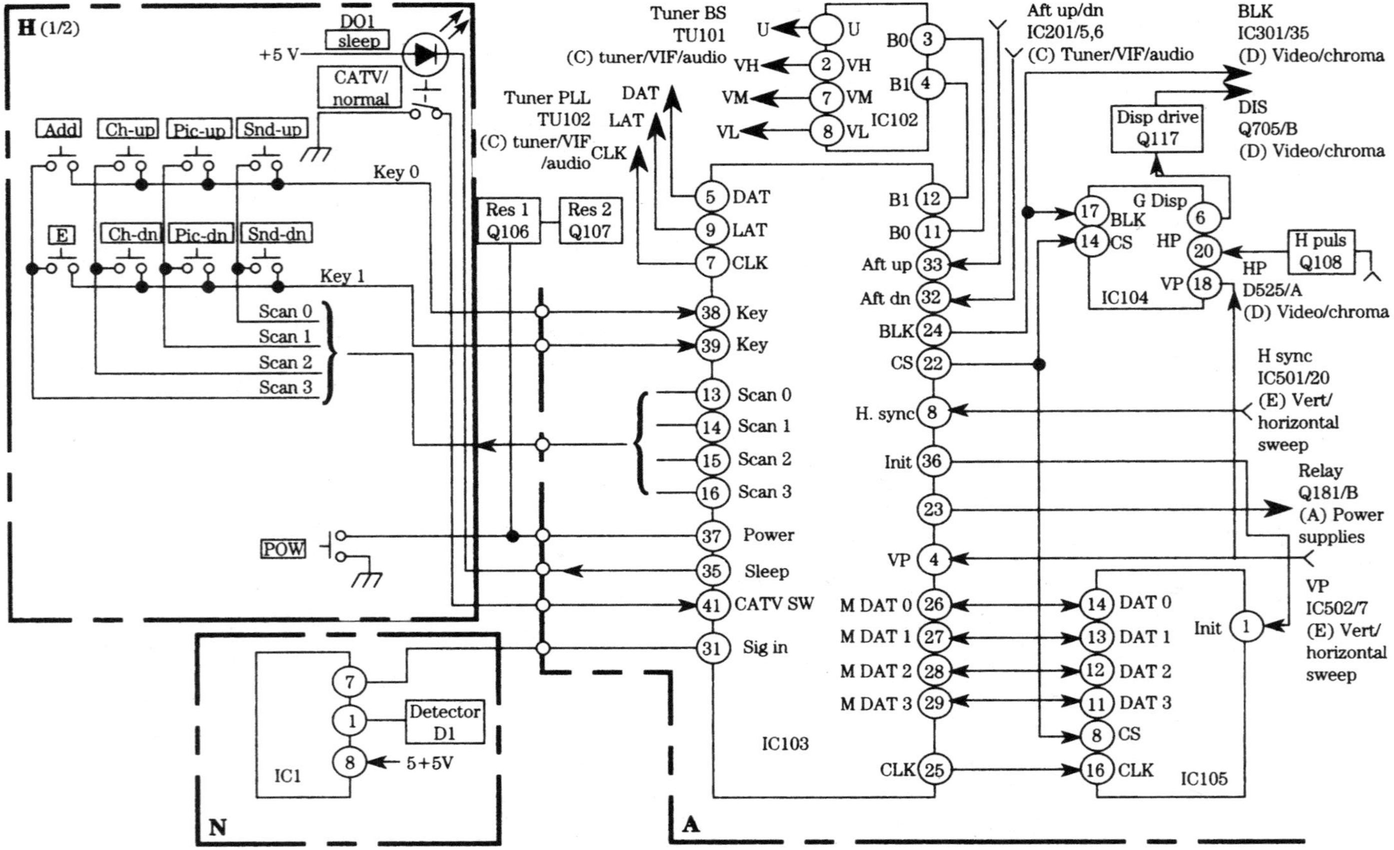

FIGURE 2.36 System-control circuits.

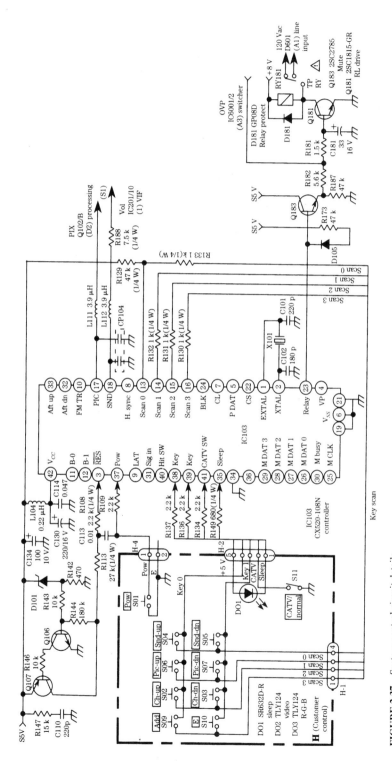

FIGURE 2.37 System-control circuit details.

2.97

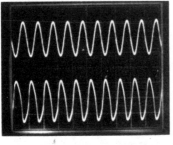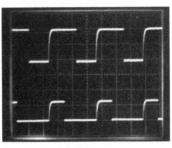

(S12)	IC103/Pin 1	EXTAL	2 V/d	0.5 μsec/d	(S1)	IC103/Pin 17	PIC	5 V/d	20 μsec/d
	IC103/Pin 2	XTAL	2 V/d	0.5 μsec/d		IC103/Pin 18	VOL	5 V/d	20 μsec/d

FIGURE 2.37   (*Continued*) System-control circuit details.

ART FPO ONLY DUP FROM FILM
PER LIST SUPPLIED

input for each switch closure. Similarly, each keyboard switch is represented by a unique combination of scanner output and signal input. The scanner signal applied to a particular input is decoded by IC_{103} and produces a specific output representing the function of the corresponding keyboard switch.

As an example, assume that the channel up (CH UP) switch S_{02} is pressed. This applies the scan-2 signal to the key-0 input at pin 38 and causes IC_{103} to apply channel-up commands to the tuner band-select IC_{102} (from pins 11 and 12 of IC_{103}, as described in Sec. 4.1). When the tuner reaches the selected channel and is properly tuned to the channel, the AFT signals at pins 32 and 33 are removed, and IC_{103} stops sending channel commands to IC_{102}. If the CH UP switch is held down, the channels are scanned continuously, going from low to high.

Front-panel CATV/Normal switch S_{11} is not scanned. Instead, S_{11} provides a high or low (open or ground) directly to pin 41 of IC_{103}. In some cases, the channel commands to the tuner band-select IC_{102} are different when cable is used (to select the additional channels available on cable).

When S_{11} is set to CATV, pin 41 goes low, instructing IC_{103} to issue the corresponding channel commands to IC_{102}. During *normal* TV operation, S_{11} is open and IC_{103} produces the corresponding band-select commands.

The sleep function is not operated by a front-panel switch, even though there is a Sleep indicator D_{01}. Instead, the sleep function is a fixed-time (1 hour) shutoff operated by the remote control. When sleep is selected on the remote transmitter, pin 35 of IC_{103} goes low, turning off the power-on relay RY_{181}.

The picture-level (PIC UP/PIC DN) and sound level (SND UP/SND DN) commands at pins 17 and 18 of IC_{103} are positive-going pulse width modulation (PWM) signals at a frequency of 15 kHz. These signals are filtered by the capacitor-coil combination, producing a corresponding dc voltage for picture control (at Q_{102}, Fig. 3.26) and sound control (at pin 10 of IC_{201}, Fig. 2.25).

The level of the dc control voltages varies, depending on the width of the 15-kHz PWM signals. As pulse width increases, the dc voltages increase, and vice versa. In turn, the width of the PWM signals is determined by the scanned picture and sound switches S_{04} through S_{07}. For example, to decrease the picture level, hold S_{07} until the desired level is reached.

2.20.5 Power On-Off Sequence

The power on-off functions are controlled by Q_{181}, Q_{183}, and relay RT_{181}, as selected by the signal at pin 23 of IC_{103}. A +8 V from the standby supply is applied to RT_{181}. When a power-on command is received by IC_{103} (from front-panel switch S_{01} or the remote), pin 23 of IC_{103} goes high. This turns on Q_{181} and Q_{183} and energizes RY_{181}. The normally open contacts of RT_{181} close, applying 120 V to the line input circuit at D_{601}. This turns the set on. If the set is on, a low at pin 37 of IC_{103} causes pin 23 to go low. This removes power and turns the set off.

The overvoltage protection (OVP) input from the switching supply is applied to the base of Q_{181}. Should an overvoltage condition occur, the OVP input goes low, turning Q_{181} off and removing power from RT_{181}. This removes power and turns the set off.

2.20.6 Steps in Troubleshooting System-Control Circuits

Troubleshooting for the system-control circuits starts with the microprocessor. Although the system-control functions are often complex, troubleshooting is not necessarily difficult because you are dealing with basic input-output and on-off functions. This is especially true when only one or two functions are absent or abnormal.

As an example, assume that the audio level does not increase when the front-panel SND UP switch S_{04} is pressed. First check to see if the sound level can be increased with the remote. If not, the problem is likely in IC_{103} or in the control circuits from pin 18 of IC_{103} to IC_{201}. If the remote does have control over the sound level (both up and down), the problem is between S_{04} and IC_{103}. (The SND UP switch applies a scan-0 output to the key-0 input at pin 38.)

The problem becomes more complex when all of the system-control functions are absent or abnormal (or are erratic). Should this occur, start with the microprocessor. The following paragraphs describe some basic approaches.

The first step is to check all power and ground connections to the microprocessor (and any other IC in system control). Make certain to check all power and ground connections to each IC because many ICs have more than one power and one ground connection. For example, as shown in Fig. 2.37, IC_{103} requires standby 5 V at pin 42 and four grounds at pins 6, 19, 21, and 34.

With all of the power and ground connections confirmed, check that all ICs are properly reset. IC_{103} is reset when the power cord is first connected and standby power is applied. Pin 3 of IC_{103} should be low (near zero) when standby power is first applied and then should rise to about 5 V. The low causes circuits within IC_{103} to reset. After reset, when pin 3 rises to about 5 V, IC_{103} remains ready to perform the various microprocessor functions.

If the IC_{103} reset function is not as described, suspect Q_{106} and Q_{107} and the associated parts. Notice that IC_{103} is not reset by temporary drops in the standby power (because of the charge on C_{110}) but should be reset once each time the standby power is removed and reapplied.

As in the case of any microprocessor, if the reset pin is open, or shorted to ground or to power, the IC will not reset (or will remain reset, low) when the power is switched on and off. So if you find a reset pin (or line) that is always high, always low, or apparently connected to nothing (floating), check the pin connections and reset circuits.

The next function to check is the clock signal at pins 1 and 2 of IC_{103}. It is possible to measure the presence of a clock signal with a scope. Obviously, if the clock is not operating, the microprocessor will not function. A frequency counter provides a more accurate clock measurement. Although crystal-controlled clocks do not usually go off frequency, it is possible that the crystal could go into a third overtone frequency, well beyond the operating capabilities of the microprocessor. This will not show up on a scope.

When you are certain that the system-control microprocessor has proper ground and power connections, and that all reset and clock signals are available, the final step is to monitor all input and output signals at microprocessor IC_{103}. Use the procedure previously described for sound- or audio-level control.

CHAPTER 3
BASIC VIDEO COLOR CIRCUITS

This chapter is devoted to the basics of video color. Although these basics apply specifically to TV sets (discrete-component, IC, and monitor), the same principles also apply to color terminals used in computer graphics, video games, etc. Television is chosen since it is the most familiar color-video device in consumer electronics. The chapter describes functions, operation, circuit theory, test and alignment procedures, and a practical troubleshooting approach for the circuits involved.

3.1 BASIC COLOR TV BROADCAST SYSTEM

As shown in Fig. 3.1, the basic color TV broadcast system consists of a transmitter, a color camera, a signal matrix, a brightness amplifier, a color amplifier, and a 3.58-MHz oscillator. The TV program sound (or audio) is broadcast through an FM transmitter. The picture (or video) portion of the program is broadcast by means of amplitude modulation, or AM. The same channels and frequencies are used for color and for black and white.

In addition to the horizontal and vertical sweeps, the color signal is made up of two components: brightness and color information. The brightness portion contains all the information pertaining to the picture details and is commonly called the Y, or luminance or *luma*, signal. The color portion is called the chrominance, or *chroma*, signal and contains the information pertaining to the hue (or tint) and saturation of the picture.

The camera in Fig. 3.1 contains three separate image pickup tubes, one tube for each of three colors (red, green, and blue). The camera tubes divide the scene being scanned into three colors (red, yellow, and blue). The relative intensity of the scene is also contained in this signal. The output signal from the camera contains the luma and chroma information.

The signal is separated in the signal matrix, and the individual components are amplified by the brightness and color amplifiers. The color information (chroma) is impressed on a 3.58-MHz subcarrier (in the form of *phase shift*) and is transmitted together with the luma signal (which is in the form of instantaneous *amplitude shift* or level). This method of color broadcast provides a compatible signal that can be reproduced on black and white as well as color sets. The black and white sets require only the luma signal (which is the equivalent of the picture information transmitted on the AM carrier of a black and white broadcast).

A compatible color signal must fit within the standard 6-MHz TV channel and have horizontal and vertical scanning rates of 15,750 and 60 Hz, plus a video bandwidth not in excess of 4.25 MHz. The color information is interleaved with the video and trans-

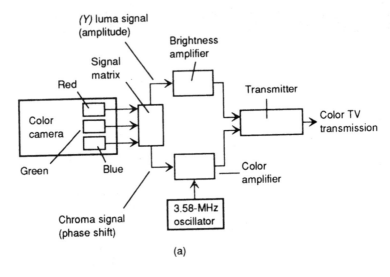

(a)

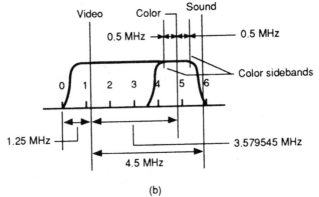

(b)

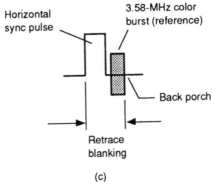

(c)

FIGURE 3.1   Basic color TV broadcast system.

mitted within the 6-MHz TV channel on the 3.58-MHz subcarrier, as shown in Fig. 3.1*b*. The color subcarrier is actually 3.579545 MHz above the video carrier since this does not interfere with reception on black and white sets.

The color signals are synchronized by transmitting about eight cycles of the 3.58-MHz subcarrier signal (of the correct phase) along with the color signal. This signal is called the *burst* sync signal (or *color burst*) and is added to the back porch of the horizontal sync pulse, as shown in Fig. 3.1*c*. The location of the burst does not interfere with the horizontal sync on black and white sets since the burst occurs during the blanking or retrace portion of the sync pulse.

3.1.1 Black and White versus Color Signals

A black and white presentation requires only the transmission of the amplitude or luma signal. A color broadcast also requires the transmission of phase-shift information (representing colors). Before going into the color circuits, let us review the nature of the NTSC color video signal.

3.1.2 History of the NTSC Color Video Signal

In 1953, the National Television Systems Committee (NTSC) of the Electronic Industries Association (EIA) established the color TV standards now in use by the TV broadcast industry in the United States and many other countries. The color system is compatible with the monochrome (black and white) system that previously existed.

The makeup of a composite video signal is dictated by NTSC specifications. These specifications include a 525-line interlaced scan operating at a horizontal scan frequency of 15,734.26 Hz and a vertical scan frequency of 59.94 Hz. Color information is contained in the 3.579545-MHz subcarrier. The phase angle of the subcarrier represents the hue or tint; the amplitude of the carrier represents the saturation.

3.1.3 Horizontal Sync (NTSC)

A line of horizontal scan is normally considered as "beginning" at the leading edge of the horizontal blanking pedestal (Fig. 3.2*a*). In a TV set, the horizontal blanking pedestal starts as the electron beam of the picture tube reaches the extreme right-hand edge of the screen (plus a little extra in most cases).

The horizontal blanking pedestal prevents illumination of the screen during retrace (until the electron beam deflection circuits are reset to the left edge of the screen and ready to start another line of video display). The entire horizontal blanking pedestal is at the blanking level or the sync-pulse level. In a TV set, the blanking and sync levels are the "blacker than black" levels that assure no illumination during retrace.

The horizontal blanking pedestal consists of three parts: the front porch, the horizontal sync pulse, and the back porch. The front porch is a 1.47-μs period at blanking level and is followed by a 4.89-μs horizontal sync pulse at a −40 IEEE units level. (An explanation of IEEE units is given in Sec. 3.1.5.) When the horizontal sync pulse is detected in a TV set, flyback is initiated, rapidly ending the horizontal scan and rapidly resetting the horizontal deflection circuit for the next line of horizontal scan.

The horizontal sync pulse is followed by a 4.40-μs back porch at the blanking level. When a color signal is being generated, 8 to 10 cycles of 3.579545-Hz color burst occur during the back porch. The color burst signal is at a specific reference phase. In a color TV set, the color oscillator is phase-locked to the color burst refer-

	H-sync	Burst	Gray 75% white	Yellow	Cyan	Green	Magenta	Red	Blue	Black
Luma IEEE units	−40	0	77	69	56	48	36	28	15	7.5
Chroma	- - -	40	- - -	62	88	82	82	88	62	- - -
Vector (from B-Y)	- - -	180°	- - -	167°	284°	241°	61°	104°	347°	- - -

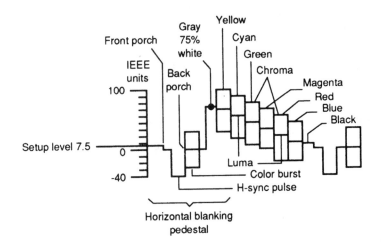

(a)

FIGURE 3.2   Composite video signal.

ence phase before starting each horizontal line of video display. When a mono signal is being generated, there is no color burst during the back porch (since the burst is in no way required for black and white operation).

3.1.4   Vertical Sync (NTSC)

A complete video image as seen on a TV screen is called a *frame*. A frame consists of two interlaced vertical fields of 262.5 lines each. The image is scanned twice at a 60-Hz rate (at 59.94 Hz, to be more precise), and the lines of field 2 are offset to fall between the lines of field 1 (the fields are interlaced) to create a frame of 525 lines at a 30-Hz repetition rate.

At the beginning of each vertical field, a period equal to several horizontal lines is used for the vertical blanking interval (Fig. 3.2*b*). In a TV set, the vertical blanking

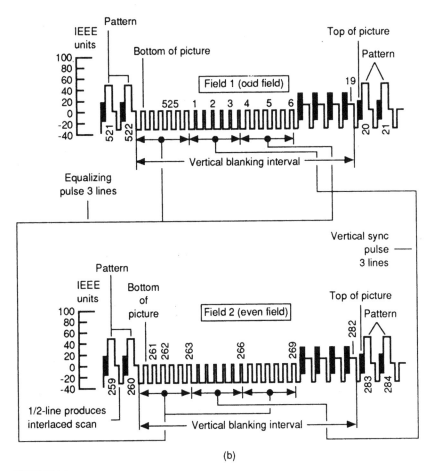

(b)

FIGURE 3.2   Composite video signal (*Continued*).

interval prevents illumination of the picture tube during the vertical retrace. The vertical sync pulse, which is within the vertical blanking interval, initiates reset of the vertical deflection circuit so that the electron beam returns to the top of the screen before video scan resumes.

The vertical blanking interval begins with the first equalizing pulse, which consists of six pulses one-half the width of horizontal sync pulses but at twice the repetition rate. The vertical sync pulse is an inverted equalizing pulse with the wide portion of the pulse at the −40 IEEE units level, and the narrow portion of the pulse at the blanking level.

A second equalizing pulse occurs after the vertical sync pulse, which is then followed by 13 lines of blanking level (no video) and horizontal sync pulses to assure adequate vertical retrace time before resuming video scan. The color-burst signal is present after the second equalizing pulse. Note that in field 1, line 522 includes a full line of video, while in field 2 line 260 contains only a half line of video. This timing relationship produces the interlace of fields 1 and 2.

3.1.5 Amplitude (NTSC)

A standard NTSC composite video signal is 1 V peak to peak (p-p), from the tip of a sync pulse to 100 percent white (Fig. 3.2). This 1-V$_{p-p}$ signal is divided into 140 equal parts called IEEE units. The zero reference level is the blanking level. The tips of the sync pulses are at -40 units, and a sync pulse is about 0.3 V$_{p-p}$. The portion of the signal that contains video information is raised to a setup level of $+7.5$ units above the blanking level.

A monochrome video signal at $+7.5$ units is at the black threshold. At $+100$ units, the signal represents 100 percent white. Levels between $+7.5$ and $+100$ units produce various shadings of gray. Even when a composite video signal is not at the 1-V$_{p-p}$ level, the ratio between the sync pulse and video must be maintained: 0.3 of the total voltage for the sync pulse and 0.7 of the total voltage for 100 percent white.

There is also a specific relationship between the amplitude of the composite video signal and the percentage of modulation of an RF carrier. A TV signal uses negative modulation, where the sync pulses (-40 units) produce the maximum peak-to-peak amplitude of the modulation envelope (100 percent modulation) and white video ($+100$ units) produces the minimum amplitude of the modulation envelope (12.5 percent modulation).

This relationship of amplitude versus modulation is very advantageous because the weakest signal condition, where noise interference can most easily cause snow, is also the white portion of the video. There is an adequate amplitude guard band so that a peak white of $+100$ units does not reduce the modulation envelope to zero.

3.1.6 Color (NTSC)

The color information in an NTSC video signal consists of three elements: luminance (or luma), hue (or tint), and saturation, as shown in Fig. 3.3.

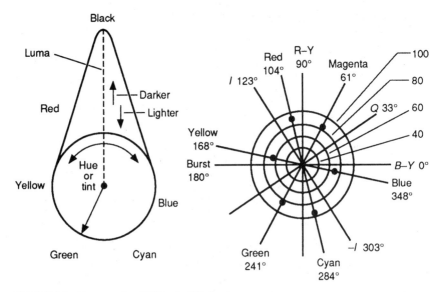

FIGURE 3.3 Elements of an NTSC color TV signal.

Luma, or brightness perceived by the eye, is represented by the amplitude of the video signal. The luma component of a color signal is also used in black and white sets, where the luma is converted to a shade of gray. Yellow is a bright color and has a high level of luminance (it is nearer to white), whereas blue is a dark color and has a low level of luminance (it is nearer to black).

Hue, or tint, is the element that distinguishes between colors (red, blue, green, etc.). White, black, and gray are not hues. Hue information is contained in the phase angle of the 3.58-MHz color carrier. The three primary video colors of red, blue, and green can be combined to create any hue. A phase shift through 360° produces every hue in the rainbow by changing the combination of red, blue, and green.

Saturation is the vividness of a hue and is determined by the amount the color is diluted by white light. Saturation is often expressed in percentages: 100 percent saturation is a hue with no white dilution and produces a very vivid shade. Low saturation percentages are highly diluted by white light and produce light pastel shades of the same hue. Saturation information is contained in the amplitude of the 3.58-MHz color carrier. Because the response of the human eye is not constant from hue to hue, the amplitude required for 100 percent saturation is not the same for all colors.

The combination of hue and saturation is known as *chroma*. This information is usually represented by a vector diagram. Saturation is indicated by the length of the vector, and hue is indicated by the phase angle of the vector. The entire color-signal representation is three dimensional, consisting of the vector diagram for chroma and a perpendicular plane to represent the amplitude of luminance.

3.2 BASIC COLOR CIRCUITS

Figure 3.4 is the block diagram of a color TV set. Again, the diagram shown is a composite of several types of color TV sets and is presented as a point of reference for troubleshooting. In the following sections we describe the basic principles of all NTSC color TV sets. The service procedures for color TV sets are provided in the remaining sections of this chapter. As shown in Fig. 3.4, the circuits of a color set are very similar to those of a black and white set, with two exceptions, the luma, or Y channel, and the chroma channel.

3.2.1 Luma Channel Operation

The luma channel consists of the detector, first video amplifier, delay line, and second video amplifier. (Note that a separate sound detector is used in the signal path prior to the video detector in many color sets. This provides good separation of the sound and video signals.)

The composite video signal is detected and amplified in the first stage of the luma channel. A portion of this signal is fed to the sync and color stages. The rest of the signal continues on to the delay line. This delay is required so that the luma information (amplitude) arrives at the picture tube at the same time as the color information (phase relationship).

The narrow bandwidth of the color circuit causes the chroma signal to take longer to reach the picture tube. Additional amplification is given to the luma signal in the second video amplifier before luma is applied to the cathodes of the picture tube.

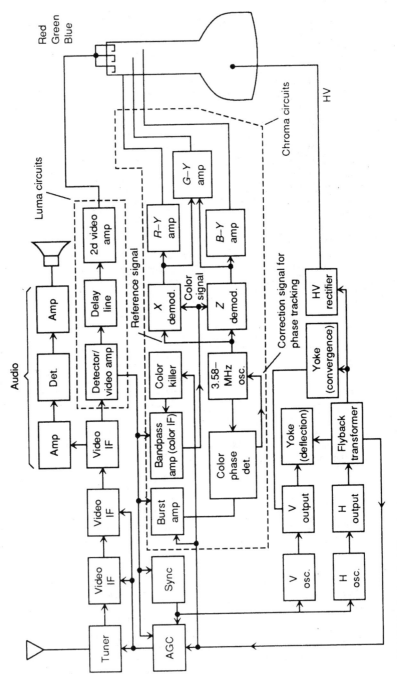

FIGURE 3.4 Basic color TV set circuits.

3.2.2 Chroma Channel Operation

The composite video signal is fed to the bandpass amplifier (sometimes called the color IF) from the first video amplifier. The chroma signal is separated from the composite color signal, amplified, and fed to the inputs of the demodulators. The burst amplifier removes the burst signal from the chroma signal and applies the burst signal to the color-phase detector.

A color-killer stage is used to prevent signals at frequencies near 3.58 MHz from passing through the color circuit during the reception of black and white broadcasts. In the absence of a color broadcast (no color-burst signals being transmitted), the color killer biases the bandpass amplifier to off, thus disabling all of the chroma channel. In most circuits, the color killer also receives pulses from the horizontal sweep circuits (usually from the flyback transformer). The horizontal pulses operate the color killer to bias the bandpass amplifier to off during the horizontal sweep retrace (so that no color information passes during the retrace period).

The 3.58-MHz oscillator provides a reference signal for demodulation of the color signal. The color-phase detector compares the 3.58-MHz oscillator signal with the burst signal and develops a correction voltage to keep the oscillator locked in phase with the burst signal.

In most circuits, the burst amplifier (or the burst control circuit) receives pulses form the horizontal sweep. These pulses gate the burst amplifier to on only during the burst signal (immediately after the horizontal sync pulses, during the retrace period, as shown in Fig. 3.1c). Thus, the 3.58-MHz oscillator is locked in phase with the burst at the beginning of each horizontal sweep.

As shown in Fig. 3.4, the X and Z demodulators detect the amplitude and phase variations of the chroma signal to recover color information. At any given instant, the X and Z demodulators receive two signals: a 3.58-MHz signal from the reference oscillator (locked in phase to the reference burst) and a 3.58-MHz signal from the bandpass amplifier (at a phase and amplitude representing the color at that instant).

The R-Y (red luma), B-Y (blue luma), and G-Y (green luma) amplifiers amplify the color signals and apply the signals to the picture-tube input (grids in this case).

3.2.3 Makeup of Colors

Color picture tubes produce most colors by mixing three colors (red, blue, and green). (Operation of the color picture tube is described in Sec. 3.2.4.) Any color can be created by the proper blending of these three colors. That is, any color can be created if the three colors are present, each at a given level of intensity or brightness. This is because of a characteristic of the human eye. When the eye sees two different colors of the same brightness level, the colors appear to be of different brightness levels. For example, the eye is more sensitive to green and yellow than to red or blue.

The combination of any two primary colors produces a third color. In color TV, the process of mixing colors is *additive* and depends on the self-illuminating properties of the picture-tube screen (not on the surrounding light source, as found with the subtractive color-mixing process used with paints and printing).

If two primary colors, such as red and green, are combined, the primary colors produce a secondary color, yellow. The combination of blue and red produces magenta. Also, a secondary color added to the complementary primary produces white.

For example, yellow + blue = white, cyan + red = white, and magenta + green = white. Thus, any color can be produced if the three colors (red, blue, and green) are mixed in the proper proportions (that is, if the amplitudes of the three signals at the picture-tube inputs are at the proper levels).

For the purposes of our discussion, it is sufficient to understand that the three picture-tube inputs receive signals of the correct amplitude proportions (identical to the proportions existing at the color camera for any given instant) provided that (1) the set 3.58-MHz oscillator is locked in phase to the burst signals and (2) the demodulators receive a 3.58-MHz color signal that is of a *phase and amplitude* corresponding to the color proportions existing at the color camera.

3.2.4 Color Picture Tube

Figure 3.5 shows a color picture tube and the related components. Remember that there are many types of color tubes. However, all color tubes have certain points in common (from a troubleshooting standpoint). Figure 3.5 shows the basic elements common to all color picture tubes.

The color picture tube is a cathode-ray tube (CRT) similar to that used for black and white sets and terminals. Like the black and white tube, the color tube has a magnetic deflection yoke which deflects the electron beam. The yoke contains two sets of coils, one for vertical deflection and one for horizontal. The vertical and horizontal sweep signals are applied to the corresponding set of coils so that the electron beam traces out a rectangular raster on the screen in the normal manner. Here, the similarity to a black and white tube ends.

The color picture tube has *three* complete electron guns. Thus, three electron beams are produced, one for each gun. The color picture tube also contains a shadow mask and a three-color phosphor-dot screen within the glass envelope.

The *screen* consists of several million phosphor dots of the primary television colors (red, green, and blue). These dots are arranged in trios and produce colored light when struck by electrons. Even though the dots are placed next to each other, they appear (to the human eye) to be superimposed so that the colors are (apparently) blended.

The *shadow mask* is a thin plate of metal with a number of small holes, each cen-

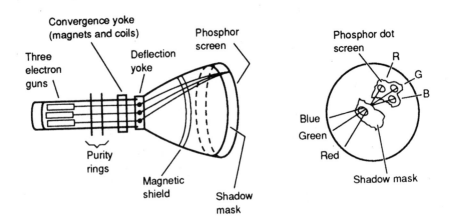

FIGURE 3.5   Color picture tube and related components.

tered over a phosphor-dot trio. This mask is positioned in front of the screen and controls the landing of the electron beams on the phosphor dots.

The *three electron guns* (one for each color) are positioned approximately 120° apart and are precisely aimed so that the individual electron beams pass through the shadow mask at different angles to strike the three corresponding phosphor dots. The grids control the amount of light leaving the gun and thus regulate the amount of colored light emitted from the phosphor screen.

3.2.5 Color Purity

For the color picture tube to reproduce color pictures correctly, the individual electron beams must strike phosphors of only one color. Each electron beam is then capable of producing a pure field of either red, blue, or green. *Purity* refers to the uniformity of the hue and the brightness over the entire area of each color field and over the entire area of the combination of all three. For example, red is pure when the red display is a uniform red with no contamination from blue or green.

Color purity is determined by where the three electron beams strike the screen and can be adjusted by means of the purity magnets and deflection yoke. Typical color purity adjustments are described in Sec. 3.4.

3.2.6 Convergence

Convergence is the registry of the three beams at the same point in the shadow mask across the entire screen. The individual aiming of the beams must be corrected so that each passes through the same holes in the mask and strikes the same phosphor-dot trio.

Static convergence is controlled by convergence magnets which correct the travel of the individual electron beams. The static magnets converge the beams in the center of the screen but cannot compensate for the curvature of the screen to converge the beam at the edges.

Dynamic convergence, as this curvature compensation is known, is provided by electromagnetic coils mounted on the convergence yoke. Sawtooth voltages from the vertical and horizontal circuits are fed to the coils, which then alter the static magnetic field in unison with the scanning. Typical color convergence adjustments are described in Sec. 3.4.

3.3 COLOR GENERATORS

Before we get into adjustment and troubleshooting for color circuits, it is essential that you understand the operation of a color generator. Such generators are essential when servicing any type of color equipment (TV, monitor, computer terminal, VCR, camcorder, video-disc player, etc.). Color generators produce color signals that are equivalent to the test signals used in TV broadcast studios.

At one time, the *keyed-rainbow generator* was popular for color TV service. Today, the NTSC color-bar generator has all but replaced the rainbow generator. For that reason, we concentrate on the many models of the NTSC generator now available, both in portable and bench versions.

3.3.1 NTSC Color-Bar Generator

An NTSC generator produces standard EIA colors at the NTSC prescribed luma level (brightness), chroma phase angle (hue), and chroma amplitude (saturation). Generally the most useful signal produced by an NTSC instrument is the standard NTSC bar pattern with an $-IWQ$ signal occupying the lower quarter of the pattern, as shown in Fig. 3.6.

Note that Fig. 3.6 shows both the pattern (as produced on a TV screen when the NTSC signal is applied to the antenna input) and the waveforms (as produced on a scope connected to various points in the TV circuits) for *one horizontal line* of the pattern. Also note that the $-IWQ$ waveform is generated for the bottom 25 percent of the display, while the color-bar waveform is on for the top 75 percent of the pattern. When using a scope, the two waveforms may be superimposed, as discussed in Sec. 3.3.3.

A typical generator also produces a number of other patterns, including various convergence, linear staircase, and full-color rasters. Use of these patterns is discussed throughout the remainder of this chapter. The following sections describe how the patterns and waveforms can be used for troubleshooting.

3.3.2 NTSC Color-Bar Pattern

The basic pattern in Fig. 3.6 is the one most often used for testing, troubleshooting, and adjusting video equipment. Analysis of the pattern shown on the screen of a TV under test, or a pattern produced by playback of a VCR or camcorder, can often localize a problem to a few specific circuits. Here are a few examples.

Overall Performance.   An overall performance test of a VCR or camcorder may be conducted by recording the NTSC color-bar pattern, then playing the pattern back on a video monitor or known-good TV set. There should be little difference between the video played back from the VCR or camcorder and an NTSC pattern applied directly to the monitor or TV.

Luma and Chroma Proportions.   In a VCR, the luma and chroma signals are generated during the recording process and recombined during playback. If luma and chroma signals are not maintained at the proper proportion when separated, color distortion will result, particularly in the vividness of colors (color saturation).

Waveforms may be examined throughout the VCR circuits for proper luma-to-chroma proportions. For that reason, our waveforms include the luma-to-chroma proportions, expressed in IEEE units for example, in Figs. 3.2 and 3.6, the yellow bar occurs when the luma is 69 units and the chroma amplitude is 62 units, at a vector of 167° (using *B-Y* as a 0° reference).

Luma and Chroma Delay.   Another problem that may be found in VCRs is a difference in delay between the luma and chroma signals because of some circuit defect. Such delays cause *fuzziness along the edges of the color bars*. The delay problem may be more pronounced along the edges of the white bar in the $-IWQ$ portion of the pattern.

RF and IF Performance.   The tuners and IF sections of VCRs and color TV sets are essentially the same. The color-bar RF and IF outputs of the generator can be used effectively to troubleshoot RF and IF sections by comparing the patterns. For example, performance of a VCR or TV should be nearly as good when using the RF signal

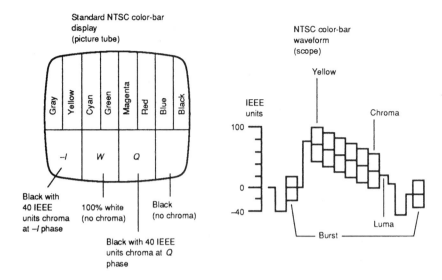

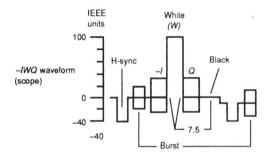

Luma amplitude	Burst	H-sync	−I	White (W)	Q	Black
	0	−40	7.5	100	7.5	7.5
Chroma amplitude	40	- - -	40	- - -	40	- - -
Vector (from B−Y)	180%	- - -	303°	- - -	33°	- - -

FIGURE 3.6   NTSC color-bar pattern with −*IWQ*.

as when applying composite video directly into the IF. If the display is substantially reduced when the NTSC generator output is applied through the tuner, as compared to when a signal is applied directly to the IF, the tuner is suspect, and you have a good starting point for troubleshooting.

Color Adjustments.   The NTSC color-bar pattern provides a standard reference for color adjustments. The pattern contains bars of the three primary colors: red, blue, and green. These are a good reference for checking 3.58-MHz phase problems. White, yellow, cyan, and magenta help define problems when the mix of colors is not in the correct proportions.

In some troubleshooting applications, it is helpful to know if the problem is chroma or luma related. Many generators are provided with a *chroma-off* function, which produces the bars and waveforms shown in Fig. 3.7a (normal bar pattern but without color).

A TV set can be checked by comparing the displays of Figs. 3.6 and 3.7a. If the display shown in Fig. 3.7a is good, but the one in Fig. 3.6 is not, the problem is in the chroma circuits. Note that in Fig. 3.7a the chroma-off waveform is generated for the top 75 percent of the pattern, and the $-IWQ$ waveform is on for the bottom 25 percent of the display. The two waveforms are superimposed on the scope display.

Phase Lock of the 3.58-MHz Oscillator.   The 3.58-MHz oscillator should be locked in phase whenever the color burst is applied. This can be checked using the *top burst-off* function available on some generators. When the top burst-off function is selected, the color is unsynchronized for the top quarter of the pattern, but there should be no delay or color distortion where the color bars start (middle of pattern) as shown in Fig. 3.7b.

Note that in Fig. 3.7b the bars with the burst-off waveform are generated for the top 25 percent of the pattern, the bar waveform is generated for the middle 50 percent, and the $-IWQ$ waveform is generated for the bottom 25 percent. The three waveforms are superimposed on the scope display.

Color-Killer Function.   Most TV sets and VCRs have a color-killer circuit (Fig. 3.4) that disables the color functions when there is no color burst present. This can be checked using the *full burst-off function* available on some generators. When the full burst-off function is selected, the bars should appear (in various shades of gray) but there is no color as shown in Fig. 3.7c. Note that in the figure the bars with burst-off waveforms (Fig. 3.7b) are generated for the top 75 percent of the pattern, and the $-IWQ$ with full burst-off waveform is generated for the bottom 25 percent. The two waveforms are superimposed on the scope display.

Audio or Sound Functions.   A 1- or 3-kHz audio signal can be added to the RF output of many generators. In a properly operating TV or VCR, this audio modulation should not affect the normal display. For example, if you select the pattern in Fig. 3.6, there should be no change in the display when the audio modulation is applied or removed. If you note some interference (typically diagonal lines across the display) when audio modulation is applied, sound is leaking into the chroma circuits. Possibly one or more of the sound traps within the TV or VCR are not properly adjusted. The 1- and 3-kHz signals can also be used to test the sound or audio circuits of the TV or VCR.

Vectorscope Applications.   The standard NTSC color-bar output (Fig. 3.6) of the generator can be used in conjunction with a vectorscope (or a conventional scope) to analyze color circuits. The procedures are discussed in Sec. 3.11.

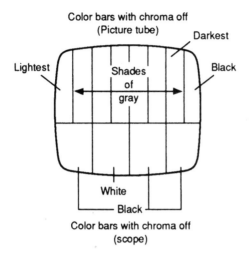

Color bars with chroma off
(Picture tube)

Lightest

Darkest

Black

Shades
of
gray

White

Black

Color bars with chroma off
(scope)

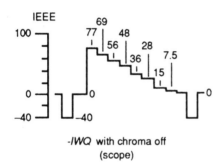

-*IWQ* with chroma off
(scope)

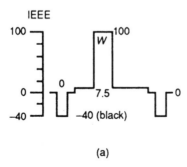

(a)

FIGURE 3.7 Normal color-bar pattern (but without color).

Color bars with top burst off (Picture tube)

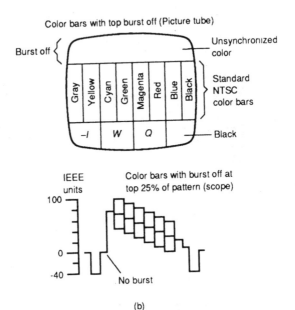

IEEE
units

Color bars with burst off at
top 25% of pattern (scope)

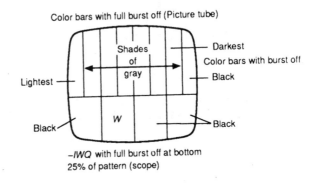

(b)

Color bars with full burst off (Picture tube)

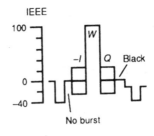

−IWQ with full burst off at bottom
25% of pattern (scope)

IEEE

(c)

FIGURE 3.7   Normal color-bar pattern (but without color) (*Continued*).

3.3.3 $-IWQ$ Patterns

The basic $-IWQ$ pattern appears on a split field with the NTSC color-bar pattern, with the $-IWQ$ pattern appearing on the lower quarter of each field, as shown in Figs. 3.6 and 3.7. When viewing horizontal lines of video (waveforms) on a scope, both the bars and the $-IWQ$ signals are superimposed, which is very handy in many applications. For example, the 7.5 percent black level (right-hand side) and the 100 percent white level (second from left) of the $-IWQ$ pattern are the key luma amplitude references.

These black- and white-level references are used for FM deviation adjustment and black- and white-clip level adjustments in VCRs. The black and white levels are also used whenever luma and chroma ratios are being adjusted or checked during troubleshooting. The $-I$ and Q signals are used primarily in video cameras and studio equipment for setting up the phase and amplitude of the $-I$ and Q signals and maintaining the proper relationship between the two.

3.3.4 Staircase Patterns

Figures 3.8 and 3.9 show the various staircase patterns and waveforms available with many generators. When the pattern in Fig. 3.9c is selected, the waveforms in Fig. 3.9a and b are generated for the top 25 percent of the pattern, and the waveform in Fig. 3.8a is generated for the bottom 75 percent. Both waveforms are superimposed on a scope display.

Amplifier Linearity Checks.   The staircase pattern contains five equal steps of increasing luminance with a constant chroma amplitude and phase. With the chroma off (Fig. 3.8c and d), only the luma steps are generated. This luma-only pattern is valuable for checking linearity in TV and VCR amplifier circuits. The amplitude of each step should be equal at the output of an amplifier or other circuit since the amplitude is equal at the input. Nonequal steps monitored at the output of a circuit represents nonlinear distortion.

Setting White-Clip Level in VCRs.   The staircase pattern is desirable for setting the white-clip levels in VCRs. Since the top step is 100 percent white level, it provides the correct reference for white-clip level adjustments. If the top step shows less amplitude than other steps, this usually indicates incorrect adjustment of the white-clip level.

Frequency-Equalization Adjustments in VCRs.   The staircase pattern is also recommended for frequency-equalization adjustment in the record amplifier of VCRs. The FM signal, which carries luma information in a VCR, is shifted to a different frequency for each step of the staircase signal. However, the record current (current applied to the tape recording heads in a VCR) should remain constant across the FM frequency band. The frequency-equalization adjustments of a VCR should be set so that the record current is *equal for all steps* of the staircase input signal.

Differential Gain and Differential Phase Lock.   The staircase patterns are most effective when checking both differential gain and differential phase. Excessive differential gain or phase can be the cause of color distortion in VCRs, color TVs, and video monitors and terminals. (Both conditions are checked often in studio equipment that processes video signals.)

Theoretically, the chroma amplitude should not change as luma is varied from 0 to

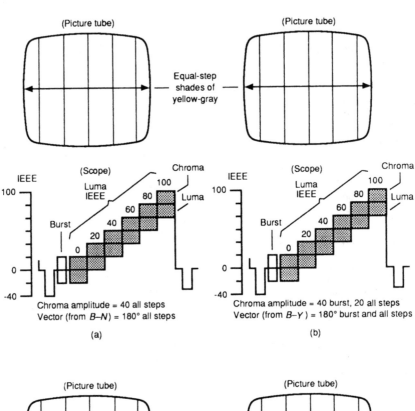

Chroma amplitude = 40 all steps
Vector (from *B–N*) = 180° all steps

(a)

Chroma amplitude = 40 burst, 20 all steps
Vector (from *B–Y*) = 180° burst and all steps

(b)

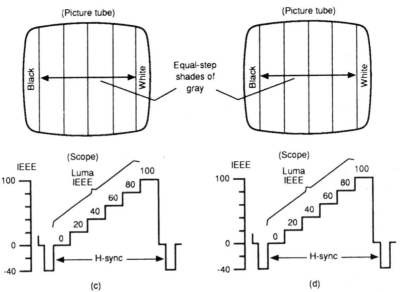

(c)

(d)

FIGURE 3.8 Staircase patterns.

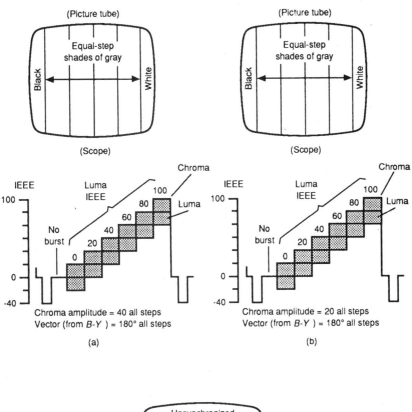

(a)

(b)

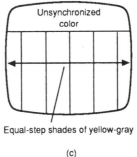

(c)

FIGURE 3.9 More staircase patterns.

100 percent. Any interaction is called *differential gain*. To check for differential gain, the output of a chroma circuit is displayed on a precision waveform monitor while the staircase-pattern input is applied from the generator. As the luma signal steps from 0 to 100 percent in 20 percent increments, any differential gain causes changes in the chroma amplitude, in synchronization with the luma steps.

Sometimes, the degree of differential gain may be affected by the peak-to-peak amplitude of the chroma signal. This characteristic can be checked by switching between the low-staircase (Figs. 3.8b, 3.8d, 3.9b) and high-staircase (Figs. 3.8a, 3.8c, 3.9a, 3.9c) patterns. The low-staircase pattern generates 20 IEEE units of chroma amplitude, whereas the high-staircase pattern generates 40 IEEE units.

3.3.5 Convergence Patterns

Figures 3.10 and 3.11 show the various convergence patterns and waveforms available with many generators. The convergence patterns are used primarily for static and

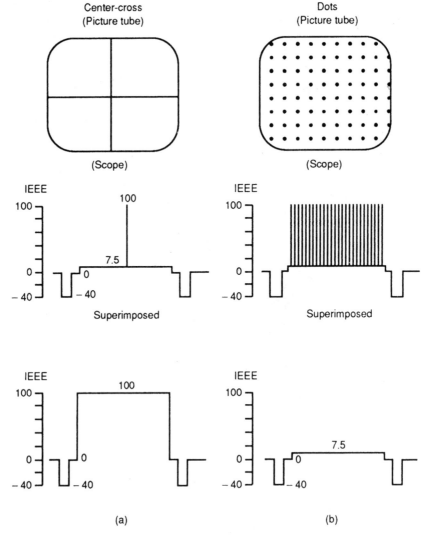

FIGURE 3.10   Center-cross and dots convergence patterns.

(Picture tube)   (Scope)

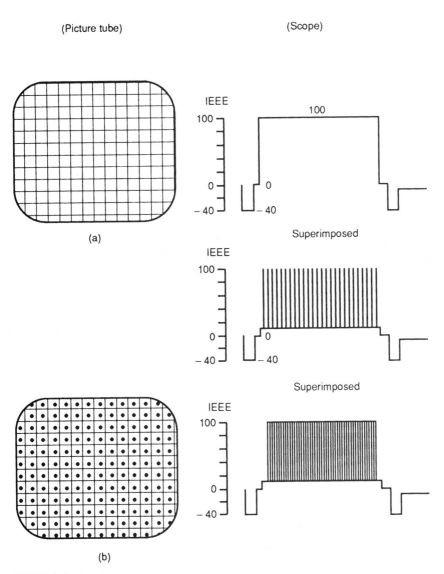

FIGURE 3.11   Crosshatch (*a*) and dot-hatch (*b*) patterns.

dynamic convergence of color TV and video monitors and terminals, as discussed in Sec. 3.4.

Center Cross.   The center-cross pattern (Fig. 3.10*a*) should intersect at the center of the screen, and there should be no tilt of the horizontal line. Improper centering indicates the need for centering adjustment or a possible deflection-circuit fault. Tilt may require repositioning of the deflection yoke for correction. The center-cross pattern also provides a good general check of vertical and horizontal sync.

Dots Pattern.   The dots pattern (Fig. 3.10*b*) is used for static convergence, usually by converging the center dot of the pattern. Most TV sets and some video monitors and terminals have *overscan,* so that all dots are not visible except possibly under low-voltage conditions. Some TV sets have a tendency toward a greater amount of over-scan than other sets. Typically, it is desirable to display at least a 17- by 13-dot pattern.

Crosshatch Pattern.   The crosshatch pattern (Fig. 3.11*a*) is normally preferred for dynamic convergence, although some technicians prefer the dots pattern for both static and dynamic convergence. The crosshatch pattern checks both vertical and horizontal linearity. Each square of the crosshatch pattern should be the same size, which is a convenient reference for making linearity adjustments. The crosshatch pattern is also used to check so-called *pincushion distortion,* which sometimes appears at the outside edges of large-screen picture tubes as bends in the lines.

Dot-Hatch Pattern.   The dot-hatch pattern (Fig. 3.11*b*) combines the dots and cross-hatch patterns for a quick overall check of static and dynamic convergence, linearity, overscan, and pincushion distortion from a single pattern. Note that there may be some slight amount of vertical jitter when monitoring the convergence patterns with some generators. This jitter is the result of interlaced scan and is normal for an NTSC generator (since the NTSC signal is interlaced). The jitter does not degrade the accuracy of the convergence adjustments and does not indicate a malfunction associated with vertical sync.

3.3.6   Raster Patterns

Figure 3.12 shows the raster patterns and waveforms available with many generators. As shown, one raster fills the entire screen, while the other raster pattern (with top burst off) fills the bottom 75 percent of the screen, with the top 25 percent occupied by unsynchronized color.

The raster patterns are valuable for checking and adjusting purity (Sec. 3.4) in TV sets, monitors, and terminals. Also, some VCR manufacturers recommend a raster pattern for setting the record current. Not only can the white raster be used in the standard manner, but the three separate guns of the picture tube may be individually adjusted (in some circuits) using a continuous chroma signal of red, blue, or green.

Analysis of hue and saturation problems may be simplified by analyzing each primary color or the yellow, cyan, and magenta hues, which are the equal mixture of two primary colors without the third one. On some generators, a "black burst" test signal is generated when none of the three primary colors is selected. The luma and chroma components for each raster color are identical to the corresponding bar from the color-bar pattern, but each color can be selected individually for analysis.

3.3.7   Sync Signals

Every pattern produced by an NTSC generator contains NTSC sync pulses. These sync pulses are the same as those produced by a TV broadcast station. The sync amplitude of 40 IEEE units is often the reference against which the remainder of the luma signal is compared. Circuits can be checked for sync clipping by monitoring the staircase pattern on a scope and checking to see if the sync-pulse amplitude remains at 0.4 compared to the 100 percent white step, which has a reference of 1.0.

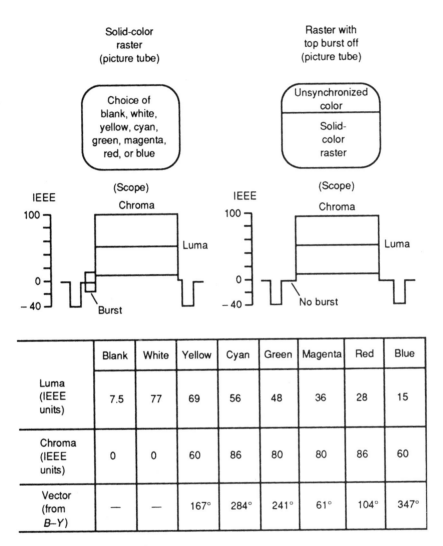

	Blank	White	Yellow	Cyan	Green	Magenta	Red	Blue
Luma (IEEE units)	7.5	77	69	56	48	36	28	15
Chroma (IEEE units)	0	0	60	86	80	80	86	60
Vector (from $B-Y$)	—	—	167°	284°	241°	61°	104°	347°

FIGURE 3.12   Raster patterns.

AGC circuits, which respond to sync-pulse amplitude, can be checked with the *composite-video output* available on many generators. Although the overall amplitude of the video is varied, the sync-pulse amplitude is a *constant percentage* of the total video amplitude.

The precise timing of the sync pulses allows proper adjustment of the servo and switching circuits in VCRs. The servo circuits control the speed of the video tape in VCRs, while the switching circuits provide for switching between video heads (to allow a continuous transition from field to field, as discussed in Chaps. 5 and 6).

3.4 PURITY, CONVERGENCE, AND LINEARITY ADJUSTMENTS

All color video equipment requires some form of purity, convergence, and linearity adjustments. These adjustments must be made in addition to the adjustments for black and white (height, width, centering, linearity, etc.). Generally, the adjustments are made when the equipment is placed in operation and should be checked when the equipment has been subjected to extensive repair, such as replacement of the picture tube and/or deflection coils. Typically, the setup procedure for color circuits consists of degaussing the color picture tube, followed by purity, linearity, and convergence adjustments.

It is recommended that the manufacturer's service instructions be followed exactly when color setup procedures are involved (when all else fails, follow instructions). However, in the absence of manufacturer's instructions, and to show what typical color setup procedures require, we describe a complete setup for color TV circuits, as recommended by the manufacturers. With the examples of the controls shown, you should be able to relate the procedures to a similar set of controls on almost any color TV (or color monitor or terminal) being serviced.

3.4.1 Black and White Adjustments

Use the following procedure for all black and white picture tubes and as a starting point on color tubes:

1. Measure from opposite diagonal corners of the picture tube screen, and mark the physical center of the screen with a grease pencil or similar marker, as shown in Fig. 3.13a.

2. Connect the generator to the antenna terminals and select a center-cross pattern (Fig. 3.10a).

3. Adjust the ion trap or centering adjustment of the picture tube so that the center cross of the test pattern coincides with the physical center of the screen. Note that if the pattern is badly distorted, the linearity adjustments (or troubleshooting and repair if necessary) should be performed before the centering adjustment is complete.

4. Switch to a crosshatch or dot-hatch pattern (Fig. 3.11). If the circuit has a width adjustment, adjust it so that the raster just fills the screen. (Some manufacturer's literature specifies the number of horizontal and vertical lines.)

5. Adjust the height control (if any) so that the raster just fills the screen. (Again, check for any recommendation in the service literature and/or generator operating instructions as to a specific number of lines.)

6. Adjust the vertical and horizontal linearity controls (if any) so that the horizontal and vertical lines are straight. Remember that any bend in horizontal or vertical lines with a crosshatch pattern is likely to show up as a bend in the picture.

7. Wipe the grease pencil centering marks from the face of the picture tube.

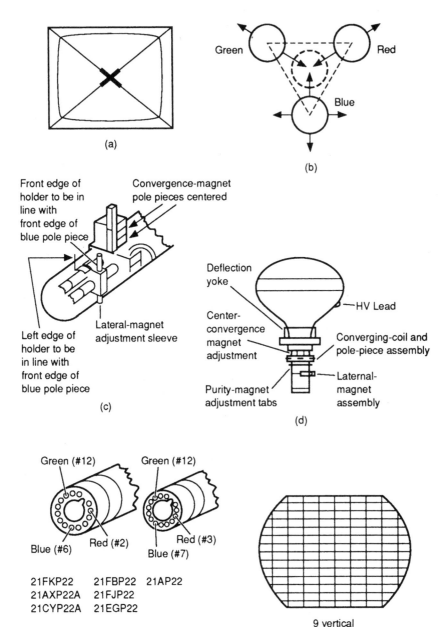

FIGURE 3.13   Center and convergence adjustments.

3.4.2 Preliminary Convergence Adjustments

Before making any convergence adjustments, check picture-tube focus. Some tubes have an electronic focus. The focus adjustment is a variable inductance in series with a winding on the horizontal output transformer and is located at the rear of the high-voltage cage.

1. Leave the generator connected to the antenna input, and select a dot pattern (Fig. 3.10*b*).

2. Adjust the red, green, and blue magnets and the lateral magnet to get proper convergence of the dots in the center of the picture tube. The direction of dot movement using the magnets is shown in Fig. 3.13*b*. Red and green movement is opposite that for blue. The red and green dots move diagonally, whereas the blue dot moves horizontally or vertically (on this particular picture tube). Location of convergence and lateral-beam magnets on a typical picture tube are shown in Fig. 3.13*c*.

3. If a greater range of adjustment is necessary, the magnets of some picture tubes may be reversed. To do this, slide the plastic magnet holder out of the metal clip and rotate the holder 180°. Replace the holder in the clip, making sure that the magnet is reinserted in the clip. If necessary, keep the picture tube in focus when making this adjustment.

3.4.3 Color Purity Adjustments

The picture tube and associated parts should be subjected to a strong demagnetizing field before any purity adjustments are made. All present-day color picture tubes have some form of demagnetizing or degaussing system, usually a coil that is energized when power is applied. However, it is possible that the degaussing system may fail or the picture tube may become contaminated. In either event, a commercial degaussing coil can be used to correct the problem.

Slowly move the coil around the picture tube and around the sides and front of the set or terminal. *Do not degauss the magnets* around the picture-tube neck. If the picture tube has a special built-in degaussing system, operate the system as described in the service literature.

1. Once the picture tube is properly degaussed, operate the generator to produce a blank raster (Fig. 3.12). If the generator is not capable of producing a blank or solid-color raster, disable the output of the video detector (short the output to ground).

2. Loosen the screw on the yoke clamp and slide the yoke as far to the rear as possible. Figure 3.13*d* shows the location of the purity magnets on a typical picture tube.

3. Shunt the blue and green picture grids to ground through individual 100-k resistors. If the generator is provided with color-gun interrupters (gun killers), connect the leads from the generator to the corresponding grid leads of the picture-tube socket. The location of color-gun leads for an assortment of color picture tubes is shown in Fig. 3.13*e*. Generally, it is not necessary to remove the socket from the color picture tube or make any direct connection to the socket pins. Most color-gun interrupter leads are provided with special alligator clips that pierce the insulation on the picture-tube lead and make contact with the wire.

4. Rotate the purity magnet around the neck of the picture tube and, at the same time, adjust the tabs on the magnet to produce a uniform red screen area (solid red spot) at the center of the picture tube.

5. Slide the yoke forward until the screen becomes completely red.

6. After checking the red field, it is helpful (but not absolutely necessary) to check the green and blue fields (individually) for purity and the white field for uniformity. The green field can be checked by shunting the red and blue picture-tube guns (or use the color-gun interrupters). The blue field can be checked by shunting the red and green fields.

With all grids shunted, there should be no color, and the white should be uniform across the screen. Note that if individual fields of red, blue, and green are present and a white field is present with all guns off, you know that the picture tube is good and is capable of producing all required colors.

7. Set the generator to select a crosshatch pattern. Check that the height, width, centering, focus, overscan, etc., are properly adjusted by comparing the crosshatch pattern with that specified for the picture tube.

8. It is sometimes possible to get good purity only by pushing the yoke too far forward. This reduces width and height such that the set does not overscan properly. The crosshatch pattern provides a convenient method for adjusting the overscan to ensure that the proper portion of the raster is extended beyond the range of the picture-tube mask.

9. The service literature for color sets and terminals usually specifies a recommended amount of overscan at the left and right and a different amount of overscan at the top and bottom. The recommended overscan varies in different models of equipment. However, the crosshatch function provides a *fixed number* of vertical and horizontal bars, so it is relatively easy to judge the amount of overscan. The appearance of the crosshatch pattern with correct overscan (9 vertical and 13 horizontal in this case) in a representative color TV set is shown in Fig. 3.13*f*.

3.4.4 Picture-Tube Temperature Adjustments

When required, picture-tube temperature adjustments can be made without an external color generator. Thus, the generator can be disconnected and/or turned off if the signal interferes with the procedure.

1. Set the picture-tube bias (sometimes called the *kine bias*) and screen controls (red, blue, and green screens) fully off.

2. Set the normal/service switch (if any) to service.

3. Advance the individual screen controls so that each control just produces a horizontal line on the picture tube. When one or more of the controls fails to produce a line, the picture-tube bias must be advanced. After the bias control has been advanced to make the missing line appear, the remaining screen controls must be adjusted to where the horizontal line just appears.

4. Return the normal/service switch to normal. This should restore the vertical sweep (which is removed in the service position).

5. Alternately adjust the blue and green video-drive controls to produce a normal black and white picture. Use the gray-scale output of the generator (Fig. 3.7*c*).

6. Check the picture from white to black at all brightness levels for proper tracking.

The compatible TV system is designed so that reception of black and white pictures is reproduced on color sets. The three picture-tube guns must be adjusted so that the telecast is reproduced as black and white within the *normal usable range* of the contrast and brightness controls.

The adjustments are called *white balance* (in some literature) as well as screen temperature, since the adjustments pertain to the reproduction of various luma values from black and white. Only a set correctly adjusted for gray-scale tracking can, in turn, reproduce proper color when tuned to a color telecast.

Alternative Screen-Temperature Adjustments. An alternative method for adjusting picture-tube temperature that sometimes provides greater accuracy involves extinguishing the horizontal line just after the line appears.

1. Set the normal/service switch to service (to collapse the vertical sweep).

2. Set the kine-bias control and all screens to minimum.

3. Adjust the red, green, and blue screens to where beams just cut off (increase kine bias only as required to support the weakest screen adjustment).

4. Reset the normal/service switch to normal.

5. Adjust the blue and green video drive for a normal black and white picture (using the gray-scale tracking pattern, Fig. 3.7c).

3.4.5 Center-Convergence Adjustments

After purity and temperature adjustments, check the convergence of dots at the picture-tube center, as described in Sec. 3.4.2. If the temperature and purity adjustments have affected convergence, repeat the convergence procedure using a dot pattern. If the convergence magnets are touched, readjust the purity and temperature controls. When the center-convergence, purity, and temperature controls require no further adjustment, proceed with the vertical and horizontal convergence adjustments.

3.4.6 Vertical Convergence Adjustments

1. Set the color generator (connected to the antenna input) to produce vertical lines. Use a crosshatch pattern if the generator does not produce separate horizontal and vertical lines. Note that some technicians prefer to use the center-cross pattern in Fig. 3.10a for all convergence adjustments.

2. Referring to the vertical line at the center of the picture-tube screen, adjust the vertical *R-G* (red-green) master *amplitude control* to converge the center line at the bottom, as shown in Fig. 3.14. Adjust the vertical *R-G* master *tilt control* to converge the center line at the top of the screen.

3. Touch up both adjustments (amplitude and tilt) for best convergence along the entire center vertical line. You may find it easier to perform this setup with the blue gun disabled. The blue gun is not necessary to the adjustment since the *R-G* controls affect only red and green.

4. Set the generator to produce horizontal lines (or a crosshatch).

5. Referring to the center line of the screen, converge the horizontal line at the bottom of the screen with the vertical *R-G differential amplitude control.* Adjust the vertical *R-G differential tilt control* to converge the top horizontal line at the center of the screen. Touch up both adjustments for the best convergence of all lines at the vertical center line.

6. Switch to a dot pattern and check convergence of the dots at the center of the screen. If necessary, readjust the convergence magnets as described in Sec. 3.4.2.

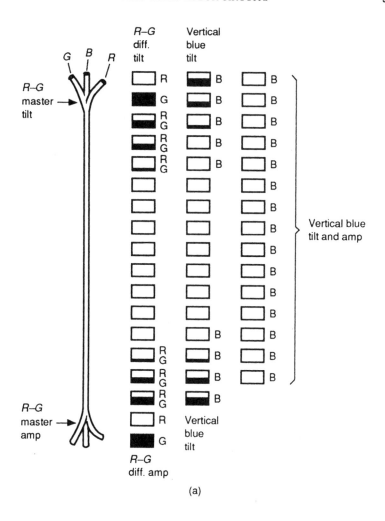

(a)

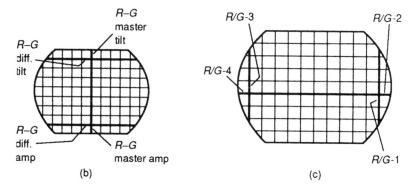

(b)   (c)

FIGURE 3.14 Vertical convergence adjustments.

7. Switch to horizontal lines (or crosshatch) and advance the vertical *blue amplitude control* to produce displacement of the lines at the top and bottom of the screen at the center line. Adjust the vertical *blue tilt control* to produce equal displacement of the lines at both top and bottom.

8. Check convergence with the generator switched to horizontal lines (or crosshatch). It may be necessary to converge the blue with the red and green using the static magnets. (Note that interaction of controls is much more severe in 90° color picture tubes.)

9. With the generator still set for horizontal lines or crosshatch, adjust the vertical *blue amplitude and tilt* controls to produce equal displacement of all lines from top to bottom of the screen along the center line.

10. With the generator set for a crosshatch pattern, check the convergence of the dots as described in Sec. 3.4.2. Reconverge if necessary (first with the dot pattern and then rechecking with a crosshatch). Retouch the vertical blue amplitude and tilt controls for best convergence along the vertical center of the screen, as necessary.

3.4.7 Horizontal Convergence Adjustments

1. With the generator set for a crosshatch pattern, check the convergence of dots at the center of the screen. Reconverge as described in Sec. 3.4.2, if necessary. Refer to Fig. 3.14c when making horizontal convergence adjustments.

2. Adjust the horizontal convergence controls to make the corresponding lines converge. For example, adjust red and green 1 to make vertical lines at the right side converge and red and green 2 to make horizontal red and green lines at the right side converge. Generally, the service literature shows some similar arrangement to those given in Fig. 3.14.

3. After completion of all vertical and horizontal convergence adjustments, check and (if necessary) repeat the picture-tube purity and temperature adjustments (Secs. 3.4.3 and 3.4.4).

3.5 COLOR SETUP PROCEDURES USING A COLOR GENERATOR

In this section, we describe how a typical color generator can be used to set up a color TV. Note that the procedures cover essentially the same areas described in Sec. 3.4 but apply specifically to a color TV, using a particular color generator. With this generator, it is possible to vary the RF level as well as the chroma level of the signal applied to the antenna. Such features are not found on all color generators.

3.5.1 Preliminary Procedures

Connect the generator output to the antenna input. Disconnect any antenna that may be connected to the set, including "rabbit ears." Set the generator RF- and chroma-level controls to midrange. Select RF output with a full color-bar pattern (Fig. 3.6). Disable any automatic fine-tuning (AFT) circuits. (This does not apply to the synthesized-tuning circuits described in Chap. 4.) Adjust all TV set controls for best picture.

3.5.2   AGC Adjustment

Use the following procedure on sets with any form of AGC:

1. Select a crosshatch pattern. Turn the set AGC control until the pattern just begins to bend or distort; then readjust the AGC to the point slightly below where distortion is eliminated. Note that the AGC control has very little effect on some sets. If so, set AGC for best picture.

2. Set the generator RF-level control to minimum. If the pattern distorts or loses sync, readjust the AGC as in step 1, setting AGC just below where the trace distorts.

3. Vary the RF-level control throughout the entire range. Readjust AGC as necessary to prevent the picture from distorting or losing sync through the widest possible range of RF levels.

3.5.3   Video Peaking

Use the following procedure on sets with any form of video peaking adjustments: Select a gray-scale pattern (Fig. 3.7c). Adjust the video peaking coil (or coils) so that the edges of the bars are sharp. Note that if you adjust the peaking control too far in one direction, "ringing" (traces of the edge repeating across the pattern) may occur. If you adjust the peaking too far in the other direction, the edges may become soft or hazy.

3.5.4   Horizontal and Vertical Centering

Use the following procedure on sets with any form of horizontal or vertical centering controls: Select a center-cross pattern (Fig. 3.10c) and adjust the centering controls so that the vertical and horizontal lines intersect at the center of the screen (Sec. 3.4.1). Then select a crosshatch pattern and check that the outer lines are equally spaced from the edge of the picture-tube mask.

3.5.5   Purity

Use the following procedure on all sets:

1. Before adjusting purity, the set may need a rough check of static convergence, as described in Sec. 3.5.10.

2. Select a blank raster. Turn off the blue and green guns or use gun-killer switches. This should produce a red screen.

3. Loosen the picture-tube yoke and move the yoke to the rear of the picture-tube neck (as far as possible without moving the purity-magnet assembly).

4. Adjust the purity magnets so that the area in the center of the screen is uniformly red with no dark or discolored areas.

5. Push the yoke forward until the entire screen is uniformly red. Tighten the yoke.

6. Turn on the blue and green guns. The raster should be uniformly white. If not, repeat the adjustment procedure.

7. If good purity adjustment cannot be obtained, part of the set may have become

magnetized and requires degaussing. Even sets with automatic degaussing can have a magnetized cabinet. If necessary, use a degaussing coil to demagnetize the affected area (Sec. 3.4.3).

3.5.6 Overscan

Use the following procedure on all sets: Select a crosshatch pattern, and adjust the width and height controls so that the raster extends beyond the edge of the picture-tube mask by the amount specified for the particular set. As discussed, the recommended amount of overscan varies but is usually specified in the service literature.

3.5.7 Linearity

Use the following procedure on all sets: Select a crosshatch pattern, and adjust the height and vertical linearity controls so that the horizontal lines of the crosshatch are equally spaced (particularly at the extreme top and bottom of the screen). Adjust the horizontal linearity controls (if any) so that the vertical lines are equally spaced.

3.5.8 Pincushion Adjustment

Use the following procedures on sets with pincushion controls: Select a crosshatch pattern (or preferably a dot hatch) and adjust the pincushion controls (top and bottom) until all horizontal lines are straight.

3.5.9 Color-Temperature Adjustment (Gray-Scale Tracking)

Correct color temperature applies to all sets (and color monitors and terminals) and is necessary to assure that all three color guns are operating at a level to cause the red, blue, and green phosphors to glow at the same level. The beam currents of the three guns must maintain the ratio throughout the excursions of the video signal and throughout the range of the brightness control. Proper color balance must be maintained from black (low lights) through the grays to white (bright lights). Use the procedure in Sec. 3.4.4 to adjust color temperature and check gray-scale tracking.

3.5.10 Convergence Adjustments

Misconvergence of a color set can be noted by observing the crosshatch or dot pattern. If the set is properly converged, the lines or dots are clear and sharp, with no *color fringing* (one or more colors appearing on the edges of the dots or crosshatch lines).

1. Before adjusting convergence, check other adjustments, especially purity, if specified in the service literature. These adjustments may include horizontal tuning, horizontal drive, high-voltage regulation, height, width, linearity, focus, etc.

2. Select a dot pattern and perform preliminary or static convergence adjustments as described in Sec. 3.4.2. Adjust the permanent magnets on the picture-tube neck to converge the red, blue, and green beams at the screen center. A "perfectly converged" pattern has pure white dots with no color fringing.

3. Select a crosshatch (or dot hatch) pattern, and perform dynamic convergence as described in Secs. 3.4.5 through 3.4.7. Adjust the dynamic-convergence controls to converge the entire screen, particularly the edge areas.

4. Repeat static and dynamic convergence procedures as required until the complete crosshatch pattern consists of crisp white lines and have the least amount of color fringing.

3.5.11 Check the Color Circuits

Use the following procedure on all color sets:

1. Select a full color-bar display. Switch out any special color controls (such as "accutint"). Adjust the hue or tint and color controls to midrange. Adjust the fine tuning, brightness, picture, and contrast controls for best color-bar pattern. The pattern should be similar to Fig. 3.6.

2. Pay particular attention to the *magenta and cyan* (bluish-green) bars. Generally, if you can get good magenta and cyan with the hue or tint control, the color circuits are operating satisfactorily (and all of the remaining colors will be good). Ideally, the hue or tint control should be near the midrange position when magenta and cyan are good and should have enough range to shift the color so that both magenta and cyan are off color. Typically, the magenta becomes red at one extreme of the hue or tint control, while the cyan goes from blue to green (or vice versa) at the other extreme.

3. If it is impossible to get a good magenta and cyan display, suspect the automatic frequency and phase controls (Sec. 3.8).

4. To check accutint or similar functions that enhance flesh tones, switch the function on, and check the color bars. The pattern should change, with several of the bars taking on a more reddish color. The range of the hue or tint control is usually more restricted in most sets (only a slight shift of color at both extremes). The color control may also be restricted so that the color cannot be taken completely out of the pattern.

5. To check color-sync lock action, turn the generator chroma-level control slowly to minimum. The color should become pale and finally disappear. Since some sets are equipped with an automatic color control (ACC) circuit, the rate of fading depends on the set. Most sets lose color sync just before the color disappears (diagonal lines run through the colors). These conditions indicate *normal operation* of the color-sync circuits. If a slight reduction of the chroma amplitude causes color to fall out of lock, however, this indicates that the color-sync ability of the set may be inadequate. As a further check, turn the RF-level control to reduce signal strength, and note the effect on color sync.

3.6 COLOR TV SET ALIGNMENT (BASIC)

The alignment procedures described in Sec. 2.7 can be used for color sets. However, color sets have a greater need for proper alignment. That is, you might be able to get by with a poorly aligned black and white set, but poor alignment in a color set is totally unacceptable. Also, color sets have additional circuits that requite alignment. For these reasons, we include a series of notes on color TV alignment. Compare these notes with the procedures found in Sec. 2.7.

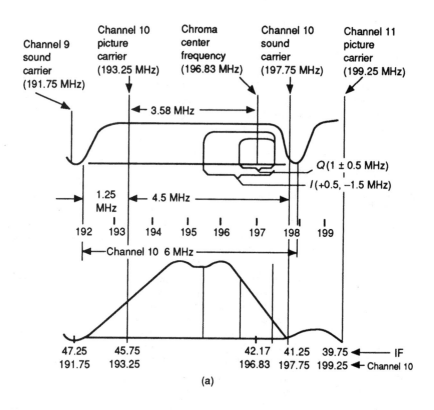

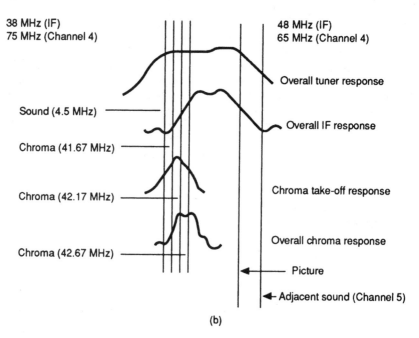

FIGURE 3.15 Simplified TV frequency spectrum of Channels 4 and 10.

3.34

3.6.1 The Transmitted Color TV Signal

Figure 3.15a shows the simplified TV frequency spectrum of Channel 10. The picture carrier is 1.25 MHz above the lower limit of the channel. Within the channel, frequencies are referenced to the picture-carrier frequency.

The I and Q signals (color signals) are centered on the chroma center frequency of 196.83 MHz. The spectrum shows that the Q-signal sidebands are symmetrical about the chroma center frequency with a distribution of 0.5 MHz. The I-signal sidebands are not symmetrical about the center frequency. Older color sets have a bandpass sufficient to pass the complete bandwidth of the I and Q signals. Present-day color sets use a narrowband chroma response (± 0.5 MHz from the color subcarrier). This means that portions of the I signal are not used.

Figure 3.15a also shows how the relative response changes when the signal is converted to IF by the RF-tuner mixer and then passes through the IF stages. The IF frequencies are indicated with the corresponding Channel 10 RF frequencies. As discussed in Chap. 2, there is a frequency inversion in the mixer process. For example, the picture carrier is at the low-frequency end of the transmitted signal (input to tuner) and at the high-frequency end in the IF stages.

As discussed, TV tuned circuits contain traps to reject unwanted signals from adjacent channels. For example, in Fig. 3.15a, notice that the sound carrier of Channel 9 (191.75 MHz) is just outside the lower end of the Channel 10 band and that the Channel 11 picture carrier (199.25 MHz) is 1.25 MHz above the upper end of Channel 10. These two frequencies are called the *adjacent channel sound carrier* and the *adjacent channel picture carrier,* respectively, of Channel 10.

3.6.2 Color TV Tuned Circuits

To process the signals properly, the RF, IF, and chroma sections must have certain bandwidth characteristics. The IF bandpass is obviously narrower than the tuner bandpass and thus contributes most to bandpass shaping. In sets with circuits similar to those in Fig. 2.22 (discrete-component and/or tuned IC), the bandwidths are determined by the number of amplifiers and associated circuits. In present-day IC sets such as shown in Fig. 2.23, the bandwidths are set by IC and filter characteristics.

3.6.3 The Need for Sweep Alignment

The IF circuits in Fig. 2.23 do not require sweep alignment. Such circuits are often found in sets with frequency-synthesis (FS) turners (Chap. 4), which also do not need sweep alignment. For these reasons, we do not emphasize sweep alignment in this book.

The IF circuits in Fig. 2.22 (or any *tuned* circuit, IF or RF) must be properly aligned. If such circuits drift or are misaligned or if the gain of one or more stages changes, the signals are affected in many ways. Signal levels may be too low, the bandwidth may be too narrow, the signals may begin to interfere with each other or if traps are misaligned, the overall performance may be degraded by interference from undesired signals, such as adjacent-channel sound or picture-carrier signals. Most of these problems can be overcome by proper alignment.

3.6.4 Recommended Sweep-Alignment Techniques

The service literature of most manufacturers recommends sweep alignment for all tuned circuits. However, the exact method and the order of steps to be performed are not standard. The following notes summarize typical recommendations.

Generally, alignment starts by injecting the sweep signal at the RF input (antenna terminals). You then monitor the IF and chroma outputs for comparison against service-literature waveform patterns. If the IF is good but the chroma is not, the problem is between the video-detector output and the bandpass-amplifier output (input to the color demodulators). If both IF and chroma outputs are abnormal, it is most likely that the IF requires a touch-up, particularly if the response is poor on the *slope affecting chroma response* (Sec. 3.6.5).

You seldom find an alignment problem in the RF portion of the tuner (unless there has been tinkering) because the bandwidth is so much greater than that of the IF section (about 6 MHz compared to 4 MHz). However, the mixer output circuit, which is located on the tuner, may require attention. This is part of the tuned matching network (called the *tuner link* or simply *link*) between the tuner and the first IF stage. A separate prealignment procedure is given for the link circuits by some manufacturers. In present-day sets, the tuner is usually a sealed package, and no alignment is possible.

3.6.5 Diagnosing the Need for Alignment

You must realize that alignment alone is not the universal cure for poor picture quality. Before attempting to diagnose the need for alignment, you must be sure that the convergence, purity, and focus (and even high-voltage regulation) are good and that the set has been properly degaussed. You should also eliminate the possibility of interference from other test equipment (caps left off RF generator outputs, etc.).

With these problems out of the way, the quickest method to determine the overall condition of the set is to use sweep alignment (with markers) as described in Sec. 2.7. As the sweep signal is processed through the tuned circuits, the signal is shaped by the gain and bandpass characteristics of the various sections (RF, IF, chroma, etc.).

Figure 3.15b shows the sweep signal with basic response curves (waveforms) of the RF tuner, IF amplifiers, and chroma bandpass circuits. The bandwidths shown are approximately to scale. These outlines are similar to the curves that you get if you monitor corresponding points in the set during sweep alignment.

Figure 3.15b includes some reference frequencies (for Channel 4) to show the importance of proper alignment. For example, note that the chroma frequencies are on the slope of the IF response curve. This area is the most critical because improper IF alignment on the slope affects the amplitude and shape of the chroma response curve. In turn, this affects color picture quality.

3.6.6 Analyzing the Response Curves

Using the procedures in Sec. 2.7 and/or the service-literature instructions, you must now determine if the curves obtained are satisfactory, if the set must be realigned. If alignment is required, to what extent? Is a touch-up required or a complete realignment?

Figure 3.16 shows typical IF and chroma bandpass response curves as the curves might appear in the service literature of a color TV set. Note that the reference marker locations are shown with given tolerances. This means that the response curves obtained may vary within these limits and still give satisfactory performance.

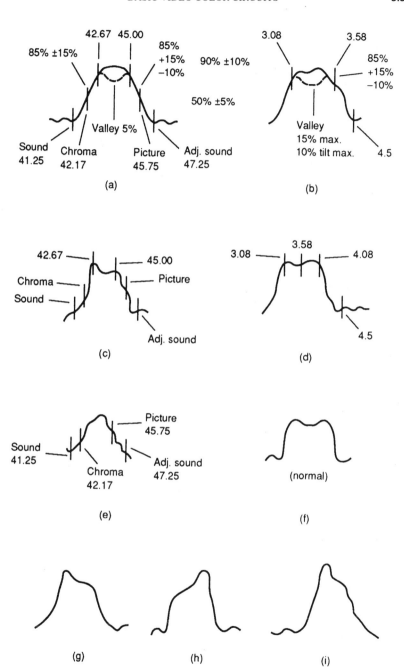

FIGURE 3.16 Typical IF and chroma bandpass response curves.

Figure 3.16c and d shows some allowable variations based on the limits of Fig. 3.16a and b. You must evaluate the response curves obtained with the allowable tolerances in mind. The areas to examine (aside from tilt across the curve top) are the areas of the trap frequencies, such as sound and adjacent sound, and the picture and chroma markers at the 50 percent reference points.

If a trap has been detuned toward the center of the response curve, the overall response is pulled downward. For example, if the sound trap is tuned near the chroma frequency, the curve response at the chroma frequency is as shown in Fig. 3.16e.

For stagger-tuned circuits, the curve is tilted if the circuit (tuned to the approximate center of the response curve) is misaligned. This is shown in Fig. 3.16g and h. Also, if one of the outside tuned circuits is tuned toward the center of the response curve, the curve peaks toward the center with reduced bandwidth and excessive gain, as shown in Fig. 3.16i.

By realizing the effect of mistuning traps and other tuned circuits, you can usually locate the mistuned circuit if the approximate alignment frequency of each circuit is given in the service literature. One way to get this information is to check the alignment procedures thoroughly for *prealignment* instructions. Then make a cross-reference between marker frequencies and the tuned circuits to be adjusted. If prealignment of traps and transformers is specified, most or all of the information related to tuned-circuit frequencies is included in that section of the service instructions.

3.6.7   Alignment Touch-Up Procedures

Touch-up procedures are used when your analysis of the curve indicates that the curves are recognizable but marginal on response limits (such as excessive tilt, abnormal peaking, etc.). If the curves fall within the limits indicated, no alignment is required. If you decide to align, the extent of alignment must be determined. If the response curves are marginal at several or all points but are still recognizable, a touch-up can be performed to correct excessive tilts and to restore response levels at various points on the response curves.

3.6.8   Complete IF and Chroma Alignment

When complete IF and chroma alignment is required, use the basic procedures in Sec. 2.7 and observe the following notes (which supplement the service-literature information).

Trap Alignment.   The standard trap frequencies are the adjacent-picture carrier (39.75 MHz), the sound carrier (41.25 MHz), and the adjacent-sound carrier (47.25 MHz), as shown in Fig. 2.22. However, several manufacturers have additional trap frequencies (such as 35.25 and 38.75 MHz found on older sets). All traps except the sound trap (41.25 MHz) are located at the input to the first IF. The sound trap is usually located in the last IF or in the output circuit of the last IF.

Prealignment.   Some alignment procedures specify a pretuning of all IF bandpass circuits, as well as the traps. Each circuit is tuned for a maximum output as indicated on the scope using modulated markers. That is, you inject the required marker signal using the generator function and then tune the specified control for a maximum trace (typically a 400-Hz trace). If prealignment is recommended, remember to use only one marker at a time and turn off the marker when alignment of a particular circuit is complete.

Marker Height.   In some service literature, the phrase "set the marker height" is used extensively. This may be misleading since actual marker height is not adjustable. (Marker height is not to be confused with marker amplitude, which is adjustable on some generators.) The phrase means that you are adjusting the IF response so that at a particular marker frequency the response is at *some percentage down from maximum*. For example, to set the height of the 42.17-MHz marker at 50 percent amplitude, you adjust the IF to alter the curve so that at 42.17 MHz the response is 50 percent of maximum.

Chroma Markers.   When aligning the chroma circuits, the 3.58-MHz oscillator within the set may produce a "marker" on the scope display. In some cases, a low-frequency beat may be produced between the set marker and the generator marker. This problem is discussed in Sec. 3.9.

3.6.9   Adjustment of Automatic Fine-Tune (AFT) Circuits

Figure 3.17 shows a typical AFT circuit. Most sets include AFT circuits to prevent the tuner oscillator from drifting in frequency. Typically, the 45.75-MHz (picture carrier) output of the last IF (or the IC IF output) is applied to the input of a tuned AFT amplifier. The amplifier is tuned to exactly 45.75 MHz and delivers an output to the AFT discriminator. The discriminator output is applied to a voltage-variable capacitor (VVC), which is connected across the resonant circuit of the tuner oscillator. (In some tuners, the base-collector junction of a transistor is sometimes used as the VVC.)

If the tuner-oscillator frequency remains fixed, the outputs of the AFT amplifier and discriminator also remain fixed, and the VVC does not vary in capacitance. Thus, the tuner oscillator circuits remain unchanged. If the oscillator should drift in frequency, the AFT amplifier and discriminator outputs change, thus changing the VVC capacitance and oscillator resonant circuits.

The FM discriminator requires the usual alignment. The following is a typical alignment procedure. The exact discriminator crossover point is obtained by injecting a 45.75-MHz carrier at the AFT input (last IF output). The crossover point represents the discriminator center frequency, as shown in Fig. 3.17b. The dc output voltage of the AFT circuit is then checked at the AFT test point using a meter.

Upon completion of the AFT alignment, perform an AFT check as described in the service literature. Usually, this involves connecting the set to an external antenna, disabling the AFT, and adjusting the fine tuning manually for best picture and sound, with minimum sound interference. Then activate the AFC and check the "pull-in" effect.

An alternative method of checking AFC operation is to adjust the fine tuning in either direction from the best-picture setting, with the AFT disabled. Then activate the AFC and check the effect. The picture should automatically be restored to the best setting obtained by manual adjustment of the fine tuning when the AFT was disabled.

Note that the frequency-synthesis tuner circuits described in Chap. 4 also contain AFT circuits. However, operation of the circuits is different, and alignment (as described here) is not required.

3.7   CHECKING AND ALIGNING AFPC CIRCUITS

All color sets have some form of automatic frequency and phase control (AFPC) circuits to lock the 3.58-MHz reference oscillator with the color-burst signal. These cir-

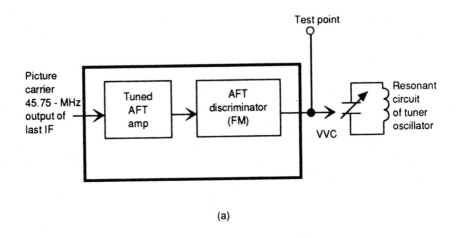

(a)

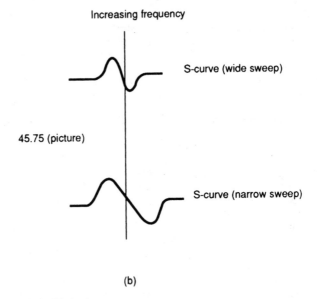

(b)

FIGURE 3.17   Typical AFT circuits.

cuits may be called APC or AFC in some literature. Usually, the circuits require some form of alignment or adjustment. The following procedures are given as a guide for AFPC alignment for the several different color-demodulation systems used in various color sets. Figure 3.18 shows a typical set of adjustment points. As always, the procedures recommended in the service literature should be used.

1. Connect the color generator to the antenna input, or inject the signal to the video stage if desired. Adjust the generator and color set for a suitable color-bar pattern. If available, use the color-bar patterns described in Sec. 3.3.2 (such as Fig. 3.7b).

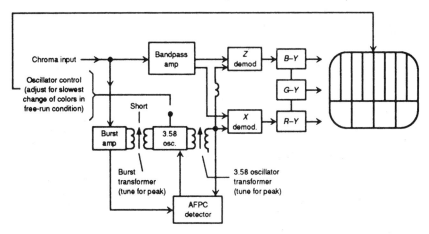

FIGURE 3.18 Typical AFPC adjustment points.

2. Turn off any accutint or other special color circuit, and turn the set color control fully on. Set the hue or tint control to midrange. Turn the color killer off (or disable it with a suitable bias).

3. Short out the burst-phase transformer, and adjust the oscillator control for a free-running "floating" pattern. The correct pattern on the picture tube will be uniform color bars, but the *bars should change color slowly* as the colors appear to drift or float from bar to bar. If the colors change rapidly, the oscillator is not properly adjusted.

4. Remove the short, and connect a scope to the burst-amplifier output. Adjust the burst-amplifier transformer to get a maximum signal condition.

5. Check for a good set of color bars on the picture tube, as described in Sec. 3.5.11. Pay particular attention to the magenta and cyan bars, as discussed.

3.8 SETTING THE 4.5-MHZ TRAP

As discussed in Sec. 2.7.3, adjustment of the 4.5-MHz trap can be quite critical for a color set. If the color generator has a 4.5-MHz signal with modulation (as is usually the case for most NTSC generators), this signal can be used for sound-trap alignment, as follows:

1. Connect the color generator to the antenna input, or inject the 4.5-MHz signal to the video stage (such as at the base of Q_1, shown later in Fig. 3.21). Adjust the generator controls to produce the 4.5-MHz signal (with modulation).

2. Monitor the video circuit with a scope at any point after the trap (such as at the picture-tube cathode shown later in Fig. 3.21). Adjust the trap until the 4.5-MHz signal (with modulation) is eliminated (or at least minimum). Try applying and removing the 4.5-MHz signal so that you can identify the pattern produced by sound modulation. You can also check the picture-tube screen for evidence of sound modulation, but the scope is more accurate for adjustment.

3.9   SERVICING VIDEO AND CHROMA AMPLIFIERS WITH VIDEO SWEEP

As discussed in Chap. 2, a sweep signal can be used to check operation of a black and white set from antenna to picture tube. This is in addition to (or in place of) an overall check with an NTSC generator. It is also possible to service the video and chroma amplifiers of a color set. For example, assume that there is a color problem and that an overall check from the mixer input to the output of the chroma amplifiers indicates a loss of response or a discrepancy in the response characteristics (but there is a video signal). At this point, the chroma and video amplifiers are suspect, with the chroma amplifiers being the most likely problem.

With a video sweep available, you can inject a signal into the last chroma output stage and observe the output with a wideband scope (or a narrowband scope and a chroma or video-detector probe). Typical connections are shown in Fig. 3.19. If all is well in the last chroma stage (before the demodulators) you should get a response curve similar to that in Fig. 3.19a. (A similar curve may be given in the service literature.)

Next, inject the sweep signal into the input of the first chroma amplifier stage (if there is more than one chroma amplifier). Again, you should get a pattern similar to Fig. 3.19a. Now inject the signal into the input of the color-takeoff point, as shown in Fig. 3.19b. You should get a response curve similar to Fig. 3.19c. This curve is the complement of the color slope of the IF bandpass curve, and the two are shown in Fig. 3.19d. Note that 42.67 and 4.08 are of equal height up on the curve, that the two curves cross over at 3.58 and 42.17, and that 3.08 and 41.67 are also equal in height. Finally, inject a sweep into the video amplifier as shown in Fig. 3.19e. This should produce a curve that is flat to about 3 MHz, as shown in Fig. 3.19e.

This procedure should pinpoint which section (video or chroma) is at fault. The problem can then be corrected by alignment and/or troubleshooting. Note that if the 3.58-MHz reference oscillator is not disabled, a 3.58-MHz "marker" may appear on the pattern.

3.10   USING THE VECTORSCOPE

A vectorscope is used in TV broadcast work to troubleshoot and adjust color demodulators. The vectorscope can also be used to judge the general condition of color TV set circuits (chroma bandpass, burst amplifier, reference oscillator, phase detectors, etc.) when used with the NTSC color-bar output of a color generator. A vectorscope measurement is often more helpful for troubleshooting than merely observing the NTSC pattern on the picture tube.

The display of an NTSC vectorscope, with the NTSC color-bar pattern applied, should be six dots located within the six boxes of the vectorscope graticule, as shown in Fig. 3.20. The pattern or signal to be displayed on the vectorscope may be probed from anywhere in the composite video or 3.58-MHz color circuits. Amplitude must be initially adjusted so that the color-burst dot aligns with the 75 percent mark on the vectorscope graticule.

If an NTSC vectorscope is not available, a good lab-type scope can be substituted. The demodulated color signals (directly from the red and blue guns) can be used as X and Y inputs to the scope, as shown in Fig. 3.20a. This connection should produce a scope display as shown in Fig. 3.20b. The scope can be set up for vectorscope operation as follows:

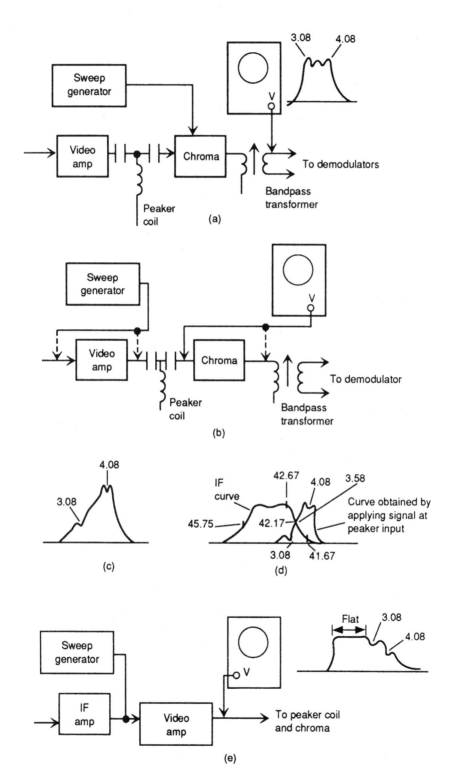

FIGURE 3.19 Typical chroma sweep adjustments.

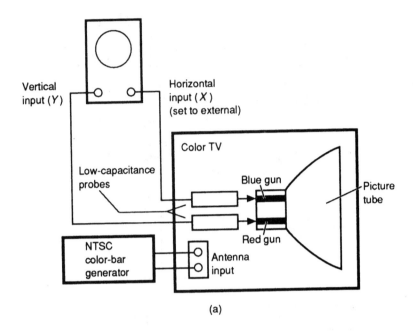

(a)

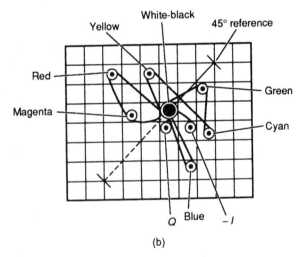

(b)

FIGURE 3.20 NTSC vectorscope connections and display.

1. Select the NTSC color-bar pattern (Fig. 3.6), and apply the generator output to the antenna input.

2. Set up the scope for X-Y operation. Adjust the position controls to center the dot on the screen with no signal input to the scope.

3. Connect both the X and Y inputs of the scope to the red gun.

4. Adjust the horizontal and vertical gain of the scope to equal amounts, which will move the spot or dot to the 45° reference position (from the center), as shown in Fig. 3.20b.

5. Now move the horizontal (X) input of the scope to the blue gun. Leave the vertical (Y) input connected to the red gun.

6. For a 90° picture tube, the display on the scope should be similar to that shown in Fig. 3.20b. The typical 105° picture tube produces a more elliptical display.

7. If desired, the various vectors may be selected one at a time, by selecting the corresponding raster color display of the generator. Raster displays are discussed in Sec. 3.3.6.

3.10.1   Basic Vectorscope Alignment Procedure

With the vectorscope connected as shown in Fig. 3.20, rotate the hue or tint control through the full range, and note the effect on the vector pattern. The pattern should turn but *should not* change in size. If necessary, adjust the 3.58-MHz reference oscillator until the pattern is the same size throughout the range of the hue or tint control.

As an alternate, set the hue or tint control to midrange and adjust the 3.58-MHz oscillator so that the color signal stays in sync as the chroma-level control on the generator is turned to minimum. The vector pattern (scope) should reduce in size but *should not rotate*. Rapid spinning of the vector pattern indicates misadjustment of the oscillator.

3.10.2   Correction for a Badly Misaligned 3.58-MHz Oscillator

This procedure is an alternate to that described in Sec. 3.10.1 and should be used when the 3.58-MHz oscillator is way off frequency.

1. Disable the correction signal to the 3.58-MHz oscillator as described in the service literature. Usually, the correction signal can be disabled by a bias or a short applied at some point in the circuit.

2. With the correction signal disabled, the oscillator is free-running, and the pattern appears to rotate or possibly appears as a blurred circle, depending on how far the oscillator is off frequency. A free-running oscillator can also be verified by changing colors on the color-bar (picture tube) pattern.

3. Adjust the 3.58-MHz oscillator until the vectorscope pattern stands still or is as close as possible to a motionless condition and the colors change slowly on the color-bar pattern. Then restore the correction signal to the oscillator.

3.10.3 Troubleshooting with Vectorscope Patterns

It is possible to check operation of the color circuits by comparing the display in Fig. 3.20*b* with the correct values shown in Figs. 3.2 and 3.3. That is, you can check each color for correct angle and amplitude. However, this is best done with a vectorscope and is not too practical with a scope (even a good lab scope).

Of course, any drastic deviations of the vectorscope pattern from that shown in Fig. 3.20*b* indicate a major problem in the color circuits. Here are some examples:

A loss of *R-Y* signal causes a loss of vertical deflection, and the vectorscope pattern changes to a straight line. If the *B-Y* signal is good, the beam is deflected along the horizontal axis. This indicates that the trouble lies in the *R-Y* demodulator, matrix, or difference amplifier, depending on the circuits. If the *R-Y* signal is weak, some deflection of the vectorscope pattern occurs, but the pattern is extremely distorted.

A loss of *B-Y* signal results in no horizontal deflection, and the vectorscope pattern changes to a straight line (a vertical line if the *R-Y* signal is good). This indicates that the problem is in the *B-Y* difference amplifier, matrix, or demodulator, depending on circuits. Again, if *B-Y* is weak, some deflection occurs, but the vectorscope pattern is extremely distorted.

If there is a complete loss of color, the vectorscope pattern usually appears as a center dot or possibly a short line in the center of the vectorscope pattern. Remember that not all circuits produce identical vectorscope patterns. The patterns discussed here are for reference only and must be considered as typical. Always consult the vectorscope instruction manual and all service literature.

3.11 VIDEO AND CHROMA PROCESSING CIRCUITS

This section is devoted to TV-set circuits between the VIF and SIF stages and the picture tube or CRT. Circuits that provide the horizontal and vertical sweep for the CRT are discussed in Sec. 2.19. Section 3.12 describes additional color circuits (such as interface circuits used in video monitors).

Video and chroma processing circuits have several functions, and there are many circuit configurations. Thus it is very difficult to present a typical video and chroma processing circuit. However, most video and chroma circuits have certain inputs and outputs that can be monitored. If the inputs are normal, but one or more of the outputs are abnormal, the problem can be localized to the video and chroma processing circuits.

3.11.1 Video and Chroma Processing Basics

Figure 3.21 shows the signal path of typical discrete-component video and chroma processing circuits between the VIF and SIF stages and the CRT. (Such circuits are often called the video-amplifier stages in older literature.) Typically, the signal input and output at Q_1 are about 1 V. This signal, generally called the *composite video signal,* actually consists of sound, video, and sync pulses, as taken from the video detector that follows the VIF. The output of Q_1 is applied to the input of Q_2, to the input of the sync-separator (Sec. 2.19), and to the input of the SIF (in this particular circuit). In

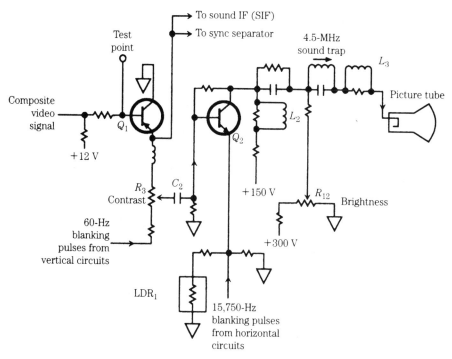

FIGURE 3.21   Signal path in discrete-component video-amplifier circuits.

the circuit of Fig. 3.21, the output of Q_1 is applied to Q_2 through contrast control R_3. In some circuits, the contrast control is part of the video-output circuit (typically in the emitter of the output transistor).

Transistor Q_2 (typically a power transistor mounted on a heat sink) amplifies the 1-V output to about 25 to 50 V, depending on CRT screen size. Brightness control R_{12} sets the voltage level on the Q_2 collector. The output of Q_2 is applied through a 4.5-MHz sound trap to the cathode of the CRT.

The sound trap, when properly adjusted, prevents any sound from entering the video circuits. The Q_2 output should contain only the video (picture information) and the horizontal and vertical retrace-blanking pulses. Because all of this information is applied to the CRT cathode, the video (picture) signal is negative, and the blanking pulses are positive. (A positive applied to the CRT cathode decreases brightness, whereas a negative increases brightness.)

In some circuits, the blanking pulses are applied to the first control grid of the CRT rather than being mixed with video. In still other circuits, the mixed video and blanking pulses are applied to the control grid. In most IC color sets, the blanking and video (both luma and chroma) are mixed in a single IC that also contains the color matrix.

In the discrete circuit of Fig. 3.21, the vertical retrace-blanking pulses from the vertical-sweep circuits (Sec. 2.19) are applied to the emitter of Q_1. Horizontal-blanking pulses are applied to the emitter of Q_2. Notice that both Q_1 and Q_2 are forward-biased during normal operation.

The bias on Q_2 is partially determined by light-dependent resistor LDR_1, connected between the emitter and ground. As the ambient lighting around the set varies, the bias on Q_2 (and thus the gain of the video amplifier) varies. This variable-gain feature provides the necessary changes in picture contrast to accommodate changing conditions of ambient light.

3.11.2 Color-Circuit Basics

The circuit of Fig. 3.21 applies primarily to discrete-component black and white TV sets. Figure 3.22 shows a matrix-demodulator circuit that is typical of discrete-component color sets. In IC sets, the circuits of Figs. 3.21 and 3.22 are often combined in a single IC (or possibly an IC with some external components).

In the circuit of Fig. 3.22, the X and Z demodulators receive chroma signals (color-phase information) from the bandpass amplifiers and a reference signal from the 3.58-MHz oscillator. These signals are amplified and applied to the red, green, and blue (RGB) CRT guns through the three matrix transistors. The relative phase of the signals at each gun depends on the phase relationship between the reference signal (3.58 MHz) and the color-phase (chroma) signals.

Notice that the circuits are not adjustable even for discrete-component sets. Thus, any failure must be corrected by replacement. Also notice that the output color signals from the matrix to the CRT are cut off by blanking pulses (during the horizontal-blanking interval). Although the matrix and demodulator are not adjustable, the CRT drive signals that follow the matrix are adjustable in many sets.

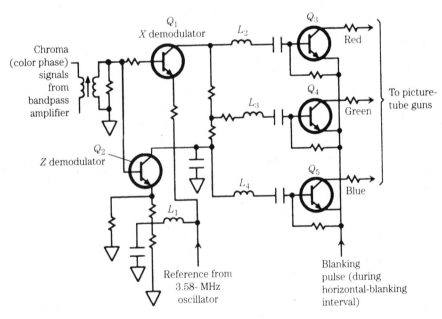

FIGURE 3.22   Typical matrix-demodulator circuit.

3.11.3   Basic IC Video and Chroma Processing

Figure 3.23 shows the stages involved in processing both video (luma) and chroma from the VIF circuits to the CRT. In this particular set (13-in Sony), composite video from the VIF (Sec. 2.18) is routed to comb-filter stages through RGB interface circuits (Sec. 3.12).

The comb filter, in conjunction with a one-horizontal-line (1H) delay, separates the chroma and video information. The chroma information is passed through T_{352} (where low-frequency video information is filtered out) and applied to the chroma-processing circuits in IC_{301} at pin 6. The video information is applied through delay line DL_{301} to the video-processing circuits in IC_{301} at pin 3. Delay line DL_{301} compensates for the inherent chroma-processing delay.

The processed video and chroma are mixed in the RGB matrix of IC_{301} and are routed to the RGB driver board on the neck of the CRT (through the RGB interface in this set). Several signals, such as automatic brightness limiter (ABL), delayed sync (D sync), blanking, and display, are input to the processing stages shown in Fig. 3.23, as discussed in the following sections.

3.11.4   Comb Filter

Figure 3.24 shows the comb-filter circuits for a typical TV set (13-in Sony). The purpose of the comb filter is to separate the luma information from the chroma information and to direct each to the respective processing stages. In the NTSC color-transmission method, the phase of the chroma signal is inverted on every other line. The comb filter takes advantage of this characteristic and operates on the assumption that luma and chroma content changes very little from line to line.

The comb filter has a 1H delay that allows the comb filter to compare a line of luma and chroma with the previous line of video and chroma information. (A 1H delay is about 63.5 µs.) Using the inversion characteristics of the NTSC signal, the chroma can be canceled out, resulting only in luma output from one terminal. Likewise, by inverting line 1 and comparing it to an inverted line 2, the luma can be canceled out, leaving only chroma at the other output.

The composite video from the VIF stages is applied to Q_{351} through the RGB interface circuits (Sec. 3.12). Transistor Q_{351} is a phase splitter that provides buffered composite video to Q_{353} and inverted composite video to Q_{352}. Transistor Q_{352} buffers the inverted composite video and applies it to one leg of the resistive bridge network at the output of delay line DL_{351}. Transistor Q_{353} inverts the composite video before application to the DL_{351} input. RV_{351} sets the level through Q_{353}.

The tuned circuit at pin 4 of DL_{351} is series-resonant at about 2.5 MHz. This effectively removes most of the lower-frequency information from the composite video at pin 4, leaving primarily high-frequency video and the chroma at the DL_{351} input. The bandpass of DL_{351} is 3.58 MHz.

The inverted chroma and high-frequency video passing through DL_{351} appears at pin 2, with the noninverted chroma and video appearing at pin 1. Because the inverter (pin 2) chroma and video is delayed by 1H, the pin-2 information arrives at the bridge simultaneously with the next line of information through Q_{352}. T_{351} adjusts the phase so that the inverted and noninverted signals are exactly $180°$ out of phase.

The upper leg of the bridge (across R_{366}) provides cancellation of the out-of-phase chroma signal, leaving only the video at Q_{356}. The lower leg of the bridge (across R_{367}) provides cancellation of the out-of-phase video information, leaving only the chroma at Q_{354}.

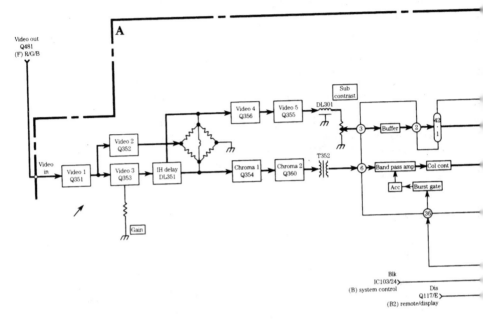

FIGURE 3.23 Video and chroma circuits.

The video or luma information is amplified by Q_{356}, Q_{355}, and Q_{369}. The resonant circuits associated with these transistors condition the video signal to restore the normal high-frequency and low-frequency content. The conditioned video is then passed through another delay line DL_{301}, which provides an approximate 0.1-ms delay. The purpose of the DL_{301} is to compensate for inherent delay encountered by the chroma signal so that the chroma and luma signals arrive at the IC_{301} matrix simultaneously. RV_{301} provides for internal contrast adjustment by setting the dc level of the luma signal.

The combed-out chroma information is buffered by Q_{354}, amplified by Q_{355}, and applied to the 3.58-MHz bandpass transformer T_{352}. This effectively removes any remaining low-frequency video from the chroma signal, presenting relatively pure chroma information to the luma/chroma (Y/C) processing stage in IC_{301}.

3.11.5 Y/C Processing

Figure 3.25 shows the circuits involved in processing the combed-out luma and chroma signals before application to the CRT output or drive circuits. Compare this to the basic circuits of Figs. 3.21 and 3.22.

The combed luma information is applied to pin 3 of IC_{301}, buffered, and then applied to a sharpness amplifier through an external video-peaking network. The sharpness amplifier enhances the high-frequency characteristics of the video signal.

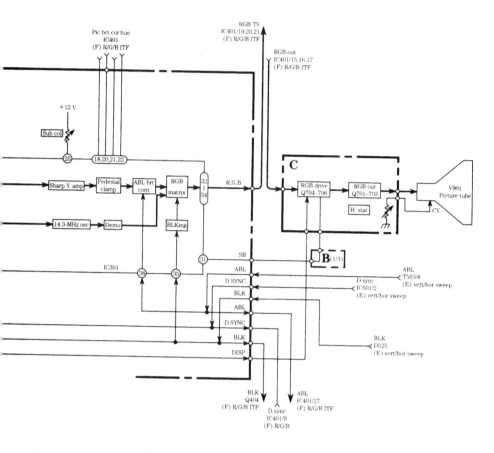

The processed and amplified video signal is then passed through a pedestal clamp for dc restoration and through an ABL brightness-control circuit to the RGB matrix.

The combed chroma information is applied to the color demodulator through pin 6 of IC_{301}, a bandpass amplifier, and a color-control circuit. The chroma information is demodulated using the 14.3-MHz oscillator signal as described in Sec. 3.11.2. The demodulated R-Y, B-Y, and G-Y signals are applied to the matrix. The resulting RGB signals from the matrix are applied to the CRT drive circuits through the RGB interface (Sec. 3.12). The reference oscillator can be adjusted to the frequency of crystal X_{301} by CV_{301}.

Several external signals are input to the Y/C processing stages. The PIX (picture) signal is a dc voltage developed in the system-control microprocessor IC_{103} through Q_{102}. This signal controls the gain of the video and color amplifiers in IC_{301}. There is also a picture control (in the RGB section) connected to pin 18.

The brightness, color, and hue inputs from the user controls on the RGB interface (Sec. 3.12) are applied to IC_{301} at pins 20, 21, and 22. These dc inputs set the brightness level for the luma signal, the luma saturation of the color signal, and the phasing or hue (also called *tint* on some sets) of the color signals. RV_{303} provides for internal adjustment of the color level.

The blanking signal at pin 35 of IC_{301} is developed in the vertical and horizontal

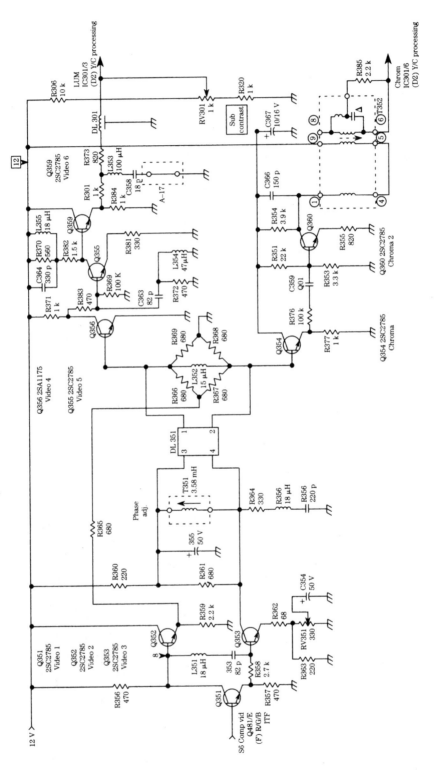

FIGURE 3.24 Comb-filter circuits.

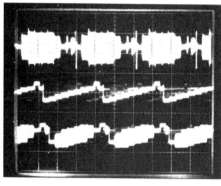

(S17)	DL351/4	CHR+HFV	0.5 V/d	20 μsec/d
	DL351/1	LUM	0.5 V/d	20 μsec/d
	DL351/2	CHR+LFV	0.5 V/d	20 μsec/d
	Color bars			

FIGURE 3.24 (*Continued*).

output stages (Sec. 2.19). These are positive-going vertical and horizontal blanking pulses used to turn off the RGB outputs during the vertical and horizontal retrace periods. The blanking signals are also applied to the dc-restoration pedestal clamp within IC_{301}. The D-sync (delayed sync) pulse at pin 36 is used to trigger the burst gate. This allows the burst amplifier to operate only when the burst signal is present (when color information is being transmitted).

The automatic brightness limiter (ABL) input at pin 38 of IC_{301} is taken from the horizontal-output or flyback transformer (FBT) T_{503} in the horizontal stages (Sec. 2.19). Beam current through the CRT is reflected in the ABL winding of T_{503}. If the beam current becomes excessive, the dc level at pin 38 decreases, lowering the brightness level. This limits the maximum beam current to prevent overdriving the CRT.

3.11.6 RGB Output

Figure 3.26 shows the circuits involved between the *Y/C* processing and the CRT. In this particular set (13-in Sony) all of the circuits are on a board (the C board) located at the neck of the CRT.

Drive and output transistors Q_{701} through Q_{706} amplify the RGB signals from pins 32 through 34 of IC_{301} (Fig. 3.25) for application to the CRT cathodes. Notice that the blue and green circuits have both a drive and a background adjustment (RV_{702} through RV_{705}), but the red circuit has only a background adjustment (RV_{701}, in this particular set).

The RGB outputs from IC_{301} are applied to the drive and output circuits of Fig. 3.26 through switches on the RGB interface boards (Sec. 3.12). The switches permit either the normal TV RGB signals from IC_{301} or external RGB signals (from computers, VCRs, video discs, etc.) to be reproduced on the CRT, thus making the set a monitor TV.

The display from the system-control microprocessor (Sec. 2.20) is applied through connector C-5 to the green driver and output circuit (Q_{702} and Q_{705}). This signal generates a channel indication and bar-graph display on the CRT.

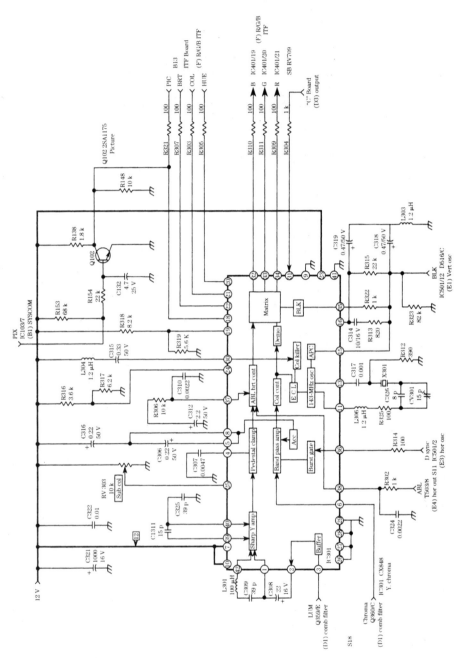

FIGURE 3.25 Y/C processing circuits.

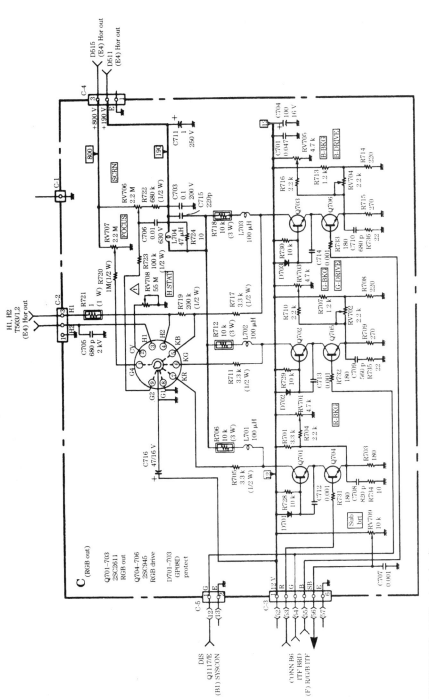

FIGURE 3.26 RGB output circuits.

Flyback transformer T_{503} in the horizontal-output stages provides the focus, screen, heater, and high voltage for the CRT. The screen and focus voltages are taken from the +800 V at pin 1 of connector C-4, with the heater voltages H1 and H2 applied through connector C-2. T_{503} also supplies the +190 V required by the CRT drive transistors Q_{701} through Q_{706}.

The CRT focus voltage at pin 1 (grid 4) is set by RV_{707}, with the screen (CRT brightness) voltage at pin 9 (grid 2) set by RV_{706}. There is also a subbrightness control RV_{709} that sets the dc voltage at pin 31 of IC_{301} (Fig. 3.25).

The convergence voltage for the internal convergence plates of the CRT is taken (within the CRT) from a high-voltage second-anode lead across a resistance network built into the CRT. The return path is from CRT pin 2 through RV_{708} to ground. RV_{708} provides for adjustment of the static convergence. (Notice that this internal static-convergence voltage is unique to Sony Trinitron.)

3.11.7   Troubleshooting Video and Chroma Processing Circuits

The recommended approach for troubleshooting video and chroma processing stages depends largely on the available test equipment. Ideally, an analyzer or NTSC generator can be used to duplicate the signals normally found at the video-detector output (composite video, as well as horizontal and vertical blanking pulses).

As a quick check, the output of the video detector can be monitored with a scope (with the set tuned to an active TV channel). If there is a signal of about 1 V at the video-processing input (and the waveform resembles that shown in the service literature), picture problems are probably in the video-processing circuits. In the circuit of Fig. 3.23, the video input is at Q_{351}. However, in this particular set, the input passes through the RGB interface circuits (Sec. 3.12) before reaching Q_{351}. So it is possible that the RGB circuits could be causing the problem, if there is a good composite-video output from IC_{201} (Fig. 2.25) but a bad input at Q_{351}.

If an analyst or NTSC generator is available, an alternate quick check is to apply a composite video signal at Q_{351} and note the response on the CRT screen. If the generator pattern is absent or abnormal on the TV screen, you have localized the problem to the video-processing circuits.

Some technicians prefer to make a square-wave test of the video-processing circuits. A square wave of about 1 V is applied to the video-processing input Q_{351} (or at the IC_{201} output), and the resultant output is monitored with a scope at all three inputs of the CRT (at pins 5, 6, and 7 of the CRT, Fig. 3.26).

This square-wave test checks the overall response of the video circuits, including the effect of contrast, brightness, and any other user controls. If square waves are passed without distortion, attenuation, ringing, and so on, the video circuits are operating properly.

3.11.8   Typical Video-Processing Troubles

The following sections discuss symptoms that could be caused by defects in the video-processing circuits.

No Raster.   A no-raster symptom is usually associated with failure of the power supply (Sec. 2.17) or horizontal circuits (Sec. 2.19). However, complete cutoff of the CRT can be produced by failure in the video-processing circuits, even though the high voltage and CRT voltages are normal.

For example, in the circuit of Fig. 3.25, if the blanking signal applied to pin 36 of IC_{301} is held positive (say, by a short in the PC wiring), the RGB outputs from IC_{301} are turned off. Likewise, if the dc level at pin 38 of IC_{301} is zero (say, by a short to ground), the CRT brightness will be reduced to zero. Even the brightness signal applied to pin 22 of IC_{301} can cut the CRT off completely (if there is a malfunction in the brightness line that simulates turning the brightness control all the way down).

If such a condition is suspected, check all of the CRT voltages (Fig. 3.26) as well as the blanking and other control voltages applied to IC_{301} (Fig. 3.25). Pay particular attention to heater and cathode voltages of the CRT. A low heater voltage can cause the heater (or filament) to glow but might not produce enough emission for a picture. If any one of the voltages is not normal, trace the particular circuit and look for such defects as shorted or leaking capacitors, breaks in PC wiring, and worn controls. If all of the voltages appear normal but there is no raster, suspect the CRT itself (or the high voltage, which is discussed in Sec. 2.19).

No Picture and No Sound. With a raster present, a no-picture and no-sound symptom is normally associated with failure of the tuner or VIF and SIF stages rather than in video. However, in the circuit of Fig. 3.21, failure of Q_1 could cut off both sound and picture even though good signals are available at the video-detector output.

No Picture but Normal Sound. With a raster present and good sound, a no-picture symptom is definitely in the video-processing circuits (unless you are servicing a monitor or other TV with a video/TV switch set to the wrong position). In the circuit of Fig. 3.23, check for a video signal at Q_{351}, corresponding luma and chroma signals (combed) at pins 3 and 6 of IC_{301}, RGB signals at pins 32–34 of IC_{301}, and RGB signals at the C-board (Fig. 3.26). If the video is good at Q_{351}, but either or both luma and chroma are bad at pins 3 and 6 of IC_{301}, suspect Q_{351}, through Q_{356}, and Q_{360}. As in the case of any amplifier-buffer circuit, total failure of these circuits is usually easy to locate. Look for such defects as an open coupling capacitor and coils, bad solder, open or worn controls, breaks in PC wiring, and the like.

If the luma-chroma signals at pins 3 and 6 of IC_{301} are good, but the RGB outputs at pins 32–34 are bad, suspect IC_{301}. Keep in mind that the blanking and other control signals applied to IC_{301} can affect the RGB output, as discussed. If the RGB outputs from IC_{301} are good, but the RGB inputs at the C-board are bad, suspect the RGB interface circuits. Finally, if the RGB inputs are good at the C-board (at Q_{704} through Q_{706}), but the picture is absent or abnormal, suspect the C-board components. (Try adjustment as described in the service literature before you throw the C-board away.)

Poor Picture Quality. When the picture quality is poor (lack of detail, smearing, poor definition, etc.), the video circuits are usually suspected, particularly if the sound is good. However, these same symptoms can be caused by problems in the tuner and/or VIF.

If you have an analyst or NTSC generator, inject a video signal at the video circuit input and check the CRT display. Suspect the video if the display is poor. As an alternate, check the video-circuit frequency response with a square wave as discussed. Look for such problems as leaking capacitors that change the resonant frequency of coils, solder splashes that short out damping resistors across coils, or in rare cases, shorted turns in a coil (such as L_{206} and L_{207} in Fig. 2.25).

Retrace Lines in Picture. In the circuit of Fig. 3.21, both vertical and horizontal blanking pulses are fed to the video amplifier to blank the retrace lines. In most IC sets, the blanking is applied to the CRT through the same IC that contains the color matrix, bandpass amplifier, etc. (For example, in the circuit of Fig. 3.25, the vertical and horizontal blanking are applied to pins 35 and 36 of IC_{301}.)

No matter what system is used, the obvious first step is to monitor the blanking pulses at the point where the pulses enter the video circuits or at the CRT. If the blanking pulses are absent, trace the particular circuit back to the pulse source.

If the blanking pulses are low in amplitude, the typical symptom is the presence of retrace lines only when the brightness control is turned up. If this is the case, check the pulse amplitude against that in the service literature. Then retrace the blanking pulses back to the source.

If you have an analyst or NTSC generator with blanking pulses, substitute the generator pulses for the normal pulses. If this corrects the retrace problems, you have localized the trouble.

3.11.9 Localizing Color Troubles

Color troubles can be grouped into four classifications: no color, wrong color, weak color, or no color-sync. The following procedures describe the basic approach to color-trouble localization, using a color-bar generator:

1. Connect the color-bar generator output to the antenna input.

2. Adjust the generator controls to produce an NTSC color-bar display.

3. Set the hue or tint control to midrange.

4. In turn, monitor each of the points shown in Fig. 3.27 with a scope. Observe the following notes for each symptom. It is not always practical to monitor each of the points in Fig. 3.27 in sets where many of the circuits are in a single IC. For example, as shown in Fig. 3.25, the burst gate, color killer, bandpass amplifier, demodulator, and matrix are all in IC_{301}. However, it is still possible to monitor the chroma (at pin 6), the reference oscillator (at pins 11 through 13), and the matrix output (at pins 32 through 34).

No Color.   Start by checking the video-detector output. If there is no color output from the video detector, no color can be expected from the chroma circuits. Likewise, with a normal color signal at the detector, a no-color symptom is definitely localized to the chroma section that follows the video amplifier and/or detector.

Start at either demodulator stage (point A) and work back to the video detector. If the input to the demodulator is normal at A, check for correct signal at B. The signal at B is generated by the reference oscillator, which is controlled by the phase detector.

The signals at points A and B are both at 3.58 MHz. (In the circuit of Fig. 3.25, the reference oscillator is at 14.3 MHz, which is 4 times the 3.58-MHz burst signal.) The signal amplitude should be shown in the service literature. If not, it is reasonable to assume that the amplitude is correct if both amplitudes (A and B) are about the same.

The signal at A is passed through the bandpass amplifier, which in turn is controlled by a bias from the color killer. If the bias is applied incorrectly (say, because of some circuit failure), the color killer cuts the bandpass amplifier off, and there is no signal at A. Always check the bias and/or color-killer operation first when there is a no-color symptom.

As discussed, it is not practical to check all of the color circuits when they are part of an IC such as IC_{301} in Fig. 3.25. However, you can check all inputs to the IC. For example, if there is a no-color symptom, look for chroma at pin 6, blanking at pins 35

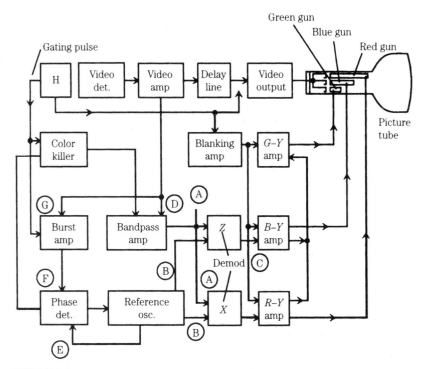

FIGURE 3.27 Localizing color troubles.

and 36, and reference-oscillator signals at pins 11 through 13. If any of these are absent or abnormal, trace back to the source. For example, if the D-sync pulse is absent, the burst gate within IC_{301} will not turn on and pass color information. If all inputs are good, suspect IC_{301} if color is totally absent. Also check the color-control signal at pin 21. If the user color control is set to one extreme, the color circuits are cut off.

Wrong Color. The demodulators are the most likely points at which to start localizing a wrong-color symptom. The procedure is essentially the same as that for a no-color symptom, except that the demodulator outputs must also be checked. These are shown in Fig. 3.27 as point C. (In the circuit of Fig. 3.25, you cannot check at the demodulator outputs, but you can check at the matrix output, pins 32 through 34 of IC_{301}.)

If the demodulator inputs are correct, but there is no output from one demodulator, the trouble is localized immediately. If the demodulator outputs are present but not phased properly, the trouble is most likely improper alignment of the demodulator stages (if the demodulators are adjustable). If the demodulator outputs are correct, the trouble then can be localized to the *B-Y, G-Y,* or *R-Y* amplifiers (also known as the matrix amplifiers) or to the CRT. (Although the demodulators and matrix of the circuit in Fig. 3.25 are not adjustable, the circuits between IC_{301} and the CRT are adjustable, as shown in Fig. 3.26.)

Weak Color. Again, the demodulator inputs are the most likely places to start. If the information is available, check the demodulator input amplitudes against those shown in the service literature. Note that the input from the reference oscillator is fixed (in amplitude), whereas the input from the bandpass amplifier can be varied by the user color control (at pin 21 of IC_{301} in Fig. 3.25).

No Color Sync. Color sync is controlled primarily by the reference oscillator and the phase detector (the automatic phase control, or APC, of Fig. 3.25). A phase difference between the oscillator and the generator (or TV station) bursts causes a loss of color sync. That is, color bars might not be in proper sequence, or the colors might be erratic (constantly changing).

The phase detector receives inputs from both the burst amplifier and reference oscillator (points E and F in Fig. 3.27). If either input is incorrect in phase, the colors will be out of sync. (Although the phase of the reference oscillator in IC_{301} cannot be adjusted, the reference-oscillator frequency can be set to the crystal X_{301} frequency with capacitor CV_{301}, using a frequency counter.)

3.12 RGB INTERFACE CIRCUITS

This section is devoted to circuits that provide interconnection or interface between the CRT/speakers and video/audio inputs. In some sets, these circuits are on a separate board. On other sets, the circuits are integrated with the circuits on various boards. Either way, the purpose of the circuits is to interface the set with VCRs, computers, video games, etc., thus making the set a monitor RV. Although called RGB, the circuits usually involve both color video and audio, possibly stereo/SAP audio.

3.12.1 RGB Basics

Figure 3.28 (see pages 3.62 and 3.63) shows the RGB interface circuits for a typical TV set (13-in Sony) where the majority of interface circuits are on a single board. Because the board circuits are primarily simple amplifiers and switching ICs, the circuits are shown in block form. In this set, the RGB interface has three modes of operation: TV, video, and RGB. Also, in the RGB mode, the circuits have the capability of processing both digital and analog RGB signals.

3.12.2 Troubleshooting Interface Circuits

Because of the circuits involved (amplifiers, buffers, and IC switches) the obvious troubleshooting approach is signal tracing. This also involves making sure that the IC switches are closed to pass the corresponding signals.

As an example, assume that the circuit of Fig. 3.28 is operated in the TV mode and that composite video is not available at the input to the video and chroma circuits at Q_{351} (Fig. 3.23). First check for video at TV OUT connector CNJ_{404}. If the video is missing at CNJ_{404}, suspect amplifier Q_{469} and buffers Q_{464}, Q_{466}, and Q_{467}.

Next, check for video at pins 8 and 9 of IC_{402}. If video is available at pin 8, but not at pin 9, suspect IC_{402}. Before pulling IC_{402}, check that pin 6 is high. If not, check back to the switch-control source. In the TV   mode, pin 6 of IC_{402} is made high through inverter Q_{470}. If there is video at pin 9 of IC_{402}, but not at Q_{351}, suspect buffer Q_{481}. Also check that there is video at Monitor Video Out connector CNJ_{403}. If not, suspect buffer Q_{465}. Finally, check for video input at pins 1 and 22 of IC_{501} (Fig. 2.31). If the video is missing at IC_{501}, but present at Q_{351} (Fig. 3.23), suspect Q_{483} and Q_{484}. (As discussed in Chap. 2, this video is used for both vertical and horizontal sync.)

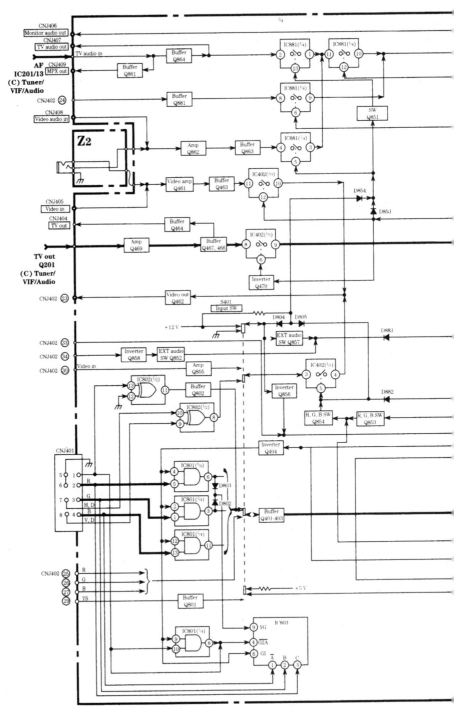

FIGURE 3.28 RGB interface circuits.

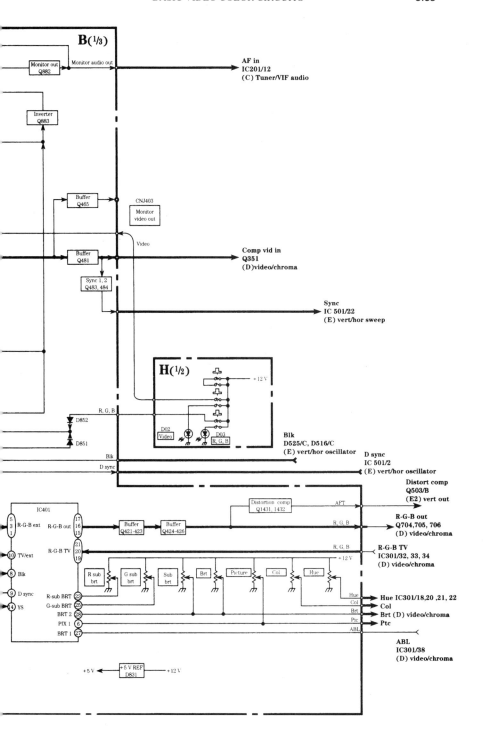

CHAPTER 4
DIGITAL VIDEO CIRCUITS

This chapter is devoted to video circuits that use some form of digital pulses. Such circuits include frequency synthesis (FS) tuning, on-screen display (OSD), infrared (IR) remote control, digital TV, and picture-in-picture circuits. Digital video discs are discussed in Chap. 9.

4.1 FS TUNING

The tuners of most modern video products (TV, VCR, etc.) use some form of *frequency synthesis*. This provides convenient push-button channel selection with automatic channel search and automatic fine-tune (AFT) capability. The FS system is also known as *digital tuning* or *quartz tuning*. The key element in any FS system is the phase-locked loop, or PLL, which controls the variable-frequency oscillator (VFO) of the tuner. Generally, the VFO is some form of voltage-controlled oscillator (VCO). Quite often, the oscillator frequency is set by a varactor diode, which in turn is controlled by a variable voltage from the AFT line.

The tuner components (oscillator, RF amplifier, and mixer) of the discrete-component set shown in Fig. 1.1 are usually part of a tuner or front-end module in IC video equipment. Figure 4.1 shows such a module. This particular module includes both the tuner and IF components. In other equipment, there are separate tuner and IF modules.

4.1.1 FS and PLL Basics

Figure 4.2 shows the basic PLL circuit. *PLL* is a term used to designate a frequency-comparison circuit in which the output of a VFO is compared in frequency and phase to the output of a very stable (usually crystal-controlled) fixed-frequency reference oscillator. PLL is not unique to video equipment; it is also found in many communications receivers.

In any PLL, if a deviation occurs between two compared frequencies or if there is any phase difference between two oscillator signals, the PLL detects the degree of frequency or phase error. Then the PLL automatically compensates by tuning the oscillator up or down in frequency until both oscillators are locked to the same frequency and phase. (The *loop* is said to be *locked* at this time.) The accuracy and frequency stability of a PLL circuit depend on the stability of the reference oscillator (and on the crystal that controls the reference oscillator).

In the basic PLL of Fig. 4.2*a*, the VCO has a desired output frequency of 1 kHz. The actual output frequency depends on the tuning voltage produced by the phase

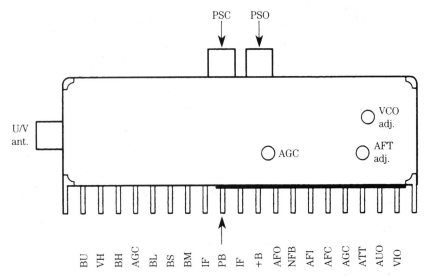

FIGURE 4.1   Typical tuner and front-end module.

comparator, which receives two input signals, both at 1 kHz. Any frequency or phase variations in the VCO cause the phase comparator to produce a correction voltage.

The magnitude of the correction voltage depends on the amount of frequency and phase deviation. The polarity of the correction voltage depends on the direction of phase and frequency deviation. The correction voltage is applied to the VCO as an increase or decrease in tuning voltage. Changes in the tuning voltage alter the VCO frequency as necessary to make the VCO output have the same frequency and phase as the reference oscillator. When this occurs, the tuning voltage stabilizes and the PLL is said to be locked in.

Figure 4.2b shows a more sophisticated PLL circuit—one capable of comparing frequencies that are not identical. This circuit includes a divide-by-10 element, which divides the VCO frequency by 10, and a low-pass filter, which acts as a buffer between the comparator and the VCO. Notice that while the inputs to the comparator remain at 1 kHz when the loop is locked, the output frequency of the VCO is 10 kHz because of the divide-by-10 element.

Figure 4.2c shows a PLL circuit similar to that found in TV and VCR, but it is far less complex. The system is generally called an *extended PLL* and holds the tuner oscillator frequency to some harmonic (or subharmonic) of the reference oscillator (3.58 MHz in this case). The fixed-divide element of Fig. 4.2b is replaced by a programmable variable divider (÷N) in Fig. 4.2c. A channel change is produced by varying the division ratio of the programmable divider with 4-bit data commands from the system-control microprocessor (which, in turn, is operated by front-panel push buttons and/or remote-control signals).

4.1.2 Pulse Swallow Control (PSC)

Most PLL tuning systems found in video equipment use some form of PSC, such as shown in Fig. 4.2d. The PSC system uses a high-speed prescaler with variable-division ratio (instead of the fixed-division ratio prescaler). The variable-division ratio

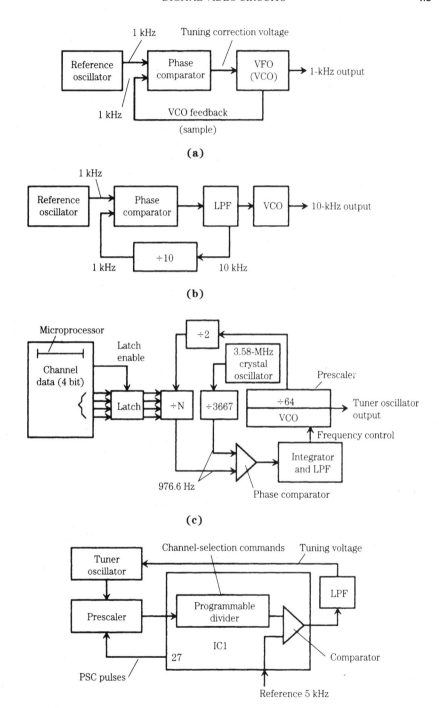

FIGURE 4.2 PLL basics.

depends on the PSC signal from the PLL IC_1. As the number of PSC pulses increases, the division ratio increases. A specific number of PSC pulses are produced by IC_1 in response to channel-selection commands.

The overall division ratio for a specific channel is the prescaler division ratio multiplied by the programmable-divider division ratio. The result of division at any channel is a precise 5-kHz output to the comparator when the tuner oscillator is locked to a given channel frequency. In effect, the programmable divider determines the basic channel frequency, and the prescaler performs the fine adjustment (AFT) to the channel frequency.

Notice that if you are troubleshooting any tuner with PSC and you cannot tune in a channel manually or with AFT, check the PSC line (pin 27 of IC_1 in this case) for pulses and check the tuning voltage to the oscillator. If the pulses are missing, the tuner cannot lock onto any channel, even with a tuning voltage present.

4.1.3 Basic TV Tuner Operation

Figure 4.3 shows the circuits of a typical TV tuner (using PLL) in block form. Notice that the PLL IC_1 receives channel commands from an IR (infrared) remote-control

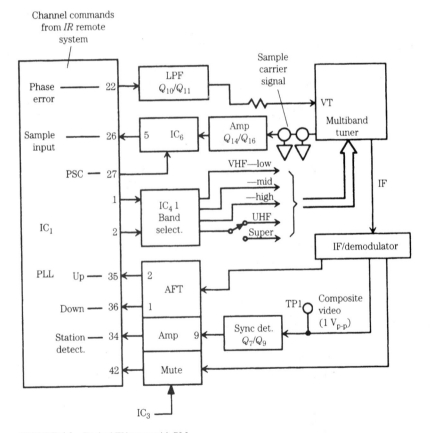

FIGURE 4.3   Typical TV tuner with PLL.

transmitter after the commands have been decoded by a remote-control receiver (Sec. 4.3).

The multiband tuner is controlled by circuits within PLL IC_1 (which, in turn, receives commands from the remote-control circuits). IC_1 monitors signals from the IF demodulator circuits to determine when a station is being received. These signals are the AFT up and down (pins 35 and 36) and the station-detect signals (pin 34).

The detected video from the output of the IF demodulator is passed to a sync amplifier and detector Q_7 and Q_9. The detected sync signal (station detect) is passed to pin 9 of IC_3, amplified, and sent to the station-detect input at pin 32 of IC_1. When a station is tuned in properly, the sync is detected from the video signal and applied as a high to pin 9 of IC_3. This applies a high to pin 34 of IC_1, indicating to IC_1 that video with sync is present.

To maintain proper tuning, IC_1 monitors the AFT up and down signals (at pins 35 and 36) from IC_3. The AFT circuit of IC_3 is a window detector, monitoring the AFT voltage and outputting a high at pins 1 or 2, depending on the magnitude and direction of the AFT voltage swings (should the tuner-oscillator frequency drift).

When a channel is selected, band-switching information is supplied from IC_1 (at pins 1 and 2) to IC_4, which develops four band-switching outputs. (In this particular circuit, one of the four outputs is switched by a normal/cable switch to provide the five bands shown.)

The tuner oscillator passes a sample carrier signal to oscillator amplifier Q_{14} and Q_{16}. The amplified oscillator signal is then passed to a prescaler IC_6. The amount of frequency division is determined by PSC pulses from IC_1. The frequency-divided output of IC_6 (pin 5) is then passed to the sample input of IC_1 (at pin 26). When a channel is selected, circuits within IC_1 produce the appropriate number of PSC pulses at pin 27. The PSC pulses are applied to IC_6 and produce the correct amount of frequency division.

The divided-down oscillator signal (sample input) at pin 26 of IC_1 is divided down again within IC_1 and compared to an internal 5-kHz reference signal. The phase error of these two signals appears at pin 22 of IC_1 and is applied to a low-pass filter (LPF) Q_{10} and Q_{11}. The dc output from the LPF is applied to the tuner oscillator. This voltage sets the tuner oscillator as necessary to get the proper frequency for the channel selected.

4.1.4 Typical Tuner Circuits

Figure 4.4 shows the circuits of a TV tuner (a 13-in Sony) where the PLL is separate from the tuner package. The purpose of this circuit is to interpret keyboard or remote inputs requesting channel changes. When such commands are received, microprocessor IC_{103} determines if a particular channel is to be tuned in or skipped (depending on the data in memory IC_{105}). IC_{103} then selects the appropriate tuning band through operation of the band-switching circuits in IC_{102}. This action tunes in the particular channel requested or, in the case of up-down tuning, tunes in the next channel from the one presently tuned. After the channel is tuned in, IC_{103} monitors the AFT up and AFT down lines to maintain precise tuning.

4.1.5 Tuner

TU_{101} is a multiband varactor-diode tuner capable of tuning channels 2 through 13, 14 through 83, and cable channels 1 through 125. TU_{101} operates from a 9-V supply and provides two prime outputs. The IF signal is output to the VIF circuits (Sec. 2.18), and a sample of the tuner local-oscillator signal is output to PLL TU_{102}. The necessary

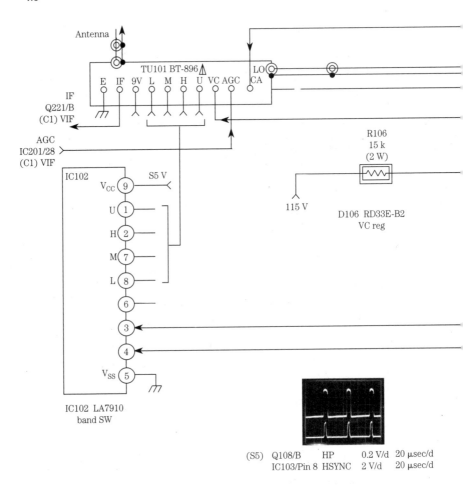

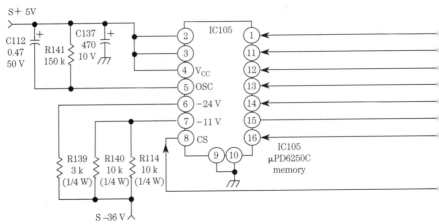

FIGURE 4.4 Tuner circuits with separate PLL.

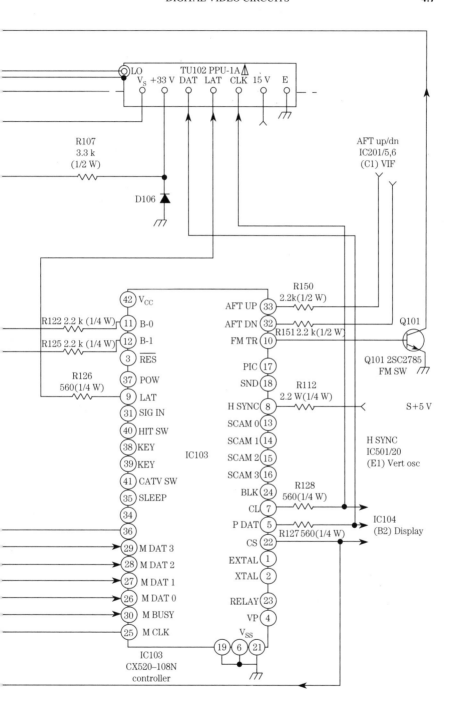

FIGURE 4.4 Tuner circuits with separate PLL.

inputs for proper operation are the band-select inputs L, M, H, and U from the band-select switch IC_{102}, and the VC tuning voltage developed in TU_{102}.

In addition to these basic inputs and outputs, there is an AGC correction voltage input from the VIF stage, and a control line (labeled CA) is used to alter the tuning characteristics when tuning certain cable channels. Band-switching for TU_{101} is done by IC_{102}. The state of the L, M, H, and U outputs from IC_{102} is determined by the binary logic applied to pins 3 and 4. In turn, the binary logic is controlled by IC_{103} outputs B-0 and B-1 (pins 1 and 12).

4.1.6 PLL Tuning Unit

PLL TU_{102} receives serial tuning data on the DAT (data) line from pin 5 of IC_{103}. This data stream is synchronized by CLK (clock) signals from IC_{103} at pin 7. At the end of each data transmission, a LAT (latch) pulse occurs at pin 9 of IC_{103}, latching the transmitted data into PLL TU_{102}. This sets the internal division ratio of TU_{102}. In turn, TU_{102} monitors and divides the local-oscillator sample input from tuner TU_{101} and develops the corresponding VC voltage output necessary to tune the particular channel requested.

4.1.7 Tuning Memory

Tuning memory IC_{105} is a nonvolatile EAROM that is enabled when the chip-select line is made high by a $\overline{CS}$ signal from pin 22 of IC_{103}. The initialize signal from pin 36 of IC_{103} is a positive-going pulse that occurs on power-up to reset the memory of IC_{105}. The clock signal from pin 25 of IC_{103} is used to synchronize the transmission of parallel data between IC_{103} and IC_{105}.

The M BUSY input from pin 13 of IC_{105} to pin 30 of IC_{103} occurs when memory operations within IC_{105} might prevent proper data transfer between IC_{103} and IC_{105}. Such data transfers occur across the directional 4-bit data-bus M DAT 3. This 4-bit bus is used to transmit commands that set the operational mode of IC_{105}, followed by data bits that are stored in IC_{105}.

In the read mode, commands are sent over the bidirectional data bus to set IC_{105} to the read mode and to select the correct address. Then the data bus is used to transfer data from IC_{105} to IC_{103}. Memory IC_{105} is used during channel up-down operations to determine if the next higher or lower channel should be skipped or selected (tuned in). This type of operation is referred to as a *skip-memory* application.

4.1.8 AFT Up-Down

The AFT up-down input signals at pins 32 and 33 of IC_{103} are developed in the VIF stages (Sec. 2.18) by IC_{201}. The signals are positive-going pulses that inform IC_{103} when it is necessary to output correction data to PLL TU_{102} for automatic fine-tuning operations.

4.1.9 FM Trap

In most modes of operation, pin 10 of IC_{103} is high and Q_{101} is turned on. When cable TV is being received and the CATV selector switch is set to the CATV position, pin 41 of IC_{103} is low. This informs IC_{103} that cable is being used. With pin 41 low, cable channels 14, 15, 16, or 17 are tuned in, and pin 10 of IC_{103} goes low, turning Q_{101} off.

This allows the CA input of TU_{101} to go high, enabling an FM trap within TU_{101}. The frequencies of cable channels 14 through 17 are close to the FM broadcast band. The FM trap prevents the FM broadcast signals from interfering with the cable signals.

4.1.10 H-Sync

The H-sync input at pin 8 of IC_{103} is developed in the sync-separator stage (Sec. 2.19) and is present only when composite video is present, indicating that a channel is properly tuned. During the normal tuning sequence, IC_{103} monitors the H-sync input to determine that the requested tuning operation has been accomplished. On receipt of the H-sync pulse, IC_{103} monitors the AFT up-down signals at pins 32 and 33 and applies any fine-tuning corrections to the PLL TU_{102}.

4.1.11 Typical Timing Sequence (Channel Surfing)

The following is a typical timing sequence for the tuner of Fig. 4.4. When the AFT up-down request is received by IC_{103}, either from the remote control (Sec. 4.3) or the keyboard input (Sec. 2.20), IC_{103} produces control signals to select the next higher or lower channel requested.

 IC_{103} first outputs a chip-select signal at pin 22 to enable the skip memory of IC_{105}. Then IC_{103} resets IC_{105} and places IC_{105} in the read mode. Finally, IC_{103} sends the address of the requested channel on the 4-bit data bus to select the next higher or lower channel.

 When the address is selected, the data bits in the address of IC_{105} are applied to IC_{103}. In turn, IC_{103} analyzes the data to determine if the next higher or lower channel is to be skipped or tuned. All of these operations on the data bus are synchronized by the clock signal applied to IC_{105} from pin 25 of IC_{103}. If IC_{105} is busy and unable to accept or transmit data on the bus, the busy signal from pin 15 of IC_{105} is applied to IC_{103}, causing IC_{103} to wait.

 If the information at the selected address tells IC_{103} that the channel is to be skipped, IC_{103} continues on to the next higher or lower channel. This operation continues until a channel that is not to be skipped is reached. When this occurs, IC_{103} outputs B-0 and B-1 signals (at pins 11 and 12) to the band-switch circuits within IC_{102}. In turn, IC_{102} applies the appropriate band-switching signals (L, M, H, and U) to tuner TU_{101} as necessary for the selected channel. IC_{103} also outputs the necessary DAT (data), CLK (clock), and LAT (latch) signals to PLL TU_{102} to hold the tuning at the selected channel.

 IC_{103} monitors the H-sync input at pin 8 to determine when the tuning operation is complete and composite video is present. When H-sync pulses are present (indicating that composite video is passing through the VIF stages), IC_{103} begins to monitor the AFT up-down signals at pins 32 and 33 to determine if fine-tuning corrections are required. If necessary, IC_{103} outputs the appropriate DAT, CLK, and LAT signals to PLL TU_{102} until the AFT up-down inputs indicate that the channel is properly tuned. IC_{103} continues to monitor the AFT up-down inputs at pins 32 and 33 and produces fine-tuning corrections as necessary.

4.1.12 Troubleshooting FS or Digital-Tuning Circuits

The most common symptoms for failure in a TV set with FS or digital tuning are a combination of no stations received, failure to lock on channels, picture snowing, audio noisy, and color dropping in and out.

I have experienced televisions with all of these symptoms. The cause was traced to poorly soldered terminals on the tuner, IF module, and PLL IC. The following steps led to the defective solder junctions.

The first troubleshooting step is to isolate the problem to the tuner and IF modules or to the PLL system. Start by checking for power to all ICs and components. For example, in the circuit of Fig. 4.4, tuner TU_{101} requires only 9 V, whereas the PLL requires both 15 and 33 V. The ICs (102, 103, and 105) require various voltages and ground connections. Once you are satisfied that all of the components have the necessary power (and grounds), start the isolation process.

Select a channel and confirm that the band-switching signal for that particular channel appears at the TU_{101} input and band-select IC_{102} output. For example, if Channel 4 is available locally, select Channel 4 and check that the VHF low-band-switching signal at pin 8 of IC_{102} is present (high) and that the high appears at the L terminal of TU_{102}.

If there is no high at pin 8 of IC_{102}, check the status of pins 3 and 4, as well as pins 11 and 12 of IC_{103}. In this particular circuit, when both B-0 and B-1 outputs from IC_{103} are high, pin L of IC_{102} (and only pin L) is high (pins M, H, and U are low). Also, there should be a tuning voltage at the VC pin of TU_{101} in the range from 2 to 22 V.

If B-0 and B-1 are not high, or do not change status when the keys are pressed (Sec. 2.20), suspect IC_{103} or the system-control circuits. If B-0 and B-1 are both high, but pin L of IC_{102} is not high (or one or more of pins M, H, and U are high), suspect IC_{102}. It is also possible that there is a problem in memory IC_{105} and that Channel 4 has been skipped. If there is no tuning voltage at the VC pin of TU_{101}, or if the voltage is not within the range of 2 to 22 V, suspect PLL TU_{102}.

Next, check for an H-sync signal at pin 8 of IC_{103}, indicating that a channel has been tuned in. Then check for AFT up-down signals at pins 32 and 33. If any of these signals are absent or abnormal, suspect the VIF circuits (Sec. 2.18). Before you condemn the VIF circuits, check that there is an IF signal at the IF output of TU_{101}. If not, suspect TU_{101} or possibly TU_{102}. Also check that there is a local oscillator signal from TU_{101} to TU_{102}.

4.2   OSD CIRCUITS

Many present-day video devices are provided with some form of OSD circuit. The on-screen channel display of TVs and the on-screen programming display of VCRs are typical examples. The OSD circuits use a *character generator* IC to produce the desired numbers and letters.

The character generator receives sync signals from the same source as the picture tube (so that the characters appear at a given position on the screen). Some OSD circuits include positioning controls. The characters to be displayed are determined by a microprocessor, which, in turn, receives commands from pushbutton controls. In some cases, the commands are hard wired into the microprocessor.

4.2.1 Typical OSD Circuit

Figure 4.5 shows the date and time OSD circuits for the electronic viewfinder (EVF) of a camcorder. These OSD circuits provide date and time information that appears on the EVF and that can be recorded on tape if desired. Figure 4.5 also shows the hard-wire and push-button command connections to produce various combinations of date and time.

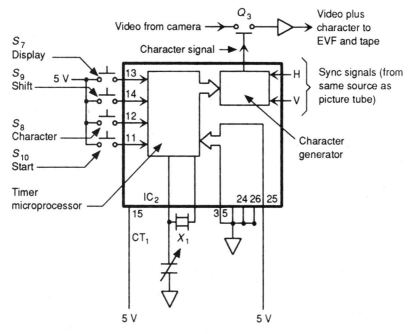

FIGURE 4.5 Typical OSD circuit.

Note that most of the OSD circuits are contained within IC_2. When 5-V power is applied to IC_2 at pin 15, crystal X_1 generates a 32.786-kHz clock pulse, and the timer-microprocessor in IC_2 starts operation. CT_1 provides a means of adjusting the clock frequency (to properly position the display).

The character generator in IC_2 receives date and time information from the timer and generates corresponding character signals synchronized with the H- and V-signals from the sync generator (the same source of H- and V-sync applied to the picture tube). The character signal is applied to the base of Q_3. When the character signal is high, Q_3 turns on, and the character is added to the video passing from the camera to the EVF and tape.

The display mode for date and time is selected by the logic at pins 3, 5, 24, 25, and 26 of IC_2. In the configuration shown, the date display is set to the month, day, year format, since pin 25 is high (5 V), while the time is set to the 12:15 A.M., 11:15 P.M. format, since pins 3, 5, 24, and 26 are low (ground or 0 V).

The date and time setting and display switches S_7 through S_{10} determine the display generated by the character generator.

When display switch S_7 is pressed, the character display appears in the EVF screen. The display mode is changed to date, time, date and time, and display-off repeatedly, every time S_7 is pressed.

When shift switch S_9 is pressed, the display blinks. The blinking display is changed to A.M. (P.M.), hour, minute, year, month, and day repeatedly, every time S_9 is pressed.

When character switch S_8 is pressed, the blinking number is incremented. In the case of the A.M. or P.M. blinking display, A.M. is changed to P.M. and vice versa every time S_8 is pressed.

Start switch S_{10} is used to start and stop the setting for time and date. When S_{10} is first pressed for setting date and time, the A.M. or P.M. blinks, and the setting can be made. When S_{10} is pressed after setting the date or time, blinking stops and date and time counting starts.

4.2.2　Troubleshooting OSD Circuits

If the EVF display is good but there is no date and time display, suspect IC_2 and/or Q_3. It is also possible that X_1 is not oscillating.

If there is a date and time display but the display is not properly positioned, try adjustment of CT_1 (as described in the service literature).

If there is a properly positioned date and time display but certain functions of the display are absent or abnormal, check the corresponding circuit. For example, if the blinking number is not incremented, suspect S_8. Check that pin 12 of IC_2 goes high (5 V) when S_8 is pressed. If so, suspect IC_2. If not, suspect S_8.

4.3　IR CIRCUITS

Many modern video products (TV, VCR, and modular home-entertainment systems) use some form of IR remote control. Similarly, most present-day remote-control systems use some form of digital position modulation, or P-M. So we concentrate on both IR and P-M remote control in this section.

4.3.1　Typical Remote-Control Transmitter Circuits

Figure 4.6 shows the circuits of a typical hand-held IR remote transmitter. The circuits shown are unique in one respect. The reference frequency of the transmitter is 255 kHz (rather than the more common 455 kHz). This variation in transmission frequency prevents the transmitter from interfering with remote operation of other video products.

The digital code representing a given function appears as a series of "bursts" from the transmitter. Each burst contains 10 pulses, of 50-μs duration each, at a frequency of 20 kHz (rather than the more common 38 kHz). The duration of each burst is 500 μs.

The *position* or distance between recurring bursts identifies the digital 1s and 0s that make up the digital code. In this case, position or distance is determined as the amount of time between the rise of the first burst and the rise of the succeeding burst. A digital or logic 0 is 2 ms, while a logic 1 is 4 ms.

The function of IC_1 is to: (1) generate an oscillator signal for the creation of scanner-signal outputs (at pins 4 through 8), which are applied to specific inputs (at pins 11 through 15) of IC_1 (through the keyboard) to identify specific functions and (2) decode the information received at the inputs and produce a digital code (at pin 17) representing the selected function.

No scanner output is available unless a key is pressed. When a given key is pressed, the oscillator in IC_1 is turned on, and it generates the appropriate scanner signals. The scanner outputs at pins 4 through 8 of IC_1 are connected to one contact of specific keyboard switches. The remaining contact of each switch is connected to a specific input at pins 11 through 15 of IC_1.

When a given key is pressed (enabling the oscillator), a scanner signal is applied to the corresponding input through the closed switch contacts. A specific scanner signal is applied to a specific input for each switch closure. Similarly, each keyboard switch is represented by a unique combination of scanner output and signal input. The scan-

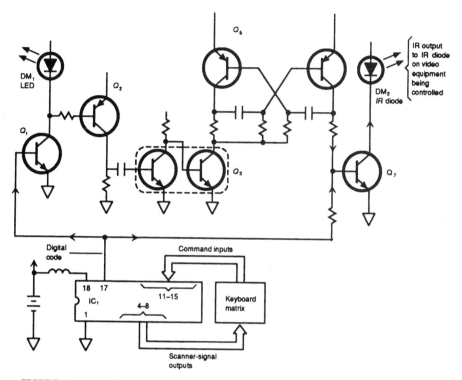

FIGURE 4.6   Typical hand-held IR remote transmitter.

ner signal applied to a particular input is decoded by IC_1 and produces a specific digital code representing the function of the corresponding keyboard switch.

The digital code (representing the selected function) at pin 17 of IC_1 is applied to the base of Q_1 and Q_7. The code signal is amplified by Q_1 and Q_2, delayed for 2.2 ms, differentiated, and applied to Q_3, which conducts for 550 μs, enabling the 40-kHz multivibrator Q_5. The output of Q_5 is applied to the base of Q_7, along with the output from IC_{1-17}.

The combination of both signals turns on Q_7, activating the IR diode DM_2. (Since two signals are required to activate DM_2, erroneous transmission is prevented.) Diode DM_2 is pulsed and the IR output is sent to the light-sensitive IR diode on the video equipment being controlled (Sec. 4.3.3). The output from DM_2 consists of a series of bursts, with the position or spacing between bursts determined by the digital code being transmitted.

4.3.2   Troubleshooting IR Transmitter Circuits

Once you have definitely pinpointed the problem to a remote-control transmitter by trying a different transmitter with the same equipment being controlled, look for weak or defective batteries. (Also look for any switch on the equipment being controlled that disables the remote-control function, as is the case on the author's TV set).

Batteries are the most common causes of trouble in any remote-control transmitter.

So start troubleshooting by putting in a known-good battery. The next most common problem is a defective switch contact. Defective switches usually show up when one or more functions are absent or abnormal but other functions are normal.

Note that when Q_1 is turned on by the output of IC_{1-17}, the remote-transmission LED DM_1 is turned on. This indicates that the transmitter is sending commands to the receiver diode. (Of course, it is possible that the transmitter is operating properly and only DM_1 is defective. However, this is not likely.) So, if DM_1 does not turn on when a key is pressed, look for a defective battery. If the battery is definitely good, suspect IC_1 or Q_1.

If the batteries and switches appear to be good but you cannot transmit any commands with the remote transmitter, press the keys while monitoring pin 17 of IC_1. Although it is not practical to determine the actual digital code being transmitted, the presence of pulse bursts usually indicates that IC_1 is good.

Next, try monitoring the pulse bursts through from Q_1 to DM_2. Also, look for 40-kHz square waves at the collectors of Q_5. If pulse bursts appear at DM_2 but the transmitter cannot transmit commands, suspect DM_2.

4.3.3   Typical Remote-Control Receiver Circuits

Figure 4.7 shows the circuits of a typical IR remote receiver. IC_1 has the dual function of decoding digital commands from the remote transmitter and converting them into

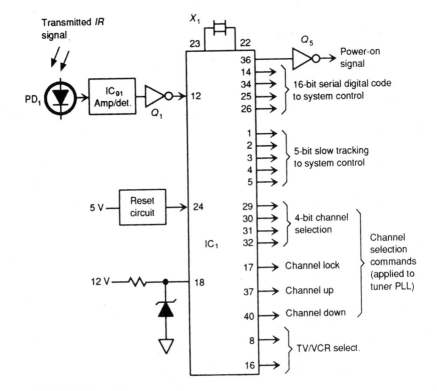

FIGURE 4.7   Typical IR remote receiver.

signals (that are applied to system-control, tuner-control, and power circuits). These signals replace or supplement the commands applied by push buttons on the equipment being controlled.

The command in IC_1 is taken from the signal transmitted by the remote-control transmitter. The transmitted IR signal is received by light-sensitive diode PD_1 and converted into an electrical signal applied to amplifier and detector IC_{91}. The detected signal from IC_{91} is amplified and inverted by Q_1 and applied to the input of IC_{1-12}.

The code transmitted as the first 16 bits of the signal indicates to IC_1 that the signal is for VCR operation. IC_1 decodes the second 16 bits to determine which function is requested and turns on the corresponding output(s).

The output at IC_{1-36} is the power-on signal, which turns on Q_5 whenever the power button on the remote-control transmitter is pressed. Transistor Q_5 is in parallel with the manual on-off push button of the VCR. The output at pins 17, 29 through 32, 37, and 40 of IC_1 are the channel-select commands. (These commands are applied to a tuner PLL such as described in Sec. 4.1.)

4.3.4 Troubleshooting Remote-Control Receiver Circuits

If you do not get proper remote-control operation, substitute a known-good remote transmitter. Next, confirm the presence of the signal at pin 12 of IC_1. This signal is a 32-bit serial data stream that cannot be intelligently monitored with a scope. That is, there is no easy way to determine which digital code is being transmitted by monitoring the signal. However, it is reasonable to assume that if a signal is present, the code is correct.

If the signal is present, check for 5 V at pin 18 of IC_1. If present, check for a 3.57-MHz clock signal at pin 23 of IC_1 and for a reset high at pin 24 of IC_1. If these inputs are correct but there is no remote-control operation (but manual operation is good), suspect IC_1.

If you get remote-control operation for some but not all functions, check the corresponding output from IC_1. For example, if you get channel-up operation but no channel-down operation (with a known-good remote transmitter), check for a channel-down output at pin 40 of IC_1. If the output is absent or abnormal, suspect IC_1. If the output is present, the problem is likely to be in the tuning system and should be checked as described in Sec. 4.1.

4.3.5 OSD and Remote Circuits

Figure 4.8 shows the OSD and remote-control circuits for a 13-in Sony TV set. The following is a brief description of these circuits.

Remote-Control Receiver. The remote-control receiver, sometimes called the *head amp,* is composed of IC_1 and IR detector D_1. The transmitted IR signal is detected by D_1 and applied to IC_1 for processing. The 40-kHz pilot signal is removed, and the negative-going square-wave remote signal (digital data stream) is applied to system-control microprocessor IC_{103} at pin 31.

This system uses the standard Sony Infrared Remote Control System (SIRCS) for all remote-control operation. The SIRCS is characterized by the ability to selectively operate only pertinent equipment. For example, a TV will respond only to a TV remote transmitter and a VCR will respond only to a VCR remote. This is done by sending a product-identifying code along with the control data.

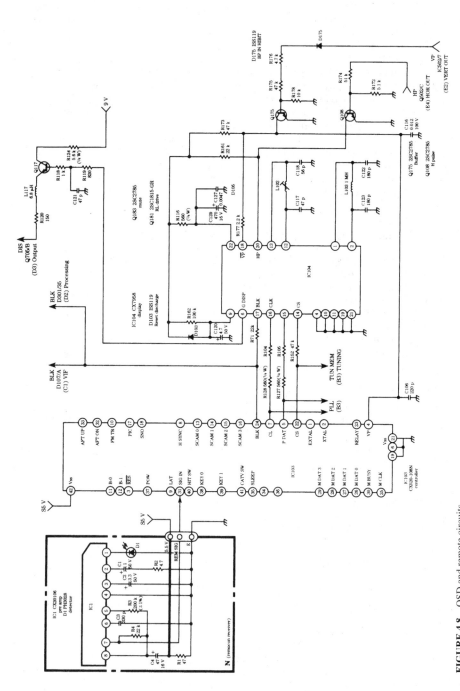

FIGURE 4.8 OSD and remote circuits.

4.16

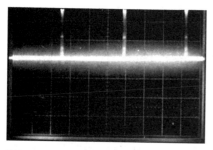

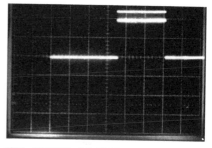

(S2) Q117/B DIS 2 V/d   5 msec/d

(S3) IC103/Pin 24 BLK 2 V/d 0.2 sec/d

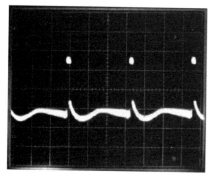

(S4) Q175/B   VP   0.2 V/d   5 msec/d

FIGURE 4.8   (*Continued*) OSD and remote circuits.

OSD Circuits.   The OSD functions are performed by IC_{104} (which corresponds to IC_2 in Fig. 4.5). However, display is provided by IC_{103} using the clock, data, and chip-select lines on pins 7, 5, and 22, respectively. Timing for the display is generated with IC_{104} using the VP (vertical sync) and HP (horizontal sync) reference signals. These signals determine the starting points of the vertical and horizontal intervals and originate from the same vertical and horizontal sweeps applied to the CRT.

The clock signals for IC_{104} are generated internally and appear at pins 1 and 13. The components at these pins determine the exact clock frequency. Notice that L_{102} is adjustable so that the clock at pin 13 can be adjusted in frequency. This permits the display to be positioned on the CRT screen.

Data bits for IC_{104} display operations are provided by IC_{103}. The data bits are serially transmitted to pin 15 of IC_{104} in sync with the clock signal at pin 16. Notice that the clock and data lines are shared by the tuning PLL (Fig. 4.4). Only the clock and data pulses occurring during the negative-going chip-select signal (pin 14 of IC_{104}) apply to the IC_{104} display operation. Also notice that the chip-select line is shared by the tuning memory IC_{105} (Fig. 4.4).

The signal developed by IC_{104}, in accordance with the data transferred from IC_{103}, is applied to the green gun of the CRT through pin 6 of IC_{104}, Q_{117}, and Q_{705} on the RGB output circuits (Fig. 3.26). As a result, all OSD characters appear as solid green on the CRT screen.

Blanking Output. During a channel-change operation, ICI_{103} outputs a positive-going pulse at pin 24. This blanking pulse mutes the audio by blanking the output to the VIF stage (Fig. 2.25) and blanks the CRT with a pulse to pin 35 of IC_{301} (Fig. 3.25). The blanking pulse is also applied at pin 17 of IC_{104} to remove the OSD signals from the green gun of the CRT. The result of this blanking action is a muting of the audio and a totally blank CRT between channel changes.

Reference Inputs. The internal clocks of IC_{104} are synchronized to the CRT vertical and horizontal rates by the VP and HP inputs applied through Q_{175} and Q_{108}, respectively. The inverted VP pulse is also applied to pin 4 of IC_{103}, telling IC_{103} when the vertical-blanking interval occurs. IC_{103} uses this information to update the OSD signals (and the PLL tuning signals) during the vertical-blanking period.

4.4 INTRODUCTION TO DIGITAL TV

The digital-TV circuits described here receive conventional TV signals and produce corresponding pictures and sound (although both video and sound are generally superior to those of conventional TV).

The key to digital TV is in analog-to-digital (A/D) and digital-to-analog (D/A) conversion. In the simplest of terms, the analog signals (composite video and sound) at the output of a TV tuner are converted to the digital equivalent by an A/D process somewhat similar to that found in compact discs. The resulting digital signals are then processed to produce the corresponding video and audio in digital form.

When the processing is complete, the digital signals are restored to analog form by D/A converters and applied to the picture-tube and audio circuits. The A/D conversion, digital processing, and D/A conversion all take place in digital ICs.

4.4.1 Advantages of Digital TV

The obvious advantage of digital processing is in the quality of the signal reproduced. As in the case of compact discs, the tuner signals to be processed are sampled (at high frequency), the sampled signals are stored digitally, and after processing, the restored signals are retrieved.

Another advantage is that a digital TV set can easily be adapted to the three basic TV systems because the sampling clock for the A/D converter is phase-locked to the broadcast color-burst frequency. Simply by changing the frequency of the clock, the system can accommodate NTSC (3.58-MHz) or PAL (4.43-MHz) color-burst systems. The same digital TV can also be used for SECAM in black and white. However, the system must be modified for color SECAM (since the PAL and SECAM color techniques are different).

For those not familiar with the three TV broadcast systems, the NTSC system (525 line) is used exclusively in North America and widely in Latin America and Japan. PAL (Phase Alternative by Line, 625 line) is used in the United Kingdom and most of western Europe, Australia, New Zealand, and South Africa. SECAM (Sequential Color with Memory or, in French, Sequential Couleur avec Memoire, 625 line) is used in France, the former Soviet Union, and eastern Europe.

Since the TV picture can be stored (in digital form), a digital TV can reduce flicker caused by interlaced scanning and can increase the *apparent resolution* of the picture. Digital TV eliminates interlace by storing all 525 lines and displaying the complete picture on the screen all at once instead of having only half the scan lines on the screen for each field of video, as is the case with conventional TV.

In addition to these obvious advantages, digital TV is generally superior because sync is checked on each horizontal line with PLL circuits, and there are a minimum of capacitors, inductors, and *RC* circuits to break down or to distort video signals.

4.4.2 Special Effects and Tricks

In addition to superior performance, the big selling feature of digital TV is the ability to "do tricks" or display special effects. The following is a brief summary of these special effects. Not all digital TVs can do all of the tricks described here. All of the effects are programmed into the digital ICs.

The *mosaic* and *paint* effects are available on some digital TV systems. With mosaic, the screen is a picture composed of small blocks. The paint effect is similar to an oil painting and is also called posterization in some literature.

The *freeze mode* freezes the current picture being viewed. Either field or frame reproduction may be selected to minimize jitter. Simultaneously, a real-time picture can be set into the lower corner of the picture-tube screen if desired.

The *preview mode* displays the still picture of nine channels in sequence, arranged in three rows and three columns on the screen. At predetermined times, the display is changed so that all channels programmed into the tuning-system memory are scanned.

The *picture-in-picture mode* inserts a $\frac{1}{9}$-normal-size, real-time picture from an external video input in the corner of the screen. The main picture and insert picture may be exchanged at any time by pressing a single button.

The *strobe mode* displays eight time-sequenced still pictures at once, while showing the real-time picture in the lower corner of the screen. The *editing mode* allows the user to change the still pictures displayed in the strobe mode.

4.4.3 Digital versus Conventional TV

Before we discuss ICs found in digital TV, it may be helpful to compare the basic functions of a digital TV to those of a conventional TV. Figure 4.6 shows the basic block diagram of a digital TV. Compare this to the block diagrams in Chaps. 2 and 3.

4.4.4 The Basic Five-Chip Digital TV

Figure 4.9 shows the basic (so-called) five-chip digital TV system in block form. If you look closely, you will find seven chips or ICs. However, the clock chip is generally not counted, and there are (typically) two identical audio-processor chips (for stereo operation).

We discuss each of these chips in the following sections. However, remember two points when studying the five-chip system. First, not all digital TVs have all five chips. Second, although analog signals are fed in and analog signals come out, all the signal processing is done digitally in the five-chip system. This is not necessarily true for some of the digital TVs that do not use the five-chip system.

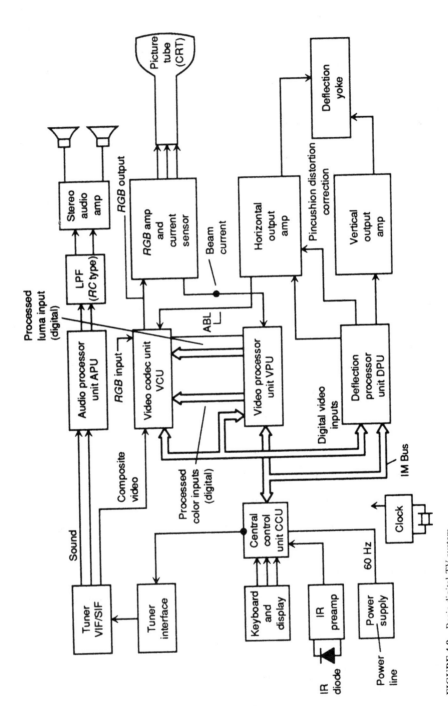

FIGURE 4.9 Basic digital-TV system.

4.20

4.5 TYPICAL DIGITAL TV CIRCUITS

This section describes the video and related circuits found in a Zenith Digital System 3 TV. The circuits described here use the full five-chip configuration. However, the five ICs used in the Zenith system are not identical to those in Sec. 4.4 (although the general circuit arrangement is as shown in Fig. 4.9.)

4.5.1 Modular Configuration

The circuits for the Digital System 3 are contained on six modules. Five of these modules include conventional circuits and/or perform TV set operations that are used on other (nondigital) Zenith sets. We do not concentrate on the five conventional modules here, except for the input-output relationship to the digital circuits.

All circuits that are unique to the digital functions are contained on one module, designated as the digital main module 9-535 and generally referred to as the *9-535* or *main* module.

4.5.2 Digital Main Module

Figure 4.10 shows the IC layout of the 9-535 module. From a troubleshooting standpoint, there are several important features to remember concerning the digital main module.

First, with one exception, all of the ICs on the 9-535 (both digital and nondigital) plug in. The exception is the ADC IC_{1404}, which is permanently installed on the module. As with any plug-in device, if all other troubleshooting steps fail, you can replace the ICs one at a time until the problem is cleared (or you can replace the entire module at unbelievable expense to the customer).

All the voltage and signals to and from the digital ICs can be measured at this module. If the voltages and/or signals are incorrect, you can trace back to the source from this module. That is the approach we use in this chapter. Before we get into the circuit details, let us see how the Zenith digital ICs compare with the ICs shown in Fig. 4.9.

4.5.3 CCU IC_{6001}

The central control unit (CCU) IC_{6001} is an 8-bit microprocessor containing a 6.5-kilobyte ROM and a 120-byte RAM. IC_{6001} has direct electrical connection to 9 of the 14 ICs on the main module. The program executed by the ROM section of IC_{6001} affects operation of four other digital ICs on the 9-535.

IC_{6001} is reset and the stored program is started each time the TV set is connected to a source of ac power. When the program is first started, IC_{6001} sets all ICs to the wait mode. Further execution of the program requires a power-on signal to IC_{6001}. No other operations can be performed until the power-on signal is received.

When the power-on signal is received, IC_{6001} distributes a reset (or initialization) signal to the ICs (see Sec. 4.6.4). After the ICs are reset, IC_{6001} sends and receives data bits to and from the remaining ICs as necessary to produce the sound and video. We describe how this is done in Secs. 4.7 through 4.9.

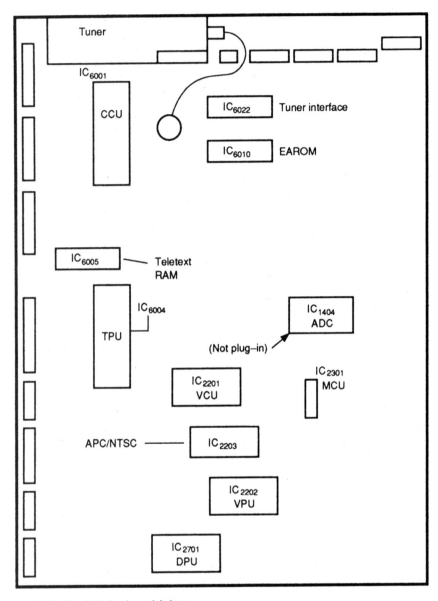

FIGURE 4.10 Digital main-module layout.

CCU IC_{6001} also responds to commands from the outside world (viewer and/or service technician) through the keyboard or remote control. All of these commands are coded digital signals (data bits) applied to IC_{6001}. No matter what the source, the commands cause changes in operation (channel changes, increase or decrease in volume, color level, black level, picture level, etc.). Similarly, the service technician (or a very ambitious tinkerer) can input data to IC_{6001} that (1) centers the picture, (2) sets picture height, and (3) sets the free-running frequency of the master (or color) oscillator.

IC_{6001} constantly monitors operation of many circuits within the set. For example, automatic black and white tracking (gray-scale tracking), automatic brightness limiting, color-oscillator drift, and so on, are constantly checked to maintain proper set operation.

The digital data bits to maintain operation are communicated between the CCU IC_{6001} and other ICs by a three-wire connection called the IM BUS. IC_{6001} can both send and receive data bits on the IM BUS (an abbreviation of Intermetall Semiconductors, a division of ITT).

4.5.4 VCU IC_{2201}

The video codec unit (VCU) IC_{2201} accepts two composite-video signals, as well as an RGB video input. (Codec is an abbreviation for encoder-decoder.) As discussed in Sec. 4.9, one video input is from the video detector in the IF and audio module, while the other video input is from a jack-pack on the rear panel of the TV set. The RGB inputs are used for (1) teletext displays, (2) on-screen displays, and (3) RGB signals for computer terminals.

Either of the two composite-video signals can be applied to an A/D converter within in IC_{2201}. The A/D converter produces a corresponding video signal in digital form. The RGB inputs are not "digitized" (the term used by Zenith) but are routed to the RGB outputs of IC_{2201}.

The VCU IC_{2201} applies digital video signals to two ICs: the DPU (Sec. 4.5.7) and the VPU (Sec. 4.5.2). After the digital composite-video signal is processed in these two ICs and the APC (Sec. 4.5.6) digital-luma and digital-color-difference signals are returned to IC_{2201}.

The luma and color-difference signals are converted to analog signals by D/A converters within IC_{2201}. The resultant analog signals are then matrixed to produce the RGB output from VCU IC_{2201}.

An *automatic brightness-limiting* (ABL) signal, developed by sensing the average total picture-tube current on the horizontal output module, is applied to IC_{2201}. Black and white operation of the set is determined by this ABL signal, and an automatic black and white tracking signal applied to the VPU.

4.5.5 VPU IC_{2202}

The video processing unit (VPU) IC_{2202} accepts a digital composite-video signal. A digital bandpass filter within IC_{2201} separates the luma and chroma information.

The digital-luma signal is output from IC_{2202} and applied to the APC (Sec. 4.5.6). (On this particular set, a binary-code converter converts the seven-lead output into an eight-lead format.)

The digital-chroma or color-difference signals R-Y and B-Y are applied to the APC. The color-difference signals are time-shared (or multiplexed) with other digital information that determines picture operation.

The other information is obtained by applying three monitor or sensing signals to IC_{2202}. These signals are: (1) signals from the output module that indicate the beam current for each picture-tube gun, (2) a signal from a photo sensor that senses the ambient light level around the TV set, and (3) input signal that represents the cut-off current for each picture-tube gun. The three monitor signals are time-shared with the chroma signals.

IC_{2202} contains a phase comparator used to synchronize the 14.3-MHz oscillator in MCU IC_{2301} (Sec. 4.5.8). The phase comparator receives the RF-burst portion of the chroma signal within IC_{2202} and the master-clock signal from IC_{2301}. The control loop for synchronization is completed by applying a digital correction signal back to MCU IC_{2301}. The correction signal is gated into the MCU by a clock signal.

4.5.6 APC and NTSC IC_{2203}

The automatic picture control (APC) IC_{2203} is also known as the NTSC IC. IC_{2203} is not part of the basic five-chip configuration but is an IC specifically developed to process NTSC video signals. Zenith includes APC and NTSC IC_{2203} to improve operation.

The outputs from IC_{2203} are corrected color and luma signals returned to the VCU IC_{2201}. These output signals are corrected automatically (by IC_{2203}) for brightness, contrast, color separation, and flesh tone.

4.5.7 DPU IC_{2701}

The deflection processor unit (DPU) IC_{2701} contains circuits that process the digital composite-video signal into correctly timed and phased horizontal and vertical drive signals.

A standard signal detector (SSD) circuit within IC_{2701} recognizes the composite-video signal as either a color or a black and white signal. In either case, both the horizontal and vertical drive signals are locked to the 14.3-MHz clock from MCU IC_{2301}.

IC_{2701} contains counters, dividers, and phase comparators to develop proper horizontal and vertical drive signals. For example, to get the proper horizontal frequency when a color signal is present, a programmable counter is set to divide the clock signal by 910. To phase the same horizontal signal properly, the counter output is phase-compared to the flyback pulse.

To get proper vertical drive signals, the horizontal signal is divided by 525 twice. To phase the same vertical signal properly, the counter output is phase-compared to the integrated vertical-pulse signal (obtained from a sync comparator within IC_{2701}).

DPU IC_{2701} also develops four keying and blanking signals. These signals are applied to the VCU, APC, and VPU to process the digital video signal.

4.5.8 MCU IC_{2301}

The master clock unit (MCU) IC_{2301} contains a VCO that generates the main timing signals for all ICs. The 14.3-MHz output of IC_{2301} is locked to the RF-burst portion of the received color signal by a phase comparator in the VPU.

The digital-correction voltage (PLL) from the phase comparator is applied to IC_{2301}, which contains a D/A converter that converts the digital-correction voltage to an analog signal used by the VCO.

The MCU IC_{2301} output is applied to seven of the ICs on the 9-535 module. The clock output is used to (1) synchronize A/D and D/A converters for both video and audio signals, (2) time horizontal and vertical sweep signals, and (3) time movement of the data from one circuit to another.

4.5.9 TPU IC$_{6004}$ and RAM IC$_{6005}$

The teletext processor unit (TPU) IC$_{6004}$ and related RAM IC$_{6005}$ (neither of which are on the basic five-chip system) are combined to (1) separate, decode, and store printable information that can be "piggybacked" on a composite-video signal during the vertical blanking period, (2) store and read out certain on-screen display information that can be superimposed on the picture, and (3) serve as an interface circuit between RGB computer inputs to the set and the VCU IC$_{2201}$.

The function of the RAM is to store data. To phase data into the RAM properly, an address bus is connected between the TPU and RAM. A read/write connection between these two ICs permits new data to be stored in, or stored data to be read out, of the RAM.

The digital circuits within TPU IC$_{6004}$ require certain input signals to perform these three functions. The input signals include (1) a digital composite-video input signal from the VCU for teletext operation, (2) horizontal and vertical blanking signals to time teletext or on-screen display information, (3) an IM BUS input to receive or send data to and from the CCU, (4) the 14.3-MHz master-clock signal for timing operations, and (5) RGB inputs for computer data.

The TPU has four signal inputs, three of which are the RGB outputs connected to the VCU. A fast blanking and switching signal is also output from the TPU to the VCU. This signal is used in the VCU to switch circuits between video information and the RGB information that produces on-screen displays.

4.5.10 EAROM IC$_{6010}$

The electrically alterable ROM (EAROM) IC$_{6010}$ is a nonvolatile, reprogrammable IC capable of storing 128 8-bit words (and is not part of the basic five-chip system). As in the case with other EAROMs, the words are retained when power is removed. The words can be changed when a programming-voltage (about 20 V) and a write signal are applied to IC$_{6010}$. When a read signal is applied, the words can be transferred out of the EAROM.

The information stored in IC$_{6010}$ consists of such factors as favorite channels, color-oscillator frequency, vertical height, and so on. IC$_{6010}$ is connected only to the CCU. All information to and from the EAROM is passed through the CCU. Data bits are transferred between the EAROM and CCU via the IM BUS.

4.5.11 Tuner-Control, or Interface IC$_{6022}$

The tuner-control, or interface IC$_{6022}$, operates with the CCU (and two transistors) to develop the frequency-control voltages for the tuner mounted on the 9-535 module. The tuning system uses frequency synthesis (FS) with PLL control (similar to that used on many Zenith TV sets and other microprocessor-based TVs, as described in Sec. 4.1).

4.6 BASIC DIGITAL TV TROUBLESHOOTING

Before we get into troubleshooting details for the Zenith Digital System 3, let us review the basic troubleshooting approach for any digital TV (and for any digital video equipment). Remember that (in this chapter) we are concerned with locating troubles caused by the digital video ICs, not with troubleshooting the remaining circuits (which are the same as those found in other nondigital TVs).

4.6.1 Plug-In IC Replacement

Because all the Zenith digital ICs plug in (except the ADC), it is practical to try correcting the problem by replacement. (This is not practical where ICs are wired in.) By replacing each IC in turn with a known-good IC and checking to see if the trouble symptoms are eliminated, you can cure about 90 percent of the problems in a digital TV (except for those problems caused by adjustment, which are covered in the service literature).

Of course, it is helpful if you start by replacing the ICs most likely to cause the problem. For example, in the Zenith set, if the problem is bad video, with good audio, start by replacing the VCU, VPU, APC, and NTSC. If the problem appears to be in deflection, either horizontal or vertical, start with the DPU. If the problem cannot be easily localized, start with the CCU.

Obviously, to make a logical choice for replacement of ICs, you must be able to *group those ICs that perform a specific function.* We do just that for the Zenith Digital System 3 in the troubleshooting discussions of Secs. 4.7 through 4.9.

Now let us consider the terrible possibility that all of the digital ICs have been replaced with known-good ICs, all adjustments have been made (in accordance with the service literature), but the problem remains.

4.6.2 Modular Replacement

At this point, if a known-good digital main module is available, you can try substitution. If this cures the problem, you can then trace the problem on the defective module. During replacement, you might find that the problem is one of poor electrical connection (such as loose connectors, dirty contacts, etc.).

Unfortunately, modular replacement may not always be practical. You may not have a replacement module readily available, or you may have only one shop-standard module (that you will not surrender at any price). A shop-standard module is recommended if you plan to service a particular model of digital TV. Of course, you may not want to invest in dozens of known-good modules for a variety of digital TVs.

If you do choose modular replacement, *take special care when reinstalling the shielding systems. Digital video signals are quite high in amplitude and frequency.* This can cause interference in nearby electronic equipment or in the picture being received. Such problems are most evident in areas where signal levels are weak, and where antennas are used (instead of cable). To eliminate problems, make certain that all shield covers are reinstalled and that locking clips are tight. All ground connections should be carefully resoldered (the author has had more problems with poor solder connections than any other trouble source).

4.6.3 Power and Ground Connections

The first step in tracing problems on a known or suspected defective module (when the problem is not corrected by substitution of known-good ICs and adjustment) is to check all power and ground connections to the ICs. Make certain to check all power and ground connections to each IC since many ICs have more than one power and one ground connection. For example, additional grounds are required for the MCU (in the Digital System 3) since the PAL and SECAM clock oscillators must be disabled (so as not to create interference). Also, look for any special power inputs. For example, IC_{6001} requires a 60-Hz sync signal (from the power line through the power-supply module).

4.6.4 Reset Signals

With all of the power and ground connections confirmed, check that all the ICs are properly reset. The reset connections are shown in Fig. 4.11a.

CCU IC_{6001} is reset when the power cord is first connected and standby power is applied. The remaining ICs in the digital system are reset (at a 4-MHz rate) from IC_{6001} (at pin 39) when a power-on signal is applied to IC_{6001}.

One simple way to check the reset function is to check for reset pulses at the appropriate pins. For example, pin 4 of IC_{6001} should be low (near zero) when standby power is first applied (because of the drop across R_{6041}) and then rise (to about 4.5 V) when C_{6013} charges through CR_{6001}.

The low applied to IC_{6001} causes the circuits within IC_{6001} to reset. After reset, when pin 4 rises to 4.5 V, IC_{6001} remains ready to perform the CCU functions, unless the voltage at pin 4 drops to a low for a prolonged period and produces a reset.

If the IC_{6001} reset is not as described, suspect C_{6013}, CR_{6001}, and R_{6041}. However, remember that IC_{6001} is not reset by temporary drops in the standby power (because of the charge on C_{6013}) but should be reset once each time the standby power is removed and reapplied.

As in the case of any digital device, if the reset line is open or shorted to ground or to power (5 or 12 V), the ICs are not reset (or remain reset) no matter what control signals are applied. This brings the entire digital operation to a halt. So if you find a reset line that is always high, always low, or apparently connected to nothing (floating), check the line carefully.

4.6.5 Clock Signals

The clock signals for the digital ICs are shown in Fig. 4.11b. Note that there are two clocks in the Digital System 3. The CCU IC_{6001} has a 4-MHz clock at pin 1. This clock signal can be monitored at the C_{6031} test point. The other clock signal (14.3 MHz) is taken from the MCU IC_{2301}. As discussed, this clock is locked to the RF-burst by a PLL phase comparator in VPU IC_{2202}. The PLL connections are at pins 5 and 6 of IC_{2301}. The 14.3-MHz master clock (for NTSC) is connected to the remaining ICs as shown in Fig. 4.11b.

It is possible to measure the presence of a clock signal with a scope or digital-logic probe. However, a frequency counter provides the most accurate measurement. Obviously, if any of the ICs do not receive the clock signal, the IC cannot function. On the other hand, if the clock is off frequency, all of the ICs may appear to have a clock signal, but the IC function can be impaired. (Note that crystal-controlled clocks do not usually drift off frequency but can go into some overtone frequency, typically a third overtone.)

If the 4-MHz clock is absent or abnormal, suspect CR_{6002} and IC_{6001}. If the 14.3-MHz clock is absent or abnormal, suspect CR_{2301}, IC_{2301}, and IC_{2202}. Also check for pulse activity on the PLL lines between IC_{2301} and IC_{2202}. Generally, the presence of pulse activity on these lines shows that the PLL circuits are *probably* good.

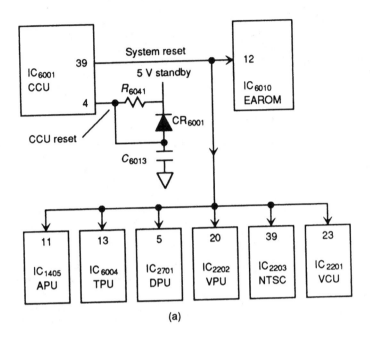

(a)

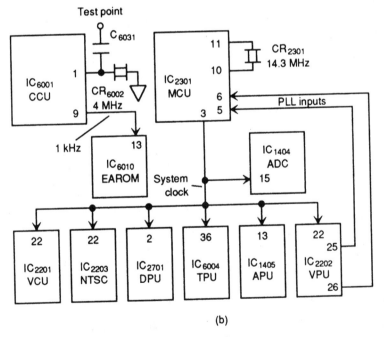

(b)

FIGURE 4.11 Reset and clock signals for digital ICs.

4.6.6 Input-Output Signals

Once you are certain that all ICs are good and have proper power and ground connections and that all reset and clock signals are available, the next step is to monitor all input and output signals at each IC. This is the approach we use in Secs. 4.7 through 4.9.

4.6.7 Logic or Digital Probe

Although you can check for the presence of input-output signals with a scope, a logic or digital probe is often more convenient (since many of the signals are digital). There are a number of probes on the market (including a Zenith Digi-Probe).

4.7 DIGITAL HORIZONTAL DRIVE CIRCUITS

As shown in Fig. 4.12a, the horizontal drive circuits originate with the DPU. When IC_{2701} is turned on, drive pulses appear at pin 31. These pulses are applied to the sweep-module circuits through Q_{2700}, Q_{2701}, pin 5 of connector 3A4, and pin 1 of connector 4A3.

The drive pulses are applied to flyback circuits on the sweep module through horizontal predriver XQ_{3202} and driver XQ_{3201}. The circuits produce flyback pulses for the picture tube and develop secondary supply voltages in the usual manner (Chap. 2). We do not discuss flyback circuits here because such circuits are essentially the same as for other nondigital Zenith sets.

4.7.1 Digital Horizontal Drive Troubleshooting

Start by checking for proper power, clock, and reset signals to IC_{2701}, as described in Sec. 4.6.

Next check for horizontal drive pulses at pin 31 of IC_{2701}. Then check for pulses at pin 1 of connector 4A3 on the sweep module.

If pulses are absent at $IC_{2701-31}$, with IC_{2701} on, suspect IC_{2701}. If pulses are present at $IC_{2701-31}$ but not at pin 5 of connector 3A4, suspect Q_{2700} or Q_{2701}.

If drive pulses are available at connector 4A3-1 but there is no horizontal sweep (or any other indication that the set is not turning on, such as lack of secondary supply voltages for the picture tube), check the sweep-module circuits, starting with XQ_{3202}.

4.8 DIGITAL VERTICAL SWEEP CIRCUITS

Figure 4.12b shows the vertical sweep circuits. Generally, there is no point in checking the vertical circuits until the horizontal circuits are proved good. The vertical sweep is produced by circuits on both the 9-535 and 9-370 modules. Note that the 9-370 module provides both power-supply and vertical signal functions (in addition to other functions not related to the video ICs).

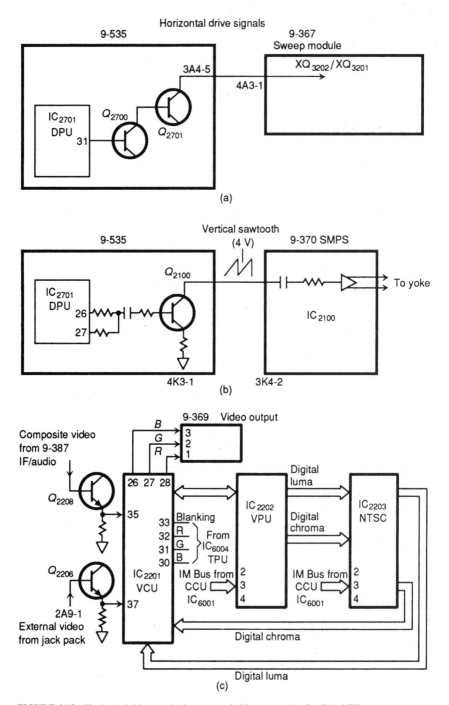

FIGURE 4.12   Horizontal drive, vertical sweep, and video processing for digital TV.

The vertical sweep signal is generated in the DPU and appears at pins 26 and 27. These outputs are combined in RC circuits to produce a typical vertical sawtooth (Chap. 2) at the emitter of Q_{2100}. The sawtooth signal is applied to vertical amplifier IC_{2100} on the sweep module. The output of IC_{2100} is applied to the vertical yoke in the usual manner.

4.8.1 Digital Vertical Sweep Troubleshooting

Troubleshooting for the vertical circuits is straightforward. Check for pulses at pins 26 and 27 of IC_{2701}, for a sawtooth at Q_{2100}, and for a sawtooth at the input of IC_{2100}. Note that IC_{2100} is an IC version of the vertical yoke drive found on many sets (and accomplishes the same functions as described for the discrete-component circuits in Sec. 2.19).

If the input to Q_{2100} is not good, suspect IC_{2701} or the RC components. If the input to IC_{2100} is not good (no sawtooth), suspect Q_{2100} or the related circuits.

The sawtooth should be about 4 V peak-to-peak. If the input to IC_{2100} is good but there is no vertical sweep, suspect IC_{2100} or the vertical yoke.

4.9 DIGITAL VIDEO CIRCUITS

Figure 4.12c shows the digital video circuits. As discussed, most of the video processing takes place in the VCU, VPU, APC, and NTSC. Likewise, most of the processing is in the form of digital signals.

In basic terms, the composite video signal is converted to digital form in VCU IC_{2201}. The digital composite video is separated into luma and chroma signals in VPU IC_{2202}. The separated luma and chroma signals are processed in APC and NTSC IC_{2203} and returned to IC_{2201}.

All of the signals in the video-processing loop are difficult to monitor. Also, both IC_{2202} and IC_{2203} are under control of IM BUS signals (from CCU IC_{6001}) that are equally difficult to monitor. You can check that the signals are present on each line but not that the signals are correct.

4.9.1 Digital Video Troubleshooting

Although troubleshooting for the video circuits appears difficult (in a digital TV), remember that the *inputs to the video loop* (pins 35 and 37 of IC_{2201}) *are conventional baseband video signals* that can be monitored and traced back to the source (IF and audio module or external jack-pack). Similarly, *the output from the loop* (at pins 26, 27, and 28 of IC_{2201}) *are conventional RGB signals* that can be monitored and traced to the video-output module.

In simple terms, if the signals applied to pins 35 and 37 of IC_{2201} are good but the signals at pins 26, 27, and 28 are bad, you have isolated video problems at IC_{2201}, IC_{2202}, or IC_{2203} or to the loop circuits (typically PC wiring and connectors). Of course, it is possible that CCU IC_{6001} is producing false information on the IM BUS, but this is not likely. In the real world, if there is a bad IM BUS input to IC_{2202} or IC_{2203}, the bus is probably at fault (shorted or stuck lines).

Note that on-screen teletext information developed in TPU IC_{6004} is added into the video-processing loop at pins 30 through 33 of IC_{2201}. If TPU IC_{6004} is suspected of causing a video problem, remove IC_{6004}. The set should operate normally, as far as video is concerned, with IC_{6004} removed. Of course, if all video functions except tele- text are normal, you have isolated the problem to IC_{6004} and the related circuits.

In addition to processing the composite video signal, the circuits of Fig. 4.12c are also the point at which viewer selections are processed. For example, if the tint button is pressed, commands are applied to the CCU through the IR remote-control system (Sec. 4.3). These commands are converted to digital signals on the IM BUS (by the CCU) and are applied to IC_{2202} and IC_{2203} (at pins 2, 3, and 4).

4.10   PICTURE-IN-PICTURE CIRCUITS

As discussed in Sec. 4.9, it is not necessary for a TV set to have all digital features (special effects) to produce the picture-in-picture (PinP) functions. It is possible to provide a conventional TV set with PinP using a few added ICs. One such circuit (found in some Hitachi models) is discussed in this section.

4.10   Basic PinP Processing

Figures 4.13 and 4.14 show the basics of PinP processing. In this basic circuit, PinP is a two-picture display (for example, the subpicture from tuner 2) located in one of the four corners of the main picture (the picture from tuner 1). The subpicture area is reduced to one-ninth or one-twelfth of the main picture. The choice of main and sub- picture is controlled by A/V switching (under direction of the remote control, in the usual manner).

The original picture (before PinP processing) is formed from the selected subpic- ture input with 49 μs (H) and 460 lines (V), as shown in Fig. 4.13. The selected origi- nal picture is written into a DRAM (in 4-bit data units or groups) using the A/D con- verter section of the PinP processor circuits. The data bits are read out from the

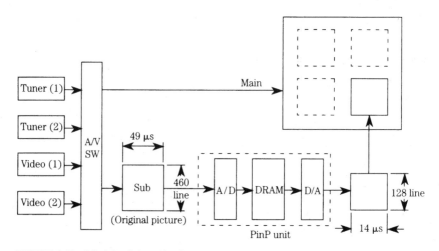

FIGURE 4.13   Principle of picture-in-picture.

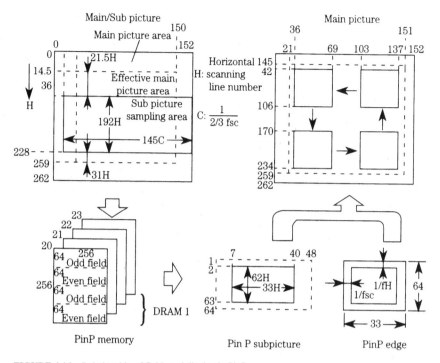

FIGURE 4.14   Relationship of fields and display in PinP.

DRAM and inserted into the main picture after conversion by the D/A portion of the PinP unit. Timing for insertion into the main picture and for digital processing is controlled by fsc (the chroma subcarrier frequency) and the sync signal.

The diagram on the upper left of Fig. 4.14 shows the original picture for one field. The vertical axis is shown as the number of horizontal-scanning lines, with the horizontal axis shown as number C (columns) by which the horizontal period (1H) is counted down with 2/3 fsc (two-thirds of the chroma subcarrier frequency).

The sampling area of the original picture used to form the subpicture is shown by the solid line. The vertical area is from 36H to 228H, with the horizontal area from 7C to 152C. The upper and lower sections of the effective picture (14.5H to 36H and 228H to 259H) are not sampled because of memory limitations.

The composite video is separated into Y and C components and is processed with a sampling frequency synchronized to fsc of the main picture. (The Y signal is approximately 2.4 MHz and the C signal is at about 0.06 MHz.) The video is converted to 6-bit units by the A/D converter.

As shown in the lower left of Fig. 4.14, the converted data bits are stored in memory (DRAM) as the PinP signal. The diagram on the lower right of Fig. 4.14 is composed from 64H (V) and 48C (H) held in memory. The DRAM memory is an IC with four memory-cell arrays. Each cell has a memory capacity of 65,536 bits (256 columns times 256 lines). Each memory-cell array has 256 vertical addresses, which are divided by 4 for a four-field subpicture writing area. As a result, the writing area is (64 vertical $\times$ 256 horizontal) $\times$ 4 fields.

The digital information of the picture (for the field that matches the scanning of the main picture in memory) is selected and is synchronized to the scanning of the main picture. The digital information is read out at a speed of 4.5 fsc. The data bits read out from memory are converted to a video signal by a D/A converter and appear in one of the four corners of the main picture (Fig. 4.13). The section 64H times 33C from the 64H times 48C of the subpicture data is displayed in the main picture.

The remaining section of data is used for two purposes. First, the information is superimposed in the main picture so as not to lose the video signal at the changeover part when scanning is changed from the main picture and vice versa. Second, the information matches the pedestal level of the signal that is demodulated into the color-difference signal. (This matches the tint of the main picture and subpicture.)

The areas immediately surrounding the subpicture are edged in white. This is done to emphasize the border of the main and subpictures. The edging line is set as 1/fH vertically and 1/fsc horizontally. When the circumference is edged, the dc-voltage equivalent to 100 percent white level is superimposed on the video signal. The position where the subpicture is inserted into the main picture can be selected (with the remote control) to one of four corners of the screen, as shown on the upper right of Fig. 4.14.

4.10.2 PinP Signal-Processing Sequence

Figure 4.15 shows the PinP signal-processing sequence. The composite video is first applied to an NTSC decoder, where the video is separated in Y, R-Y, and B-Y components. This is done so that brightness (Y) and color difference (R-Y and B-Y) components can be sampled individually. If the composite video was sampled as is, the correct phase could not be obtained because the sampling rates are changed to reduce the subpicture to 1/9 size. (To reduce the original subpicture to one-ninth, the sampling rate of the horizontal axis to the read-out time is set at 1:3, with the number of vertical-axis samplings set to one-third of the total scanning lines.)

After decoding, the bands of the Y, R-Y, and B-Y signals are limited by an LPF and applied to the multiplexer. The purpose of the LPF is to make the maximum frequency of each signal one-half (or less) of the sampling frequency. (The sampling frequency for Y is about 2.4 MHz, or fsc times 4/6, with the sampling frequency for R-Y and B-Y about 0.6 MHz, or fsc/6.)

Because the sampling frequency of the color-difference signals can be one-fourth of that of the Y component, and to save memory capacity, a multiplexer is used following the LPF. The multiplexer samples the signals on a time-share basis. The Y signal is sampled once per three horizontal-scanning lines. Then, to sample the chroma signals from the following scanning line, the B-Y and R-Y signals are alternately sampled by time sharing at 1/6 fsc. This processing method is known as the *line sequence method* and is shown in Fig. 4.16.

The A/D that follows the multiplexer converts the multiplex output to a 6-bit digital signal and supplies the signal to a gate array before being written into the PinP memory. This process is known as *quantization*. The gate arrays arrange the 6-bit signal to a 4-bit signal (which can be written more easily).

When the desired picture-insertion position is reached, the digital Y, R-Y, and B-Y components read from the PinP memory are separated from each other by the demultiplexer and are converted to 6-bit signals (from the 4-bit signals) by the 6-bit decoder. Then, the components are converted back to the original Y and C signals by the D/A converter. Because the 6-bit decoder outputs B-Y and R-Y color-difference signals, a color encoder is necessary to restore the original C (or chroma) signal.

To frame the subpicture correctly in the specified portion of the main picture, and to get the same tint, the sync and burst signals (separated from the video used for the

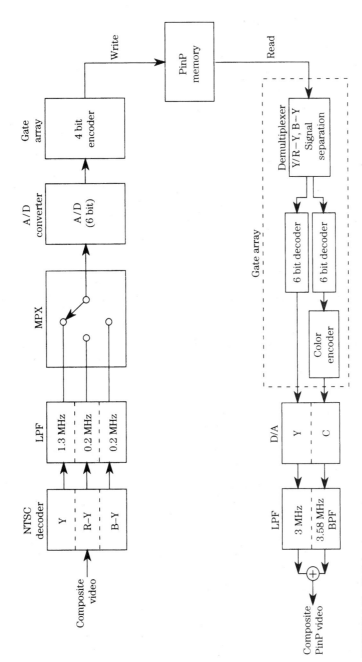

FIGURE 4.15 PinP signal-processing sequence.

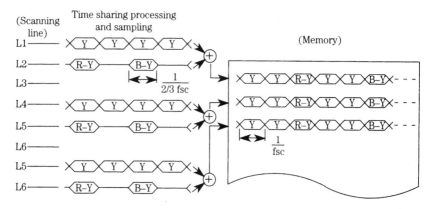

FIGURE 4.16   Principle of line-sequence method.

main picture) are used for timing in the PinP memory and for producing the chroma subcarrier signal required for NTSC modulation. All of these processes are executed according to a procedure programmed in the PinP gate array.

4.10.3   Typical PinP Circuits

Figure 4.17 shows the circuit configuration of the PinP unit. Notice that there are 10 ICs involved, including the PinP memory IC_{10}. The following is a brief description of the IC functions.

This PinP unit uses a digital-video process, with a 16-MHz (4.5-fsc) high-amplitude pulse signal. Because such a pulse signal can cause interference with nearby circuits, the PinP unit is housed in a shielded case. As discussed in Sec. 4.6.2, always make sure that all shields are in place, and properly grounded, to prevent interference.

The video signal is input and output to and from the PinP unit through the PB connector. The control signals for the PinP functions are input through the PA connector. The PinP function is on when pin 2 of PA is high and is off when pin 2 is low. Pin 1 of PA is the input terminal for moving the position of the subpicture (to each of the four corners of the main picture, in turn, as shown in Fig. 4.14). Each time pin 1 of PA is made high, the subpicture position shifts in the counterclockwise direction.

Main and Subvideo Input-Output Switch IC_1.   The input switch of IC_1 interchanges the video signals of the main and subpictures. However, in this configuration, the function is performed at a stage prior to the PinP unit, so the switch is not used. The output switch of IC_1 determines whether the subpicture is to be inserted into the main picture. If PinP is selected, the output switch also adds the edging signal (equivalent to 100 percent white) to emphasize the border of the main and subpictures. The timing control for these functions is performed by the PinP controller IC_6.

NTSC Demodulator, Sync Separator, Y/C Multiplexer IC_2.   This IC contains the 3.58-MHz (fsc) generator or VCO, sync separator, H/V pulse generator, and Y/C multiplexer.

The Y/C multiplexer multiplexes the video signal in the subpicture. The H/V sync pulses generated after the main sync separator are read to controller IC_6 and are used

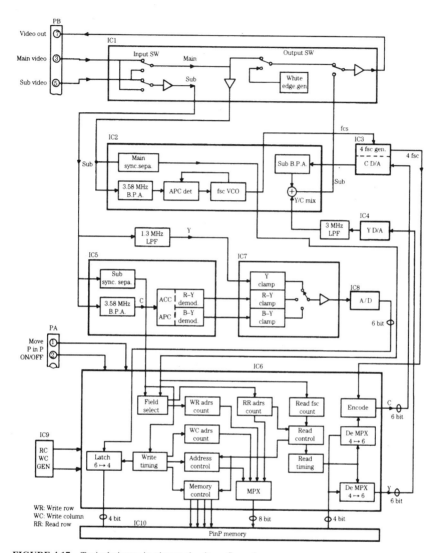

FIGURE 4.17 Typical picture-in-picture circuit configuration.

to determine the subpicture display position. The pulses also control the timing of write-read data to and from the PinP memory.

NTSC Demodulator, Sync Separator IC$_5$.   This IC performs NTSC demodulation of the subpicture video signal and feeds the signal to multiplexer IC$_7$. The IC also performs sync separation and delivers H/V sync pulses to controller IC$_6$.

Clamper, Multiplexer IC$_7$.   This IC sets the amplitude of the *Y, B-Y,* and *R-Y* signals before multiplexing the three signals into a serial data stream (on a time-share basis).

The (3N + 1) H scanning line of the Y signal is time-shared at 2/3 fsc. The color-difference signals of R-Y and B-Y are alternately time-shared by 1/3 fsc. This means that the sampling frequency of the Y signal is 2/3 fsc and the sampling frequency of the color-difference signal is 1/3 fsc. Both the clamping time of the clamper and the time-sharing of the multiplexer are controlled by IC_6.

A/D Converter IC_8.   This IC converts the subpicture video signal (separated into R-Y and B-Y signals with a maximum amplitude of 1 V_{p-p}) to a 6-bit quantized digital signal. The clock signal needed for A/D conversion is supplied from IC_6.

PinP Controller IC_6.   Control operation of IC_6 is performed by two control signals (PinP and PinP position) from the tuner or channel-selector microprocessor (Chap. 2). The functions of IC_6 are as follows:

Selects the video signals of the main and subpictures when the PinP mode is set

Determines where the subpicture is to be inserted

Generates the control signal for the multiplexer

Determines the clamp timing of the color-difference signals

Generates the clock signal for A/D conversion

Receives the sync signals separated from the video of the main and subpictures and generates the control signals for writing and reading to and from the PinP memory IC_{10}

Arranges the 6-bit data into 4-bit data units and performs writing and reading to and from IC_{10}

Converts the digital subpicture data (read from IC_{10}) to a Y signal and converts color-difference signals (also read from IC_{10}) to a C signal. This includes encoding the color-difference signals to restore the original chroma, or C, signal.

PinP Memory IC_{10}.   This memory has a capacity of 262,144 bits and can store data for four fields. Writing and reading is performed in real time (or apparent real time).

Y-Signal D/A Converter IC_4.   This IC converts a 6-bit quantized digital Y signal to an analog signal. The analog output is shaped by an LPF and is limited within the band specified by NTSC.

Chroma Signal D/A Converter IC_3.   This IC converts the 6-bit quantized digital chroma (C) signal to an analog chroma signal. The output becomes a chroma signal (within the NTSC band) through processing by a BPF in IC_2. The 4-fsc VCO (generator) and phase detector are part of IC_3. Phase control of the 4-fsc VCO is locked to the fsc of the main-picture video from IC_2. The 4-fsc signal is supplied to IC_6 as the reference clock for digital encoding.

Clock Signal Generator IC_9.   This IC generates the clock signals required for writing and reading to and from PinP memory IC_{10}. The writing frequency is 14.3 MHz (4 fsc), and the reading frequency is 16.1 MHz (4.5 fsc).

4.10.4   PinP Circuit Troubleshooting

Although troubleshooting for the PinP circuits appears difficult, remember that the inputs to the PinP loop (plus 3 and 5 of the PB connector, Fig. 4.17) are conventional video signals that can be monitored and traced back to the source (typically VIF or external-video circuits). Similarly, the outputs from the loop (at pin 7 of the PB connector) are conventional video signals that can be monitored and traced to the video-output circuits.

In simple terms, if the signals applied to pins 3 and 5 are good, but the signal at pin 7 is bad, you have isolated the trouble to the PinP unit (which includes the 10 ICs shown in Fig. 4.17). In some cases, individual ICs can be replaced. In other cases, the PinP unit must be replaced as a package. Either way, it is a case of digital troubleshooting, such as described in *Lenk's Digital Handbook* (McGraw-Hill, 1993).

Before you condemn the PinP circuits, make certain that they are turned on. Pin 2 of the PA connector must be high for the PinP circuits to operate. If pin 2 is not high, trace back to the source (which is probably a signal from system control in most sets but could be from a user switch).

If the PinP circuits are operating, but the subpicture does not move from corner to corner, check that pin 1 of the PA connector goes high each time the appropriate button is pushed. If not, trace back to the source (probably from the remote transmitter through system control).

CHAPTER 5
VHS VIDEO CIRCUITS

This chapter is devoted to the video circuits found in VHS VCRs. (Beta VCRs are covered in Chap. 6.) The chapter also covers other portions of the VCRs that are directly related to the video circuits (such as the system control and servo system for the tape-drive mechanism). Because of the complexity of the subject, it is not practical (if not impossible) to provide full details for all aspects of VCRs (even if limited to VHS) in the space allotted here. Full coverage requires several books (such as the author's many best-selling VCR books) and cannot be reduced to one or two chapters.

Instead of trying to provide such coverage, this chapter gives the video technician practical information that can be put to immediate use when servicing VCRs. In addition to providing a summary of VHS-circuit operation (including HQ, S-VHS, and VHS-C), the chapter includes troubleshooting, adjustment, and service notes for all types of VHS video circuits.

5.1 INTRODUCTION TO VCRs

This section is devoted to the basics of videocassette recorders (VCRs) and is included for those readers totally unfamiliar with VCRs and those who need a refresher.

5.1.1 VCR Basics

Figure 5.1 shows the basic functional sections of a VCR, which include a tuner, an RF section, a timer section, and a mechanical section (including tape transport, stationary audio heads, stationary control head, and rotating video heads). Note that the heads used for record are also used for playback.

On early-model VCRs, there are two rotating heads used for video record and playback. On present-day VCRs, the trend is toward four heads or possibly five heads (or six heads for the high-fidelity, or hi-fi, and S-VHS VCRs described at the end of this chapter). The multihead models (which are available in more than one head configuration—90°, 180°, etc.) provide additional features, particularly with regard to high-speed recording and still-frame or slow-motion playback.

In basic VCRs there is one audio head and one control head. Usually both heads are combined in one head stack, with the audio head at the top. In modern VCRs, the heads provide two audio channels for stereo or two-channel independent operation. Virtually all VCRs have a full-track erase head which erases audio, video, and control information from the tape. Some VCRs also have an audio-track erase head which erases either or both audio channels without disturbing the video or control informa-

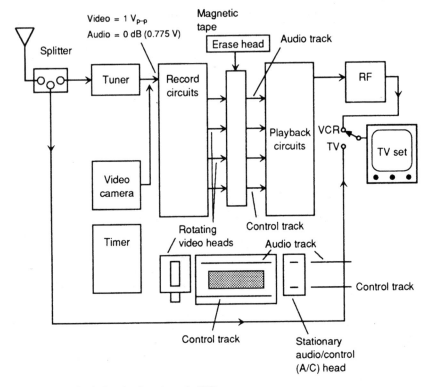

FIGURE 5.1   Basic functional sections of a VCR.

tion. This permits dubbing of audio information onto tapes recorded with video and control information.

The *tuner* shown in Fig. 5.1 is similar to tuners found in TV sets and functions to convert broadcast signals to formats suitable for the VCR. TV Channels 2 through 83 are covered by all VCRs, with additional cable-TV channels available on many modern VCRs. The tuners found in early-model VCRs use mixer and oscillator circuits. The tuners of modern VCRs have some form of frequency synthesis, using PLL circuits (Sec. 4.1). Typically, tuner output to the record circuit is 1 V (peak to peak) for video and 0 dB (0.775 V) for audio (compatible for NTSC).

The *record circuits* convert tuner output into electrical signals used by the heads to record the corresponding information on the magnetic tape along tracks. Note that the audio and control tracks are parallel to the tape, whereas the video track is diagonal. The video track is recorded diagonally to increase tape writing speed and thus to increase the frequency range necessary to record video signals.

The *playback circuits* convert information recorded on the tracks and picked up by the heads into electrical signals used to modulate the RF section. In essence, the RF section is a miniature TV broadcast station operated on an unused TV channel (typically 3 or 4). The output of the RF section is applied to the TV set.

During record operation, you select the channel you wish to record using the VCR tuning controls. This need not be the channel being watched on the TV set. Similarly, the TV set need not be on while recording with the VCR. You turn on the timer and the program or programs are recorded.

During playback operation, you select the correct TV channel (3 or 4) using the TV set channel controls. Then you turn on the VCR and play back the program, using the TV set as a display device. Many modern VCRs have monitor outputs (audio and video) which can be applied directly to a monitor TV (or another VCR), bypassing the RF section.

5.1.2 Record and Playback Heads

Figure 5.2 shows the relationship of the recording and playback heads on a typical two-head VCR. The same principles apply to multihead models. Instead of moving the tape at a high speed, the video heads are rotated to produce a *high relative speed* between the head and tape. This increases the frequency range necessary for video (4 to 5 MHz). The actual tape speed is in the range of 2 cm/s, and the video heads are rotated at 1800 rpm. This results in a relative speed in the range of 5 to 7 m/s (225 to 275 in/s).

The video frequency range is also increased by the use of FM for video recording and playback and by micro head-gaps for the video heads. Audio is recorded with AM, except for true hi-fi–stereo VCRs, which use FM for both audio and video. The control track is recorded with AM.

Note that the video heads rotate in a horizontal plane (on a *drum, cylinder,* or *scanner,* depending on the literature you read), while the tape passes the heads diagonally. This is known as *helical scan* and produces *slant tracks* or *diagonal tracks* for video recording. The audio head and control track head (mounted one above the other) are stationary and separate from the video heads, as is the erase head.

5.1.3 VCR Video Fields and Frames

Figure 5.3*a* shows the basic relationship between the video heads and video tracks recorded on tape. Video heads A and B are positioned 180° apart on the drum or cylinder, which rotates at a rate of 30 times a second (1800 rpm). The tape is wrapped around the drum to form an omega (Ω) shape. The tape then passes diagonally across the drum surface to produce the helical scan. Since there are two heads on the drum (which is rotating at 30 rps), each head contacts the tape once each $\frac{1}{60}$ s. Each head completes one rotation in $\frac{1}{30}$ s, and one slant track is recorded on the tape during half a rotation ($\frac{1}{60}$ s).

Since the tape is moving, after the first head has completed one track on tape, the second head records another track immediately behind the first one, as shown in Fig. 5.3*a*. If head A records during the first $\frac{1}{60}$ s, head B records during the second $\frac{1}{60}$ s. The recording continues in the pattern A-B-A, and so on. During playback, the same sequence occurs (the heads trace the tracks recorded on the tape and pick up the signal, producing an FM signal that corresponds to the recorded video signal).

Figure 5.3*b* shows the theoretical relationship among tracks, fields, frames, and the TV vertical sync pulses (Chap. 2). Since there are two heads, 60 diagonal tracks are recorded every second. One field of the video signal is recorded as on the track on the tape, and two fields (adjacent tracks A and B) make up one frame. In actual practice, there is some overlap between the two tracks. For example, the video signal recorded by head A (just leaving the tape) is simultaneously applied to head B (just starting the

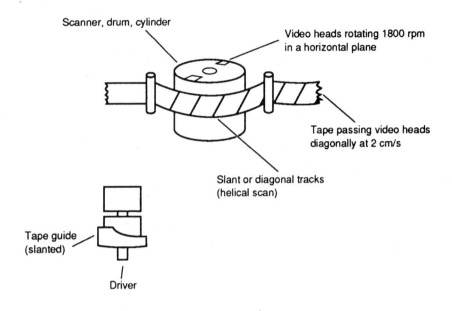

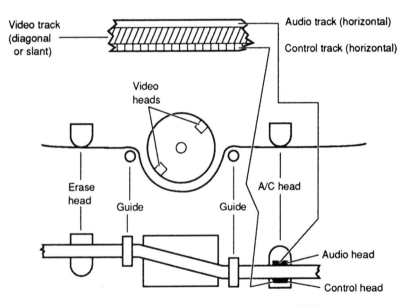

FIGURE 5.2 Relationship of recording and playback heads on a two-head VCR.

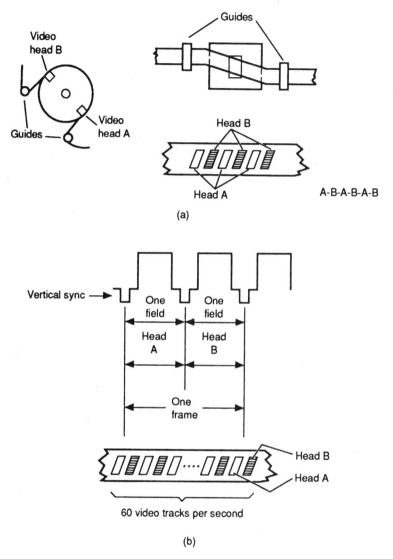

FIGURE 5.3 VCR video fields, frames, tracks, and sync pulses.

track). During playback, this overlap is eliminated by electronic switching so that the output from the two heads appears as a continuous signal, as described in Sec. 5.3.2.

5.1.4 The Basic VCR Servo System

It is obvious that no matter how precisely the tracks are recorded, the picture cannot be reproduced if these tracks are not accurately traced by the rotating heads during playback. In addition to mechanical precision, VCRs use an automatic self-governing arrangement called the *servo system*. Early-model VCRs generally use a single motor that drives both the head scanner and tape capstan through belts and gears. Present-day VCRs use direct-drive (DD) motors for scanner and capstan. The DD motors are *electronically synchronized as to phase and frequency* by the servo system ICs, as discussed in Sec. 5.3.5. Here, we consider the basic servo system.

Figure 5.4*a* shows operation of a basic servo system for a typical two-head VCR. The vertical sync pulses of the TV-broadcast signal are used to synchronize the rotating heads with the tape movement. The TV-sync pulses are converted to a 30-Hz control signal (often referred to as the CTL signal). This CTL signal is recorded on the tape by the separate stationary control-track head.

One major function of the servo is to rotate the cylinder at precisely 30 Hz during record. Note that 30 Hz is one-half the vertical sync frequency (60 Hz) of the input video signal. With a 30-Hz speed, the vertical blanking period can be recorded at any desired point on each video track. In TV, the vertical blanking occurs at the bottom of the screen, where blanking does not interfere with the picture. For this reason, the vertical sync is recorded at the bottom (or start) of each video track, as shown in Fig. 5.4*b*.

In the system in Fig. 5.4, there are two heads (channel 1 and channel 2), and each head traces one track for each field. Two adjacent tracks or fields make up one complete frame. To ensure that there are no blanks in the picture, the information recorded on tape overlaps at the changeover point (from one head to another). This changeover point must also occur at the bottom of the screen, where the changeover does not interfere with the picture.

To ensure proper changeover, the vertical sync signal is recorded precisely in a position 6.5H from the changeover time of the channel 1 and channel 2 tracks. (The term *1H* refers to the period of time required to produce one horizontal line on the TV screen, or about 63.5 μs, and is sometimes referred to as *line period*. 6.5H is 6.5 times 1H.)

The precise timing requires that the *speed and phase* of both the cylinder motor and capstan motor be controlled (since the cylinder motor determines the position of the heads at any given instant, while the capstan motor determines the position of the tape). In older VCRs, the synchronization is achieved by driving both capstan and cylinder from a common motor through belts. In a modern VCR servo system, such as shown in Fig. 5.4*c*, five separate (but interrelated) signals are used to get the precise timing. The following sections describe each of these signals and, in general terms, how the signals are used.

Cylinder FG Pulses.   The cylinder FG pulses are developed by a generator in the video-head cylinder. In the system in Fig. 5.4, the generator consists of an eight-pole magnet installed in the cylinder rotor and a detection coil in the stator. When the cylinder rotates at 30 rps, the stator coil detects the moving magnetic fields and produces the cylinder FG pulses at a frequency of 120 Hz.

Capstan FG Pulses.   The capstan FG pulses are developed by a generator in the tape capstan and are applied to the capstan speed-control circuits, as well as the capstan phase-control circuits (through a divider) during a record. In Fig. 5.4, the generator consists of a 240-pole magnet installed in the lower part of the capstan shaft and a detection coil in the stator. (In many modern VCRs, all of the servo-control pulses—cylinder, capstan, tach, reference, and control—are developed by Hall-effect generators.)

When the capstan rotates, the stator coil detects the moving magnetic fields and

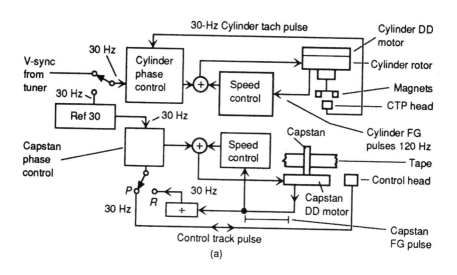

(a)

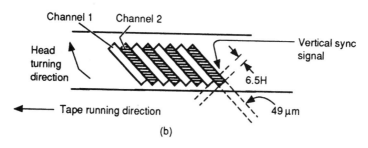

(b)

Pulse	Frequency	
Cylinder FG	120	
Capstan FG	SP (2 hour)	720
	LP (4 hour)	360
	EP (6 hour)	240
	Slow (slow motion)	120
	Quick (fast motion)	720
	Search	2160
Cylinder tach	30	
Ref 30	30	
Control track	30	

(c)

FIGURE 5.4   Basic VCR servo system.

produces the capstan FG pulses. The frequency of the capstan FG pulses depends on the capstan speed (which also controls tape speed). The tables in Fig. 5.4c show some typical capstan FG pulse frequencies for various playing times and tape speeds. Note that not all VCRs have the same six play modes or tape speeds shown in Fig. 5.4.

Cylinder Tach Pulse. The cylinder tach pulses (CTPs) are developed by another generator in the cylinder and are applied to the cylinder phase-control circuits. In some VCRs, the generator consists of a pair of magnets installed symmetrically in a disk in the lower part of the cylinder shaft and a stationary pickup head. Hall-effect generators are used in other VCRs.

With the system in Fig. 5.4, the CTP pickup head detects the moving magnetic fields when the cylinder rotates. The pulse frequency is a constant 30 Hz. In effect, the tach pulse indicates video-head channel switching and is used as a comparison signal in the cylinder phase-control circuits during both record and playback.

REF$_{30}$ Pulse. The reference signal for the phase-control system of both the capstan and cylinder motors is taken from a crystal oscillator with a frequency of 32.765 kHz. A frequency of 30 Hz is obtained when the crystal oscillator signal is divided. The REF$_{30}$ pulse is used for the cylinder phase control only during playback. During record, the cylinder phase control receives V-sync pulses from the tuner.

Control Track Pulse. The 30-Hz control-track pulses are the broadcast V-sync pulses recorded on tape during record. At playback, the pulses are picked up by the control head and applied to the capstan phase-control circuit.

5.1.5 Relationship of Luma and Chroma Signals in a VCR

Figure 5.5a shows the typical sequence in recording and playback of the VCR luma signal.

During record, the entire luma signal (from sync tips to white peaks) is amplified and converted to an FM signal that varies in frequency from about 3.4 to 4.4 MHz for VHS (3.5 to 4.8 for Beta, Chap. 6). During playback, the FM signal is demodulated back to a replica of the original luma signal. Note that this provides an FM luma bandwidth on tape of about 1 MHz for VHS (1.3 MHz for Beta).

As discussed in Chap. 3, color TV information is transmitted on the 3.58-MHz chroma subcarrier. Color at any point on the TV screen depends on the instantaneous amplitude and phase of the 3.58-MHz signal. In VCRs, the 3.58-MHz subcarrier is *down-converted* to a frequency of 629 kHz for VHS (688 kHz for Beta) and recorded directly (AM, not FM) on tape. Figure 5.5b shows the typical sequence for such down conversion (known as a *color-under* system, since the color-signal frequency is always well below the luma signal frequency). No bias is needed to record the chroma signal, since the FM luma signal is recorded together with the chroma (on the same video head), and the FM luma signal acts as a bias. Note that 629 kHz is 40 times the H-sync frequency of 15,750 Hz (actually 15,734.26 Hz during a color broadcast).

5.1.6 The VHS Recording System

Figure 5.6 shows the basic VHS recording system. VHS uses high-density recording to get the maximum amount of program information on a given amount of tape. This involves *zero guard band* recording and results in a *crosstalk* problem. Figure 5.6a

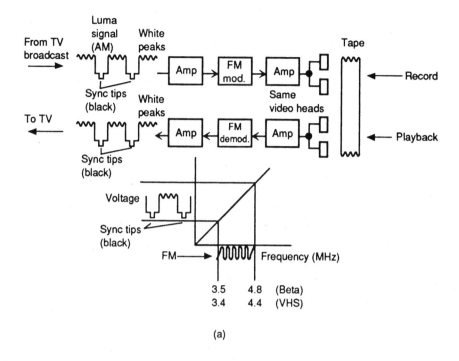

(a)

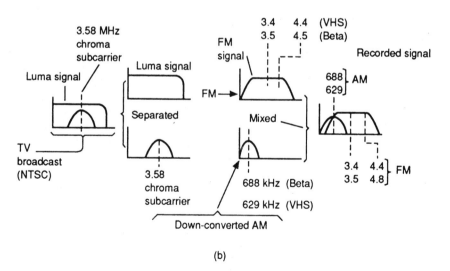

(b)

FIGURE 5.5 Relationship of luma and chroma signals.

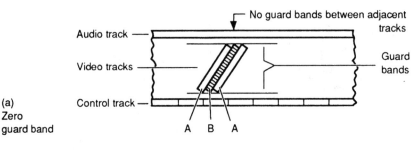

(a)
Zero
guard band

(b)
Azimuth
recording

(c)
Phase
inversion
(PI)

Line	1	2	3	4	5	6
Head A	0°	90°	180°	270°	0°	90°
Head B	0°	270°	180°	90°	0°	270°

(d)
Crosstalk cancellation
through 1H delay line

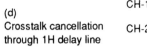
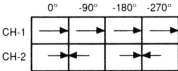

FIGURE 5.6   Basic VHS recording system.

shows a comparison of tapes with and without guard bands. Both VHS and Beta use *azimuth recording* and *phase inversion* (PI) to eliminate the crosstalk problem. However, the two systems are different.

Azimuth Recording in VHS.   Figure 5.6*b* shows how the azimuth recording principle is used to minimize crosstalk between adjacent tracks. The two video heads (or pairs of video heads on four-head VCRs) are mounted so that one head is at a different angle from the other. The angle for one head (head A) is +6° from the reference point (typically at right angles to tape movement), whereas head B is −6° from the reference. This produces a 12° difference between heads and, during playback, a strong signal is picked up only when head A traces over track A. If head B runs over track A for any reason, the track B signal is weak and does not produce interference or crosstalk.

Phase-Inversion Recording in VHS.   The phase of the 629-kHz color signal being recorded on head A is advanced in increments of 90° at each successive horizontal line. At the end of four lines, the 629-kHz signal is back to the original phase. For example, as shown in Fig. 5.6*c,* line 1, 2, 3, and 4 are shifted 0°, +90°, +180°, and 270° in succession. When head B is recording, the 629-kHz color signal is shifted in phase (retarded) in the opposite direction (0°, 270°, 180°, 90°). Thus, recorded phase shifts for odd-numbered lines (1, 3, 5) are the same but are opposite for even-numbered lines (2, 4, 6).

When the 629-kHz color signal is played back, the 4.2-MHz signal is again phase-inverted and mixed with the 629-kHz signal to restore the 3.58-MHz chroma signal. When both the playback 629-kHz and reference 4.2-kHz signals are phase-shifted in the same direction, the effect in the mixer is to restore the 3.58-MHz signal to normal phase.

When the playback 629-kHz and reference 4.2-MHz signals are phase-shifted in opposite directions, the phase of the 3.58-MHz chroma signal is shifted on every other line. When such a signal is phased through a 1H delay line (as shown in Fig. 5.6*d*), the crosstalk component is canceled out and the normal chroma signal component is double in amplitude.

5.2   VCR TEST EQUIPMENT AND TOOLS

If you have a good set of test equipment suitable for TV and audio work, you can probably service VCRs. That is, most service procedures are performed using meters, signal generators, scopes, frequency counters, power supplies and assorted clips, patch cords, and so on. Tools are quite another problem since VCRs have elaborate mechanical sections for tape transport and cassette load and unload mechanisms.

5.2.1   VCR Test Equipment

The following test equipment is recommended for VCR service.

Signal generator: An NTSC color generator (Chap. 3) is essential for VCR service. The two signals most often required for VCR work are the standard NTSC color-bar pattern with a *−IWQ* signal and an assortment of linear staircase patterns. Section 5.4 describes use of the NTSC generator during adjustment.

Receivers and monitors: If you are planning to go into VCR service on a full-scale basis, you should consider a receiver or monitor such as used in studio or industrial video work. These receivers and monitors are essentially TV receivers but with video and audio inputs and outputs brought out to some accessible point (usually on the front panel). There are also monitor-TV sets designed specifically for VCRs, video-disc players, video games, etc.

The output connections make it possible to monitor broadcast video and audio signals as the signals appear at the output of a TV IF section (the so-called *baseband signals,* generally in the range of 0 to 4.5 MHz, at 1 V peak to peak for video and 0 dB, or 0.775 V for audio). These output signals from the receiver or monitor can be injected into the VCR at some point in the signal flow past the tuner and IF. (Note that monitor-TV sets do not generally provide baseband outputs).

The input connections on either a receiver or monitor or a monitor-TV make it possible to inject video and audio signals from the VCR (before the signals are applied to the RF unit) and monitor the display. Thus, the baseband output of the VCR can be checked independently from the RF unit.

If you do not want to go to the expense of buying an industrial receiver or monitor or a monitor-TV set, you can use a standard TV to monitor the VCR. Of course, with a TV set, the VCR video signals are used to modulate the VCR RF unit. The output of the RF unit is then fed to the TV antenna input. Under these conditions, it is difficult to tell if faults are present in the VCR video or in the VCR RF unit. Similarly, if you use an NTSC generator for a video source, the generator output is at an RF or IF frequency but not always at the baseband video frequencies (on some NTSC generators).

If you use a TV set as a monitor, adjust the vertical height control to *underscan the picture.* This makes it easier to see the video switching point in relation to the start of the vertical blanking.

5.2.2 Tools and Fixtures for VCR Service

Figure 5.7 shows some typical tools and fixtures recommended for field service of VCRs. These tools are available from the VCR manufacturer. In some cases, complete tool kits are made available. There are other tools and fixtures used by the manufacturer for both assembly and service of VCRs. These factory tools are not available for field service (not even to factory service centers, in some cases). This is the manufacturer's subtle way of telling service technicians that they should not attempt any adjustments, electrical or mechanical, not recommended in the service literature. The author strongly recommends that you take this subtle hint. Most VCR troubles are the result of tinkering with mechanical adjustments. One effective way to avoid this problem is to use only recommended factory tools and perform only recommended adjustment procedures.

Alignment Tapes. Many VCR manufacturers provide an alignment tape as part of the recommended tools. An alignment tape is housed in a standard cassette and has several very useful signals recorded at the factory using very precise test equipment and signal sources. Although there is no standardization, a typical alignment tape contains audio signals at low and high frequencies, such as 333 Hz and 7 kHz, an RF sweep, a black and white pattern, and NTSC color-bar signals. If you intend to service one type of VCR extensively, you would do well to invest in the recommended alignment tape.

A typical use for the audio signals recorded on the alignment tape is to check the

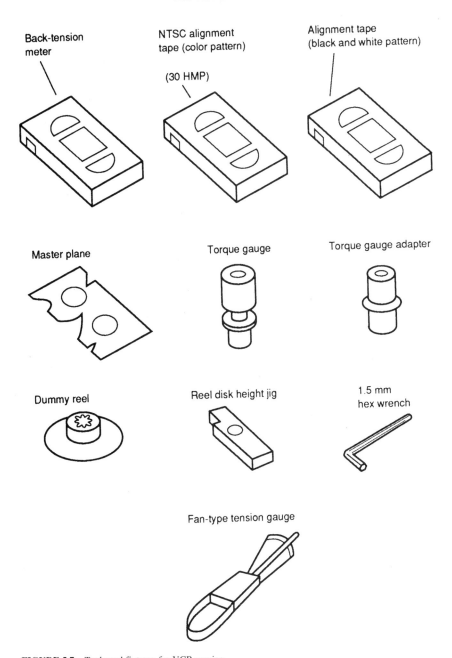

Back-tension meter

NTSC alignment tape (color pattern)

(30 HMP)

Alignment tape (black and white pattern)

Master plane

Torque gauge

Torque gauge adapter

Dummy reel

Reel disk height jig

1.5 mm hex wrench

Fan-type tension gauge

FIGURE 5.7 Tools and fixtures for VCR service.

overall operation of the servo speed and phase control systems. For example, if the frequency of an audio playback is exactly the same as recorded (or within a given tolerance) and remains so for the entire audio portion of the recording (as checked on a frequency counter), the servo control system (both speed and phase) must be functioning normally. If there are any mechanical variations (or variations in servo control) that produce wow, flutter, jitter, and so on, the audio playback varies from the recorded frequency.

If you do not want to invest in a factory alignment tape or if you do not want to wear out an expensive factory tape for routine adjustments (alignment tape deteriorates with continued use), you can make up your own alignment, or "work," tape using a blank cassette. The TV stations in most areas broadcast color bars before or after regular programming. (Use the VCR timer for convenience.)

The color bars must be recorded using a VCR known to be in good operating condition. Any stationary color pattern with vertical lines is especially helpful. If you have access to a factory tape, you can duplicate the recorded material on your own work tape. Of course, make certain to use a *known-good* VCR when making the duplicate.

Miscellaneous Tools.   In addition to the special tool shown in Fig. 5.7, most VCRs can be disassembled, adjusted, and reassembled (to the extent described in service literature) with common hand tools such as wrenches and screwdrivers. Remember that most VCRs are manufactured to Japanese metric standards, and your tools must match. For example, you will need metric-sized allen wrenches and phillips screwdrivers with Japanese metric points.

Since VCRs require periodic cleaning and lubrication, you will also need tools and applicators to apply the solvents and lubricants (cleaner sticks for the video heads, etc.). Always use the recommended cleaners, lubricants, and applicators, as discussed in Sec. 5.4.

Cleaning Cassettes.   While on the subject of cleaning, you should be aware of a special cleaning cassette, also known as a *lapping cassette.* Such cassettes contain a nonmagnetic tape coated with an abrasive. The idea is to load the lapping cassette and run the abrasive tape through the normal tape path (across the video heads, around the tape guides, etc.) *for a few seconds.* This cleans the entire tape path (especially the video heads) quite thoroughly.

Prolonged use of a lapping tape can result in damage (especially to the video heads). The author has no recommendation regarding cleaning tapes. If lapping cassettes are used, always follow the manufacturer's recommendations, and never use any cleaning tape for more than a few seconds.

5.3   TYPICAL VHS VCR CIRCUITS

This section describes the luma and chroma video circuits, as well as the related servo and system-control circuits, in a typical VHS VCR. Again, the circuits described here are composites of many VCRs. Also, in many present-day VCRs, some of the discrete-component circuits are combined into one or two ICs. The discrete-component versions are described to help understand the many functions performed by VCR ICs.

5.3.1   VHS Luma (During Record)

As shown in Fig. 5.8, the video signal from the tuner is fed to LPF_4 where the 3.58-MHz color signal is attenuated. The video is then fed to an AGC circuit (to maintain

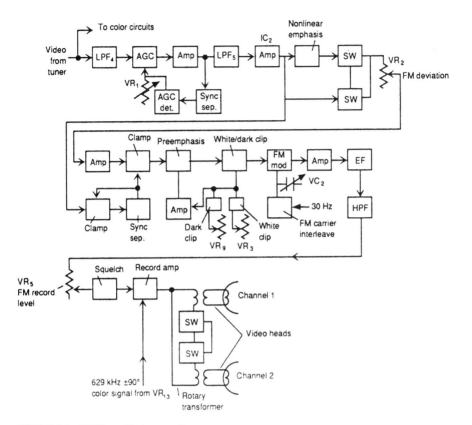

FIGURE 5.8   VHS luma (during record).

the output level constant) and to LPF$_5$ where the 3.58-MHz color is completely removed. The pure video signal is amplified by IC$_2$ and fed to the nonlinear-emphasis circuit, which emphasizes the luma frequencies by different amounts, depending on playing time (2 or 6 hours). The selection of frequency emphasis is made by a pulse applied to the switches that follow the emphasis circuit.

The emphasis circuit output is applied through VR$_2$, which sets the FM deviation. Note that the emphasis network is completely bypassed on the 2-hour playing mode. In all modes, the signal from the emphasis circuit is amplified by the video amplifier (part of IC$_2$) and is fed to a clamp where the dc voltage of the video sync tip remains constant regardless of the video signal. This keeps the sync tip at 3.4 MHz (Fig. 5.5a).

The clamped signal is fed to the preemphasis network where the high-frequency signals are emphasized to improve the signal to noise (S/N) ratio. The white and dark clip circuits are included to prevent overshoot and are adjusted by VR$_3$ and VR$_9$, respectively. Output from the clip circuits is applied to the FM modulator, which operates at 3.4 MHz for the sync tips and 4.4 MHz for white peaks (Fig. 5.5a). The FM modulator also receives 30-Hz pulses (Sec. 5.4) through the FM carrier-interleave circuits.

The FM luma signal is amplified and passed through an emitter-follower and a high-pass filter, which attenuates the lower end of the FM so as not to interfere with the 629-kHz chroma signal added later in the signal path. The output is applied to a squelch circuit through VR$_5$, which sets the FM luma record level.

The squelch circuit prevents the signal from being fed to the record amplifier for about 1.5 seconds after completion of cassette loading. This prevents the recorded signal from being erased if the tape runs near the drum in the middle of loading. The signal passing through the squelch circuit is amplified and applied to the video heads through a rotary transformer. Note that the record amplifier also receives the 629-kHz chroma signal from the color system (Sec. 5.3.3).

5.3.2   VHS Luma (During Playback)

As shown in Fig. 5.9, the reproduced signal from the video heads is applied to the preamps through rotary transformers. Switch circuits process the signals from the two channels to remove any overlap, as shown in Fig. 5.10. The composite signal is amplified and applied to the video amplifier through VR_4. The reproduced signal is also applied to the color circuits (Sec. 5.3.4) through VR_4.

After amplification, the signal is passed through a high-pass filter to extract only the luma FM and is applied to a drop-out cancel (DOC) circuit. The DOC prevents picture deterioration by supplying a 1H preceding signal if the FM signal is partially missed (dropout) for any reason (dirt on video heads, flaws in video tape, etc.). The DOC also provides a pulse to the AFC circuits (Sec. 5.3.6).

The FM signal is then fed to a double limiter (HPF, first limiter, LPF, mixer, and second limiter), which removes the AM from the FM. The luma signal is fed to the FM demodulation circuit. The demodulated signal is amplified and only the video signal is taken from LPF_2. The video signal is deemphasized (in reverse to the preemphasis at recording, Sec. 5.3.1). An edge-noise canceler removes the noise, and the video signal is applied to a compensator.

During the 6-hour playing mode, the compensator (together with the nonlinear deemphasis and feedback amplifier) returns the nonlinear emphasis supplied during record (Sec. 5.3.1). Note that the nonlinear deemphasis network is turned on during the 6-hour mode by the 6H signal. The feedback amplifier output is applied to a video amplifier through a low-pass filter (during a color broadcast). For black and white, the LPF is bypassed by the color/black and white switch. The LPF removes any noise that may arise where the video signal overlaps the demodulated chroma signal.

The signal is then sent to a noise-canceler circuit which suppresses any noise in the video signal. The luma (*Y*) and chroma (*C*) signals are combined in the *Y/C* mixer and then applied to the E-E/V-V switch through a clamp-mute circuit.

The term *E-E,* or electric to electric, can be explained as follows. During record, the VCR record-output circuit is connected to the playback input so that the video signal to be recorded can be monitored on a TV, if desired. Since the magnetic components (heads, tape, etc.) have nothing to do with this signal (the signal is passed directly from one electrical circuit to another), the function is called the E-E mode. When the heads and tape are involved in the normal record and playback cycle, the term *V-V,* or video to video, is sometimes used.

The E-E/V-V switch in Fig. 5.9 is applied to the RF unit (Fig. 5.1), which produces NTSC signals on Channels 3 or 4. The E-E amplifier and E-E/V-V switch combination permits a signal being recorded to be monitored on a TV set, if desired.

5.3.3   VHS Chroma (During Record)

As shown by Fig. 5.11, video from the tuner is passed through a BPF to remove only the chroma (3.58 MHz ± 500 Hz) and is amplified. The color signal is then passed

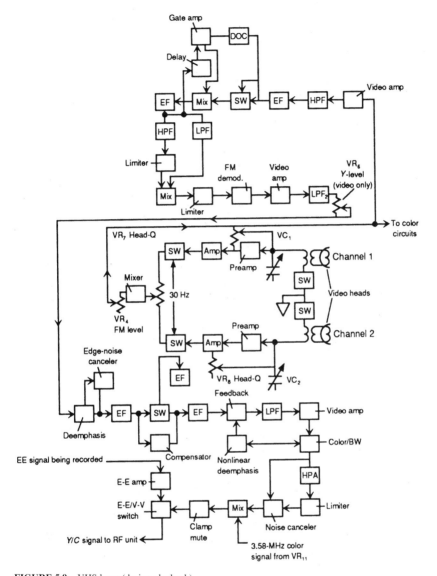

FIGURE 5.9 VHS luma (during playback).

through an expander to boost the burst signal by about 6 dB. The burst is fed to the automatic color control (ACC) through an emitter-follower.

The signal at pin 11 of IC_6 is applied to the color-control detector through a switch and a gate. The color-signal peak is detected and produces a control voltage which is applied to the ACC circuit. The detected voltage controls the ACC to maintain the color signal at a certain voltage level. The color signal is then applied to the main converter and mixed with a reference signal to produce the desired 629 kHz for recording on tape (with the luma signal, Sec. 5.3.1).

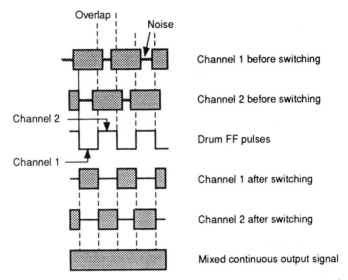

FIGURE 5.10 Switching and mixing process to produce a continuous signal from the video heads.

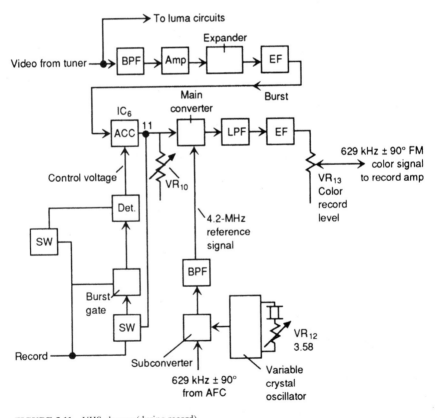

FIGURE 5.11 VHS chroma (during record).

The reference signal applied to the main converter is developed by mixing a 3.58-MHz signal from the variable crystal oscillator and a 629-kHz ±90° signal from the AFC circuit (Sec. 5.3.6). (Note that the term *±90°* applied to a signal in VHS means that the signal is rotated or shifted in phase every 1H period.)

The 3.58-MHz and 629-kHz signals are combined in the subconverter to produce a 4.2-MHz signal. This 4.2-MHz signal is passed through a BPF to the main converter, where the signal is combined with the 3.58-MHz chroma signal to produce a 629-kHz ±90° signal. The resultant signal is then fed to the record amplifier (with the luma signal) through the LPF, emitter-follower, and color record-level control VR_{13}.

5.3.4 VHS Chroma (During Playback)

As shown in Fig. 5.12, the recorded signal applied through the FM level control VR_4 (Sec. 5.3.2) is amplified. At this point, the signal contains both luma and chroma. Only the 629-kHz ±90° chroma signal is passed by the LPF. This signal is amplified and applied to the ACC.

The chroma signal is maintained at a constant level by the ACC circuit and is mixed with a 4.2-MHz ±90° signal from the APC circuit (Sec. 5.3.5). The resultant 3.58-MHz signal is amplified after passing through the BPF.

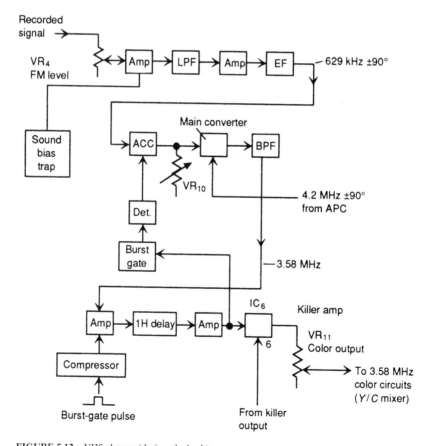

FIGURE 5.12   VHS chroma (during playback).

The compressor circuit operates whenever there is a burst-gate pulse to reduce gain in the amplifier. This is necessary to restore the burst signal to a normal level at playback. As discussed in Sec. 5.3.3, the burst signal is increased in amplitude by about 6 dB during record.

The restored chroma signal is passed through a 1H-delay to remove crosstalk. The 1H-delay output is amplified and fed to the killer amplifier. This killer amplifier is a switch that allows the signal to pass only when there is a color signal present in the video.

When playback is black and white (no color carrier), the color killer prevents a color signal from being applied to VR_{11}. This function eliminates noise components from the chroma circuit being applied during a black and white playback. The color killer is operated by signals from the APC (Sec. 5.3.5). When pin 6 of IC_6 is high, the color passes through VR_{11} and is superimposed on the luma signal (Sec. 5.3.2).

5.3.5 Typical VHS (Automatic Phase Control, APC) Circuit

As shown in Fig. 5.13, the variable crystal oscillator controlled by VR_{12} operates as a fixed 3.58-MHz oscillator *during record*. The output is mixed with the 629-kHz $\pm 90°$ signal from the AFC (Sec. 5.3.6) in the subconverter to form a 4.2-MHz reference voltage, as discussed in Sec. 5.3.3. The oscillator output is also applied to the phase-detector killer through diode switches.

During playback, the oscillator operates as a phase-locked 3.58-MHz oscillator. The phase is controlled by an error voltage from the phase-detector APC, which compares two 3.58-MHz inputs. One input is the playback color burst (which includes any phase shifts caused by jitter in the tape travel), while the other input is from a fixed 3.58-MHz oscillator. If there are any phase differences between the two inputs, the error voltage produced by the phase detector shifts the phase of the variable oscillator to correct the condition. Since the playback color-burst phase is controlled by the variable oscillator, any phase shift in the playback color signal is eliminated.

Note that the term $\Delta \Phi$ (delta phi) shown on Fig. 5.13 means that the signal has been shifted in phase or is of differing phase. In the case of the 3.58-MHz signal applied to the phase detector in VHS, the term means that the signal contains any possible jitter which could shift the phase.

The killer circuits in Fig. 5.13 have two functions. First, the circuits prevent a color signal from being passed when the signal is black and white only (to eliminate color-circuit noise from being mixed with the black and white signal). Second, the killer circuits provide an identification pulse (ID) which is used in the AFC circuit to prevent 180° out-of-phase lockup (Sec. 5.3.6).

During color operation the 3.58-MHz color-burst signal is passed through the burst gate to the phase detector killer, which also receives a 3.58-MHz signal from either the fixed oscillator (during playback) or the variable oscillator (during record). The two signals are compared (in phase) by the phase-detector killer. If both signals are of the same phase, the output of the killer detector goes low and the killer output goes high. This high is applied to pin 6 of IC_6. As discussed in Sec. 5.3.4, with a high at pin 6 of IC_6, the color signal is passed.

During black and white operation there is no 3.58-MHz color-burst signal, so the phase-detector killer sees only one signal. The output of the killer detector goes high, and the killer output goes low. This low cuts off the killer amplifier and prevents passage of color signals (or color noise).

Also during color operation, if the 3.58-MHz color burst is exactly 180° out of phase with the fixed 3.58-MHz oscillator (locked in phase but 180° out), the phase-detector killer develops a burst ID pulse, which is applied through D_9 to the AFC.

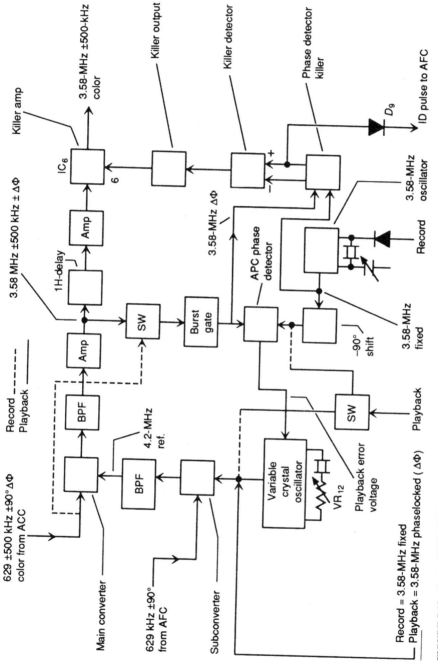

FIGURE 5.13 Typical VHS APC circuit.

5.21

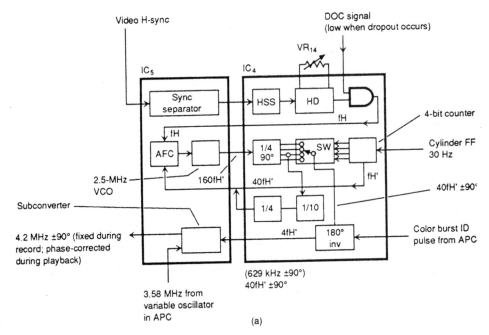

FIGURE 5.14 Typical VHS AFC circuit.

5.3.6 Typical VHS (Automatic Frequency Control, AFC) Circuit

Figure 5.14 shows typical VHS video AFC circuits. Note that the AFC system operates in much the same way for both record and playback. However, during record, the AFC uses H-sync pulses in the video signal from the tuner. During playback, the AFC uses the H-sync signals recorded on tape.

The AFC system has five inputs and one output. The five inputs include the video H-sync pulses, a dropout pulse from the DOC (Sec. 5.3.2), a 30-Hz cylinder signal from the servo (Sec. 5.1.4), a color-burst ID pulse from the APC (Sec. 5.3.5), and a fixed or phase-corrected signal from the variable oscillator in the APC circuits. The output of the AFC circuit is a 4.2-MHz ±90° signal (fixed reference during record or phase corrected during playback).

The video signal (from tuner or playback) is applied to a sync separator where only the vertical and horizontal sync signals (V- and H-sync) are passed. The resultant signal is then applied to the H-sync separator (HSS) where only H-sync is passed. The H-sync signals (or fH, as the signals are referred to in most VCR literature) are shaped into a 2-μs pulse by a horizontal drive (HD) circuit. The output from the HD circuit is adjusted to exactly 2 μs by VR_{14} and is applied to an AND gate. The other AND gate input is normally high so that the 2-μs fH pulses can pass. However, if there is a dropout (Sec. 5.3.2), the other AND gate input goes low, preventing the fH pulses from passing.

The output of the AND gate is applied to an AFC circuit within IC_5. This AFC also receives an fH prime (fH′) signal developed by a 2.5-MHz VCO. Note that the actual frequency of the 2.5-MHz oscillator is 160 times the H-sync frequency of 15,750 Hz (for black and white) or 160 times 15,734.26 Hz (for color). Note that the term *prime*

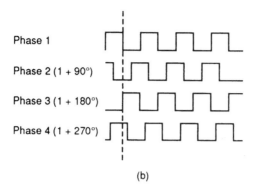

(b)

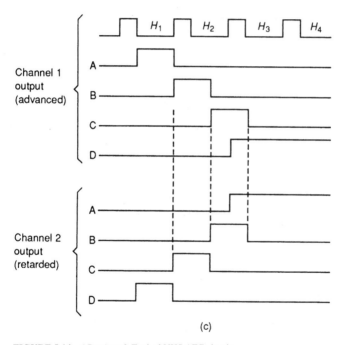

(c)

FIGURE 5.14   (*Continued*) Typical VHS AFC circuit.

applied to a signal here means that the signal has been locked in frequency to some other signal (to the H-sync signals in this case).

The 160fH′ from the VCO is divided by 4 into 40fH′ through operation of a one-quarter 90°-shift or switch circuit which is operated by a 4-bit counter. The 40fH′ output of this circuit is further divided by 10 to produce 4fH′ and by one-fourth to produce 1fH′ (or simply fH′). This fH′ is fed back to the AFC. If there is any difference in frequency between the fH signal coming from the AND gate and the fH′ signal originating at the VCO, the VCO is shifted in frequency by an error-correction voltage developed in the AFC circuit, locking the VCO precisely onto the H-sync frequency.

The 4-bit counter (operated by the fH′ and cylinder pulses) produces switch signals which select each of four signal from the one-quarter 90° shift circuit in sequence. Each of the four signals is shifted by 90° from the previous signal, as shown. In effect, the switch can be thought of as a rotary switch where the rotor direction and speed are determined by the 4-bit counter. The counter supplies the pulses to the switch each time an H-sync pulse is applied.

As shown in Fig. 5.14c, the channel 1 signals are advanced in phase by 90°, whereas the channel 2 signals are retarded or delayed in phase by 90°. These signals (40fH′ ±90°) are applied through the 180° inverter to be mixed in the subconverter with the 3.58-MHz signal from the variable oscillator in the APC and result in a 4.2-MHz ±90° signal that is locked precisely to the VCO.

The 180° inverter is operated by the burst-ID pulse from the APC. As discussed in Sec. 5.3.5, the ID pulse occurs only when the 3.58-MHz color burst is 180° out of phase with the 3.58-MHz oscillator (locked in phase but 180° out). The inverter normally passes the 40fH′ ±90° signal without change. However, if the ID pulse is present (indicating an undesired 180° lockup), the inverter reverses the phase of the 40fH ±90° signal to correct the condition.

5.3.7 Video Circuit Operation versus Troubleshooting

It can be seen by the discussion thus far that VCR video circuits are very complex. However, the video circuits are not necessarily complex to troubleshoot. This is especially true in modern VCRs, where most of the video circuits are contained within a very few ICs. It is relatively easy to trace signals through the circuits (it usually boils down to tracing a few inputs and outputs between ICs). Likewise, 90 percent of the video-circuit problems are cured by proper adjustment, followed by replacement of a few ICs, as necessary. Unfortunately, this is not true for the mechanical (tape transport, system control, servo, etc.) functions in VCRs, where most problems arise. That is why we concentrate on mechanical-control functions throughout this book.

5.3.8 VHS Tape Loading and Tape Transport

Figure 5.15a shows the tape-loading system for early-model VHS VCRs. Modern VHS VCRs often use a loading ring similar to that shown in Fig. 5.15b. Note that the term *tape loading* applies primarily to threading of the tape and is not to be confused with *cassette loading,* which involves inserting and locking the tape cassette within the VCR. However, the term *loading* is used interchangeably in VCR literature.

Figure 5.15b shows the major components of a typical front-load video tape transport. The tape-transport deck performs all the mechanical functions related to video tape handling and movement. When the front-load cassette is locked into position, the tape transport withdraws the tape from the cassette and routes the tape through a system of precision alignment posts and guides, precisely aligning the tape for application to the erase, video, audio, and control-track heads.

The tape transport also supplies the drive necessary to pull the tape from the cassette, through the transport assembly, and to return the tape back into the cassette. Throughout the process, electronic sensors continually monitor the condition of the tape and freedom of movement throughout the transport.

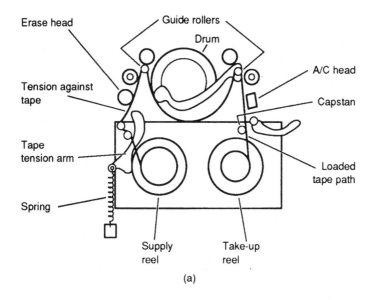

(a)

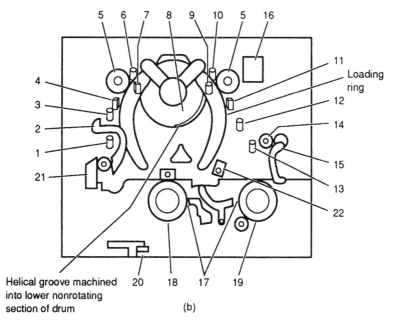

(b)

FIGURE 5.15 VHS tape loading and tape transport.

All of these functions are under control of microprocessors in the system-control circuits (Secs. 5.3 and 5.11 through 5.13). The following is a brief description of the operating sequence for all VHS tape transports (except the one you are working on). Refer to the index numbers in Fig. 5.15*b*.

Loading Ring.   All active transport elements are positioned by a loading ring. The use of a loading ring for VHS is one feature that distinguishes most modern VCRs from early-model VHS units. (Beta has used loading rings for some time.) The loading ring is responsible for the initial withdrawal of the tape from the cassette, as well as directing the movable guides to the final positions.

Tape Path.   Exiting the cassette, the tape is initially positioned by a fixed guide post (1), contacts the supply-tension arm (2), and is further positioned by a larger-diameter supply guide post (3) with upper and lower positioning flanges. The tape then moves past the full-erase head (4) and pressure idler (5) which supplies the tension necessary to maintain tape contact with the input guide roller (6). The final tape position is refined by the input slant-guide post (7) to provide the proper approach angle to the drum assembly (8). The lower edge of the tape is guided around a portion of the drum by a precision helical groove machined into the lower, nonrotating section of the drum.

The exit slant-guide post (9) positions the tape for departure from the drum, and the tape is applied to the takeup guide roller (10), past the audio and control head (11) and fixed takeup guide post (12) to the capstan drive shaft (13), where the tape is held against the shaft by the capstan pinch roller (14). The capstan drive shaft provides the momentum necessary to move the tape through the transport at a consistent predetermined speed.

Upon leaving the transport deck, the tape is prepared for reentry into the cassette by the takeup tension arm (15). The loading motor (16) supplies the power to drive the loading ring, located underneath the transport plate, via a belt-and-pulley coupling.

Reel Motors and Brakes.   Additional elements of the video tape transport include mechanical reel brakes (17), activated when the VCR is in the stop mode, and the electronically controlled supply reel (18) and takeup reel (19) with the associated drive motors.

Status Switches.   Many present-day VCR video-tape transports have status switches that inform the mechanical-control microprocessor in system control (Secs. 5.3 and 5.11 through 5.13) of the progress of certain cassette and tape-handling procedures. For example, during tape loading or unloading, the play/record (P/R) switch informs system control that video-tape loading or unloading is complete, terminating drive to the tape-loading motor.

The record-inhibit switch (20) prevents recording on a cassette where the *erasure-prevention tab* has been removed. When the tab is removed, the system control ignores record commands (from the front panel or remote control) and the original recording is preserved. If you wish to record on a cassette with the tab removed, cover the hole on the cassette with tape. This actuates the record-inhibit switch, and the record function is normal. During service, it is sometimes necessary to operate the VCR with a cassette installed (for instance, to observe rotation of the tape drive motor). In this case, use tape to hold the switch in place, just as if actuated by the tab.

Sensors.   All video-tape transports have sensors. Many, but not all, sensors operate through the system control. For example, the supply-tension sensor (21) detects tension on the tape just as the tape leaves the supply reel. A sensor (21) controls the supply-reel motor voltage (directly, not through system control), thus varying the back tension as required for uniform tape tension throughout the transport path.

Sensor inputs to system control have priority over all other inputs (on most VCRs). For example, the cassette lamp (22) works in conjunction with the tape-end sensors (Sec. 5.3.13) to detect the transparent leader at the beginning and end of the VHS tape, placing the VCR in the stop mode. In certain cases, this is followed by the rewind mode.

5.3.9 Typical VHS Cylinder-Servo Operation

Figure 5.16 shows the overall operation for the cylinder-servo system of a typical VHS VCR. Figure 5.17 is the diagram for the phase and speed loops of the cylinder servo. Note that the phase-loop circuit is contained in IC_1, while the speed-loop portion is in IC_2. The outputs of the phase and speed loops are added together to determine the drive voltage to the cylinder motor-drive IC (Fig. 5.4). The phase-loop portion of the cylinder servo is a pulse width modulation (PWM) system.

During playback, the reference signal for the phase loop is taken from a 3.58-MHz signal at IC_{1-25}. This is divided down to 30 Hz. The comparison signal is derived from the 30-Hz cylinder input at IC_{1-14}. This signal is delayed by two multivibrators (MVs).

The time constant of the two MVs is controlled by the two shift adjustments at pins 12 and 13 of IC_1 (often called PG shift adjustments). One output from the MVs is the 30-Hz head-switching signal (Sec. 5.7). The other output from the MVs is applied to the PWM circuit, which also receives the 30-Hz record or playback reference signal. The two signals are compared in the PWM circuit to determine the correction necessary for proper phase.

During record, the reference signal is the divided-down vertical sync signal at IC_{1-17} (instead of the 3.58-MHz signal). The output of the PWM at IC_{1-29} is a 1784-Hz

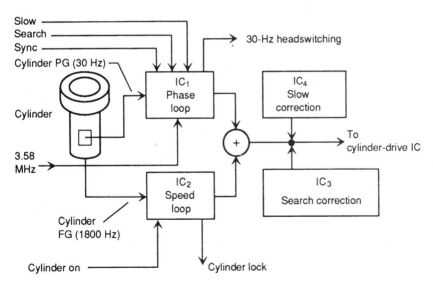

FIGURE 5.16 VHS cylinder servo.

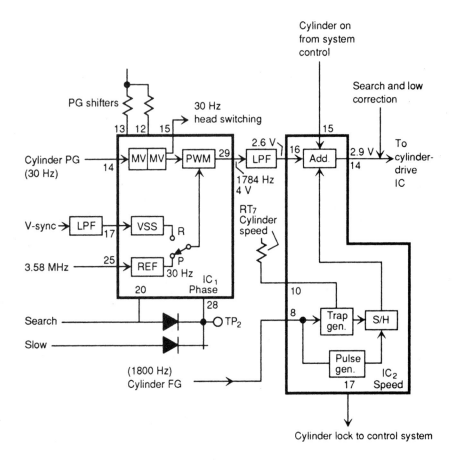

FIGURE 5.17   Phase and speed loops for VHS cylinder servo.

signal, which is passed through an LPF to remove the ac component. The resultant dc output (about 2.6 V) from the LPF is applied to pin 16 of the speed-loop IC_2.

During the search and slow modes, the system-control circuits apply a high at IC_{1-28}. This high, which can be monitored at TP_2, places the output of the PWM system in a fixed 50 percent duty cycle, inhibiting any phase shift of the cylinder motor.

IC_2 is essentially the coarse speed control for the cylinder-servo system. IC_2 receives a primary input from the cylinder 1800-Hz signal at IC_{2-8}. The signal applied to pin 16 of IC_2 from IC_1 is a form of vernier speed (or phase) control voltage.

The speed-loop circuits in IC_2 are of the sample-and-hold (S/H) type. The 1800-Hz signal at IC_{2-8} is processed and passed to the S/H circuit, which develops a correction voltage. This voltage is passed to the adder circuit, along with the phase-control voltage. The adder passes the combined control voltage to the cylinder motor-drive IC through IC_{2-14}. Note that the cylinder motor-drive signal is active only when system control applies a cylinder-on signal (a low at IC_{2-15}). This enables the adder circuit to output the motor-drive voltage.

During initial start-up of the cylinder servo system, the cylinder motor is at zero rotation. So when the cylinder-on command occurs, the cylinder-lock signal at IC_{2-17} goes high for a short period of time (until rotation of the cylinder motor reaches the proper value) and then returns to low. The cylinder lock tells system control that the cylinder motor is not up to speed.

When IC_{2-17} is high, system control places the VCR in stop. However, system control does not monitor the cylinder-lock signal during load and initial start-up. After proper loading and when the VCR is in play mode, system control then monitors the cylinder-lock signal for proper level. If the signal is low, the VCR remains in play. If high, system control places the VCR in stop.

Cylinder-Servo Troubleshooting. If the picture is out of horizontal sync (the most common symptom for trouble in the cylinder phase and speed circuits), play back a known-good tape or a test tape. Connect 5 V to TP_2, locking the PWM output of IC_1 to the 50 percent duty cycle. Check for the presence of a signal at IC_{1-29}. The signal should be about 4 V at 1784 Hz, with a 50 percent duty cycle.

If the signal is absent or abnormal, suspect IC_1. If the signal is correct, look for 2.6 V at IC_{2-16}. If this voltage is absent or abnormal, suspect the LPF. If the signal at IC_{2-16} is normal, check for a low at IC_{2-15}.

If IC_{2-15} is high, indicating that the cylinder is locked, check the system-control circuits. If IC_{2-15} is low and there are about 2.6 V at IC_{2-16}, try monitoring the dc voltage at IC_{2-14} while adjusting the cylinder-speed control RT_7. Adjust RT_7 for about 2.9 V at IC_{2-14}, and check that the TV picture is in horizontal sync.

Note that if the phase control IC_1 is disabled by 5 V at pin 28, there may be some noise bars floating through the picture, even if the picture is in sync. Remove the 5 V, and recheck for noise bars and horizontal sync. If you cannot get proper horizontal sync by adjustment of RT_7, with all inputs to IC_2 correct, suspect IC_2.

5.3.10 Typical VHS Capstan-Servo Operation

Figure 5.18 shows the overall operation for the capstan-servo system of a typical VHS VCR. Note that the capstan servo is functionally similar, but not identical, to the cylinder servo described in Sec. 5.3.9. Both use PWM for speed and phase control.

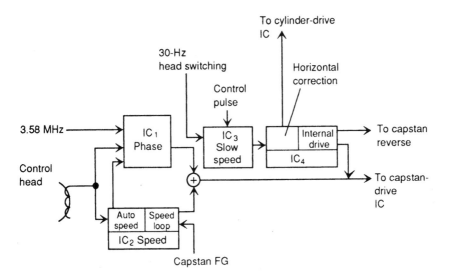

FIGURE 5.18 VHS capstan servo.

The use of PWM is typical for most modern VCRs (although some VCRs use sample-and-hold circuits). The servos of most modern VCRs use three-phase motor drives (direct drive, or DD) with Hall-effect control.

As shown in Fig. 5.18, the phase-loop portion of the capstan servo is in IC_1, and the speed-loop portion is in IC_2. During playback, the auto-speed select circuits in IC_2 monitor the control track signal (picked up by the control head) and determine the proper speed for the capstan motor. The correction voltage from the phase loop and speed loop are added together and passed to the capstan motor-drive IC.

During slow motion and special effects, correction signals developed from IC_3 and IC_4 are passed to the capstan motor-drive IC along with the speed-loop correction voltage. During slow-motion operation, the phase-loop portion of the servo (IC_1) is inhibited, and phase correction is done by IC_3.

Capstan Phase-Loop Operation.   As shown in Fig. 5.19, the capstan phase loop uses a 3.58-MHz signal at IC_{1-25} as a reference signal for the PWM comparator. During playback, the comparison signal is derived from the control-track signal recorded on tape. During record, the comparison signal is taken from the capstan signal (the frequency of which depends on playing time).

The control-track signal and the capstan signal are passed through the programmable divider in IC_3, which *does not* divide the signal in normal play or record modes but does divide in the search modes. This division is necessary since capstan speed is increased but phase lock must be maintained.

To accommodate the increased frequency of the control-track and capstan signals (SP is increased by 5; SLP, by 15), the signals are divided down to normal frequencies so that both the phase and speed loops can control the capstan motor. The control-track signal is passed from IC_{3-3} to IC_{4-3}, where the signal is amplified and applied to the phase-loop IC_{1-21}. The capstan signal at IC_{3-4} is passed through a divider within IC_2 and applied to IC_{1-6}.

During playback, the control-track signal at IC_{1-21} is applied to the input of the PWM counter-latch circuit. During record, the capstan signal is divided down to 30 Hz and applied to the PWM system. The capstan servo system adjusts capstan speed (and thus tape speed), thereby maintaining the divided-down capstan signal at 30 Hz.

To get proper tracking, the 3.58-MHz reference signal at IC_{1-25} is divided down and delayed by an MV. The time constant of the tracking MV is determined by the resistance and capacitance at IC_{1-19}. The front-panel *tracking control* VR_1 and preset controls RT_{13} and RT_{14} determine the amount of delay to the reference signal to get proper phase of the control-track signal, minimizing noise in the video signal.

The output of the PWM system is a 437-Hz signal appearing at IC_{1-2}. The *duty cycle or pulse width* of the PWM output signal is proportional to the phase error of the system. The output is passed to an external LPF, which removes the ac component and passes the dc component to the speed loop. The dc component of the PWM signal is about 2.6 V.

During special effects, slow motion, pause, or cylinder-lock condition, a high is applied to IC_{1-3}. This disables the PWM system and generates a fixed 50 percent duty cycle for the 437-Hz signal at IC_{1-2}. In effect, the capstan phase control is removed from the circuit.

Capstan Phase-Loop Troubleshooting.   The most common symptom of problems in the capstan phase loop is a *noise bar* (or bars) floating through the picture but with the picture properly synchronized. (If the capstan speed loop is malfunctioning, you get excessive audio wow and flutter, along with picture instability, out of sync, etc.). If you suspect problems in the capstan phase loop, set the VCR to pause, and check the

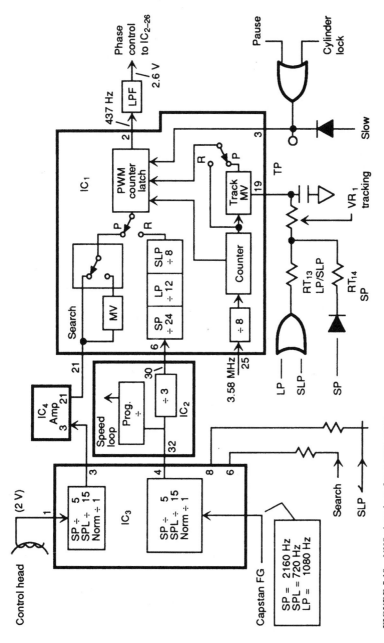

FIGURE 5.19 VHS phase loop for capstan servo.

437-Hz PWM signal at IC_{1-2} for a 50 percent duty cycle. If absent or abnormal, suspect IC_1.

If the PWM signal is good, return the VCR to play, and play back a known-good tape or a test tape. Check for the presence of the control-track signal at IC_{3-1}. If the control-track signal is absent or abnormal, suspect a defective control-track head, connector cable, or possible tape drive-path problem (tape not moving past the control head properly).

If the control-track signal is good, check for a signal at IC_{3-3}. If it is absent or abnormal, suspect IC_3. Then check for a signal at IC_{4-21} and IC_{1-21}. If the signal is absent or abnormal, suspect IC_4. If the signal at IC_{1-21} is good, check the 3.58-MHz reference signal at IC_{1-25}.

If both the comparison signal (IC_{1-21}) and reference signal (IC_{1-25}) are normal but the capstan phase loop appears to have no control of the capstan (noise bars, etc.), check that IC_{1-3} is low (normal phase-loop operation). If IC_{1-3} is low, suspect IC_1. If IC_{1-3} is high (in normal play), check the pause, slow, and cylinder-lock lines from system control.

Capstan Speed-Loop Operation.   As shown in Fig. 5.20, the capstan speed loop develops a correction voltage resulting from a difference in frequency between the capstan signal and the divided-down 3.58-MHz signal. The capstan signal is passed through the programmable divider in IC_3 to IC_{2-32}. IC_3 divides the capstan signal by 5 in the SP search mode and by 15 in the SPL search mode. The capstan signal is not divided by IC_3 in normal playback. The capstan signal is further divided within IC_2 and passed to a digital counter, which is part of the PWM system of IC_2.

The 3.58-MHz reference signal is first divided by 8 in IC_1 and passed to IC_{2-35} as a 477-kHz signal. The divided-down signal is then processed and applied to a counter within IC_2 for comparison with the divided-down capstan signal. The PWM circuit develops a dc output voltage corresponding to the difference in frequency between the capstan signal and the 3.58-MHz reference.

Capstan speed control RT_8 at IC_{2-27} adjusts the level of the dc correction voltage for the speed loop. The correction voltage is combined with the capstan phase-control signal by an adder in IC_2. The combined speed and phase-loop voltages are applied to a variable amplifier. The amplifier gain is determined by the auto-speed-select circuit in IC_2. The output at IC_{2-28} is passed to the capstan motor-drive IC.

During slow motion, the capstan servo speed loop is disabled by a high at IC_{2-31}. The output drive voltage is also pulled up to about 3.5 V by turning on Q_{11}. In still operation, the output drive voltage is pulled to ground by IC_4.

Capstan Speed-Loop Troubleshooting.   The most common symptoms of problems in the capstan speed loop are excessive audio wow and flutter, picture instability (out of sync), or both. The first step is to disable the phase-loop circuits by connecting TP_1 (IC_{1-3}) to B + (usually about 5 V). This disables the phase-loop operation and sets the output of the PWM system to 50 percent. Check for about 2.6 V at IC_{2-26}. If absent or abnormal, suspect a problem with the 50 percent duty cycle square-wave output from the phase loop or in the LPF between phase loop and speed loop (Fig. 5.19).

If the input from the phase loop at IC_{2-26} is normal, monitor the capstan free-run waveform while slowly rotating capstan speed control RT_8. Adjust RT_8 to get the proper sampling pulse and trapezoid lockup. (The service literature describes exact procedures. Follow them.)

If the capstan speed loop can be locked in (picture stable, minimum, or no wow and flutter) by adjustment, with the phase-loop disabled, suspect the phase-loop com-

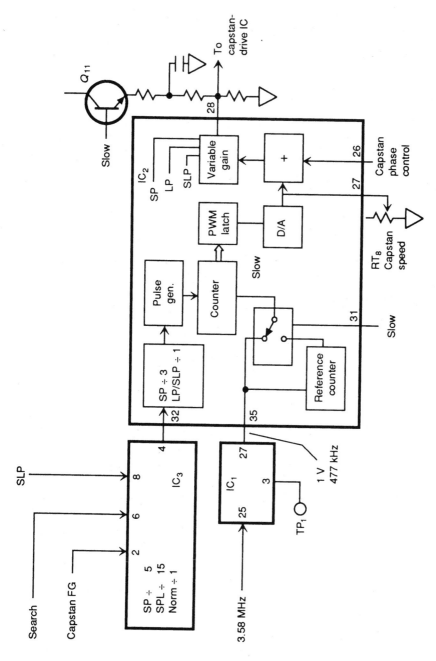

FIGURE 5.20 VHS speed loop for capstan servo.

ponents. If you cannot correct capstan speed-loop problems by adjustment of RT_8, look for the correct inputs to the capstan speed loop at IC_{2-32} and IC_{2-35}.

If the input signals appear normal, suspect IC_2. If the signals are absent or abnormal, look for a 3.58-MHz reference at IC_{1-25} and for a capstan signal at IC_{3-2}. If either signal is absent or abnormal, trace the 3.58-MHz reference line and/or capstan line. If the inputs at IC_1 and IC_3 are good, suspect IC_1 and IC_3.

5.3.11 System-Control Overview

Figure 5.21 shows overall operation for the system-control circuits of a typical VHS VCR. Microprocessor-A IC_1 receives the various input commands from the tuner and remote IC (through a 16-bit serial data bus), the camera and remote input, and/or the front-panel operating buttons. IC_1 also controls the power on-off circuits, which latch the power relay. IC_1 communicates to microprocessor-B IC_2 (through a 4-bit data bus).

Microprocessor-B IC_2 drives the mechanical systems (Secs. 5.3 and 5.10) of the VCR deck and determines the condition of the deck (tape loading, cassette loading, etc.) in all modes of operation. IC_2 also drives the LCD counter-display module and monitors the various trouble sensors (Sec. 5.3.13) which protect the system should a malfunction occur.

IC_1 and IC_2 function together as one microprocessor (and are physically mounted on the same system-control board in this VCR). As discussed, it is very difficult (nearly impossible) to monitor the data communications between IC_1 and IC_2. So the manufacturer generally recommends that both microprocessors be replaced as a package, along with the system-control board. This is a practical approach since both ICs are flat-pack types and very difficult to replace on an individual basis.

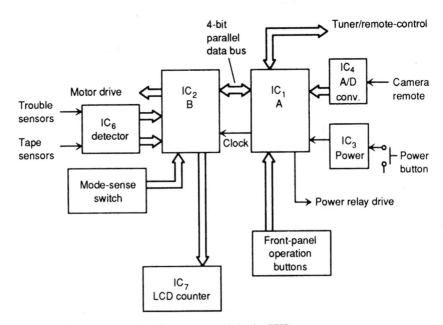

FIGURE 5.21 Overall operation for system-control circuits (VHS).

Even though the microprocessors are replaced as a package, it is still helpful in troubleshooting to know what inputs are applied to which microprocessor and how both microprocessors produce outputs in response to these inputs. The following paragraphs summarize some typical functions that affect the video circuits of a VCR.

5.3.12 System-Control Input Operation

Figure 5.22 shows the system-control input circuits for a typical VHS VCR. The operating mode of the video circuits is selected from three different sources: the front-panel keyboard (push buttons), the tuner and remote IC_{41}, or the camera. Microprocessor-A IC_1 monitors these three input sources.

Keyboard operation occurs when IC_1 generates phase 1 scanning pulses at pin 35. These pulses are amplified by Q_5 and applied to the keyboard switches. IC_1 also generates phase 0 pulses, which are output at pin 34 and amplified by Q_6 to drive the LED displays on the front panel. The return signals from the keyboard and LED displays are routed to eight key-in and display-drive input pins on IC_1.

IC_1 multiplexes the function of keyboard input and display drive. During the time when phase 0 is active, LED-drive operation is being performed. When phase 1 is active, keyboard scanning occurs, and IC_1 monitors for a command or function input.

IC_1 also monitors for input from remote and tuner IC_{41} (over the 16-bit serial data bus and handshaking lines) at pins 44 through 47. The ready 1 (pin 46) and ready 2 (pin 47) lines inform the microprocessors that one or the other is ready to send data. After receipt of the ready signal from the transmitting microprocessor, the receiving microprocessor outputs an acknowledge signal on the acknowledge 1.2 line. The

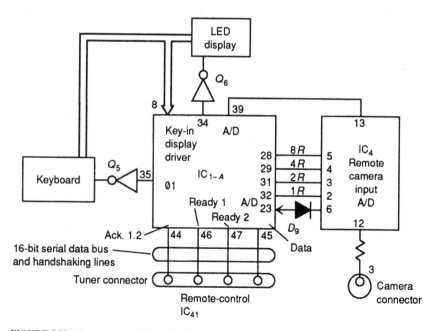

FIGURE 5.22 System-control input circuits (VHS).

transmitting microprocessor then acknowledges this signal and outputs serial data on the data line (pin 45). This sequence occurs for each 16-bit word.

The third source of input is applied through pin 3 of the external camera connector to IC_4 (the camera input A/D converter). IC_1 passes four scanning pulses (pins 28, 29, 31, 32) to IC_4 (pins 2, 3, 4, 5) and monitors the A/D signal at pin 23 (through D_9). The camera connects a predetermined value of resistance from pin 3 to ground for the mode required. A constant-current source in IC_4 is passed through the camera resistance, developing a voltage at pin 3 proportional to the resistance. IC_4 compares this voltage to the voltage developed from the combination of 8R, 4R, 2R, and 1R outputs from IC_1.

IC_1 monitors the signal at pin 23 for an indication of when the voltage developed from the 4-bit signals matches the voltage developed from the external resistor (at pin 3 of the camera connector). The 4-bit signal is then decoded by IC_1 to determine the function requested.

You can see from this that simply monitoring the levels on the 4-bit bus with a scope or probe is of little value in troubleshooting. You need a logic analyzer to monitor the 4-bit coding between IC_1 and IC_4. It is even more difficult to monitor the 16-bit serial coding between IC_1 and IC_{41}. However, there is a way around this troubleshooting problem in any VCR, no matter what data-bus and coding arrangement is used.

System-Control Input Troubleshooting.   If you get good remote operation but no manual operation, the problem is probably in the keyboard circuits or Q_5. (In this particular VCR, you must replace the entire system-control board, increasing your profits, in that event).

If you have good manual operation but no remote, look for signal activity on the ready 1 and ready 2 inputs of IC_1, using a scope or probe. (You should see evidence of rapidly changing pulses on both ready lines if normal communication is being passed between IC_1 and IC_{41}.) If pulse activity is missing, suspect IC_{41}, the remote-control circuits, or the hand-held IR remote unit (Chap. 4).

If you get pulse activity on the ready lines, look for similar activity on the acknowledge 1.2 line. If it is missing, replace the system-control board.

If you get pulse activity on the acknowledge 1.2 line, look for similar activity on the data line. If it is missing, suspect the remote IC_{41} (or IR remote unit).

If you get pulse activity on the data line, try substituting the system-control board. Of course, it is possible that either IC_{41} or the IR remote unit is producing some unintelligible code which cannot be interpreted by IC_1, but this is a long shot.

If you get good manual and remote operation but no camera function, try a different camera. If this is not practical, try substituting various resistors between pin 3 of the camera connector and ground. The resistor values are given in the service literature schematic.

If you do not get the correct mode with a resistor mounted at pin 3, suspect IC_4 or IC_1 (in that order). Either way, you must replace the entire system-control board.

If you do get the desired mode of operation with the correct resistance connected at pin 3 (for example, this VCR should go into slow motion when 15.5 k is placed between pin 3 and ground), look for problems in the camera cable or in the camera circuits (if you feel courageous enough to tackle a video camera repair job).

5.3.13   VCR Trouble Sensors

Figure 5.23 shows the system-control trouble-sensor circuits for a typical VHS VCR. IC_2 monitors the various trouble sensors and detectors: dew sensors and rewind or forward, tape-end reel-rotation, cassette-up, and cylinder-lock detectors.

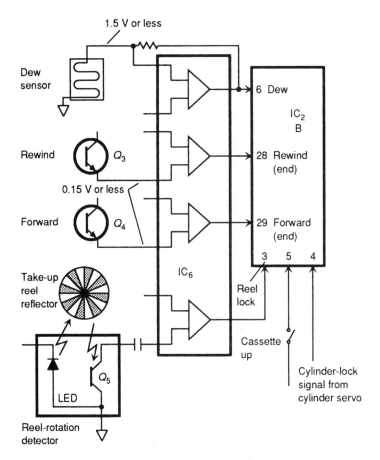

FIGURE 5.23   VHS system-control trouble sensors.

Dew Sensor.   If there is low moisture in the VCR, the resistance of the dew sensor is high. This produces a high at pin 6 of IC_2, which responds by doing nothing. However, if there is considerable moisture within the VCR, the resistance of the dew sensor drops, producing a low at pin 6 of IC_2, and the VCR is placed in the stop mode.

Rewind or Forward Tape-End Detectors.   The rewind tape-end detector Q_3 senses the *infrared signal* that is passed through the clear leader of the VHS tape when the end of the tape is reached. This generates a high at pin 28 of IC_2, which responds by placing the VCR in stop.

Note that most modern VHS VCRs use infrared (IR) end-of-tape sensors and light sources. This is not true of early-model VCRs, which use ordinary light and photodetectors. The use of conventional light can be a problem during service. If you remove the tape or remove some covers that expose the sensors to outside light, the VCR goes into stop, ready or not. So you must cap the sensors. *Do not* remove the sensor lamp or light source. On many VCRs this triggers the VCR into stop. These problems do not exist on Beta VCRs, which use a metal foil at the end of the tape (Chap. 6).

The forward end-of-tape detector Q_4 senses the IR signal passed through the leader at the forward end of tape. This generates a high at pin 29 of IC$_2$, which responds by placing the VCR in stop.

Reel-Rotation Detector. Q_5, a combined LED and light sensor, is located under the takeup reel. A reflector disk with reflective and nonreflective areas is located at the bottom of the takeup reel. As the reel rotates, pulses of light are generated when the LED light is alternately reflected and not reflected onto the light sensor. The light pulses are converted to electrical pulses by the sensor and applied to pin 3 of IC$_2$ through a capacitor and amplifier. If the reel stops rotating, the pulses are removed and the IC places the VCR in stop. (In this VCR, the pulses are divided in IC$_2$ and are also applied to the front-panel tape counter.)

Cassette-Up Switch.   When a cassette is loaded into the holder (front load in this case) and the holder is moved to the down position, a high is applied to pin 5 of IC$_2$. If the cassette-up switch is actuated (even momentarily) for any reason during normal operation (play, rewind, slow, etc.), the high is removed from pin 5 and IC$_2$ places the VCR in stop.

Cylinder-Lock Detector.   The cylinder-lock signal is supplied from the cylinder servo (Fig. 5.17) and applied to pin 4 of IC$_2$. During normal operation, if the cylinder motor speed decreases, the cylinder servo generates a high at pin 4 of IC$_2$, causing IC$_2$ to place the VCR in stop.

Trouble-Sensor Troubleshooting.   If the VCR does not go into play, check the inputs from all trouble sensors. Look for a voltage of less than 1.5 V at the dew sensor and less than 0.15 V at end-of-tape sensors Q_3 and Q_4. Also look for a high at pin 5 of IC$_2$ from the cassette-up switch (make sure that there is a cassette in place).

If any of the trouble-sensor inputs are incorrect (voltages substantially higher than 1.5 for the dew sensor and higher than about 0.15 V for the end-of-tape sensors), trace the signal back to the source. If all signals are normal, replace IC$_2$ (or the board).

Note that if the VCR does not make any attempt to load tape from a cassette (after the cassette is pulled in and down) the reel-rotation detector and cylinder-lock detector are *probably not* at fault. These two trouble sensors place the VCR in stop only after the cassette is loaded, the tape is loaded from the cassette, and the tape starts to move.

If the VCR loads tape from the cassette *and then immediately* unloads the tape and stops, look for pulses from Q_5 (about 1 V) at the time when the tape just starts to move (takeup reel rotating). If the pulses are missing, suspect Q_5, reel-motor problems, or capstan-servo problems. If the pulses are present, check for a cylinder-lock signal at pin 4 of IC$_2$. The cylinder-lock line should be high while the tape is being loaded from the cassette and then go low at completion of the tape load (when the tape starts to move). If not, suspect cylinder-servo problems (Sec. 5.3.9).

5.4 TYPICAL VHS ADJUSTMENTS, CLEANING, LUBRICATION, AND MAINTENANCE

This section describes typical adjustments, cleaning, lubrication, and maintenance procedures for VHS VCRs. Remember that these specific procedures apply directly to specific VCRs. When servicing other VCRs, you must follow the service literature exactly. Using these examples, you should be able to relate the procedures to a similar set of adjustment points on most VCRs (except for the one you just got in today).

5.4.1 Typical Electrical Adjustments

Figure 5.24 shows the waveforms involved for typical VHS electrical adjustments.

Switching-Point Adjustment.   The purpose of this adjustment is to set the video-head switching point to almost the center where the channel 1 and channel 2 envelopes overlap each other during playback. If the adjustment is not correct, the vertical sync signal is degraded and vertical jitter occurs. Also, you may get switching noise bars in the lower part of the picture.

1. Connect the scope to the video out jack and trigger the scope with SW_{30} (switching 30-Hz pulses). Play an NTSC alignment tape. Set the tracking control to center (detent) position.

2. Turn the head-switching adjustment control until the vertical sync signal is 6.5H ±0.5H from the trailing edge (trigger point) of the SW_{30} pulse, as shown in Fig. 5.24a.

Tracking Preset Adjustment.   The purpose of this adjustment is to optimize tracking when playing back a tape recorded by the same VCR. If the adjustment is not correct, noise occurs even when the tracking control is centered, and noise cannot be removed by turning the tracking control.

1. Connect the scope to the video out jack and trigger the scope with control-track pulses. Connect an NTSC color-bar generator to the video in jack. Set the tracking control to center.

2. Record and play back a color-bar signal (using a blank tape) on the same VCR. Use the SP mode for both record and playback.

3. Turn the tracking-preset adjustment control until the control-track pulse and the vertical sync pulse are matched in phase, as shown in Fig. 5.24b.

Slow Tracking Preset Adjustment.   The purpose of this adjustment is to set timing so that a brake pulse is generated for the capstan motor during slow play and to minimize noise. If the adjustment is not correct, noise appears in the slow-motion picture. Both the EP and SP tracking-preset adjustments must be used.

1. Connect a monitor-TV to the video out jack. Connect an NTSC color-bar generator to the video in jack. Center the slow tracking control.

2. Record and play back a color-bar signal (using a blank tape) on the same VCR. Record in the EP mode, but play back in the slow mode. Turn the EP tracking-pre-set adjustment control until the noise is at the bottom of the monitor-TV screen.

3. Now record in the SP mode and play back in the slow mode. Turn the SP tracking-preset control until the noise is at the bottom of the monitor-TV screen.

Y-Output Level Adjustment.   The purpose of this adjustment is to set the luma-signal output to a specified level. If the adjustment is not correct, proper brightness cannot be obtained.

1. Connect a scope to the Y output. Connect an NTSC color-bar generator to the video in jack.

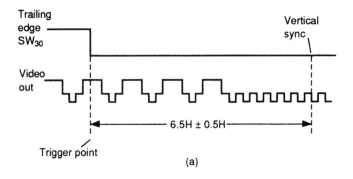

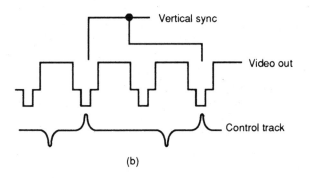

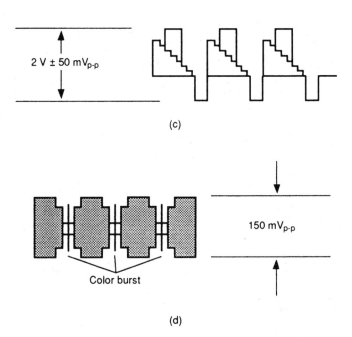

FIGURE 5.24 Waveforms for typical VHS electrical adjustments. (*a*) Switching point. (*b*) Tracking preset. (*c*) *Y*-output level. (*d*) Chroma record level.

2. While receiving a color-bar signal, adjust the amplitude of the Y signal to 2 V, as shown in Fig. 5.24c. Note that this separate Y-output level adjustment applies primarily to VCRs with S-VHS capabilities (Sec. 5.8).

Output Level Adjustment. The purpose of this adjustment is to set the video signal output to a specified level. If the adjustment is not correct, the proper brightness cannot be obtained.

1. Connect a scope to the video out jack. Connect an NTSC color-bar generator to the video in jack.
2. While receiving a color-bar signal, adjust the amplitude of the video signal to 1 V. If the VCR is capable of S-VHS operation, this adjustment should be made in the S-VHS mode.

Chroma Record-Level Adjustment. The purpose of this adjustment is to set the chroma record level to an optimum value. If the adjustment is not correct, the color is poor and diamond-shaped beats may appear in the picture.

1. Connect a scope to the test point that monitors the playback FM (such as at VR_4 in Fig. 5.9). Connect an NTSC color-bar generator to the video in jack.
2. While receiving a color-bar signal, adjust the amplitude of the color burst to 150 mV (peak to peak) as shown in Fig. 5.24d.

Additional Electrical Adjustments. There are many more electrical adjustments required for VHS VCRs, both old and new. Fortunately, most service literature describes the adjustments for the specific VCR fully (often in boring detail). Section 5.6 describes trouble symptoms related to adjustment.

5.4.2 Typical Mechanical Adjustments

Figures 5.25 through 5.29 show the components involved for typical VHS mechanical adjustments. These adjustments apply primarily to the video tape-transport mechanism. The tape-transport system is the path from the supply reel passing through the video heads to the takeup reel (Fig. 5.15a). The transport-system parts, especially the parts which come into direct contact with the video tape, should be kept clean (without scratches, dust, oil, etc.).

The tape-transport system is adjusted before the VCR is shipped from the factory. When parts are replaced in the tape transport, the recommended adjustments should be performed (or at least checked) as described in the service literature. However, if there is no trouble with the VCR, the author recommends (strongly) that you do not make any adjustment to the mechanism. (The expression "If it ain't broke, don't fix it" was invented for VCR tape transports.)

Reel-Disk Height Adjustment. The purpose of this adjustment is to set the reels of the cassette to a specified height, thus determining the height of the tape. If the adjustment is not correct, the video tape does not make proper contact with the heads (video, audio, control track, etc.).

1. Remove power from the VCR, and remove the cassette loading mechanism. Mount the master plane and place the reel-disk height jig on the plane (Fig. 5.7), as shown in Fig. 5.25a.

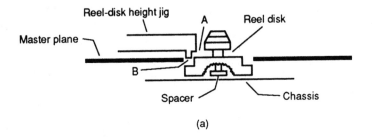

(a)

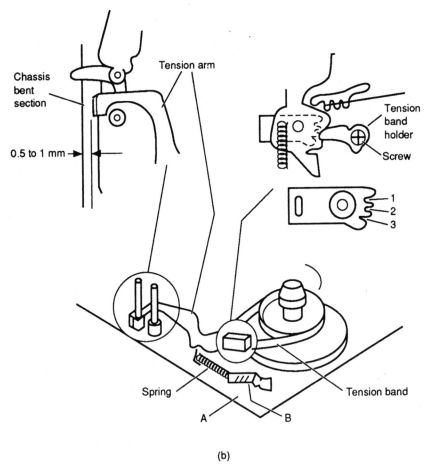

(b)

FIGURE 5.25   VHS reel-disk height, tension-pole position, and tension adjustments.

2. Check that the reel-disk top is between sections A and B of the reel-disk height jig. If not, add or remove spacers in the reel disk.

3. If the reel-disk is removed to add or replace spacers, the tension arm and tension band must also be removed (on virtually all VHS VCRs). Both the arm and band must be adjusted when replaced, as described next.

Tension Pole-Position and Tension Adjustment. The purpose of this adjustment is to make the tape tension constant so that the contact between video heads and tape is stable throughout the entire tape run. If the adjustment is not correct, the tape tension can change as the supply reel gets smaller and the takeup reel gets larger.

1. Set the VCR to the loading state without inserting a cassette. Generally, this involves removing power, disabling the power supply and takeup end sensors (Sec. 5.3.13), and then reapplying power. Always follow the service literature instructions.

2. Check the position of the tension pole as shown in Fig. 5.25b. The gap between the tension pole and the chassis should be 0.5 to 1 mm. If not, loosen the tension-band retaining screw, insert the tension-band holder into one of the grooves (1, 2, or 3), and set the gap. Tighten the retaining screw and recheck the tension-pole position.

3. Install the tension meter (Fig. 5.7) in the cassette holder. (The tension meter is actually a cassette.) Set the VCR to play, and check that the tension meter reads 34 to 44 g/cm. If not, move the spring in direction A or B as required. If the tension-meter reading is higher than 44 g/cm, move the spring in direction A (to decrease tension on the spring as shown in Fig. 5.25b). If the reading is lower than 34 g/cm, move the spring toward B. If it is necessary to change the tension more than about 6 g/cm, recheck the tension-pole position as described in step 2. Readjust the tension-pole position if necessary.

Guide-Pole Height Adjustment. The purpose of this adjustment is to regulate tape height. If the adjustment is not correct, the video tape does not make good contact with the heads.

1. Remove power from the VCR, and remove the cassette loading mechanism. Mount the master plane and place the reel-disk height jig on the plane (Fig. 5.7), as shown in Fig. 5.26a. (This is the same jig used for reel-disk height adjustment but is now used for adjustment of the nuts on the supply and takeup guide poles.)

2. Set the clearance between the bottom of the guide-pole upper flange and the top of the reel-disk jig to 0 to 0.2 mm as shown in Fig. 5.26a.

3. Load and play a blank cassette. Check that the tape does not ride over the upper and lower flanges of the guide pole. If the tape rides over the upper flange, turn the height-adjustment nut counterclockwise. If the tape rides over the lower flange, turn the nut clockwise.

Guide-Roller Height Adjustment. The purpose of this adjustment is to regulate the tape height so that the bottom of the tape runs along the *tape-guide line on the cylinder.* If this adjustment is not correct, the tape travel (and picture) becomes erratic.

1. Remove power from the VCR, and remove the cassette loading mechanism. Mount the master plane and place the reel-disk height jig on the plane (Fig. 5.7), as shown in Fig. 5.26b. (Again, the same reel-disk jig is used but for adjustment of the screws on the supply and takeup guide rollers.)

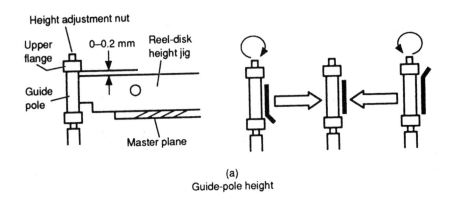

(a)
Guide-pole height

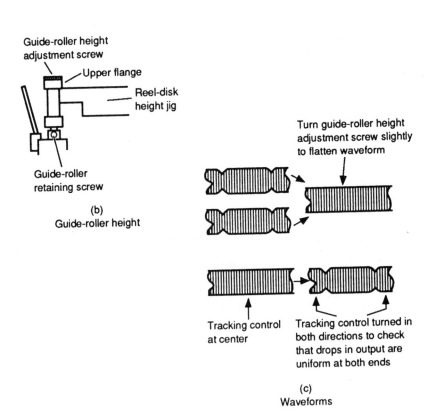

(b)
Guide-roller height

(c)
Waveforms

FIGURE 5.26 VHS guide-pole height and guide-roller height adjustments.

2. Loosen the guide-roller retaining screw (so that the guide roller does not turn during loading, unloading, and play). Align the bottom of the guide-roller upper flange and the top of the reel-disk height jig.

3. Install an alignment tape (30 HMP, Fig. 5.7) in the cassette holder. Connect a scope to monitor the playback FM (channel 1) and 30-Hz switching pulses (channel 2). Trigger the scope with channel 2. The waveforms should appear as shown in Fig. 5.26c, when the alignment tape is played.

4. Set the tracking control to center (detent) position. (If this adjustment is performed after a cylinder is replaced, set the tracking control so that the FM output is maximum.)

5. Turn the guide-roller height adjustment screw to flatten the waveforms as shown. Then move the tracking control left and right to check that the drops in FM output are uniform at the start and end. When the FM output is flat and drops are uniform, tighten the guide-roller retaining screw (using the hex wrench, Fig. 5.7).

Audio and Control (A/C) Head Adjustment. The purpose of this adjustment is to keep contact between head and tape even so that the same track is recorded and played back. If this adjustment is not correct, the picture becomes erratic.

1. Remove power from the VCR, and remove the cassette loading mechanism. Mount the master plane and place the reel-disk height jig on the plane (Fig. 5.7), as shown in Fig. 5.27.

2. Check that the spring section of the A/C head-retaining screw protrudes 6.3 mm over the top of the head base (1), as shown in Fig. 5.27a. If not, turn the A/C head-retaining screw as necessary to get the correct distance.

3. Check that head bases (1) and (2) are parallel, as shown in Fig. 5.27b. If not, turn the tilt-adjustment and azimuth-adjustment screws as necessary until the head bases are parallel.

4. Check that the clearance between the master plane and head base (1) is about 1.25 mm, as shown in Fig. 5.27c. If not, turn the height-adjustment nut as necessary to get the correct distance.

5. Remove the master plane and adjustment jigs, load a blank tape, and set the VCR to play.

6. Check that there is no excessive curling and/or overriding of the tape around the A/C head. If necessary, turn the tilt, azimuth, and height adjustments (but not the retaining screw) to eliminate curling. The ideal height of the A/C head is where the bottom edge of the tape is 0.1 to 0.15 mm from the bottom edge of the A/C head, as shown in Fig. 5.27d.

7. Remove the blank tape and install a tape with a 7-kHz segment. Connect a scope to the audio out jack, and check the 7-kHz waveform, as shown in Fig. 5.27e. If necessary, turn the tilt, azimuth, and height adjustments (but not the retaining screw) until the audio output is maximum and flat, as shown. This should require only slight adjustment (and possibly no adjustment).

X-Value Adjustment. The purpose of this adjustment is to make the VCR compatible with tapes recorded on other VCRs. If the X-value adjustment is not correct, the VCR can have interchange problems, discussed in Sec. 5.5.9. Always perform the tracking-preset adjustment (Fig. 5.24b) before making the X-value adjustment. Note that X-value is important for all VCRs but is critical for the S-VHS VCRs described in Sec. 5.8.

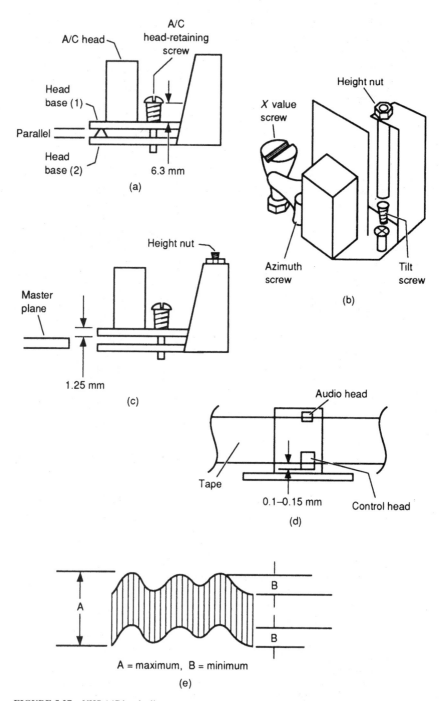

FIGURE 5.27 VHS A/C head adjustments.

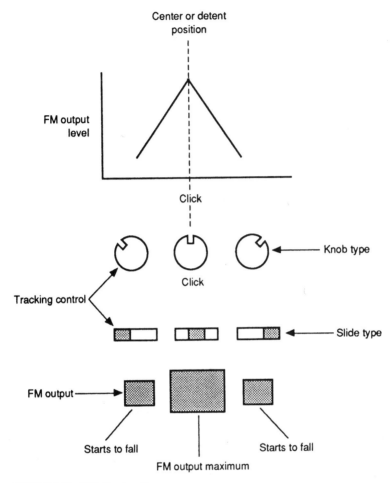

FIGURE 5.28   VHS *X*-value adjustments.

1. Install an alignment tape (30 HMP, Fig. 5.7) in the cassette holder. Connect a scope to monitor the playback FM (channel 1) and 30-Hz switching pulses (channel 2). Trigger the scope with channel 2. The FM waveforms should appear as shown in Fig. 5.28, when the alignment tape is played.

2. Set the tracking control to center (detent) position. (If this adjustment is performed after a cylinder is replaced, set the tracking control so that the FM output is maximum.)

3. Check that the FM output is maximum when the tracking control is at the center position. If so, the *X*-value adjustment is good. If not, check in which direction the FM output is maximum when the tracking control is moved from center. If FM is maximum when tracking is to the right, center the tracking control, and turn the *X*-value screw counterclockwise (until the FM is maximum with tracking centered). If FM is maximum with tracking to the left, center tracking, and turn the *X*-value screw clockwise until FM is maximum.

Adjustments after Replacing Cylinder (Video Heads). The purpose of this adjustment is to correct any change in tape height, or change in X-value, after replacing the cylinder. The procedure is essentially a repeat of the guide-roller, switching-point, tracking-preset, and X-value checks. If the VCR performs properly, without further adjustment, after the cylinder has been removed and reinstalled for any reason (usually for video-head replacement), count yourself as lucky. If any one of the checks does not produce the correct indications, after cylinder replacement, *repeat all the adjustments.*

Tension and Torque Checks. The purpose of these checks is to measure tension, torque, and compression force at the tape-travel components. The checks can be performed at any time but should be performed if the tape travel appears not to be smooth or if tape speed is abnormal.

1. Connect the torque gauge and torque-gauge adapter (Fig. 5.7) to the corresponding reel-disk as shown in Fig. 5.29.
2. Set the VCR to each operating mode, without inserting a cassette.
3. Measure all of the values shown in Fig. 5.29.
4. If any of the values are not within tolerance, try correcting the problem with adjustments (particularly the tension pole-position and tension adjustments, Fig. 5.25).

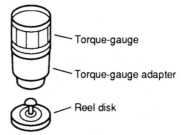

Torque measured	Operating mode	Measured reel	Measured value
Main brake	Stop [1]	Supply & takeup	170 g/cm or more
Slack removal	Unloading	Supply	90–230 g/cm
Fast forward	Fast forward	Takeup	400 g/cm or more
Rewind	Rewind	Supply	400 g/cm or more
Takeup	Play	Takeup	80–170 g/cm
Back tension	Fast forward	Supply	4–25 g/cm
	Rewind	Takeup	

[1] Value measured when VCR is shifted in the unloading direction from fast forward or rewind, and quick breaking is applied to both disks.

FIGURE 5.29 VHS tension and torque checks.

5.4.3 Mechanical-Section Checks and Maintenance Schedules

Schedules for maintenance and inspection of any section in a VCR are not fixed. One reason for this is that the need for maintenance varies greatly according to the way in which the customer uses the VCR. Equally important, most customers regard a VCR in the same way as they do any other appliance. That is, the customer brings in the VCR when it breaks down (a rule that is observed religiously).

As a general guideline, you should get a good picture if the VCR (especially the mechanical section) is inspected, cleaned, and lubricated after each 1000 hours of operation. Thus, if the customer uses the VCR about 1 hour a day, the VCR should be brought in for service once every 3 years. If use is increased to 3 hours a day (who counts?), maintenance is required every year. You can pass this along to your customers (who will promptly ignore everything you say). However, it is always wise to ask the customer how long it has been since the VCR was in for service (if you believe the customer).

Preliminary Checks.   Before you launch into a full maintenance routine, it is a good idea to make certain preliminary checks as to lubrication and cleanliness of the mechanical section. These checks, together with information from the customer regarding possible trouble symptoms, can give you a good basis for troubleshooting the mechanism.

Figure 5.30 shows the location of major points to be checked, as well as cleaning and lubrication points for the mechanism. The following describes the relationship between dirty or worn parts and trouble symptoms:

If there is jitter (*vertical or horizontal*) in the picture, look for dirt on the video heads or at any point in the tape path.

If there is no color or the S/N ratio is poor (noisy picture), look for dirt on the video heads. If the symptom persists after cleaning, look for worn video heads.

If there are color beats in the picture, look for dirt on the full-erase head.

If the audio volume is low or the sound is distorted, look for dirt on the A/C head. If the symptom persists, look for a worn A/C head.

If fast-forward or rewind is slow, look for dirt on the reel belt (if any). Again, replace the belt if there is any sign of stretching or distortion.

Maintenance Tools and Kits.   Always check the service literature for recommended tools and kits to perform mechanical-section inspection, cleaning, and lubrication. As a guideline, you should have a head-cleaning kit, a VCR oil kit, alcohol (or Freon), and gauze.

Methyl alcohol does the best cleaning job but can be a health hazard (especially if you drink it). Isopropyl alcohol is usually satisfactory for most VCR mechanism cleaning (don't drink that either).

Although no special tools are required for routine maintenance (generally), many special tools are required for mechanical adjustment and test procedures, as well as for replacement of mechanical components (including tools for disassembly, reassembly, and after-service mechanical adjustment). Always check the service literature for recommended tools. As a minimum, you will probably need a *screwdriver* to check the *X*-value and *lock paint* to fix adjustment screws when adjustment is complete.

Top view

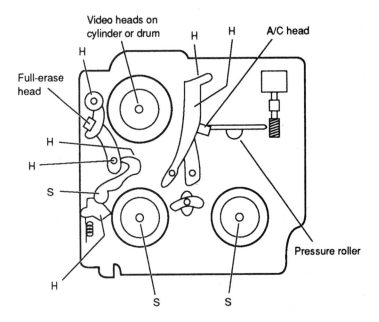

Bottom view

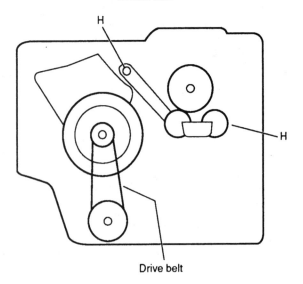

FIGURE 5.30 Location of major points to be checked, cleaned, and lubricated on VHS mechanism.

5.4.4 Mechanical-Section Cleaning and Lubrication

The following paragraphs describe typical VCR mechanical-section cleaning and lubrication procedures. Compare these procedures to those you find in the service literature for the VCR you are servicing.

As a guideline, never clean or lubricate any part not recommended in the service literature. Most VCR mechanisms use many sealed bearings that do not require either cleaning or lubrication. A drop of oil in the wrong places can cause problems, even possible damage. Clean off any excess or spilled oil using gauze soaked in alcohol.

In the absence of specific recommendations, use a light machine oil (such as sewing-machine oil) and a medium grease or lubricating paste. *Lubricate the rotary portions* of the mechanism (tape transport) with one or two drops of oil. *Apply grease to the sliding portions* of the mechanism. *Be very careful to keep oil or grease from getting on belts, flywheels, or pinch rollers.* Oil or grease on these surfaces can cause slippage (resulting in erratic tape drive and poor picture and audio problems).

Cleaning Video Heads. Some manufacturers recommend that the video heads (and the entire tape path) be cleaned first with a cleaning tape or cleaning cassette (Sec. 5.2.2). If dirt is too stubborn to remove with a cleaning tape, use a head-cleaning kit. Other manufacturers recommend using the head-cleaning kit first (or as the *only* cleaning method).

Figure 5.31 shows the procedures for cleaning the video heads. It is assumed that you are using a head-cleaning kit with a wand or cleaning stick. However, you can usually get by with Q-tips (although they are often too small and tend to tear easily).

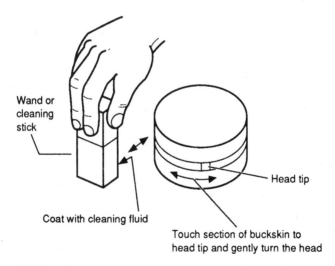

Wand or cleaning stick

Coat with cleaning fluid

Head tip

Touch section of buckskin to head tip and gently turn the head

FIGURE 5.31   Cleaning video heads.

Coat the cleaning stick with cleaning fluid (probably alcohol) as shown in Fig. 5.31. Touch the stick to the head tip and gently turn the head (by rotating the cylinder) to the right and left. *Do not move the stick vertically (or across the tape path).* Make sure that only the buckskin on the stick comes into contact with the head. Otherwise, the head may be damaged. You can also damage the video heads with Q-tips if the head comes into contact with the plastic or wooden stick.

Cleaning the Tape Transport and Tape Drive.   The drive system consists of those parts that run the tape (reels, capstan). The tape transport is any part that comes into contact with the running tape, including the full-erase and A/C heads (and capstan).

Unless otherwise directed by the service literature, clean the cylinder surface and each surface where the tape passes (except the full-erase and A/C heads) with a soft cloth moistened with alcohol or cleaner. When cleaning the cylinder surface, be careful not to touch the video heads with the cleaning cloth. If necessary, rotate the drum to move the heads away from the spot to be cleaned.

To clean the full-erase and A/C heads, moisten the cleaning stick with alcohol or cleaner, press the stick (gently) against each head surface, and clean the heads by moving the stick in the same direction as the tape path (never across the path).

Lubrication Guidelines.   In the absence of specific instructions, here are some very general guidelines for lubrication of any VCR mechanism. Figure 5.30 shows all of the points that normally require either oil or grease (in a VHS mechanism). The points marked with an S require Sonic Slidas Oil (#1600), while the points with an H require Hitazol (M0138) grease. Pan motor oil (X10W40) is also recommended for high-speed rotating sections. Likewise, Froil (GB-TS-1) grease can be used to lubricate metal or molded sections under a light load.

Use the oiler (supplied with many VCR lubrication kits) to apply one or two drops of oil. Again, make sure not to use too much oil. The oil may spill over or leak out and come in contact with rotating parts. This can cause slippage and other problems. If too much oil is applied, wipe the part clean with alcohol or Freon. Apply oil at the locations shown in Fig. 5.30 every 1000 hours (again, who counts?).

Apply grease with a stick or brush (usually supplied with the kit). *Never* apply grease with the video-head cleaning stick. Do not use excess grease. Just as in the case with oil, if excess grease comes into contact with the tape-transport or drive systems, wipe off any excess and clean with gauze dipped in alcohol. Apply grease at the locations shown in Fig. 5.30 every 5000 hours (sure you will).

Finally, whether inspecting, adjusting, cleaning, oiling, or greasing the mechanism, do not touch the tape-transport or drive system with screwdriver tips, hacksaw blades, or soldering tools. Likewise, never apply force to any part of the mechanism that might cause damage or even slight deforming. To do so can bring even the best VCR to an untimely end.

5.5 BASIC VCR TROUBLESHOOTING PROCEDURES

The notes in this section summarize practical suggestions for troubleshooting all types of VCRs. They review some simple, obvious, but often overlooked steps to be performed before you rip into the VCR with pick axe and dynamite. These basic steps involve such things as checking for proper connections, adjusting the TV, operating the controls in the proper sequence, and so on. The section concludes with video-circuit (chroma and luma) troubleshooting approaches. Section 5.6 describes trouble symptoms that are related directly to improper adjustment.

5.5.1 Basic VCR Diagnostic Procedure

If the video playback (*TV picture*) *is bad,* set the program-select switch on the VCR to TV. Check picture quality for each TV channel (using the TV channel selector). If the picture quality is still bad, check for defective antenna connections. For example, the antenna (or cable broadcast) may be defective, the VHF and UHF connections may be reversed, the F-type connector plug may be improperly connected, the center wire in the coax cable may be broken, or the TV 75/300 switch may be in the wrong position. Also check the TV fine tuning.

If the TV picture is good when the program-select switch is set to TV but video playback is not good, set the program-select switch to VCR, tune the TV to the inactive channel (3 or 4), and check reception on each channel using the VCR channel selector. If picture quality is bad or there is no picture on all channels, it is possible that the TV fine tuning is not properly adjusted. If the problem appears only on certain channels, the VCR fine tuning is suspect, as is the VCR tuner.

If picture quality is good only when viewing a TV broadcast through the VCR (*E-E operation*), it is possible that the video heads are dirty (head gaps are slightly clogged). If there is sound but no picture, the video-head gaps may be badly clogged.

If the playback is unstable with a new TV set (never previously used with the VCR), it is possible that the AFC circuits of the TV are not compatible with the VCR. This problem is discussed further in Sec. 5.5.3.

5.5.2 Operational Checklist

The following checklist describes symptoms and possible causes for some very basic VCR troubles (often reported to the technician by the customer over the telephone).

Record button cannot be pressed: Check that there is a cassette installed and that the safety tab has not been removed from the cassette. If necessary, cover the tab hole with tape.

No E-E picture, no picture and sound: Check that the VCR program-select switch is in the correct position. Check the fine tuning on the TV inactive channel.

No color or very poor color: If there is no color on playback, check the fine tuning on the TV inactive channel. Note that if the VCR fine tuning is bad during record, color may appear normal while recording but may not be good during playback. Always check fine tuning of both the VCR and TV as a first step when there are color problems.

Playback picture is unstable: If you have periodic problems with picture instability, before tearing into the VCR, check the following points: Has the VCR been operated in an area having a different line frequency? While recording, is it possible that a fringe-area signal was weak (intermittently) so that the sync signal was not properly recorded? During record, could there have been some interference or large fluctuations in the power-supply voltage? Could the cassette tape be defective? Could the tracking control have been improperly adjusted?

Snow appears on the picture during playback only: Check the tracking control.

Sound but no picture and excessive black and white snow noise: Check for very dirty video heads.

Upper part of picture is twisted or entire picture is unstable: The time constant of the AFC circuits in the TV is not compatible with the VCR. Refer to Sec. 5.5.3.

Tape stops during rewind: Is the memory counter switch on? If the memory switch is on, the tape stops automatically at 9999 or 0000 during rewind (on most VCRs).

Rewind and fast-forward buttons cannot be operated: Is the cassette tape at either end of its travel? If the tape is at the beginning, rewind does not function. Fast-forward does not function if the tape is at the end.

Cassette will not eject: Is the power on?

Acoustic feedback (whistle-like sound) when recording on video tape with camera and microphone: Keep the microphone away from the TV. Turn down the TV volume.

Noise band in playback picture or picture unstable, with a too-high or too-low pitched sound: In some VCRs, the tape is automatically locked to the correct speed by the servo. However, other VCRs also require some form of manual switching (to select SP, LP, and EP or SLP). Always check the operating controls for such a possibility, especially when you get sound that is *consistently* too high or too low in pitch.

5.5.3 TV AFC Compatibility Problems

If the AFC circuits of the TV are not compatible with the VCR, *skewing* may result. Generally, the terms *skew* and *skewing* are applied when the upper part of the repro- duced picture is bent or distorted by *incorrect back tension on the tape.* However, the same effect can be produced when the time constant of the TV AFC circuits cannot follow the VCR playback output.

This AFC-incompatibility problem is *very rare* in present-day TV sets (which are designed with VCRs and video-disc players in mind) and appears in less than 1 per- cent of older TV sets (and almost never when the VCR and TV are produced by the same manufacturer). So, do not change the TV AFC circuits unless you are absolutely certain that there is a problem. Try the VCR with a different TV. Then try the TV with a different VCR. Then check for proper tension on the tape (Secs. 5.4 and 5.6).

5.5.4 Tuner and RF-Unit Problems

When a VCR is first connected to a TV, it is likely that the unused channel (3 or 4) of the TV is not properly fine tuned. When fine tuning the TV, operate the VCR in the playback mode using a known-good cassette, preferably with a good color program.

Note that if you try to fine tune the TV in the record or E-E mode, both the VCR and TV tuners are connected in the circuit, and the picture is affected by either or both tuners. With playback, the picture depends only on the TV tuner. Once the normally unused channel of the TV is fine tuned for the best picture, the VCR tuner can be fine tuned as necessary.

Replacing the RF Unit.   In virtually all VCRs, the RF unit (also known as the RF modulator or RF converter) must be replaced in the event of failure. No adjustment is possible, and internal parts cannot be replaced on an individual basis. This is because the RF unit is essentially a miniature TV transmitter and must be type-accepted using very specialized test equipment, as are other transmitters. You must replace the RF unit as a package if you suspect failure. As an example, if you have found proper audio and video inputs (and power source) to the RF unit but there is no output (or low output) at the unused channel, the problem is likely to be in the RF unit.

As a point of reference, a typical RF unit produces 1 V into a 75-ohm load (or 2 V into a 300-ohm load) on the selected channel. Signals outside the channel frequency by more than 3.5 MHz are reduced at least 35 dB.

Replacing the Tuner. In many VCRs, the tuner is also replaced as a unit in the event of failure. However, some manufacturers supply replacement parts for tuners. Also, some manufacturers provide for tuner adjustment as part of service. As a point of reference, a typical VCR tuner (including the IF) produces 1 V peak to peak of video into a 75-ohm load. Typically, audio output from the tuner is in the −10- to −20-dB range.

5.5.5   Copy Problems

It is possible to copy a video tape using two VCRs. One VCR plays the cassette to be copied, while the other makes the copy. Keep two points in mind when making such copies. First, if the cassette being copied contains any copyrighted material, you may be doing something illegal. Second, a copy is never as good as the original, and copies of copies are usually terrible. Even with professional recording and copying equipment, the quality of a copy (particularly the color) deteriorates with each copying.

The quality of a first copy (second generation) can be acceptable, provided that the original is of very good quality. However, a second copy (third generation) is probably of unacceptable quality. Forget fourth generation (or beyond) copies. So if you are called in to service a VCR that "will not make good copies of other cassettes" patiently explain that the problem probably has no cure.

5.5.6   Video Camera Sync and Interlace Problems

If you are to service a VCR that operates properly in all modes except when used with a known-good video camera, the problem may be one of incompatibility. The cameras recommended for use with a VCR (generally of the same manufacturer) should certainly produce good cassette recordings. In general, most cameras designed for use with VCRs, even though of a different manufacturer, are compatible with any VCR. Such cameras are usually designated as having a 2:1 interlace.

Essentially, a 2:1 interlace means that both the vertical and horizontal sync circuits of the camera are locked to the same frequency source (possibly the power line) by a definite ratio. When operating a camera, the sync signal (normally supplied by the TV broadcast) is obtained from the camera and recorded on the control track of the VCR tape.

Some older (and inexpensive) cameras, particularly those used in surveillance work, have a *random interlace* where the horizontal and vertical syncs are not locked together. The playback of a recording made with a random interlace camera often produces a *strong beat pattern* (*herringbone effect*).

One way to confirm a random-interlace condition is to watch the playback while observing the last horizontal line above the vertical blanking bar. Operate the TV vertical hold control as necessary to roll the picture so that the blanking bar is visible. If the end of the last horizontal line is stationary, the camera has 2:1 interlace and should be compatible. If the last horizontal line is moving on a camera playback, the camera is not providing the necessary sync and probably has random interlace.

5.5.7 Wow and Flutter Problems

VCRs are subject to wow and flutter, as are most audio recorders. Wow and flutter are tape-transport speed fluctuations that may cause a regularly occurring instability in the picture and a quivering or wavering effect in the sound during record and playback. The longer fluctuations (below about 3 Hz) are called wow; shorter fluctuations (typically 3 to 20 Hz) are called flutter.

Wow and flutter can be caused by mechanical problems in the tape transport or by the servo system. Wow and flutter are almost always present in all VCRs, but it is only when they go beyond a certain tolerance that wow and flutter are objectionable.

If you are to service a VCR where the complaint appears to be excessive wow and flutter, first check the actual amount. This is done using the low-frequency tone recorded on an alignment tape and a frequency counter connected to the audio line at some convenient point. Typically, the low-frequency tone is about 333 Hz, and an acceptable tolerance is 0.03 percent. If necessary, operate the frequency counter in the period mode to increase resolution.

5.5.8 System-Control Problems

It is difficult to generalize about system-control problems. In most modern VCRs, system-control is performed by microprocessors. In the simplest of terms, you press front-panel buttons to initiate a given operating mode, and the microprocessors produce the necessary control signals (to operate relays, motors, and so on). Each VCR has unique system-control functions. You must learn these functions.

Automatic-Stop Functions.   Virtually all VCRs have automatic stop functions operated by the system-control microprocessor (Sec. 5.3). These functions, such as end-of-tape stop, dew sensor, and so on, must be accounted for in service. For example, when checking any system control, make certain that all the automatic-stop functions are capable of working.

Equally important, make certain that an automatic function has not worked at the wrong time (end-of-tape stop has occurred in the middle of the tape). Also, make certain that a normal automatic stop is not the cause of an imaginary trouble. (Do not expect the tape to keep moving when the end-of-tape stop has occurred.)

It is often necessary to disable or override the automatic-stop functions during service. The following notes describe some generalized procedures for checking and testing system control:

Slack-tape sensors can be checked by visual inspection and by pressing on the tape with your fingers to simulate slack tape. If the slack-tape sensor includes a microswitch (not usually on VHS but often on Beta), the sensor circuit can be disabled by forcing a match or cardboard against the sensor to keep the microswitch from triggering.

End-of-tape sensors can be checked by simulating the end-of-tape condition. For VHS, this involves exposing a photosensor to light (to simulate the clear plastic tape leader) to trigger automatic stop. To disable the sensor, cover the photosensor with opaque tape or a cap. Do not remove the light source for the end-of-tape sensor on a VHS machine, as discussed in Sec. 5.3.13. Note that even with the photosensor covered, stray light may trigger the auto-stop condition (unless the sensor is IR).

If the VCR has a switch that is actuated when a cassette is in place, locate the mechanism that actuates the switch and hold the mechanism in place with tape. In many (but not all) cases, it is possible to operate the VCR through all modes without a cassette installed, if the cassette switch can be actuated manually.

Takeup reel detectors can be checked by holding the takeup reel. This causes the takeup reel clutch to slip (to prevent damage), but the detector senses that the reel is not rotating and produces automatic stop.

Always check that all automatic-stop functions work and that all bypasses and simulations are removed after any service work.

5.5.9 Interchange Problems

When a VCR can play back its own recording with good quality but playback of tapes recorded on other machines (or commercial tapes) is poor, the VCR is said to have interchange problems. Such problems are almost always located in the mechanical section, usually in the tape path. Often interchange problems are the result of improper adjustment. Manufacturers sometimes include interchange adjustments as part of the overall electrical and mechanical adjustments.

The simplest way to make interchange adjustments is to monitor the output from the video heads during playback and adjust elements of the tape path to produce a maximum, uniform output from a factory alignment tape. On some VCRs, this is called the X-value adjustment (Sec. 5.4.2). Generally, the output is measured at a point after head switching so that all heads are monitored. Always follow the manufacturer's adjustment procedures exactly. This is especially true of S-VHS VCRs (Sec. 5.8), which must be adjusted more precisely than a conventional VHS VCR.

5.5.10 Servo Problems

Total failures in the servo system are usually easy to find. If a servo motor fails to operate, check that the power is applied to the motor at the appropriate time. If power is present but the motor does not operate, the motor is at fault (burned out or open windings, and so on). If the power is absent, trace the power back to the source. (Is the microprocessor delivering the necessary control signal or power to the relay or IC, and so on?)

The problem is not so easy to locate when the servo fails to lock on either (or both) record or playback or locks up at the wrong time (causing the heads to mistrack even slightly). Obviously, if the control signal is not recorded (or is improperly recorded) on the control track during record, the servo cannot lock properly during playback. So your first step is to see if the servo can play back a properly recorded tape.

Servo Failure Symptoms.   There are usually some obvious symptoms when the servo is not locking properly. (There is a horizontal band of noise that moves vertically through the picture if the servo is out of sync during playback.) The picture may appear normal at times, possibly leading you to think that you have an intermittent condition. However, with a true out-of-sync condition, the noise band appears regularly, sometimes covering the entire picture. Remember that the out-of-sync condition during playback can be the result of servo failure or the fact that the sync signals (control signals) are not properly recorded on the tape control track.

To find out if the servo is capable of locking properly, play back a known-good

tape. If the playback is out of sync with a good tape, you definitely have a servo problem. The symptoms for failure of the servo to lock during record are about the same as
during playback, with one major exception. During record, the head-switching point
(which appears as a break in the horizontal noise band) appears to move vertically
through the picture in a random fashion.

Checking Servo Lock with Fluorescent Light.   Another way to check to see if the
servo is locking on either record or playback involves looking at some point on the
rotating scanner or video-head assembly under fluorescent light. When the servo is
locked, the fluorescent light produces a (blurred) pattern on the rotating scanner that
appears almost stationary. When the servo is unlocked, the pattern appears to spin.
Note that this check is not possible on all VCRs. In many cases, there is no point on
the scanner that will reflect the fluorescent light.

Try checking the scanner of a known-good VCR under fluorescent light. Stop and
start the VCR in the record mode. Note that the blurred pattern appears to spin when
the scanner first starts but settles down to almost stationary when the servo locks.
Repeat this several times until you become familiar with the appearance of a locked
and unlocked servo under fluorescent light.

Analyzing Servo Failure Symptoms.   Once you have studied the symptoms and
checked the servo playback with a known-good tape, you can use the results to localize the trouble in the servo. For example, if the servo remains locked during playback
of a known-good tape, it is reasonable to assume that the circuits between the control
head and servo motors are good.

In early-model VCRs, such circuits include the playback amplifiers, tracking delay
network, sample-and-hold circuits, power amplifier, servo motors and brakes, feedback pulses, and ramp generators. In modern VCRs, the servo circuits include both
speed and phase ICs, where it is generally more difficult to localize trouble.

Using Adjustments to Find Servo Failures.   Servo troubles may be mechanical or
electrical and may be the result of either improper adjustment or component failure or
both. As a general guideline, if you suspect a servo problem, start by making the electrical adjustments that apply to the servo (using the service literature procedures).

Simply making the electrical adjustments for the servo system may cure the problem. If not, the adjustment procedure tells you (at the very least) if all the servo-control signals (such as cylinder and capstan control pulses) are available at the appropriate points in the circuits. If one or more of the signals are found to be missing or
abnormal during adjustment, you have an excellent starting point for troubleshooting.

Solving Servo-Lockup Problems.   Two points are sometimes overlooked when troubleshooting a servo that fails to lock up. First, the free-running speed of the servo may
be so far from normal that the servo simply cannot lock up. This problem usually
shows up during adjustment.

Second, on those VCRs that use rubber belts to drive servo motors (typically the
early-model units rather than modern VCRs), the rubber may have stretched (or be
otherwise damaged). If you have replacement belts available, compare the used VCR
belts for size and conformation. Hold a new and used belt on your finger under no
strain. If the used belt is larger and does not conform to the new belt, install the new
belt and recheck the servo for proper lockup.

5.5.11   Basic Luma-Video Troubleshooting

The first step in troubleshooting luma circuits is to play back a known-good tape or an alignment tape. (This is not a bad idea when troubleshooting any VCR circuit.) The tape playback identifies the problem as playback or record or both. Next, run through the electrical adjustments. Keep the following points in mind when checking performance and making adjustments:

If the playback has excessive snow (electrical noise), try adjustment of the tracking control. Mistracking can cause snow noise. Next, try cleaning the video heads before making any extensive adjustments. (Cleaning the video heads clears up about 50 percent of all noise or snow problems.) Remember that snow noise can result from mistracking caused by a mechanical problem. For example, if there is any misadjustment in the tape path, snow can result. So if you have an excessive noise problem that cannot be corrected by tracking adjustment, head cleaning, or electrical adjustment, try mechanical adjustments, starting with the tape path.

If playback from a known-good tape has poor resolution (picture lacks sharpness), look for an improperly adjusted noise canceler and for bad response in the video-head preamps. When making adjustments, study the stair-step or color-bar signals for any transients at the leading edges of the white bars. Note that in many modern VCRs the video-head circuits are in ICs and are not adjustable.

If playback of a known-good tape produces smudges on the leading edge of the white parts of a test pattern (from an alignment tape) or on a picture, the problem is often in the preamps or in the adjustments that match the heads to the preamps. The head and preamp combination is not reproducing the high end (near 5 MHz) of the video heads. The adjustment procedures (if any) usually show the head and preamp characteristics.

If you get a herringbone (*beat pattern*) *in the playback of a known-good tape,* look for carrier leak. There is probably some unbalance condition in the FM demodulators or limiters, allowing the original carrier to pass through the demodulation process. If a very excessive carrier passes through the demodulators, you may get a negative picture (blacks are white and vice versa). Recheck all carrier-leak adjustments (if any).

Most luma adjustment procedures include a check of the video output level (typically 1 V peak to peak). If the VCR produces the correct output level when playing back an alignment tape but not from a tape recorded on the VCR, you may have a problem in the record circuits. As an example, the record current may be low. (One common symptom of low record current is *snow or excessive noise.*) Another common problem in the record circuits is the *white-clip adjustment.* Some luma adjustment procedures include both a white-clip adjustment and record-current check.

To sum up luma-video troubleshooting, if you play back an alignment tape or a known-good tape and follow this with head cleaning and a check of the recommended alignment procedures, you should have no difficulty in locating most black and white picture problems.

5.5.12 Basic Chroma-Video Troubleshooting

As in the case of luma-video circuits, the color circuits of a VCR are very complex but not necessarily difficult to troubleshoot (nor do the circuits fail as frequently as the mechanical section). Again, the first step is to play back an alignment tape, followed by a check of all adjustments pertaining to color. As with the luma circuits, when performing the adjustments, you are tracing the signals through the color circuits.

There are two points to remember when making the checks. First, most color circuits are contained within ICs, possibly the same ICs as the luma circuits. Similarly, the color and luma circuits are interrelated. If you find correct inputs and power to an IC but an absent or abnormal output, you must replace the IC. A possible exception in the color circuits are the various filters and traps located outside the IC on many VCRs.

Second, in most VCRs, the fixed input to the color converters comes from the same source for both record and playback (from crystal-controlled oscillators). If you get good color on playback but not on record, the problem is definitely in the record circuits. However, if you get no color on playback of a known-good tape, the problem can be in either the color-playback circuits or in the common fixed-signal source. So a good place to start color-circuit signal tracing is to check any common-source signal. For example, check any AFC circuit signals (629 or 688 kHz) and any APC signals (3.57 or 3.58 MHz). If any of these signals are absent or abnormal, the color will be absent or abnormal.

The following notes describe some typical VCR color-circuit failure symptoms, together with some possible causes:

If you get a "barber pole" effect, indicating a loss of color lock, the AFC circuits are probably at fault. Check that the AFC circuit is receiving the H-sync pulses and that the VCO is nearly on frequency, even without the correction circuit.

If the hue or tint control of the TV must be reset when playing back a tape that has just been recorded, check the color subcarrier frequency (using a frequency counter).

If you get bands of color several lines wide on saturated colors (such as alternate blue and magenta bands on the magenta bar of a color-bar signal), check the APC circuits as well as the 3.58-MHz oscillator frequency.

If you get the herringbone (beat) pattern during color playback, try turning the color control of the TV down to produce a black and white picture. If the herringbone pattern is removed on black and white but reappears when the color control is turned back up, look for leakage in both the color and luma circuits. For example, there could be a carrier leak from the FM luma section beating with the color signals, or there could be leakage of the 4.2-MHz reference signal into the output video.

If you get severe flickering of the color during playback, you could be losing color on every other field. (This is a problem with Beta but not usually VHS.) This can occur if the phase of the 4.2-MHz reference signal is not shifted 180° at the H-sync rate when one head is making a pass. The opposite head works normally, making the picture appear at a 30-Hz rate. (In VHS, this problem is usually the result of one bad head.)

If you get minor flickering of the color during playback, look for problems in the ACC system. It is also possible that one video head is bad (or going bad) or that the preamps are not balanced, but such conditions show up as a problem in black and white operation.

If you lose color after a noticeable dropout, look for problems in the burst-ID circuit (Figs. 5.13 and 5.14). It is possible that the phase-reversal circuits have locked up on the wrong mode after a dropout. In that case, the color signals have the wrong phase relation from line to line, and the comb filter is canceling all color signals. Check both inputs (3.58-MHz input from the reference oscillator and the video-input signal) applied to the burst-ID circuit. If the two signals are present, check that the burst-ID pulses are applied to the switching flip-flops (FFs).

Again, remember that all of the color-circuit functions discussed here may be contained within one or two ICs and cannot be checked individually. So you must check inputs, outputs, and power sources to the IC and then may end up replacing the IC.

5.6 TROUBLE SYMPTOMS RELATED TO ADJUSTMENT

The following notes describe trouble symptoms that can be caused by improper adjustment. Remember that these same symptoms can be caused by circuit failures other than adjustment.

Vertical sync signal is degraded and vertical jitter occurs. Switching noise occurs in the lower part of the picture: This problem is often the result of improper switching-point adjustment. The video-head switching point must be set to almost the center where the channel 1 and channel 2 envelopes overlap each other during playback (Fig. 5.10). The basic procedure involves playing an alignment tape, with the tracking control centered and adjusting the head-switching control until the vertical sync signal occurs 6.5-H ± 0.5H from the trailing edge of the SW_{30}-Hz pulse (Fig. 5.24).

Noise occurs even with the tracking control centered. Noise cannot be removed by turning the tracking control: This problem is usually caused by improper tracking-preset adjustment. Tracking must be optimized when playing back a tape recorded by the VCR. The basic procedure involves recording and playing back a color-bar signal (on the same VCR), with the tracking control centered, and adjusting the preset control to match the phase of the control-track pulse and the vertical sync signal.

Noise appears in the slow-motion picture (if any): This problem (only on VCRs with slow-motion display) is usually caused by improper slow-tracking preset adjustment. It is necessary to adjust the timing to generate a brake pulse for the capstan motor during slow-motion play (on most VHS VCRs) and to minimize noise during slow motion (which is not always easy to see). The basic procedure involves recording and playing back a color-bar signal (on the same VCR), with the slow-tracking control centered, and adjusting the EP and SP controls to move the noise to the bottom of the screen.

Noise occurs when switching between normal play and digital play: This problem (only on VCRs with digital functions such as picture-in-picture, Chap. 4) is usually caused by improper *fsc adjustment.* It is necessary to synchronize the digital-memory timing with the servo. The basic procedure involves stopping the VCR and adjusting the fsc control until the digital memory and servo signals are synchronized.

Correct brightness cannot be obtained: Assuming that the monitor TV is properly adjusted, this problem is usually caused by improper adjustment of the VCR *luma or video-output levels.* Both signals must be set to the correct level. The basic procedure involves receiving a color-bar signal from the line input and adjusting the luma and chroma controls to provide the correct level. Typical levels are 2 V ± 50 mV$_{p-p}$ for luma and 1 V ± 0.2 V$_{p-p}$ for chroma.

Moire pattern occurs on TV screen. Colors smear: On an S-VHS VCR (Sec. 5.8), this problem is usually caused by improper adjustment of the *Y/C* separation filter or the chroma filter. It is necessary to separate the luma (*Y*) signal from the chroma (*C*) signal. The basic procedure involves receiving a color-bar signal from the line input and adjusting the *Y/C* and chroma filters to minimize chroma components in the luma signal (and vice versa). Typically, there is a variable resistance and variable inductance in both filters.

Brightness is abnormal in an S-VHS VCR: Assuming that the monitor TV is properly adjusted, this problem is usually caused by improper adjustment of the luma level. It is necessary to align the luma level of the input signal with that passed through the chroma filter. Generally, this should be done before the chroma filter is adjusted. The basic procedure involves receiving a color-bar signal from the line input and adjusting the luma signal to the same level as the chroma signal.

Frequency-synthesis tuning is abnormal. No sync-signal detection in the FS circuits: On some VCRs, this problem is caused by improper free-run adjustment of the 15.7-kHz sync in the FS circuits. It is necessary to set the free-run frequency applied to the sync-detection circuit in the channel-select circuits to 15.7 kHz. The basic procedure involves operating the VCR in stop and adjusting the free-run sync to 15.73 ± 0.1 kHz. (This is sometimes called the AFC adjustment.)

Coloring is poor. Beats in picture: If you get a combination of poor color and diamond-shaped beats (with a good monitor or TV), the problem can be improper adjustment of the chroma record level. First try playing back a known-good tape. If this playback is good but color is poor (and there are beats in the picture) when a tape is recorded and played back on the same VCR, try adjusting the chroma record level. The basic procedure involves receiving a color-bar signal from the line input and adjusting the chroma record level to a given value. A typical level for the chroma color burst is 150 mV$_{p-p}$ (when measured at a test point in the chroma record circuits).

The on-screen display (OSD) characters appear to flow and do not stand still (Chap. 4):   This problem is often the result of improper AFC or horizontal sync adjustment. The procedure is essentially the same as for setting the free-run frequency of the FS sync. The basic procedure involves operating the VCR in stop and adjusting the free-run AFC or horizontal sync to 15.73 ± 0.1 kHz.

The OSD-display position is not correct: If the OSD display is in the wrong position but not moving or flowing, the problem is most likely one of improper character-position adjustment. It is necessary to set the positions of the characters displayed on the screen. The basic procedure involves receiving a color-bar signal from the line input, operating the controls (often on the VCR remote unit) to display the tape-counter display on screen, and adjusting the character-position control until one of the zeros in the tape-counter display is aligned at the desired position.

Picture displayed in the line following head switching: This problem, often called *skew error* or *skew,* is often the result of improper back-tension adjustment.

Tape stretched, broken, loose in cassette, or spilled from cassette: These problems can often be corrected by proper cleaning of the brakes. However, if the symptoms remain after cleaning, it may be necessary to replace the brakes.

Tape creased or frilled at the edges: This problem can usually be corrected by tape-guide or tape-path adjustments.

No color, no color sync, or improper tint or hue: Assuming a known-good monitor TV (properly fine tuned if necessary), these problems are usually the result of an improperly adjusted subcarrier (3.58-MHz) oscillator but can also be caused by chroma carrier-level adjustments.

Horizontal instability and/or picture noise: This problem can be the result of improper cylinder motor-speed adjustment.

Vertical instability and/or picture noise: This problem can be the result of improper capstan motor-speed adjustment.

Horizontal and vertical instability: This problem can be the result of improper 30-Hz reference-frequency adjustment.

Vertical jitter (head-switching noise):   This problem can be the result of improper PG shifter adjustment.

Noise in picture but picture stable: This problem is usually the result of improper tracking-preset adjustment but can also be caused by the user tracking control.

Picture overloaded or washed out: This problem can be the result of improper record luma-level adjustment.

5.7 VIDEO HEAD CONFIGURATIONS AND SWITCHING CIRCUITS

This section describes the various configurations for video heads and the related switching circuits. As discussed, modern VCRs use four-, five-, and six-head configurations (and even more heads) instead of the two-head systems common in early-model (and less expensive) VCRs.

There are three general configurations (but many minor variations) for the four- and five-head models: four head with 90° spacing, the 4X system, and five head. The six-head models are essentially four-head models with two additional heads for audio hi-fi (which is recorded as FM in addition to the AM audio recorded on the A/C head, as discussed in Sec. 5.8).

5.7.1 Concepts

The terms *head switching* and *head selection* are often interchanged in VCR literature. In this book, head switching refers to switching between two heads in a pair or set of heads, typically at a 30-Hz rate, using signals from the drum or cylinder servo. Head selection refers to selection of a particular pair or set of heads to accommodate a given tape speed or playing time, such as SP, LP, and SLP (also known as 2H, 4H, and 6H). Head selection is controlled by relays (or analog switches) and is used only on VCRs with more than two video heads.

Basic Head Switching.   To prevent crosstalk between adjacent recorded video tracks during playback, VCRs use video heads with dissimilar azimuths ($\pm 7°$ for Beta, $\pm 6°$

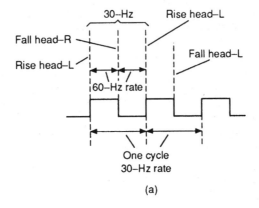

(a)

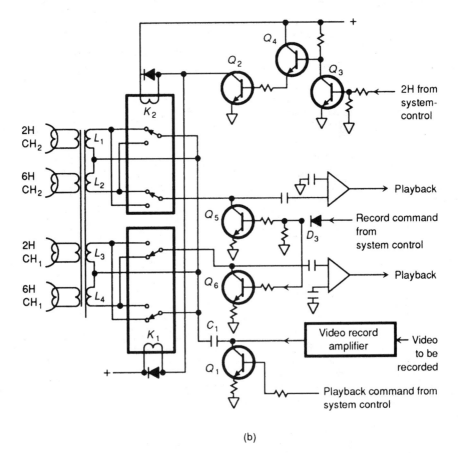

(b)

FIGURE 5.32   Basic four-head selection and switching (VHS).

for VHS). In addition to dissimilar azimuths, tracking errors are corrected by the capstan phase-control system, using the control pulses (CTL) placed on the tape during record.

In addition to keeping the heads scanning the track with the right azimuth, it is necessary that each head scan the entire recorded track and maintain consistency between tracks. This is the job of the drum phase-control system, which provides a highly stable reference frequency, phase-locked to the vertical sync frequency (and appearing at the vertical frame rate of the TV, or 30 Hz).

As shown in Fig. 5.32a, one complete drum or cylinder cycle, occurring at a 30-Hz rate, consists of a rise and fall of the pulse. Note that head L (left) is selected by the rise of each pulse, and head R (right) is selected by the fall of each pulse. So each head is selected alternately during each complete cycle, at twice the frame rate, or 60 Hz (which is the TV vertical field rate). Viewed individually, each head is selected only once for each cycle. If consideration is given only to heads R or L, selection occurs at a 30-Hz rate. As a result, the head-switching process yields one vertical field per head, or one complete vertical frame, when the head pairs are combined.

Basic Head Selection. Early-model VHS VCRs use two video heads, each with identical gap widths (typically 30 μm). Later two-head VCRs evolved to the *dissimilar-head concept.* Not only are the azimuths different, but each head has a *different gap width* (typically 26 and 31 or 28 and 32 μm). This makes it possible to record in the 6H mode (special long play, extra long play, etc.) without guard bands and to have still or freeze-frame pictures (sometimes referred to as special effects or trick play).

Unfortunately, still pictures in 2H are a problem using any two-head system. If tape movement is stopped on a 2H recording (with guard bands), the heads typically scan on a track of information (with an output) or a guard band (without an output), producing noise in the picture. The problem of noisy still pictures in 2H is overcome by the four-head concept.

Basic Four-Head Selection Circuit. Figure 5.32b shows the video-head selection circuits for a typical four-head VCR. Since all four heads are not used simultaneously, a means is provided to select a pair of heads (to be used at a particular speed).

2H-Mode Head Selection. In 2H, Q_3 is turned on by a 2H command from system control. This disables Q_2 and Q_4 and deenergizes K_1 and K_2, closing the K_2 lower contacts and the K_1 upper contacts. Head windings L_2 and L_4 are shorted, the 6H heads are cut off, and only the 2H heads are used.

In 2H playback, the returns for coils L_1 and L_3 are provided through C_1 and Q_1 (turned on by a playback command from system control). The outputs of the 2H heads are applied through L_1 and L_3 to the channel 1 and channel 2 amplifiers of IC_1.

In 2H record, Q_1 is turned off by the absence of a playback command. This permits the output of the record amplifier to be applied through C_1 and L_1 and L_3 to the 2H heads. One side of the L_1 and L_3 coils is returned to ground through Q_5 and Q_6, which are turned on when a record command from the system control is applied through D_3.

5.7.2 Four-Head VCR with 90° Spacing

Figure 5.33 shows the video-head circuits and configuration for a four-head VCR with 90° head spacing. The channel 1 and channel 2 heads are used during normal playback and record in all three speeds. In trick play (still, double speed, search) a second pair of heads (both channel 2′) are used. The bottom edge of the channel 2′ heads uses the

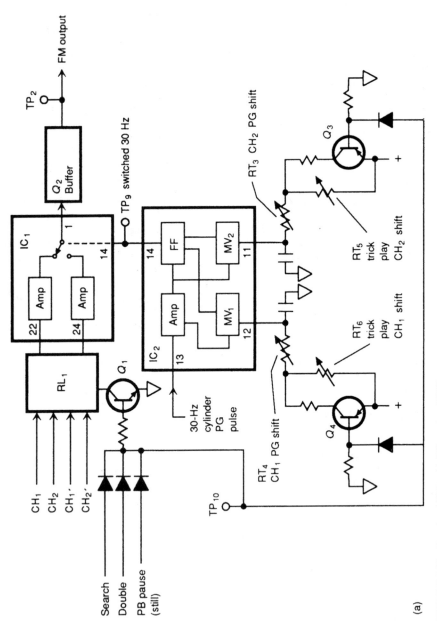

FIGURE 5.33 Four-head VCR with 90° spacing (VHS).

(a)

5.66

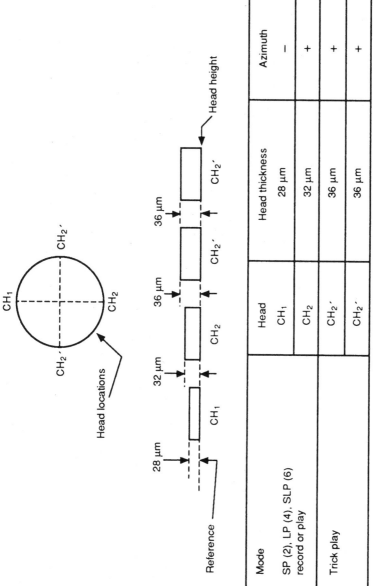

Mode	Head	Head thickness	Azimuth
SP (2), LP (4), SLP (6) record or play	CH₁	28 μm	−
	CH₂	32 μm	+
Trick play	CH₂′	36 μm	+
	CH₂′	36 μm	+

(b)

FIGURE 5.33 (*Continued*) Four-head VCR with 90° spacing (VHS).

5.67

same reference as the normal channel 1 and 2 heads. This concept is different from other video-head arrangements where the *center of the heads* is used as a reference.

Head-switching relay RL_1 is energized by turning on Q_1 during trick play with a high from system control (for search, double speed, still, or pause). With RL_1 turned on, the channel 2′ signal is passed to pins 22 and 24 of IC_1. The switched 30-Hz signal at $IC_{1\text{-}14}$ (produced by IC_2) causes IC_1 to switch between the channel 2′ heads. The $IC_{1\text{-}14}$ signal determines which FM head signal is amplified and applied to buffer Q_2.

The 30-Hz head-switching signal at $IC_{2\text{-}14}$ is developed by the cylinder-motor generator input at $IC_{2\text{-}13}$. The cylinder-motor input is processed by two multivibrators in IC_2. The time constant of the multivibrators is controlled by the shifter controls at pins 11 and 12 of IC_2. These controls set the *delay time* that occurs between cylinder pulses (sometimes called the drum FF pulses) and the rising and falling edges of the 30-Hz head-switching signal (Fig. 5.32a).

During normal playback, the inputs of Q_3 and Q_4 are low, turning on both transistors. This applies power directly to the shifter controls RT_3 and RT_4, allowing for adjustment of the delay time to accommodate normal playback.

During any one of the three trick-play modes, Q_3 and Q_4 are turned off by a high at the inputs. This allows trick-play shifter controls RT_5 and RT_6 to be connected in series with the existing normal-playback controls RT_3 and RT_4. The additional resistance changes the time constant of each multivibrator, causing a relative time shift of 90° to occur in the 30-Hz head-switching signal. With the combination of RL_1 being energized and the 30-Hz switching signal being shifted, the FM output at TP_2 contains the signal from the appropriate pair of video heads (for the particular mode of operation selected).

Four-Head VCR with 90° Spacing Troubleshooting. *If you get noisy video playback,* check for proper logic signals applied at Q_1, Q_3, and Q_4. The logic should be low for normal play and high for trick play. If the signals are absent or abnormal, suspect system control. If the signals are correct, check that Q_1, Q_3, and Q_4 are responding properly. (Q_1 should turn on and Q_3 and Q_4 should turn off during trick play.) If it becomes necessary to replace Q_3 and Q_4, readjust all four shifter controls as described in the service literature.

If you get no video playback or the playback appears choppy, check for an FM signal at TP_2 and a 30-Hz signal at $IC_{2\text{-}13}$ and $IC_{2\text{-}14}$ (or at TP_9 if more convenient). If the 30-Hz head-switching signal is absent or abnormal at $IC_{2\text{-}13}$, suspect the cylinder servo. If the head-switching signal is good at $IC_{2\text{-}13}$ but not at $IC_{2\text{-}14}$ (or TP_9), suspect IC_2. If you get good head-switching signals at $IC_{1\text{-}14}$ but the FM is absent or abnormal at $IC_{1\text{-}1}$, suspect IC_1. If you get good FM at $IC_{1\text{-}1}$ but not at TP_2, suspect buffer Q_2.

You can also check for signals from the heads at $IC_{1\text{-}22}$ and $IC_{1\text{-}24}$. However, video-head signals are usually low (in any VCR) and difficult to measure. If you get any inputs from RL_1 to IC_1 and the inputs are equal in amplitude, the heads and RL_1 are *probably good.*

5.7.3 4X Video-Head System

Figure 5.34a shows the head configuration for the 4X system. Note that four heads are used and that spacing between individual heads from an active head pair is 180°. However, an *adjacent-spacing* configuration is used in 4X.

In 4X, additional video heads R and L are positioned adjacent to primary video heads R′ and L′. The *adjacent heads* are located one horizontal scan line apart (about 12°). The letter designation of L or R denotes left or right azimuth. The prime notation (after the letter designation) denotes the head pair used primarily in the 2H mode.

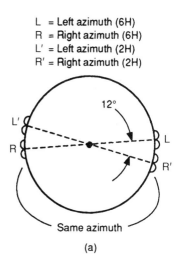

L = Left azimuth (6H)
R = Right azimuth (6H)
L' = Left azimuth (2H)
R' = Right azimuth (2H)

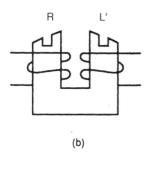

(a)

(b)

Mode → Speed ↓	PB/REC	Speed search	Still	Slow
2H	R' L'	R' L' R L	L' L	L' L R'
6H/4H	R L	R L	R R'	R R' L

(c)

FIGURE 5.34 A 4X video-head system (VHS).

Only two heads are used during normal playback and record at a given speed. Since the two heads of each pair retain conventional 180° spacing, head-switching techniques described thus far remain essentially unchanged. The R′ and L′ video heads are used exclusively for 2H, while R and L heads are used for both 4H and 6H.

Because of the close proximity of the heads (12°), adjacent heads are contained in the same head assembly, as shown in Fig. 5.34*b*. Note that the gap widths of the 2H heads (R′ and L′) are smaller than the gap widths used in many other four-head systems. With smaller gap widths, some guard bands are produced in 2H. However, the 4X system overcomes the potential noise problem by a unique combination of the four heads, as shown in Fig. 5.34*c*.

In play and record at the 2H speed, only the 2H heads are used; at 4H and 6H speeds only the 6H heads are used. *Speed search* (SS) uses all four heads, while 6H speed search uses only two heads (since there are no guard bands to make noise). In *slow motion,* three heads are used. In *still,* two heads are used, but both heads scan the same recorded track (or vertical field), as in the system in Sec. 5.7.2 (90°). This minimizes both horizontal and vertical jitter.

Figure 5.34*d* shows the head-selection and switching circuits for a VCR using the

$R = 32\,\mu m$ (6H)
$L = 30\,\mu m$ (6H)
$R' = 32\,\mu m$ (2H)
$L' = 45\,\mu m$ (2H)

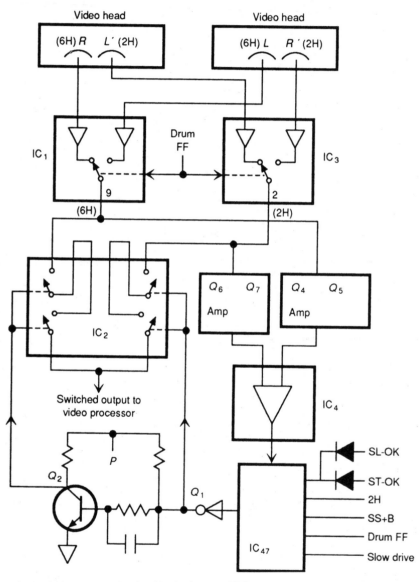

FIGURE 5.34   (*Continued*) A 4X video-head system (VHS).

4X system. Although the circuits are far more complex, troubleshooting for the 4X system is essentially the same as for the four-head system with 90° spacing. That is, you check for proper control logic and head-switching signals and monitor the FM video output to the video-processing circuits.

If the logic-circuit signals are absent or abnormal, suspect the system control. If the logic signals are good, look for problems in gate array IC_{47} and Q_1 and Q_2. If head-switching signals are not correct, suspect the drum servo.

If the logic and head-switching signals are all good but there is considerable video noise, suspect IC_2. If the video is choppy (as monitored at IC_{1-9} and/or IC_{3-2}), suspect IC_1 and IC_3. Again, it is difficult to monitor the outputs directly from the video heads.

If you have problems (video noise, bad picture) only in the 2H mode, suspect Q_4 and Q_5, Q_6 and Q_7, IC_4, and (of course) the 2H heads.

5.7.4 Five-Head System

The fifth head in a five-head system is used primarily for slow motion or still (freeze-frame) operation. Figure 5.35a shows the slow or still playback head-selection circuit for a five-head VCR, while Fig. 5.35b shows the relationship of the heads.

The cylinder has five heads: two SP heads, two LP and SLP heads, and a fifth still-field head. The channel 1 and channel 2 SP heads are used in normal playback and record in the SP mode. In LP and SLP, the channel 1 LP and SLP head and the channel 2 LP and SLP head are used for normal playback and record. The SP heads are 75 μm thick, while the LP and SLP pair of heads are 28 μm thick. The fifth still-field head is 32 μm thick.

The fifth still-field head is located one horizontal line away from the channel 1 LP and SLP head, with the same azimuth as the channel 2 LP and SLP head. During still operation, the tape mechanism stops the tape so the channel 2 LP and SLP head and the still-field head play back the *same video track*. By using the two video heads at about 180° apart and at the same azimuth, playback of two identical fields per frame is obtained.

Five-Head Slow or Still Troubleshooting. If you get normal playback at all speeds *but noise in the still-field mode,* first check that IC_{5-8} is high. If not, check the capstan servo. If IC_{5-8} is high, check IC_{5-1} for still-field playback signals. (Note that these signals are low in amplitude and may be difficult to measure. However, you should be able to detect the presence of signals with a sensitive scope.) Compare the signals at IC_{5-1} (and IC_{4-6}) with those at IC_{4-4}. If the signals at IC_{5-1} are absent or abnormal, suspect IC_5 or the still-field head.

If the signals at IC_{5-1} are normal, check for FM at TP_1 and a 30-Hz head-switching signal at IC_{4-1}. If the 30-Hz signal is absent or abnormal, suspect the cylinder servo. If you get good head-switching signals at IC_{4-1} but the FM is absent or abnormal at IC_{4-15}, suspect IC_4. If you get good FM at IC_{4-15} but not at TP_1, suspect Q_2 or Q_9.

If you get normal playback at all speeds *but noise in the slow-motion mode,* first check for noise in the still field as described. If the still field is good, check that IC_{5-8} is pulsed low by the signal from the capstan servo. If the low signal is absent or abnormal, suspect the capstan servo. If the signal at IC_{5-8} is normal but there is noise in slow motion, suspect IC_5.

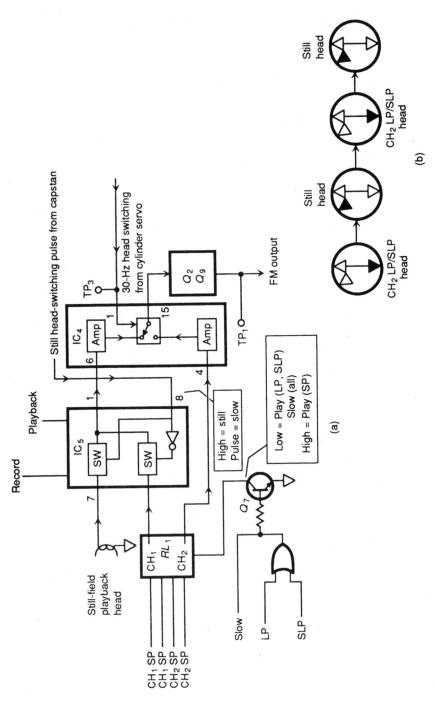

FIGURE 5.35 Five-head system (VHS).

5.72

5.8 S-VHS CIRCUITS

This section is devoted to the basics of S-VHS and hi-fi VCRs. Such units are similar to (and partially compatible with) conventional VCRs. However, there are major differences between standard VCRs and the S-VHS and hi-fi equipment described in this section.

Note that virtually all S-VHS VCRs are also hi-fi. However, the reverse is not necessarily true. Many hi-fi VCRs do not have the S-VHS feature. With hi-fi VCR, the audio is recorded and played back through stereo heads located on the same cylinder as the video heads and in FM (in addition to the conventional audio record and playback in AM by the A/C head). In VHS, separate heads are used for audio and video. In Beta, the same heads are used for audio and video.

5.8.1 S-VHS VCRs

With S-VHS (super-VHS), the VCR is capable of delivering over 400 horizontal lines of resolution in either SP or SLP playback. A conventional VHS produces 240 horizontal lines. The improved resolution with S-VHS is produced by raising the top end (the sync tip) of the video recording frequency to 5.4 MHz and increasing the video-frequency deviation to 1.6 MHz, as shown in Fig. 5.36a.

The S-VHS video spectrum shown in Fig. 5.36a increases the luma bandwidth by 0.6 MHz and eliminates much of the interference with the familiar 629-kHz chroma (used in both S-VHS and VHS).

A conventional TV-broadcast signal has a horizontal resolution of about 330 lines, and a conventional TV set has a similar resolution. Thus, S-VHS improves the picture playback quality by about 90 horizontal lines of resolution (provided a good TV or monitor is used).

To get the full benefit from an S-VHS VCR, you must have (1) a high-resolution video source, (2) an S-Video connector on the monitor, and (3) an S-VHS tape. The high-resolution source can be either an S-VHS camcorder or a prerecorded S-VHS tape.

The S-Video connector, such as that shown in Fig. 5.36b, is actually a pair of four-pin connectors that keep the luma and chroma signals separate. (Note the pair of four-pin connectors shown passing signals to and from the luma and chroma switching circuit in Fig. 5.37.)

S-VHS tape uses a very fine-grained iron oxide and was developed to get the maximum high-resolution recording and playback offered by the S-VHS system.

A typical S-VHS VCR can operate in either the VHS or S-VHS mode, depending on the status of an S-VHS *cassette switch* in the mechanical section and the S-VHS *auto switch* on the front panel.

As shown in Fig. 5.36c, the S-VHS tape is the same size as the VHS tape. However, the S-VHS tape has an ID *hole* that causes the S-VHS cassette switch to remain closed when an S-VHS tape is inserted in the VCR. When a normal VHS tape (without the ID hole) is inserted in the VCR, the cassette switch is opened, and the VCR is placed in the VHS mode. During record, the S-VHS auto switch can override the cassette switch. For example, when an S-VHS tape is inserted into the VCR and the auto switch is set to off, the VCR records in the VHS mode.

Since the sync-tip frequency and deviation for S-VHS differ from standard VHS, the two are not totally compatible in all playback applications. For example:

1. A VHS tape can be recorded in VHS and played back in VHS on an S-VHS VCR.

S-VHS spectrum

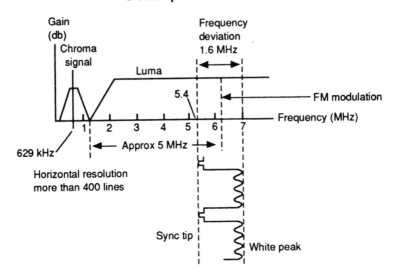

Convential VHS spectrum

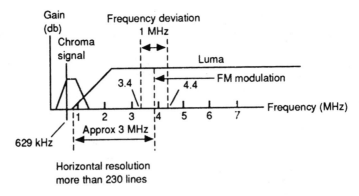

FIGURE 5.36   S-VHS spectrum, connectors, and cassette ID hole.

2. A VHS prerecorded tape can be played back in VHS on an S-VHS VCR.

3. An S-VHS tape can be recorded in VHS and played back in VHS on a VHS VCR.

4. An S-VHS prerecorded tape cannot be played back on a VHS VCR.

Note that the connectors shown in Figs. 5.36b and 5.37 are called the S-connectors or possibly the S Y/C connectors in some literature. No matter what the S-VHS con-

In – S-Video – Out

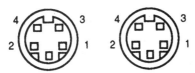

Pin 1 = Luma ground
Pin 2 = Chroma ground
Pin 3 = Luma in/out (1 V_{p-p}, 75 Ω, unbalanced, negative sync)
Pin 4 = Chroma in/out (burst, 0.286 V_{p-p}, 75 Ω, unbalanced)

(b)

S-VHS cassette
(Bottom view)

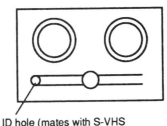

ID hole (mates with S-VHS
cassette switch in mechanical section)

(c)

FIGURE 5.36 (*Continued*) S-VHS spectrum, connectors, and cassette ID hole.

nectors are called, the luma signal at the connectors (either input to or output from the VCR) is a 1-V_{p-p} signal (with negative sync). The chroma signal has a burst level of 0.286 V_{p-p}. Both luma and chroma have a characteristic impedance of 75 ohms (at the connectors).

As shown in Fig. 5.36a, the improved tape construction allows significantly higher frequencies to be recorded. The S-VHS system takes advantage of this capability by changing the dark and white clip levels. The change in clip level improves the pulse response and permits high-definition edge reproduction.

Since FM is more susceptible to noise at high frequencies, the low-level, high-frequency components are emphasized during record and are restored during playback to improve the S/N ratio. This nonlinear-subemphasis circuit is used exclusively for S-VHS and is used in both the SP and SLP modes. The dynamic-emphasis circuit is used in the VHS SLP mode and is not used in S-VHS.

Basic S-VHS Record Signal Path. As shown in Fig. 5.37, the record signal path for an S-VHS VCR is substantially different from that of the conventional VHS VCR

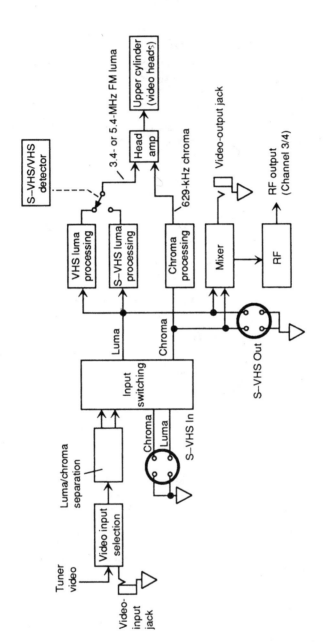

FIGURE 5.37 Basic S-VHS signal paths.

(a) Record

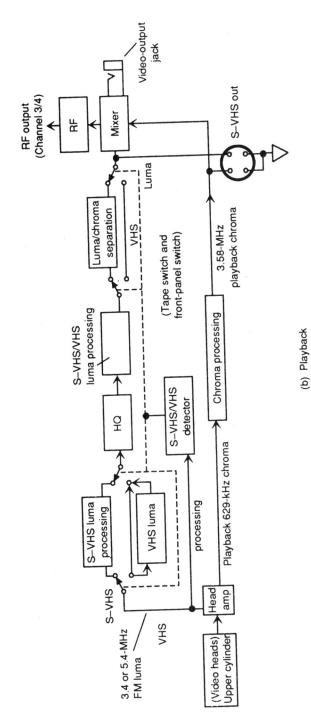

FIGURE 5.37 (*Continued*) Basic S-VHS signal paths.

5.77

(even though many of the remaining circuits are the same). In S-VHS, video IF from the FS tuner is applied to the tuner and line switch. Video from either the FS tuner or the video input jack is selected by the tuner and line switch for recording. This video goes to the luma and chroma switch where the luma and chroma signals *are separated* and sent to luma and chroma processing circuits.

Luma and chroma from the S-VHS input connector (the S or S-video connector) are applied to the luma and chroma switch and then to the processing circuits *if the S-VHS input is selected.*

Processing for the luma and chroma are the same as for normal VHS when the VCR is in the VHS mode. However, in the S-VHS mode, the luma sync-tip frequency is 5.4 MHz instead of 3.4 MHz. In either case, the 3.4- or 5.4-MHz luma carrier and the 629-kHz chroma carrier are applied to the head amp, where the signals are mixed, amplified, and sent to the rotating video heads on the cylinder.

Basic S-VHS Playback Signal Path. As shown in Fig. 5.37, the playback signal path is also different for an S-VHS VCR. During playback, the signal from the video heads goes to the head amp and is amplified, and the luma and chroma carriers are partially separated. These carriers are fed to the luma and chroma process circuits where the carriers are filtered even more. Then the luma and chroma signals *are reassembled* from carrier information and applied to the luma and chroma switch.

Separate luma and chroma signals are applied to the S-VHS output connector when the VCR is in the S-VHS mode. During either mode (VHS or S-VHS), the luma and chroma signals are mixed and applied to the RF modulator and video-output jack.

When the VCR is not in the play mode, the E-E video is separated, and the luma signal is applied to an AGC circuit in the luma-process circuits. After AGC processing, the luma and chroma signals are mixed and applied to the RF modulator and video-output jack.

Note that the luma and chroma process circuits also produce the 3.58-MHz reference (color burst) and vertical sync signals needed by the servo-control circuits to control the capstan and cylinder motors. These signals and circuits are essentially the same as for normal VHS VCRs.

5.8.2 Test Equipment for S-VHS Service

With one possible exception, you can probably service all circuits of an S-VHS and hi-fi VCR if you have a good set of test equipment suitable for conventional TV and VCR work. The exception is an S-VHS monitor.

The monitor-type TV sets designed specifically for S-VHS provide the only practical means of monitoring an S-VHS video signal. If you monitor an S-VHS tape with a conventional monitor TV, you get an improved picture, but you do not get the full benefits of S-VHS (400-line resolution) unless you connect the S-video output of the S-VHS VCR to the S-video input of an S-VHS monitor.

As a practical matter, if you can find an S-VHS monitor with built-in dual speakers and stereo amplifier, you can then check both the S-VHS and hi-fi functions on a shop standard. No matter what you use as the monitor, make certain that both the audio and video circuits are properly adjusted. (If you adjust a VCR to produce good colors on an improperly adjusted monitor or TV, you are in trouble.)

5.8.3 Basic S-VHS Record System

Figure 5.38a shows typical S-VHS record circuits. Note that the blocks with a double border represent those areas that are different from the standard VHS system. Also note that (in this VCR) the rear-panel S connector is connected mechanically to a switch that determines when separate luma and chroma signals are routed to the VCR. When the S connector is used, the switches are in the position shown in Fig. 5.38a.

When a composite video signal is routed from either the tuner or external video input, two comb filters are used to minimize the interaction between the luma and chroma portions of the signal. These filters are not needed when the S connector is used (in S-VHS mode). The luma signal is routed through a 5-MHz low-pass filter into a sub-preemphasis circuit.

The sub-preemphasis circuit is used exclusively for S-VHS to perform nonlinear emphasis. That is, the amount of emphasis varies with the input level. Figure 5.38b shows the S-VHS sub-preemphasis characteristics. Since the FM signal is more susceptible to noise at high frequencies than at low frequencies, low-level, high-frequency components are emphasized during record. Then the original level is restored during playback. This function improves the S/N ratio. Note that the familiar dynamic emphasis used in the VHS mode is not used in S-VHS.

The main preemphasis circuit is a linear circuit similar to the standard VHS emphasis circuit. The signal from the main-preemphasis circuit is routed to the dark and white clip circuit. Note that this clip circuit performs the same function as in standard VHS, but the point at which the clipping occurs is changed to take advantage of the increased bandwidth in S-VHS. Figure 5.38c shows the relationship of the dark and white clip levels for the three formats (normal VHS, VHS HQ and S-VHS). VHS high-quality (HQ) circuits are discussed in Sec. 5.8.6.

The output of the dark and white clip circuit is applied to the FM modulator. In S-VHS, the carrier frequency of the luma signal is increased so that the sync-tip is at 5.4 MHz and the peak white is at 7 MHz (Fig. 5.36). The output of the FM modulator is routed to the Y/C mixer, where the FM luma signal is added to the down-converted chroma signal. The output of the mixer is routed to the record amplifier and on to the video heads.

The chroma signal is unchanged in the S-VHS mode. The separate chroma input signal is routed through a 3.58 BPF to the ACC circuit, which monitors the burst level to keep the color level constant (as in the case of most modern VHS VCRs). After ACC, the signal is routed to a balanced modulator where the chroma signal is down converted to the familiar 629-kHz subcarrier.

The down-converted chroma is passed through a burst preemphasis circuit and color-killer circuit. These circuits are the same as found in most standard VHS VCRs. That is, the burst preemphasis raises the burst signal level by 6 dB to improve the S/N ratio and reduce phase distortion, while the color-killer circuit disables the chroma signal output to prevent false chroma triggering.

The chroma signal is passed through a 1.3-MHz LPF to the Y/C mixer. Both the down-converted chroma and luma are added in the Y/C mixer and sent to the video heads.

S-VHS Luma Record Circuits.   Figure 5.39 shows typical S-VHS luma record circuits. These circuits process the luma portion of the video signal and pass the signal through the FM modulator to prepare the signal for recording on tape. The circuit in

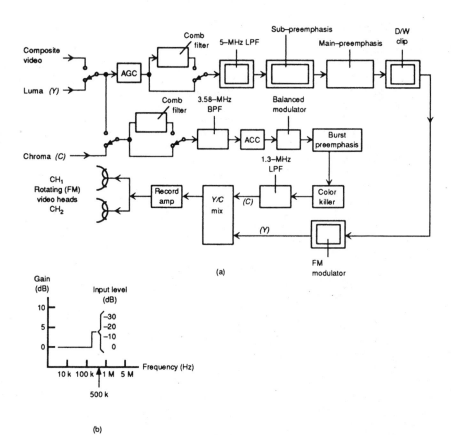

(a)

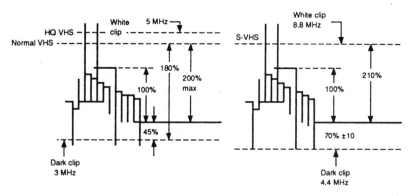

FIGURE 5.38   Simplified S-VHS record circuits.

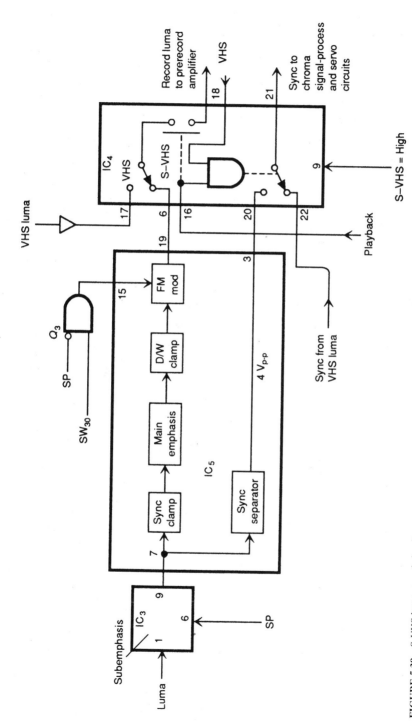

FIGURE 5.39 S-VHS luma record circuits.

Fig. 5.39 covers only the S-VHS signal path. The VHS signal processing is similar to standard VHS VCR circuits.

The luma signal enters the subemphasis circuit at pin 1 of IC_3. The subemphasis is used only in S-VHS and provides nonlinear emphasis to the high-frequency portion of the luma signal, as discussed. The amount of subemphasis is reduced in the SP mode. The SP input from system control is used to select the range of subemphasis.

The output of the subemphasis circuit is routed to the S-VHS luma processing circuits in IC_5. The luma signal enters at pin 7 and is routed to the sync-clamp and sync-separator circuits. A 4-V_{p-p} positive composite-sync signal exits IC_5 at pin 3 and is routed to a switch in IC_4. The switch is used to select the sync source from either the S-VHS or standard VHS luma circuit.

The position of the pin 20 and 22 switch in IC_4 is determined by logic in IC_4. For VHS, the sync signal is generated by the luma circuit ahead of IC_3. The sync signal from IC_5 is used only in playback of an S-VHS tape, when no other sync signal is available. The selected sync signal is sent to the chroma signal-processing circuit, as well as the cylinder servo circuit (where the sync is used in essentially the same way as for any VHS VCR).

The luma signal at pin 7 of IC_5 is also applied to a sync clamp where the dc level of the sync tip is fixed, independently of any variation in video level. The clamp output is routed to an emphasis circuit, which provides fixed emphasis to improve S/N ratio.

The emphasized luma signal over- and undershoots at the rising and falling edges of black-to-white transitions. The dark and white (D/W) clip circuit prevents overmodulation of the FM luma signal by clipping the over- and undershoots at a preset level (Fig. 5.38c).

Overmodulation can produce black and white inversion of the signal. The improved tape performance of the S-VHS tape allows the dark and white clip points to be increased so that the white clip occurs at 210 percent and dark clip at 70 percent (versus a white clip of 200 percent and a dark clip of 45 percent for standard VHS).

The output of the D/W clip circuit is routed to the FM modulator, which frequency modulates the luma signal and performs carrier interleaving in the SLP mode. The signal is modulated such that the sync tip is 5.4 MHz and white peak is 7.0 MHz (compared to the standard 3.4-MHz sync tip and 4.4-MHz white peak of standard VHS), as shown in Fig. 5.36a.

In the SLP mode, the signal is recorded on tape with no guard bands between adjacent tracks of video. The FM carrier frequency is offset by $\frac{1}{2}fH$ for every field to prevent beat interference. In the SP mode, the adjacent tracks do not overlap, so interleaving is not necessary. The FM carrier interleaving is performed by the SW_{30}-Hz signal applied to pin 15 of IC_5. Transistor Q_3 is used to disable this input in the SP mode and allows the SW_{30}-Hz signal to pass in the SLP mode.

The FM modulator output exits IC_5 at pin 19 and is passed to the VHS and S-VHS mode-select circuit IC_4, which contains switches to select either the S-VHS or standard VHS FM luma signal. The selected FM luma output is supplied to the record amplifier at pin 18 of IC_4. The second switch in IC_4 (in series with the pin 6 and 17 switch) is used during playback to prevent the playback signal from leaking to the record-amplifier circuit, producing noise in the playback picture.

5.8.4 Basic S-VHS Playback System

Figure 5.40 shows typical S-VHS playback circuits. Again, the blocks with a double border represent those areas that are different from the standard VHS system. The

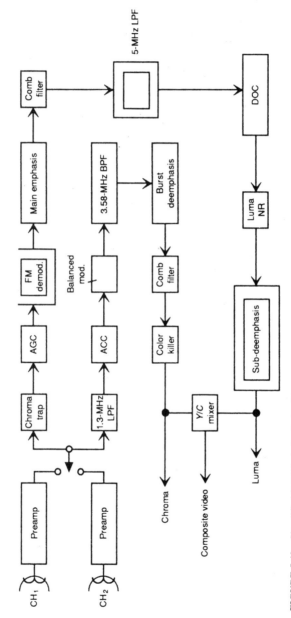

FIGURE 5.40 Simplified S-VHS playback circuits.

playback mode of S-VHS is similar to standard VHS. Chroma processing is identical for both VHS and S-VHS. *The major change in luma processing is an increase* in the luma FM carrier frequency. The head-switching and preamplifier circuits operate with the video heads to recover the video signal from the tape. The amplified signal is then split by a filter into separate luma and chroma signals.

The luma signal is passed through an AGC circuit to maintain a fixed amplitude for the FM luma information. The signal is then passed to an FM demodulator to recover the baseband luma signal. After recovery, the signal is routed through the main deemphasis circuit, comb filter, and 5-MHz LPF to the dropout compensator (DOC) circuit.

The main deemphasis circuit is used to restore the signal to the original level. The comb and 5-MHz filters are used to remove noise from the baseband luma. The DOC and luma noise-reduction (NR) circuits are combined. The DOC repeats a horizontal line of video whenever a loss of the FM carrier is detected. The delay line used for DOC is also used for luma noise reduction.

The luma NR is similar to a comb filter in that adjacent lines of video are compared, and the noise components tend to cancel. The output of the NR circuit is sent to the sub-deemphasis circuit to remove the variable emphasis used in the S-VHS process (Sec. 5.8.3). The baseband luma signal is split. The signal goes to the rear-panel S connector as luma output and to the *Y/C* mixer to become composite video.

The chroma signal is unchanged from the standard VHS system. The signal recovered from tape is passed through an ACC circuit to maintain a constant color level. The signal is routed to the balanced modulator where the chroma information is converted back to the NTSC standard 3.58-MHz color carrier.

The modulator output is passed through the burst deemphasis circuit to restore the level of burst and through a comb filter to remove any noise from the chroma signal. The color-killer circuit is used to prevent any output from the chroma circuits whenever the burst signal is missing.

The recovered chroma signal is finally routed to the rear-panel S connector and to the mixer circuit where chroma and luma are added. The combined *Y/C*, or luma and chroma, signals become the composite video signal.

S-VHS Luma Playback Circuits. Figure 5.41 shows typical S-VHS luma playback circuits. These circuits recover the luma portion of the playback FM signal and route the signal to the video-out circuits. The circuits of Fig. 5.41 are in parallel with the standard VHS playback circuit. The output of the VHS and S-VHS mode-select circuits is used to select the proper luma playback mode.

The playback FM signal is passed through a 5-MHz peaking circuit in IC_6, an 8.3-MHz peaking circuit in CP_3, and chroma trap L_{15} and C_{45}. The two peaking circuits act as a playback-equalization network to compensate for the frequency response of the video heads. The preamplifier in the video-head circuit has a flat frequency response, and the equalization circuit raises the level of the components close to the limit of the FM carrier frequency. The chroma trap is used to remove the 629-kHz chroma subcarrier.

The output of the playback equalization circuit is applied to the AGC dropout detector at pin 20 of IC_5. The AGC circuit compensates for differences in the output levels of the heads and for changes in signal level. The output of the AGC circuit is applied to the dropout compensator.

The dropout detector consists of a limiter, rectifier, and low-pass filter. The LPF removes the ac component from the AGC output that has passed the limiter. A drop in the level of the signal because of a dropout causes a change in the dc level at the output of the LPF. This dc voltage is used as a switching pulse that triggers the dropout compensator in IC_2. When a dropout is detected, the original signal is replaced with the 1H-delayed signal from the 1H-delay line.

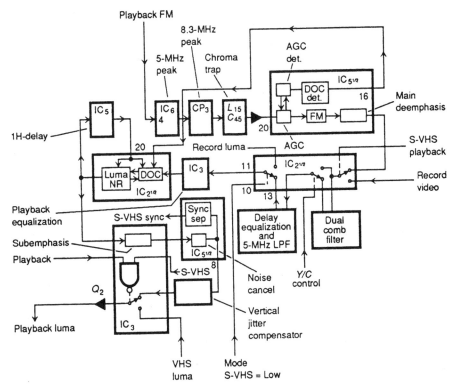

FIGURE 5.41 S-VHS luma playback circuits.

The FM AGC output is also applied to the FM demodulator in IC_5. The demodulator removes the FM carrier and recovers the luma signal, which is passed through the main-deemphasis circuit (the reverse of the main-emphasis process). The demodulated signal is sent through a comb filter to remove noise and any chroma information from the signal. The luma is then passed through a delay equalization network and 5-MHz LPF to match the delay of the chroma signal-processing circuit. The switch in IC_2 that routes the signal through the comb filter is controlled by the Y/C control input developed.

The comb filter and equalization networks are used both in record and playback, and IC_2 contains switches to route the signal to the appropriate circuit. The playback luma signal is routed from IC_2 to the playback-equalization network in IC_3. This circuit raises the signal near 6 MHz for S-VHS playback. The signal then goes to the luma noise-reduction and dropout compensator circuit in IC_2.

Part of the luma noise reduction is used as the dropout compensator. If the AGC circuit in IC_5 detects a dropout, pin 16 goes high. This high is sent to pin 20 of IC_2, causing the 1H-delay signal to be substituted for the original signal to compensate for the dropout. The output of the noise-reduction circuit is applied to the subemphasis circuit in IC_3.

The subemphasis circuit restores the original level of the signal which is emphasized during record. The output of the subemphasis circuit is routed to a noise canceler in IC_5. The noise canceler removes small-amplitude, high-frequency components that are usually noise components. The circuit also clips high-amplitude, high-frequency components, which correspond to the edges of the picture. As a result, the noise is reduced, but the edge definition of the picture is not affected.

The output of the noise canceler is routed to a sync separator to provide composite sync to the chroma and servo circuits. The output of the noise canceler also exits IC_5 at pin 8 and is routed to a vertical jitter compensator network. The vertical-jitter compensator uses the sync signal to clamp the sync tip. This prevents variations in the sync signal from affecting the luma signal and reduces vertical jitter. The output of the vertical jitter compensator is routed to IC_3.

The switch in IC_3 selects either the S-VHS playback-luma signal or the VHS signal. The selected output is passed through buffer Q_2 to become the playback-luma signal. This signal is mixed with the chroma signal and routed to the video-out circuit (Fig. 5.40).

5.8.5   Basic S-VHS Troubleshooting Approach

This section describes the basic troubleshooting approach for the S-VHS circuits covered in this chapter.

Preliminary Checks.   The first step in checking any VCR is to play back a known-good tape. There is no point in checking the record functions unless you are sure that the playback is good (since record and playback use many of the same functions in both VHS and S-VHS). On an S-VHS VCR, play back *both* VHS and S-VHS tapes, and check the results. If the playback of either tape is bad, use the procedure for playback troubleshooting in the following paragraphs.

If playback of both tapes is good, make a recording in *both* the S-VHS and VHS modes, using an S-VHS tape. Then play back both recordings. If playback of either recording is bad, the problem is likely in the record functions (since playback is good with the known-good tape). Use the procedures for record troubleshooting in the following paragraphs.

Note that if an S-VHS recording is played in the VHS mode, the *picture may streak in areas of high brightness.* This same problem can occur if the S-VHS and VHS *detector circuits are not working properly.* Start by checking those detector circuits controlled by the front-panel S-VHS and VHS switch and the S-VHS cassette switch (located on the mechanism and actuated by the ID hole on an S-VHS cassette), as shown in Fig. 5.37.

If a known-good S-VHS VCR is available, play back the tape recorded on the VCR being serviced. If playback is bad, the problem is likely in the record circuits of the VCR being serviced. If the playback is good, the record circuits of the VCR under service are *probably* good.

S-VHS Luma Record Troubleshooting.   If VHS record is good but video is missing in S-VHS, set the front-panel switch to S-VHS, and place the VCR in the record mode with an S-VHS tape. Check for a high at pin 9 of IC_4 (Fig. 5.39). If it is missing, suspect the front-panel switch.

If pin 9 of IC_4 is high, check for FM luma at IC_{5-19}. If it is present but there is no record luma at pin 18 of IC_4, suspect IC_4.

If there is no signal at pin 19 of IC_5, check for a signal at pin 9 of IC_3. If it is present, suspect IC_5. If it is missing, check for video at pin 1 of IC_3. If it is missing, suspect the video-input circuits (Fig. 5.38*a*).

S-VHS Luma Playback Troubleshooting. Standard video signal tracing is the best approach for troubleshooting the S-VHS luma playback circuits. As in the case of record, follow the luma signal through the switching networks and ICs. Make certain that all switch actions are checked. Start by checking for the presence of FM luma at pin 4 of IC_6. If the signal is missing, suspect the head-switching and preamp circuits (Figs. 5.40 and 5.41).

If the FM luma is present at IC_{6-4}, monitor the luma waveform at various points in the circuit to determine where the luma is absent or abnormal (using the service literature waveforms).

If the signal is missing at the output of a switch, check the switch-control signal. For example, if the signal is present at IC_{2-13} but not at IC_{2-11}, check the VHS mode signal at IC_{2-10} (which should be low during S-VHS in this particular VCR).

5.8.6 HQ Circuits

High quality (HQ) circuits are found on many present-day VCRs, including conventional VHS VCRs. The HQ circuits improve (or enhance, as the sales literature says) the picture by extending the luma-signal frequency spectrum, as shown in Fig. 5.38c. Most S-VHS VCRs have HQ circuits, but the circuits are not necessarily used in the S-VHS mode.

Figure 5.42 shows typical HQ circuits used during playback. In this particular VCR, either the S-VHS or VHS luma signal is applied to the HQ circuits at IC_{10-32}. The output from IC_{10} is applied to IC_9 through Q_{71}, Q_{70}, Q_{69}, and C_{70}. After the luma signal is enhanced in IC_9, the signal is buffered by Q_6 and applied to the luma-processing circuits as shown in Fig. 5.37. Troubleshooting the HQ circuits is a straightforward signal-tracing function, using the service-literature waveforms.

5.9 VHS-C CIRCUITS

When a VHS-C cassette is used in a VHS VCR, the circuit operation remains the same, as does the troubleshooting approach. VHS-C is a compact version of the VHS format and is used in compact VHS camcorders (Chap. 7).

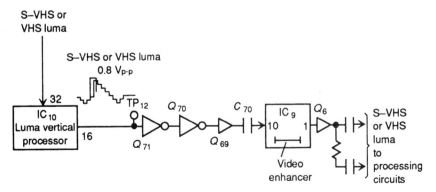

FIGURE 5.42 VHS HQ circuits.

When the VHS-C cassette is removed from the camcorder, the small cassette is placed within an *adapter,* and the adapter is installed within the larger (standard) VHS transport. At that point, the VCR sees a tape recorded in the conventional VHS format (as described in this chapter).

From a troubleshooting standpoint, if a VHS-C cassette can be played back in a VHS-C camcorder but not with an adapter in a known-good VHS VCR, suspect the adapter.

CHAPTER 6
BETA VIDEO CIRCUITS

This chapter is devoted to the video circuits found in Beta VCRs. (VHS VCRs are covered in Chap. 5.) It also covers other portions of Beta VCRs that are directly related to the video circuits (such as the system control and servo system for the tape-drive mechanism). Here we show the video technician the major differences between VHS and Beta. All the general troubleshooting and service notes for VHS also apply to Beta. Of course, such procedures as electrical and mechanical adjustments are different. However, these procedures are given in Beta service literature and need not be duplicated here. To sum up, if you can service VHS (using the information of Chap. 5 and the service literature) and understand the *practical circuit differences* discussed here, you should have no trouble with Beta.

6.1 INTRODUCTION TO THE BETA SYSTEM

It is assumed that you have digested the VHS basics of Chap. 5, particularly Secs. 5.1 through 5.3. (If not, read them.) The illustrations shown in Figs. 5.5 and 5.6 also apply to Beta. That is, Beta records and plays back video without guard bands on the tape and uses azimuth recording, down conversion, and phase inversion to prevent crosstalk between video tracks.

The video-head azimuth for Beta is ±7° (rather than the ±6° of VHS). Beta color is down-converted to 688 kHz (instead of the 629 kHz for VHS). Beta also uses phase inversion (PI) to minimize crosstalk (as does VHS), but the Beta system is substantially different. So let us start our discussion of the differences between Beta and VHS with PI color recording.

6.1.1 Phase-Inversion Color Recording for Beta

In the simplest of terms, the chroma signal to be recorded on track A (Fig. 5.6) is phase-inverted by 180° with every line period (1H), while the chroma signal recorded on track B remains continuously in the same phase. At playback, both track A and track B signals are restored to the same phase relationship.

The line-by-line phase inversion of the chroma signal is done during the conversion process from 3.58 MHz to 688 kHz and applied to head A. The 3.58-MHz chroma signal is mixed with a 4.27-MHz reference signal (which is phase-inverted each time that head A is in contact with the tape). The resultant 688-kHz signal is amplified and applied to the video heads (head A inverted, head B not inverted).

During playback, the 688-kHz signal is converted back to 3.58 MHz. In an exact counterpart of the recording process, the head A signal is again phase-inverted using the same 4.27-MHz signal used during record. Adding the recovered line signal (head A) to the adjacent line signal (head B) restores the signal back to normal. However, the phase of the crosstalk component in the 3.58-MHz playback chroma signal remains phase-inverted at every other line.

The playback chroma signal is then passed through a comb filter using a 1H delay line and a resistive matrix, as shown in Fig. 6.1*a*. Both the delayed and nondelayed signals are added together in the resistive circuit, with the result that the crosstalk component is canceled out and the normal chroma signal component is *double in amplitude.*

In addition to canceling crosstalk, the PI color recording system also minimizes the effect of mechanical jitter or flutter in the drum servo. Such jitter causes a phase shift and results in poor picture quality. Jitter effects are eliminated by locking the frequen-

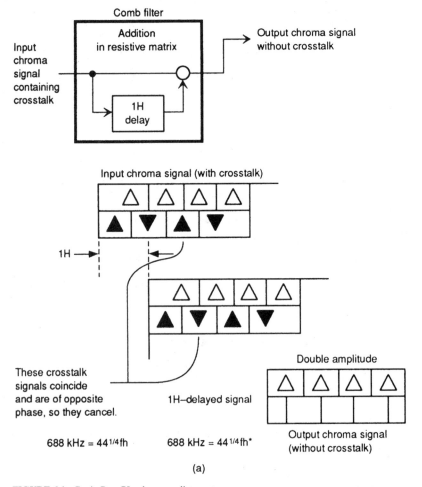

(a)

FIGURE 6.1   Basic Beta PI color-recording system.

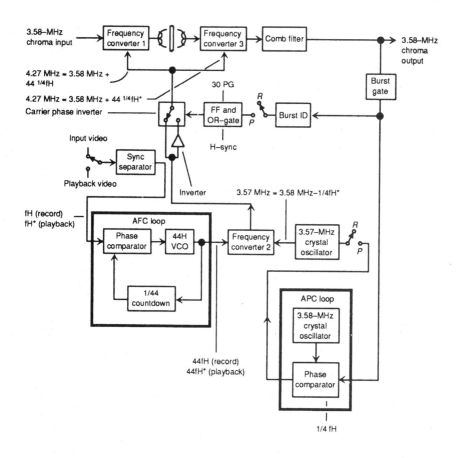

(b)

FIGURE 6.1 *(Continued).*

cy and phase of the 4.27-MHz reference signal to the TV horizontal sync signal (known as fH in Beta) during record.

At playback, the 4.27-MHz reference signal is locked to the recorded H-sync signal (known as fH* in Beta). If there is any jitter component (from any cause, mechanical or electrical), the 4.27-MHz reference is also locked to the jitter, eliminating the jitter effect. This feature is similar to locking the drum speed to the TV vertical sync during record and to the recorded V-sync during playback, as described for VHS in Chap. 5. However, operation of the PI color-recording circuits is far more complex.

6.1.2 Basic Beta PI Color-Recording Circuit

As shown in Fig. 6.1*b,* there are two phase-locked loops, or PLLs, involved in Beta PI color recording. Operation of the PLLs is similar to that used by the FS tuner PLLs described in Sec. 4.1.

One PLL is known as the AFC loop and produces a signal at a frequency 44 times fH, or about 693 khz. The AFC loop receives the fH input from the TV video signal during record and an input fH* from the recorded video signal during playback. The AFC loop output (either 44fH or 44fH*) is combined with the output of a 3.57-MHz crystal oscillator in frequency converter 2.

The 3.57 MHz is free-running during record but is locked in phase to the chroma 3.58-MHz signal during playback. (Note that 3.57 MHz is equal to 3.58 MHz, less than $\frac{1}{4}$fH.) Phase lock of the 3.57-MHz oscillator during playback is done by the APC loop. In either playback or record, the 3.57-MHz signal is combined with the 44fH signal to produce a 4.27-MHz signal which, in turn, is applied to frequency converters 1 and 3 through a *carrier-phase inverter.*

During both playback and record, the 4.27-MHz signal is passed by the carrier-phase inverter (usually a center-tapped transformer). The phase inverter is operated by an FF and OR gate that receives both cylinder tach pulses (called 30 PG pulses in Beta) and H-sync pulses. Both signals are required since the carrier-phase inverter serves to phase-invert the 4.27-MHz-signal with H-sync only when track A is being made. The 30PG pulse overrides the H-sync pulses when track B is being traced by head B.

During record, the 3.58-MHz chroma signal to be recorded is applied to frequency converter 1, where the 4.27-MHz signal is added, resulting in a difference frequency of 688 kHz. Since the 4.27-MHz signal is locked to the H-sync signal, the 688-kHz signal to be recorded is also locked to H-sync.

During playback, the 688-kHz chroma signal from the head is applied to frequency converter 3, where the 4.27-MHz signal is again added, resulting in a difference frequency of 3.58 MHz. The 3.58-MHz chroma output signal is compared with the APC loop oscillator (also 3.58 MHz).

Any phase variations because of jitter are used to shift the 3.57-MHz oscillator signal (free-running during record). Since the phase of the 4.27-MHz signal is controlled by the 3.57-MHz oscillator, any phase shift in the 3.58-MHz chroma output signal is eliminated.

Even with the AFC circuit, there is still a possibility that the 3.58-MHz chroma playback signal burst can lock up on a wrong phase of the 4.27-MHz signal (locked in but 180° out of phase). This condition is prevented by the burst ID circuit that compares the phase of the 4.27-MHz reference signal with the 3.58-MHz chroma signal during playback.

If the APC system has locked up on the wrong phase, the carrier-phase inverter FF circuit is switched by a trigger pulse developed in the burst ID circuit. The burst ID compares the phase of the 3.58-MHz chroma playback signal for each horizontal line and produces the corrective pulse whenever the phase-invert FF switch has locked on the incorrect phase.

6.1.3 Automatic Tape Loading for Beta

Figure 6.2 shows the autoloading system for Beta. Compare this to the VHS system shown in Fig. 5.15. The Beta system uses a so-called U loading or threading system (the tape appears to form the letter U when fully threaded). When the cassette is inserted into the cassette holder, a loading ring picks up the tape as shown and then threads the tape around the drum in about 3 seconds.

When the eject button is pressed, the loading ring turns in the reverse direction, and the excess tape is taken up by the takeup reel. When the loading ring returns to the

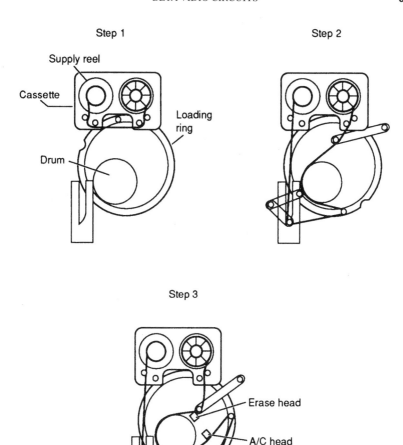

FIGURE 6.2   Beta autoloading system.

original position and the tape is all back inside the cassette, the cassette automatically rises and is ejected. All of these functions are operated by system-control microprocessors as discussed in Sec. 6.5.

6.2 INTRODUCTION TO BETA CIRCUITS

This section describes the overall operation of a Beta VCR during both record and playback. Section 6.3 describes operation of the Beta video circuits.

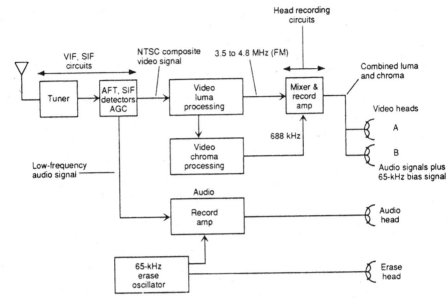

FIGURE 6.3   Basic Beta record functions.

6.2.1   Basic Beta Record Functions

Figure 6.3 shows the overall block diagram of a typical Beta VCR during record.

Tuner (VIF/SIF) detection: Beta tuners are similar to VHS tuners (and TV tuners). That is, present-day tuners use FS tuning (Sec. 4.1) to produce the standard NTSC composite video signal, as well as the audio signal. Older Beta VCRs use the RF-mixer-IF combination found in early-model TV sets.

Luma signal record: The luma and chroma signals are separated by circuits under AGC control (to maintain constant NTSC signal levels). Then, to convert the wide-frequency band (0 to 4.2 MHz) luma signals into an easily recordable FM signal, a clamp (among many other circuits) matches up the level of the luma signal, which is then processed to reduce crosstalk and improve the S/N ratio. An FM modulator converts the luma signal to a 3.5- to 4.8-MHz signal.

Chroma signal record: The separated chroma signal (the 3.58-MHz color burst) is mixed with a 4.27-MHz reference signal and converted to 688 kHz. Azimuth recording and PI are used to eliminate crosstalk between tracks, as discussed in Sec. 6.1.2.

Head-recording amplifier: The FM-modulated luma signal (3.5 to 4.8 MHz) and the chroma signal (688 kHz) are combined and amplified for optimum recording performance before application to video heads A and B.

Audio recording: Because of the nonlinear response of magnetic tape, preemphasis and deemphasis, respectively, are applied during record and playback of audio. A 65-kHz bias signal is superimposed on the audio signal during record to raise efficiency and reduce distortion (as is typical for most audio tape recorders). The same 65-kHz oscillator output is used to erase previously recorded signals on the magnetic tape.

6.2.2 Basic Beta Playback Functions

Figure 6.4 shows the overall block diagram of a typical Beta VCR during playback.

FM playback amplifier: The FM luma signal and the 688-kHz chroma signal picked up by the two video heads are amplified by an amplifier (with high S/N ratio), and frequency response is corrected. Then the two playback outputs are mixed to form a single continuous signal.

Luma signal playback: An HPF is used to separate the FM component of the playback signal. A DOC operates if dropout is present. After passing the DOC circuit, the luma signal is frequency corrected and passed through a noise canceler before emerging as a proper video signal.

Chroma signal playback: The 688-kHz chroma subcarrier is separated from the playback signal by an LPF and combined with a 4.27-MHz signal. The combined signals are demodulated as a 3.58-MHz chroma signal. At the same time, a comb filter is used to eliminate crosstalk between the two video tracks. As discussed in Sec. 6.1.2, the 4.27-MHz signal is processed by APC/AFC circuits to remove time-base errors (which is an engineering phrase for jitter) that may appear during record and playback.

Y/C mixing: The luma (*Y*) and chroma (*C*) signals are mixed to form an NTSC composite color video signal.

Audio playback: The audio signal picked up by the A/C head is corrected for frequency response and S/N ratio and then amplified to a suitable level.

RF unit: As in the case of VHS, the NTSC composite video signal and the audio signal are converted to Channel 3 or 4 so that the VCR output can be viewed on a conventional color TV. Most Beta VCRs also provide for video and audio outputs to a monitor-TV (bypassing the RF unit).

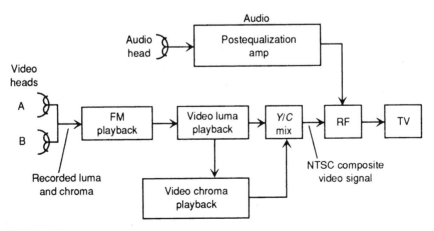

FIGURE 6.4   Basic Beta playback functions.

6.2.3   Beta II and Beta III

Most present-day Beta VCRs are capable of operating in either (or both) Beta II (Beta 2) and Beta III (Beta 3) tape formats. The formats are essentially the same except in video track pitch and tape speed. (The slower tape speed permits longer playing time for a given amount of tape.) Beta II uses a track pitch of 29.2 μm and a speed of 2 cm/s. Beta III has a track pitch of 19.5 μm and a speed of 1.22 cm/s. Because of the greater recording density, Beta III is more affected by jitter than Beta II. For this reason, the Beta III color burst is increased by 6 dB (in the 688-kHz stage) and then restored after the burst is used by the APC circuits.

From a troubleshooting standpoint, Beta II and III should present no particular problem, unless a customer tries to play a Beta II tape on a VCR that is set to Beta III, or vice versa. Always check the Beta II-III switch settings first.

6.3   BETA VIDEO CIRCUITS

This section describes the basic operation of the Beta video circuits during both record and playback. These descriptions can be used as a guide in troubleshooting. However, it is essential that you check the service-literature schematics and waveforms when performing stage-by-stage signal tracing.

6.3.1   Typical Beta Video Signal Flow During Record

As shown in Fig. 6.5, the AGC section maintains a constant output level, regardless of picture brightness. The LPF removes the SIF signal (4.5 MHz) and other unnecessary high-frequency components. The ATT (attenuator) reduces the signal amplitude to a suitable level. The *Y/C* separator separates the luma (*Y*) and chroma (*C*) components of the color input.

The *sync-tip clamp* lines up the sync-signal level. The dc component of the video signal is normally removed by capacitive coupling in the signal path. Clamping is used to match up the sync signal since the sync tip becomes the low-frequency (3.5-MHz) reference for the FM signal (as shown in Fig. 5.5).

The *E-E trap* ensures that the direct E-E output (for a monitor-TV) switches to the black and white mode when signal conditions are bad. Under such bad conditions, the chroma signal level of the input video signal is low, and the VCR color killer (ACK, or automatic color killer) circuit operates, switching the recording mode to black and white. (Always check this circuit function if you are troubleshooting a "no color but good picture" symptom.)

The *noise-reduction system* includes both preemphasis and compressor functions to produce a nonlinear emphasis of the video signal during record. During playback, deemphasis (having the opposite characteristics of the emphasis applied during record) is used to reproduce the original signal. (This noise-reduction system is found in virtually all Sony VCRs.)

The *H-step cancel circuit* corrects nonuniformity in H-sync spacing. In areas where reception is bad, the H-sync signal may not be uniform, resulting in possible

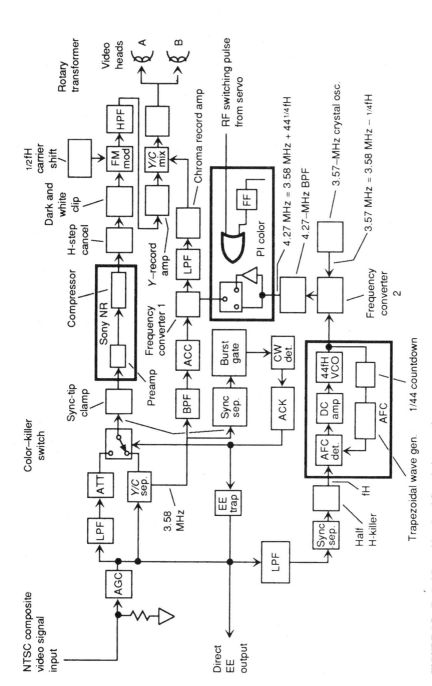

FIGURE 6.5 Typical Beta video signal flow during record.

6.9

crosstalk problems during playback. The H-step cancel circuit maintains constant H-sync spacing.

The *dark and white clip circuit* clips unwanted high-level pulse components (arising from preemphasis) to help stabilize the picture. Dark clipping cuts off excess sync excursions, while white clipping prevents a whiter-than-white level (overmodulation).

The *HPF* differentiates the FM-modulator output and passes the differentiated peaks. Differential recording is used because of undesirable self-demagnetization by high-frequency components when the luma signal (FM) is recorded.

The $\frac{1}{2}fH$ *carrier-shift* circuit produces a $\frac{1}{2}fH$ difference between the carrier frequencies and the head-A period and the head-B period. Without such a circuit, there is a possibility of "beating" because of crosstalk between adjacent tracks during playback. The $\frac{1}{2}fH$ carrier-shift circuit shifts the beat frequency by a $\frac{1}{2}$-offset relationship, making the beat unnoticeable on the TV picture.

6.3.2 Typical Beta Video Signal Flow During Playback

As shown in Fig. 6.6, the head preamps amplify the video signals at the heads. The equalizer corrects the frequency response of the head signals and adjusts channel balance so that both head signal levels are equal.

The *switcher and mixer* combines the signal of both channels and removes any overlap to provide a composite signal output. Since the tape typically wraps the drum more than 180°, pulses from the servo are used to remove the overlap and produce a continuous output signal, as shown in Fig. 5.10.

The *attenuator* matches the level of black and white reproduction with that of color. The HPF passes only the luma FM component and rejects the 688-kHz chroma signal. In the absence of a color signal during playback, the ACK circuit switches the video signal through the HPF to remove any chroma signal or noise that might interfere with black and white reproduction.

The DOC senses any dropout and compensates by using the preceding horizontal line (1H) signal. The output of the LPF is applied to the chroma playback circuit, consisting of the frequency converter, comb filter, AFC, APC, PI, and burst ID (Sec. 6.1.2).

The DOC output is applied to the limiter, which limits amplitude of the FM signal and removes amplitude fluctuations from the playback signal. The FM demodulator and LPF combination changes the FM signal back into the original video signal.

Most video FM playback demodulators use a multivibrator-multiplier circuit rather than the familiar balanced FM discriminator. No matter what form of video FM demodulator circuit is used, the circuit is part of an IC and is neither accessible nor adjustable (which also applies to most signal circuits in Beta VCRs).

The demodulated video signal is applied through a *noise-reduction system* similar to that described for record (Sec. 6.3.1). Since nonlinear emphasis is applied during record, the opposite process takes place during playback.

The $\frac{1}{2}fH$ *carrier-shift return circuit* restores the $\frac{1}{2}fH$ carrier shift produced during record. If this carrier shift is not restored, the dc component fluctuates when the FM signal is demodulated.

In the noise canceler, the noise component (high frequency) of the video signal is removed by an HPF. The phase is then inverted and, with reverse phase and the same amplitude, the processed video signal is added back to the original video signal, thus canceling the noise.

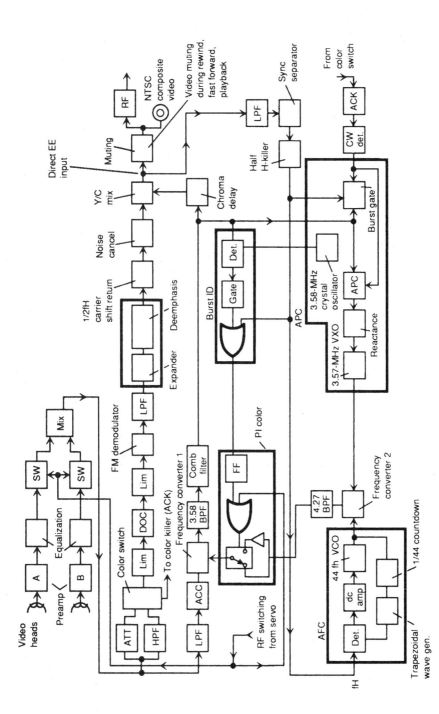

FIGURE 6.6 Typical Beta video signal flow during playback.

During rewind, fast forward, and playback-servo startup, a *muting signal from system control* prevents the video signal from being applied to the RF modulator (or to the video output connectors). Always check the muting signal when troubleshooting a "no video" symptom. It is possible that the muting signal is present at the wrong time (or at all times and in all modes).

6.4   BETA SERVO CIRCUITS

Figures 6.7 and 6.8 show the basic Beta servo circuits. The following sections summarize operation of the servo. Compare this to the VHS servo circuits described in Sec. 5.3. As with VHS, most Beta servo functions take place within ICs. The basic circuit includes a drum servo, a capstan servo, a picture-search circuit (which is part of the drum servo), and a loading and unloading circuit (part of the capstan servo).

6.4.1   Beta Drum Servo

The drum-servo signals (Fig. 6.7) are processed mainly by the circuits of IC_4. These circuits provide *error signals* to the drum motor driver. The circuits also provide switching and control signals to the capstan motor-control IC_{16} (Fig. 6.8).

The signals generated by the drum PG coils (A and B) and applied to pin 1 of IC_4 are used for both speed and phase control of the drum motor. The signals at IC_{4-18} (the drum error signal) contains both speed and phase information. The speed information is taken from the PG signals, whereas the phase information is based on a comparison of the PG-signal phase with V-sync signals (during record) or a servo reference signal (during playback).

The servo reference signal is obtained by a countdown of a 31.485-kHz signal. Either the V-sync or servo reference signals are applied at pin 12 of IC_4. The *picture-search* circuit (pin 24 of IC_4) varies drum speed so that the H-sync signal of the playback video is set to become 15.734 kHz during the picture-search mode.

6.4.2   Beta Capstan Servo

The capstan-servo signals (Fig. 6.8) are processed by the circuits of IC_{16}, as well as by several discrete-component circuits. The capstan servo has three major functions: (1) keeping the tape speed stable, (2) selecting the tape speed for either BII or BIII, and (3) operating as a tracking servo during playback.

The tape is held against the capstan by a pinch roller, and the tape is pulled from one cassette to another. The cassette reel motor provides for rewind, fast forward, and takeup of the tape onto the cassette and is controlled by the system-control circuit (Sec. 6.5). The tape is also driven by the reel motor during picture search (with tape speed set by system control). The *loading and unloading* functions are controlled, in part, by the capstan servo and by the system-control circuits.

6.5   BETA SYSTEM CONTROL

Figure 6.9 shows the basic Beta system-control circuits. The following sections summarize operation of the system-control functions. Compare this to the VHS system-

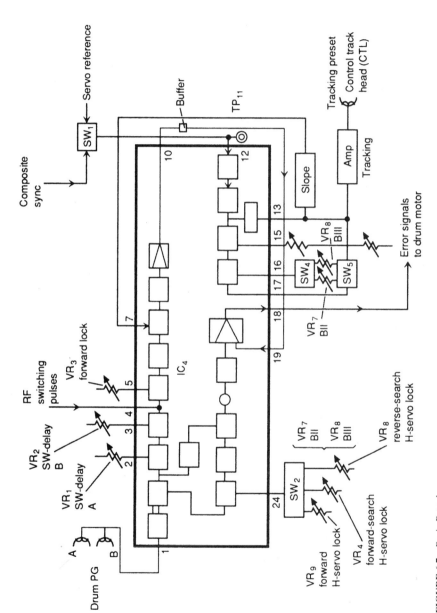

FIGURE 6.7 Basic Beta drum servo.

6.13

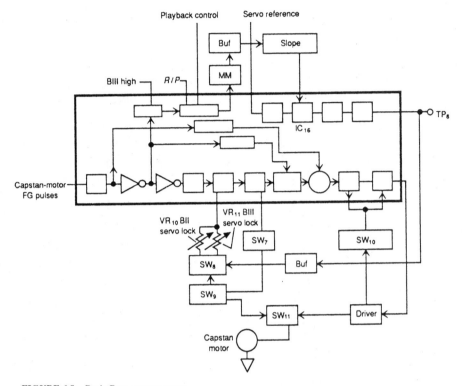

FIGURE 6.8   Basic Beta capstan servo.

control circuits described in Sec. 5.3. As with VHS, most Beta system-control operation is determined primarily by microprocessors. The microprocessors accept logic
control signals from the VCR operating controls (typically feather-touch push buttons
on a Beta VCR) and from various sensors. In turn, the microprocessor sends control
signals to video, audio, servo, and power supply circuits, as well as drive signals to
solenoids and motors.

6.5.1   Keyboard and Operating Modes

The keyboard shown in Fig. 6.9 contains the circuits for 10 operating-control push
buttons. When the buttons are pushed, control signals are applied to the micróproces-
sor. In turn, the microprocessor sends control signals to the various circuits.

The microprocessor in this particular Beta VCR performs the functions in 13
modes. The stop, rec, play, audio dub, pause, still, F-search, R-search, F fwd, and rew
modes are selected by one of the corresponding 10 push buttons. The loading, unloading, and cassette-up modes are selected (by the corresponding switches) when the cassette is loaded or ejected.

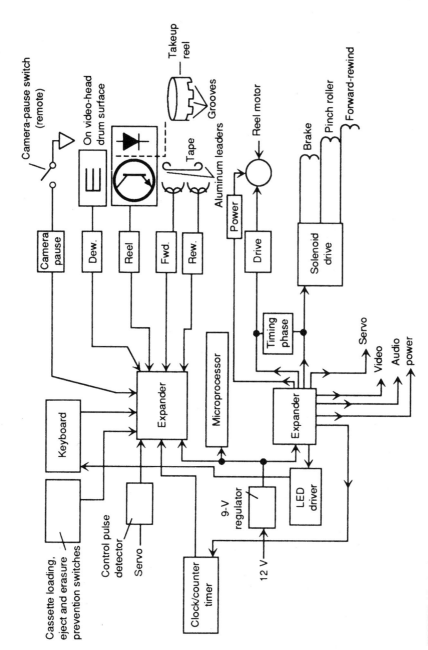

FIGURE 6.9 Basic Beta system control.

6.5.2 Summary of System-Control Functions

The *camera-pause circuit generates* pulses at the on-time and off-time of the remote pause switch. These pulses are applied to the microprocessor instead of the control signal during the pause mode.

The *dew-sensor circuit* senses when dew or moisture forms on the video-head drum surface. The dew-sensor resistance (part of the feedback network of an oscillator in this Beta unit) decreases when moisture is present, causing the oscillator to produce signals. These oscillator signals cause the microprocessor to place the VCR in stop mode.

The *reel-base sensor circuit* detects the pulses generated when grooves cut into the takeup reel intermittently shut off a photocoupler. The absence of pulses indicates that the reel base has stopped rotating. (Compare this to the rotating reflector disk used in VHS, Sec. 5.3.)

The *forward- and rewind-sensor coils* are part of oscillator circuits. When nonmagnetic (aluminum) trailer tape at both ends of Beta tape approaches the sensors, the Q of the sensor coil decreases as does the oscillator output. This signals the microprocessor that the tape is at the end (rewind or start, depending on which sensor output changes). Compare this to the clear leader used on each end of VHS tape, Sec. 5.3.

The *reel-motor drive circuit* supplies voltage to the reel motor according to the respective mode. The rec, play, and audio dub modes, a voltage of about 3 V is supplied to the reel motor (for this particular VCR). About 8 V is supplied in the F fwd, rec, loading, and unloading modes.

The *solenoid-drive circuit* controls on and off of the three solenoids (part of the mechanical functions described in Sec. 6.1.3). The three solenoids are the pinch-roller solenoid, which pushes the pinch roller against the capstan; the brake solenoid, which pushes a brake against the supply reel; and the roller solenoid, which switches tape travel between forward and rewind (on this particular VCR).

The *timing-phase circuit* drives the pinch-roller solenoid and the reel motor in synchronization with the drum motor. This is necessary so that the control pulses are recorded in line on the control track (when recording stops temporarily and starts again).

6.6 SUPER BETA AND BETA HIFI

Figure 6.10 shows the frequency spectrum of both Super Beta and Beta Hifi (which are essentially the Beta versions of S-VHS and hi-fi described in Sec. 5.8). The following sections summarize operation of Super Beta and Beta Hifi.

6.6.1 Beta Hifi

Beta Hifi permits the VCR to record left- and right-channel audio (stereo), using the video heads, in addition to the A/C head audio record and playback. As in the case of S-VHS and hi-fi, this provides much higher fidelity.

In Beta Hifi, the left- and right-channel audio signals are FM modulated and divided into four *pilot audio carriers.* Four carriers are required to maintain separation between the left- and right-channel audio, as well as to reduce the crosstalk between

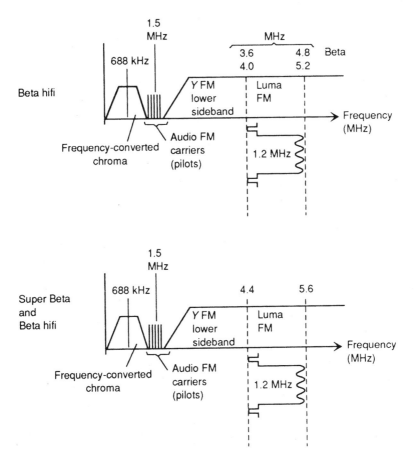

FIGURE 6.10   Super Beta and Beta Hifi spectrums.

adjacent tracks of the video information. Unlike S-VHS and hi-fi, which uses separate audio heads on the same drum as the video heads, Beta Hifi uses the same heads for video and audio. The four audio carriers are centered about 1.5 MHz and are mixed with the luma and chroma information to be recorded by the video heads on tape.

The addition of the four audio FM carriers requires that the FM luma signal be shifted upward by 0.4 MHz to make room between the chroma and luma information. The high-frequency limitations of the video heads results in a loss of some FM sidebands (because of this shift in luma frequency). This causes a slight reduction of resolution in the picture produced by some Beta Hifi VCRs.

6.6.2 Super Beta

Super Beta overcomes this reduction of resolution problem by narrowing the video-head gaps and shifting the FM luma carrier up by 0.8 MHz, resulting in a larger luma FM sideband (shown as the *Y* FM lower sideband in Fig. 6.10). The increased bandwidth of the total luma signal results in resolution greater than achieved by both Beta Hifi and conventional Beta VCRs.

6.6.3 Super Beta Record

Figure 6.11 shows the Super Beta record circuits in block form. Compare this to Figs. 6.3 and 6.5. Composite video from the tuner or the rear-panel input is selected for the AGC stage. Composite video leaves the AGC and takes two paths. One is through the E-E video-output line for an external monitor to display the recorded picture. The other path is into a comb filter which separates the luma and chroma components of the composite video signal.

The chroma down-converting process is conventional and need not be discussed. However, the edit function is unique. When receiving composite video from another VCR, the edit switch can be operated to negate the effects of the comb filter. This permits individual bandpass filters to separate the composite video for a clearer edited recording.

The LPF restores an overall flat frequency response to the luma signal (that was altered by the comb filter). The luma signal is then applied through a noise-cancel stage to a series of selected traps. Note that when the edit switch is operated, less noise cancellation occurs, permitting the full bandwidth of the playback signal to pass during the record-edit mode.

The upper frequency limit of the luma signal is fixed by the trap that follows the noise canceler. During high-band (HB) or Super Beta recording, a higher-frequency trap is selected (by front-panel controls), thus providing a wider bandwidth for the recorded luma.

The preemphasis stage provides the usual increase in gain for the high-frequency end of the luma signal (which is reduced during playback by deemphasis circuits). This boost in record and reduction in playback reduces the loss of detail during the record and playback process.

The FM modulator changes the AM luma signal to an FM signal containing upper

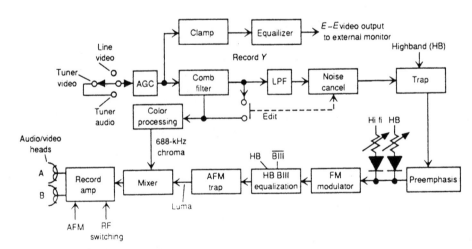

FIGURE 6.11   Super Beta record circuits.

and lower sidebands. Beta Hifi and Super Beta modes add a voltage to the incoming luma signal to shift the carrier frequency of the FM modulator (0.8 MHz for Super Beta).

The equalization stage rolls off the upper sideband (that could contribute noise to the signal) and balances the level of the lower sideband (lower frequencies) to the luma FM carrier signal (high frequencies). This prevents *black streaks* in the playback video (commonly called *overmodulation noise*).

The Super Beta switch and the BIII switch connect to the equalization stage. The extended bandwidth in Super Beta requires that the equalization emphasis be changed to maintain frequency-spectrum balance. However, this is only necessary in Super Beta operated at BII speed. The normal high-frequency losses at the slower Super Beta BIII speed maintain proper balance without the Super Beta equalization emphasis. Therefore, the Super Beta emphasis is not used in Super Beta BIII.

The AFM (audio FM) trap removes any noise in the frequency and between the chroma and luma lower sidebands so that the four audio FM carriers can be inserted. The luma signal (that has been controlled in amplitude and bandwidth, boosted, modulated, and cleaned) is mixed with the chroma. The combined luma and chroma signal is then mixed with the audio FM and recorded on tape.

6.6.4 Super Beta Playback

Figure 6.12 shows the Super Beta playback circuits in block form. Compare this to Figs. 6.4 and 6.6. Signals from the tape are picked up by the video heads, amplified by the head amplifier, and mixed together. The luma and chroma signals are separated using BPFs. The chroma-process path is not affected by Super Beta and Beta Hifi modes.

The luma path is through an equalization amplifier which provides flat frequency response across the entire FM luma carrier range. The equalization amplifier is necessary since the video heads cannot provide a linear response across the entire luma spectrum (lower FM sidebands through the luma FM at 5.6 MHz).

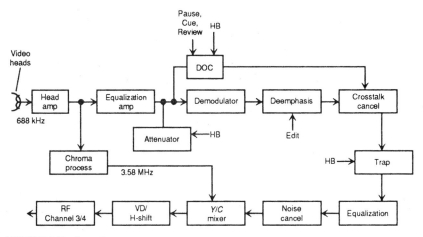

FIGURE 6.12   Super Beta playback circuits.

During high band (HB), the luma playback bandwidth is wider. The balance between the lower frequencies and extended higher frequencies is maintained with an attenuator (turned on by the HB signal).

The RF luma signal divides into two paths after the equalization and attenuation stages. One path is into the DOC stage, which produces an output pulse if the input luma signal falls below a threshold level. During Super Beta, the short time-constant, high-frequency signals may not reach the DOC threshold and thus produce an incorrect dropout signal (which calls for dropout control when none is needed).

During Super Beta, the DOC *sensitivity is reduced* to prevent false dropout control. Likewise, the DOC circuit is *completely disabled* during pause, cue, or review modes. Keep this in mind when troubleshooting a "the dropout circuit doesn't seem to work" symptom.

The main luma playback path is through the demodulator. During record, the low-level, high-frequency components of the luma signal are boosted. These components are attenuated in playback by the deemphasis stage that follows the demodulator. This reduces the low-level high-frequency losses that occur in any type of magnetic-tape recording.

The crosstalk-cancellation stage compares the past and present horizontal lines of the signal, produces a difference of the two signals, and subtracts the difference from the present horizontal line. The crosstalk-cancellation stage also receives signals from the DOC, which produces a control signal (high) when a loss of RF from the tape is detected. The DOC control signal is applied to the crosstalk-cancellation stage which then selects the last active line of horizontal luma from a 1H delay line. This insertion of lines continues until the crosstalk-cancellation stage is turned off by a low from the DOC.

After the crosstalk-cancellation stage, a trap-frequency stage is used to set a maximum frequency limit for the playback signal. This is necessary to eliminate the high-frequency noise *outside the luma bandpass*. In Super Beta, a higher trap frequency is selected (by the HB signal) because of the wider bandwidth of the playback signal. An equalization circuit follows the trap-frequency stage.

The noise-cancel stage removes high-frequency noise *inside the luma spectrum*. This completes processing of the luma signal, from the signal played back by the video heads through demodulation, frequency-response equalization, and noise reduction.

The processed luma and chroma are combined in the *Y/C* mixer. The composite video output is acted upon by a VD and H-shift stage during special effects modes to correct for possible distortion of the H- and V-sync signals. The playback video signal (with both chroma and luma) is then applied to the TV set through the modulator in the normal manner.

Note that some Super Beta VCRs have four drum heads, two of which are for special effects. This is similar to the 4X system of S-VHS and hi-fi (Fig. 5.34). However, in Super Beta, both sets of drum heads process both video and audio, unlike VHS and hi-fi.

CHAPTER 7

CAMERA AND CAMCORDER VIDEO CIRCUITS (VHS AND 8 MM)

This chapter is devoted to the video circuits found in video cameras and camcorders. Because a camcorder is essentially a video camera combined with a VCR, the examples here are based on the circuits found in camcorders (both VHS and 8 mm), including Newvicon, Saticon, MOS, and CCD video-camera circuits. Because of the complexity of the subject, it is not practical (if not impossible) to provide full details for all aspects of camcorders in the space allotted here. Instead of trying to provide such coverage, this chapter gives the video technician practical information that can be put to immediate use when troubleshooting camera and camcorder equipment.

7.1 INTRODUCTION TO CAMERA AND CAMCORDER VIDEO CIRCUITS (VHS AND 8 MM)

This section is devoted to the basics of VHS and 8-mm camcorders (camera-recorders). A camcorder is a combination color video camera and VCR using VHS or 8-mm cassette tape. Camcorders contain conventional video-camera and VCR signal-processing circuits and a single-speed VCR mechanism. However, the physical size of the mechanism is greatly reduced from that found in conventional VCRs. The major difference between the camcorder mechanism and conventional VCRs is a reduction in size of the cylinder (or scanner or drum, whichever you prefer).

7.1.1 The Basic Camcorder

Figure 7.1 shows a simplified block diagram of a typical or generalized camcorder. Before we get into the operation of camcorders, let us review some of the basics, such as differences between the camcorder and standard VCR mechanisms, the principles of color cameras, and the 8-mm format.

7.1.2 Comparison of Camcorder and Standard VHS Mechanism

Figure 7.2 shows the basic mechanical difference between a standard VHS camcorder and a VHS VCR. With the VHS VCR (two head) the tape is wrapped approximately

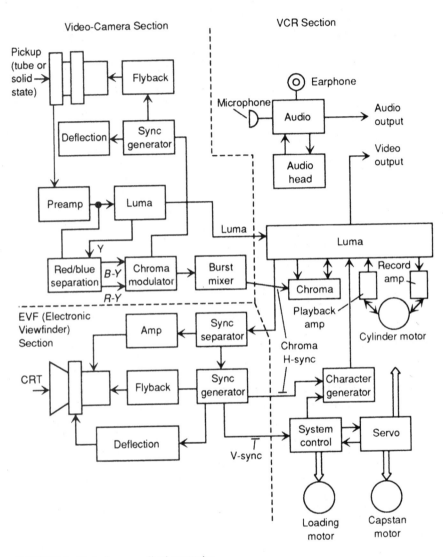

FIGURE 7.1 Typical or generalized camcorder.

180° around a 62-mm cylinder that rotates at 1800 rpm. With a VHS camcorder, four heads are used, and the tape is wrapped 270° around a 41.3-mm cylinder that rotates at 2700 rpm.

The smaller cylinder is used to make the VCR portion of the camcorder more compact. The four-head configuration and 270° wrap are used to make the camcorder tape-recording pattern compatible with that of conventional VHS-format recording patterns. (This is somewhat similar to the way that the adapter used in VHS-C makes the smaller VHS-C cassette compatible with standard VHS in the VCR.) Since the cam-

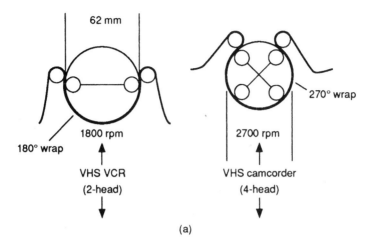

(a)

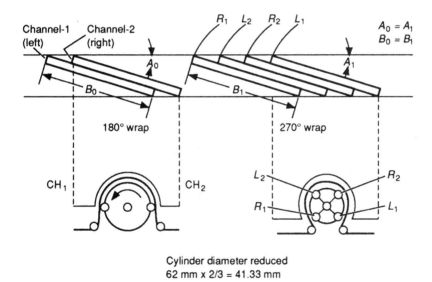

Cylinder diameter reduced
62 mm x 2/3 = 41.33 mm

(b)

FIGURE 7.2 Comparison of camcorder and standard VHS mechanisms.

corder cylinder is exactly two-thirds that of a standard VHS cylinder, three-fourths (or 270°) of the cylinder-wrap is required to keep the same track length.

Figure 7.2b shows how the four-head configuration and 270° tape wrap produce a record pattern that is equal to that of a two-head VHS VCR pattern. As shown, the four video heads record (and play back) in sequence. The typical sequence is L_1 (left-head number 1), R_2, L_2, and R_1. This produces the same pattern as the conventional VHS left-right pattern.

In the conventional VHS format, where the heads are 180° apart, one head is leaving the tape just as the next head touches it. This produces some overlap (about 1.3 ms), which is eliminated during playback by circuits following the head-switching functions.

In the four-head camcorder configuration, at least three of the heads touch the tape at the same time. As a result, only one head must be turned on during both record and playback. This head-selection process is done by the head-switching circuits.

Head Switching during Record. Figure 7.3a shows typical or generalized head-switching circuits during record. Selection among the four heads is performed by the SW_1 through SW_4 switching signals. These signals actuate gate circuits that control video signals from the camera to the recording heads.

The switching signals hold the gate open for 17.3 ms (or longer) and permit camera signals to be applied at the corresponding head for this period. Note that there is some overlap (about 1.3 ms) between L_1 and R_2, R_2 and L_2, L_2 and R_1, and R_1 and L_1. This is the same overlap that occurs on a conventional VHS VCR between the left and right head switching.

Head Switching during Playback. Figure 7.3b shows typical or generalized head-switching circuits during playback. Again, at least three of the heads pick up recorded information from the tape simultaneously, and all four heads are controlled by the head-switching circuits. For example, during period 1, only SW_1 is high, so Q_1 turns off and Q_2 through Q_4 turn on. Under these conditions, the signal picked up by head L_1 is applied through amplifier 1 to the head-switching circuit. The signals at R_1, L_2, and R_2 are shorted to ground through Q_2 through Q_4. During period 2, only SW_2 is high, so only the R_2 head output is passed. During period 3 only the L_2 output is passed. During period 4, only the R_1 output is passed.

The amplified head-output signals are applied to the playback circuits. The head-switching circuits are controlled by 16.7-ms pulses. This eliminates any overlap between head signals. The head-select circuits are controlled by pulses of twice that during playback, so the signals of only one head pass at a time. Head switching is discussed further in Sec. 7.6.

Camcorder Tape Path and Drive. Figure 7.4 shows a comparison of the tape path and drive for a VHS camcorder and a VHS VCR. Note that the 270° wrap requires a somewhat different path (and much more twisting) than does that for the VHS VCR. Also note that for most VHS camcorders, the tape supply reel and takeup reel are rotated by the capstan motor through an idler. A friction clutch reduces the speed so that the reels are rotated at the correct speed in relation to the tape capstan.

7.1.3 8-mm Format

The 8-mm format is substantially different from that of VHS or Beta. For that reason, all of Chap. 8 is devoted to the differences between the 8-mm format and other camcorder formats.

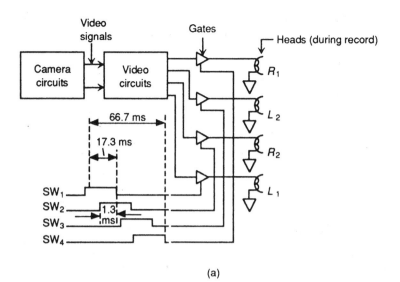

(a)

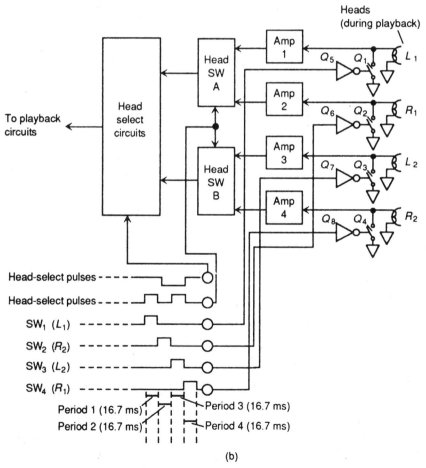

(b)

FIGURE 7.3 Typical VHS camcorder head switching.

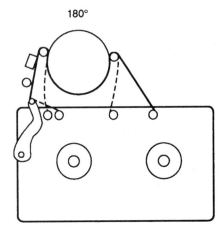

VHS camcorder

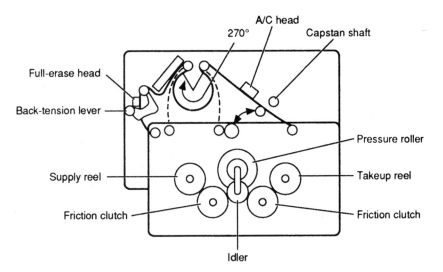

FIGURE 7.4 Comparison of tape path and drive for VHS camcorder and VHS VCR.

7.1.4 Color Camera Basics

The color video cameras used in camcorders are some of the most complex pieces of electronic equipment used in the consumer electronics field. To service a color video camera of any kind, you must understand both optical technology and the operation of the color-processing circuits. For this reason, all of Sec. 7.4 is devoted to color video cameras. For now, let us review the basics, such as the principles of light, forming images with electromagnetic deflection, and processing of colors.

Principles of Light and Color.   As shown in Fig. 7.5*a,* color may be produced by either the *subtractive* or the *additive* process. The subtractive process is used when working with paints. In the case of light and color video cameras, the additive process is used exclusively.

The three *primary colors* in the additive process are *red, green,* and *blue.* Using the additive process (color cameras), any desired color can be obtained by varying the *intensity* of the three primary colors.

In video cameras, the most important factor in the reproduction of color is the type of light source. Even if the same object is photographed (or "shot") by a color camera with different light sources, the reproduced color varies according to the light source. That is why the color of an object appears quite different when photographed under incandescent and fluorescent lamps or when photographed on sunny or rainy days. Light sources are classified as to *color temperature,* which is measured in kelvins and refers to the color of light given off by carbon at different temperatures.

To create a natural color in the color camera, regardless of the light source, camcorders (and most color video cameras) are equipped with a *white-balance circuit.* Most camcorders have both an automatic white-balance circuit (for normal color balance) and a manual override (for special color effects).

The Color Pickup.   Most older camcorders use some form of *color pickup tube* (Newvicon, Saticon, etc.) in the camera. Later-model camcorders use solid-state MOS or CCD pickups, discussed in Secs. 7.1.5 and 7.1.6, respectively.

Figure 7.5*b* shows the relationship of the object to be photographed and the pickup. As in the case of any camera, the visual image shot by the camera is presented to the pickup (and ultimately to the viewer) through an *optical lens.* The lens focuses the image on the photosensitive surface of the pickup, which, in turn, converts the image to electrical signals.

The Basic Black and White Camera.   Figure 7.5*c* shows the deflection system for a typical video camera pickup tube (the classic Newvicon). Note that this deflection system is quite similar to that of a TV picture tube. That is, electrostatic deflection is used when focusing the beam, while electromagnetic deflection is used to form the screen raster. This similarity results from the fact that a video camera pickup tube is essentially the reverse of a TV picture tube.

With a TV picture tube, an electron beam strikes light-sensitive material on the inside of the tube surface. The amount of light produced where the beam strikes the tube surface (the *target point*) depends on the intensity of the beam. In turn, the beam intensity is determined by the video signal.

In a video camera, the amount of light at the target point determines the intensity of the signal produced by the camera pickup.

With both TV picture tubes and video camera pickups, the beam is deflected to produce a raster (typically the EIA standard of 525 lines, 60 fields, and 30 frames) on the tube surface. As a result, the amount of light at any given point on the camera pickup surface produces a corresponding amount of light at the same point on the TV picture-tube surface.

In the Newvicon tube in Fig. 7.5*c,* the electron beam from the tube gun is accelerated by grid G_2 and then passes through the beam-limiting aperture to generate fine-diameter beams. These beams are then focused by the electrostatic lens composed of G_3, G_4, and G_5. Grids G_5 and G_6 form a collimating lens through which the beams are deflected so that the beams always hit the target at right angles.

Figure 7.5*d* shows how the video signal is formed and varied by light in a Newvicon tube. The electron beam scans across the target area, which is coated with a

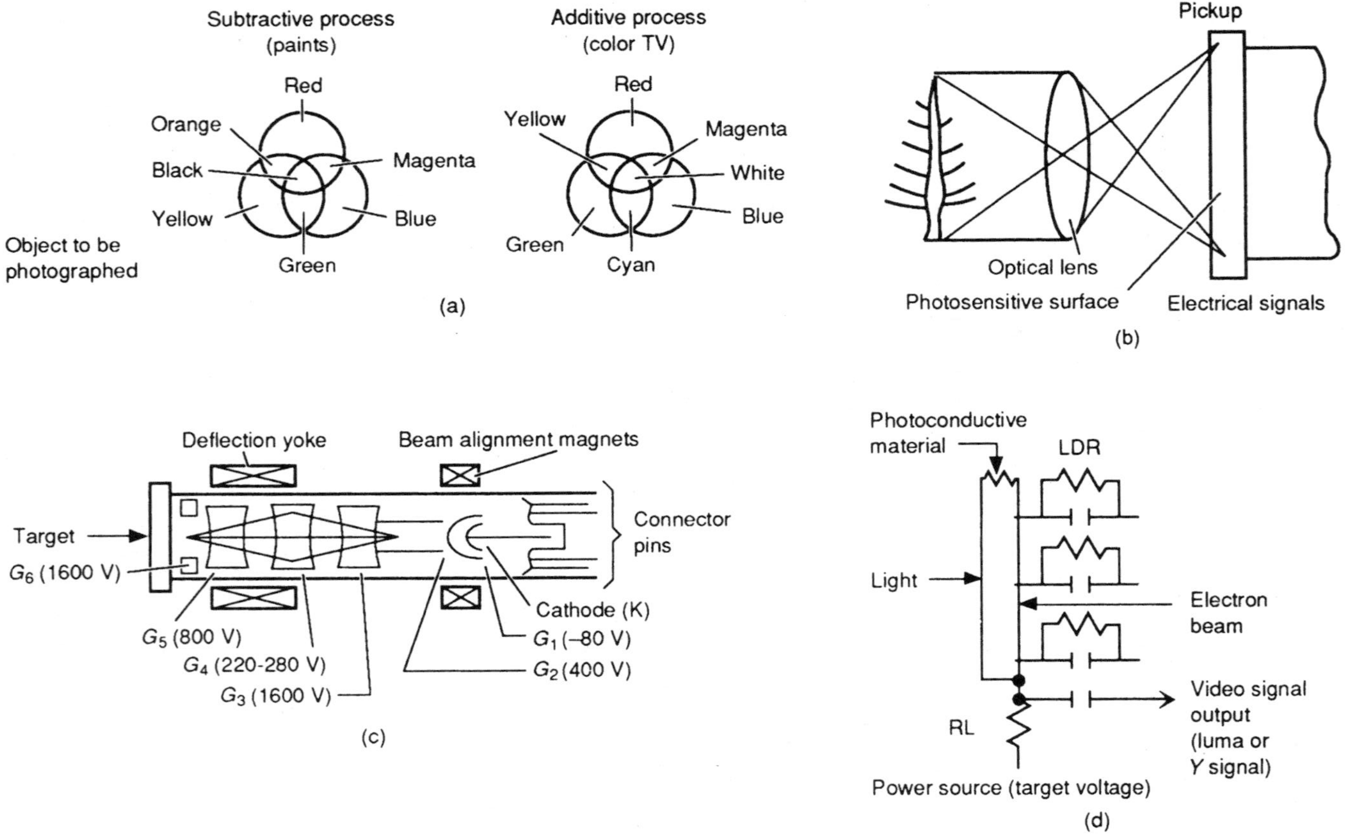

FIGURE 7.5 Video camera basics.

7.8

photoconductive material. This produces a raster on the material (which forms an inner layer on the tube surface).

The photoconductive layer creates a number of elements (or the electrical equivalent of those elements). In effect, electrostatic "capacitors" in parallel with light-dependent resistors (LDRs) are formed by the layer materials. All these elements are connected and produce a video signal output when subjected to variations in light.

When there is no light striking the face of the pickup tube, the LDRs create a high resistance. Whenever light hits the face of the target area, the resistance drops at that point. The level of the drop depends on the intensity of the light.

When the beam first scans the target area, each "capacitor" is charged through the circuit loop formed by the beam, load resistance RL, power source, and photoconductive materials. When the beam is not in contact with the element, the capacitor slowly discharges through the LDR connected across the capacitor.

Since the LDR resistance varies with changes in light, the capacitor recharges back to the target potential on each scan of the beam. This produces a corresponding charging current at each point. The detection of this charging current produces the video signal. As a result, the video signal intensity corresponds to the light intensity at any given point on the raster. As in the case of a TV picture tube, the black and white video signal is referred to as the luma, or Y, signal.

The Basic Color Camera. In addition to a luma signal, a color video camera must also generate a chroma signal to represent the color at any given point on the raster. In the simplest of terms, this is done by separating the colors from each other at the pickup, forming signals for each separated primary color and then recombining the color signals with the luma (black and white) signal to produce a color video signal (that is equivalent to the familiar NTSC broadcast signal).

Separation of the colors is done by means of a *stripe filter* located between the incoming light and the target surface. Figure 7.6a shows the composition of a stripe filter as well as how the signals are separated. Note that the filter, composed of *yellow* and *cyan* stripes, separates the R- (red), B- (blue), and Y- (luma) signals from each other in the signal output of the pickup tube. The following is a brief description of how this separation is done. Refer to Sec. 7.4 for a more detailed discussion of color video camera functions.

To see how the colors are separated, consider the signals produced by the pickup when reading out two consecutive lines, identified as N and $N + 1$ in Fig. 7.6a. When the beam crosses the transparent, diamond-shaped space, a white signal is developed (since white is considered as equal parts of red, green, and blue). In effect, all colors pass and produce corresponding signals.

When the beam crosses the yellow stripe, a red-plus-green signal is produced, but blue is not passed. When the beam crosses the cyan stripe, a blue-plus-green signal is produced, but red is not passed. When the beam crosses an area where the yellow and cyan stripes cross, the yellow stripe blocks blue, while the cyan stripe blocks red. As a result, only green passes.

Note that the two consecutive lines, N and $N + 1$, have a *phase difference* of 90°. This is important to remember since the next step in producing a color signal is to combine the signals from all lines in consecutive order.

As shown in Fig. 7.6b, signals from the pickup are applied through a preamp to an LPF and a BPF. The LPF separates the luma signals, identified as YH (or Y-high, high-frequency luma) at this point, while the BPF passes the color signals. (As you may already know, different colors produce signals at different frequencies.) The color signals are applied to a 90° phase-shift network and a 1H delay network.

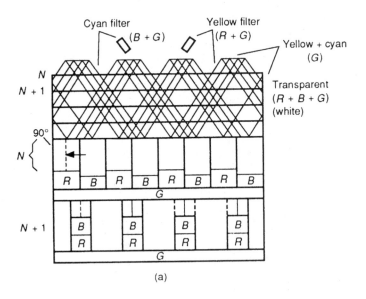

(a)

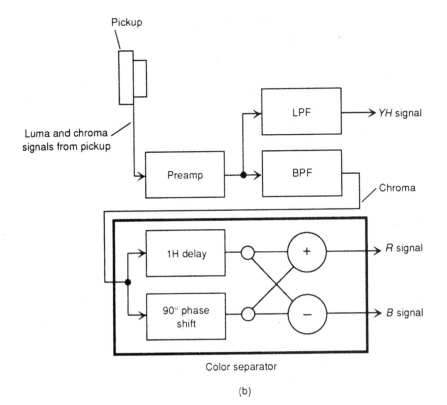

(b)

FIGURE 7.6   The basic color camera.

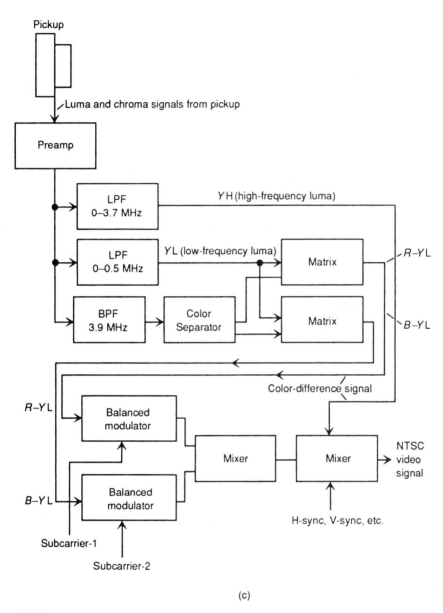

(c)

FIGURE 7.6   (*Continued*) The basic color camera.

The red, or R, signal is obtained by adding the 1H and $90°$ signals. The blue, or B, signal is obtained by subtracting the 1H and $90°$ signals. The red, blue, and luma signals are then combined to produce the equivalent of an NTSC signal.

Producing an NTSC Signal from Camera Signals. Figure 7.6c shows how the basic color camera signals (produced by the pickup, with either tube or solid-state) are combined to produce the video signal (that is then recorded on tape by the VCR portion of the camcorder).

In this system, the brightness of the video signal (at any given point on the raster) is determined by the level of the luma signal. The color information is converted to a *color-difference signal* (R-YL and B-YL).

Note that the luma signal is split into two signals, YH and YL. YH (Y-high) is in the frequency range from zero to 3.7 MHz, while YL (Y-low) is in the range from zero to 0.5 MHz. The YL signal is combined with the R and B color signals to produce R-YL and B-YL. In turn, R-YL and B-YL are combined with the YH signal to produce the composite NTSC signal.

As shown in Fig. 7.6c, after the R and B signals are passed through the BPF and color separator (Fig. 7.6b), the R and B signals are combined with YL in a matrix. The outputs from the matrix are the color-difference signals R-YL and B-YL. The R-YL and B-YL signals are amplitude-modulated by two 3.58-MHz subcarriers. (The color subcarriers are similar to the 3.58-MHz color-burst signal in TV and are $90°$ apart.)

The modulated R-YL and B-YL signals are then combined in a mixer. This process is known as *quadrature modulation*. The output of the mixer is further combined with the YH signal in another mixer. Both horizontal and vertical sync signals are added in this second mixer to produce the composite NTSC signal.

7.1.5 MOS Color Image Sensor

Figure 7.7 shows the solid-state structure and basic circuit configuration of a typical metal-oxide semiconductor (MOS) color image sensor or pickup. The solid-state structure suppresses "blooming" of the picture (a condition common in camcorders with pickup tubes). A three-layer NPN structure provides good color representation across the color spectrum. The MOS sensor is not sensitive to infrared.

A *color-resolution filter* (that produces four basic colors) is made by arranging complementary white, yellow, cyan, and green color filters in a mosaic. The picture elements (or *pixels*) are driven by signals from a drive pulse generator (as discussed in Sec. 7.4). The four color signals—W, Ye, Cy, and G—are read out on four output lines. This provides for interlaced scanning with high resolution and without residual image.

The relationship between scanning and picture elements to produce the color signals is as follows. When light falls on a picture element, or pixel, an electron and hole pair is generated inside the $n+$ layer and $p+$ layer, which form the photodiode. As the electron in the $p+$ layer flows out to the $n+$ layer, a hole remains. As the hole in the $n+$ layer flows out to the $p+$ layer, an electron remains. The resultant photoelectrons are stored in the $n+$ layer as a photoelectronic conversion signal.

When the vertical scanning pulses open the gates of the TVs and the horizontal scanning pulses open the gates of the corresponding THs of the same pixel (in sequence from left to right), the photoelectrons of the photodiodes of two horizontal lines are output in sequence. When the gate opened by the vertical scanning circuit is changed in sequence and repeated, the photoelectrons of all photodiodes are output in sequence, and all pixels are scanned.

The four signals (Ye, Cy, G, and W) are output simultaneously by a 5.43-MHz sampling signal from the drive-pulse generator (Sec. 7.4). The signals are mixed in matrix after preamplification to produce the luma and chroma signals.

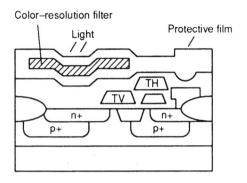

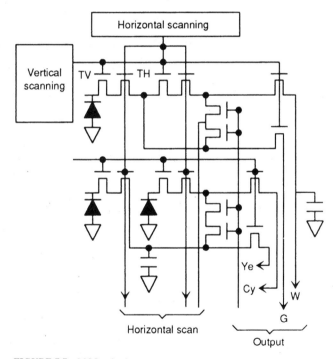

FIGURE 7.7   MOS color image sensor.

7.1.6   CCD Color Image Sensor

Figure 7.8*a* shows the arrangement of photodiodes in a typical charge coupled device (CCD) color image sensor (also known as CCD imager). The photodiodes convert light energy into electrical energy. Shift registers move the electrical image out of the CCD. Resolution of the CCD is determined by the number of photodiodes concentrated in the image area. The CCD in Fig. 7.8 uses 422 vertical columns and 489 horizontal rows, arranged in a 3 by 4 *aspect ratio*. This amounts to 206,358 photodiodes (or pixels).

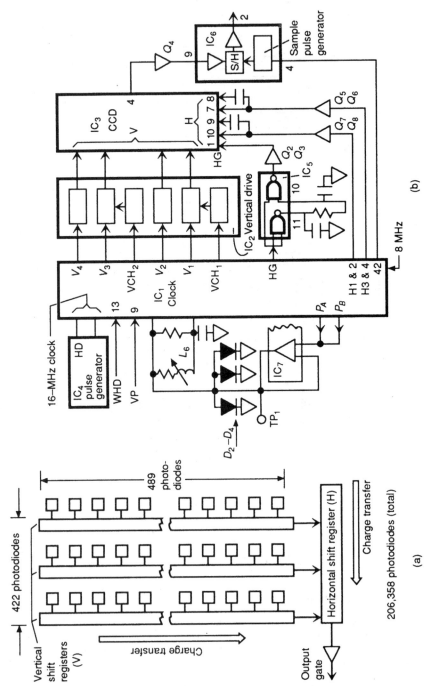

FIGURE 7.8 CCD color image sensor.

7.14

Each photo diode is negatively charged during the *integration period* by the light striking the sensor. Then the negative charge is transferred to the vertical shift registers. (The integration period is the time of one field, less the vertical blanking period.) Adjacent negative charges are mixed in the vertical shift registers to form 244.5 horizontal lines for one field of video information.

Each horizontal line is shifted into the horizontal shift register, one at a time. Between the time each line is shifted down, the horizontal shift register shifts the line out to the output amplifier in serial form. All 244.5 horizontal lines are shifted out of the vertical and horizontal shift registers before the next field begins.

Since the CCD is not a vacuum tube (Sec. 7.1.4), many of the undesired characteristics inherent in a vacuum tube are eliminated. *Image lag* is one characteristic eliminated because of the instantaneous charge-transfer of a CCD. If a vacuum tube pickup is aimed at a bright object, such as the sun, permanent damage can result in the form of *image burn*. CCD (and MOS) pickups are far less prone to image burn. Also occurring in a vacuum tube, but not a CCD, are geometric distortions such as pincushion effects, beam-deflection nonlinearities, and deflection-field nonuniformities. Another advantage of a CCD over a vacuum tube is low power consumption.

CCD image sensors use both digital and analog techniques to produce video signals. The charge developed in a photodiode is an analog function of the light falling on the diode (more light, more charge). Moving the charge from the photodiodes to the shift registers and then out of the CCD, is a digital technique.

CCD Circuits.   Figure 7.8*b* shows typical CCD circuits in simplified form. The V- and H-clocking signals are produced by IC_1 and IC_4. V- and H-sync of IC_1 is controlled by VD and WHD signals, applied at pins 9 and 13, respectively.

The wide horizontal drive (WHD) signal is produced by the system sync generator (Sec. 7.4). The vertical pulse (VP) is taken from the vertical sync signal (VSS) which, in turn, is produced by the sync generator. Internal timing for IC_1 is set by the 16-MHz oscillator (adjusted by L_6).

IC_1 develops two correction voltages, P_A and P_B, at pins 2 and 3. These correction voltages are combined and applied to varactor diodes D_2 through D_4 through IC_7. If the 16-MHz oscillator drifts off frequency, the correction voltages change the varactor-diode capacitance as necessary to bring the oscillator back to 16 MHz. (In this circuit, L_6 is adjusted until the reading at TP_1 is 3 V.)

IC_2 produces the vertical clock signals V_1 through V_4. Each signal is amplified by a driver in IC_2. Signals V_2 through V_4 require two states, low (-7 V) and medium (+4 V), whereas signals V_1 and V_3 require three states, low (-7 V) and medium (+4 V), and high (+9 V). IC_2 uses trilevel drivers for signals V_1 and V_3.

The trilevel drivers produce either a low or medium signal when driven by the V_1 and V_3 signals from IC_1 and a high signal when gated by the VCH_1 and VCH_2 signals from IC_1. Signals V_1 through V_4 are applied to the CCD (IC_3) as shown.

The 8-MHz H-clock signals H_1 through H_4 are produced by IC_1. H_1 and H_2 are inverted by Q_7 and Q_8 and then applied to pins 9 and 10 of CCD IC_3. H_3 and H_4 are inverted by Q_5 and Q_6 and applied to pins 7 and 8 of IC_3. An 8-MHz HG signal is buffered by the NAND gates in IC_5. The RC-network between pins 10 and 11 of IC_5 offers a brief delay before pin 8 of IC_5 goes low. Q_2 and Q_3 invert the signal before application to pin 1 of CCD IC_3. The CCD uses the HG signal to control an internal output gate and permits charges to exit the CCD only when the gate is open.

During each horizontal scan period, one horizontal line is shifted out of the CCD, amplified by Q_4, and applied to pin 9 of the sampling IC_6. The function of IC_6 is to remove unwanted noise generated by the high-speed switching of the CCD.

The 8-MHz control signal at pin 42 of IC_1 is applied to the sample pulse generator in IC_6. The output of this generator controls the sample and hold circuit which, in turn, reduces the switching noise. After the noise is removed, the video signal is applied to the process circuits at pin 2 of IC_6.

7.1.7 Typical Camcorder Features

Figure 7.9 shows features found on a typical camcorder. It is essential that you be aware of these features to understand the video circuits used to provide them. Remember that not all camcorders have all the features described here.

General Information. As discussed, a camcorder combines the functions of a video camera and a VCR in one lightweight, easy-to-use instrument that eliminates interconnecting cables and simplifies operation. A typical camcorder can record directly from the built-in video camera or, with an optional *audio-video input adapter,* from an audio video source such as a VCR. The audio-video played back by the camcorder or fed directly from the camera portion of the camcorder may be viewed and heard on a TV receiver or monitor by feeding the camcorder audio-video to the TV inputs through an optional *audio-video connector cable* or by using an optional RF adapter.

Because a VHS camcorder uses standard VHS video cassettes, the camcorder tapes can be played back on the camcorder or on any VHS VCR. An exception is where the camcorder records in S-VHS. In that case, the cassette must be played back on a S-VHS VCR and S-VHS TV or monitor (to get the full benefit of S-VHS) as discussed in Chap. 5.

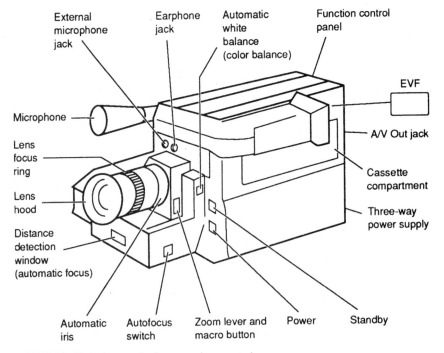

FIGURE 7.9 Typical camcorder features and user controls.

If the camcorder records in VHS-C, the output of the camcorder can be played directly on a TV or monitor (using the RF modulator, which is part of the camcorder). However, an *adapter* must be used to play a VHS-C cassette on a VHS VCR. (There are no VHS-C VCRs.)

In the case of an 8-mm camcorder, the output of the camcorder must be played directly on a TV or monitor (using the RF modulator). There are no 8-mm VCRs (at least none readily available in the consumer electronics market). At the present time, 8-mm cassettes are often played back and rerecorded on a VHS VCR so that a permanent VHS tape can be made.

In addition to being easy to use, lightweight, and compatible with other recording means, a typical camcorder is equipped with autofocus, power zoom, automatic white balance, electronic viewfinder, built-in microphone, and quick review. Let us go through some of these features in more detail.

Pickup (Camera).   Most early-model camcorders have a $\frac{1}{2}$-in pickup tube (Saticon, Newvicon, etc.). Later-model instruments use a $\frac{2}{3}$-in solid-state MOS or CCD pickup. With any pickup, the camcorder can be operated indoors or out with light levels as low as about 7 to 10 lux (less for solid state). The optimum light level for a typical VHS camcorder is about 1500 lux (10 lux is about equal to 1 footcandle).

Electronic Viewfinder.   An electronic viewfinder (EVF) displays (in black and white) exactly what is recorded on the tape (in color). The EVF also doubles as a black and white monitor for viewing "instant replays" after taping. On most EVFs, an LED signals when the camcorder is in the record mode and when battery level is low.

Some EVFs have an *adjustable diopter* that allows EVF focus to be changed without affecting lens focus. This is for the convenience of users who wear eyeglasses. Many EVFs also include a number of displays or graphic indicators. Typical display indicators include indoor-outdoor light setting, time and date, tape counter, memory-on, low-light warning, tape warning, and dew warning.

Lens.   A typical VHS camcorder has an f1.2 lens with a 6-to-1 *zoom ratio*. The zoom can be manual or motorized. A hand-grip lets you zoom in for close-ups or zoom out for panoramic shots. The lens stops at the desired perspective when the hand control is released. The zoom ratios can also be adjusted manually by rotating the lens ring.

Most camcorder lenses also include a *macro* function that permits you to obtain sharp images close up (from about $\frac{3}{8}$ to 1 in). This gives a sharp, enlarged image for shooting small objects without loss of detail.

Automatic White Balance.   As discussed in Sec. 7.1.4, white balance is *critical* to get proper colors. Most camcorders have circuits that continuously adjust for proper white balance (also called *color balance* in some camcorder literature). A manual white-balance (or color-balance) control is included for creating special visual effects or for unusual lighting conditions.

Automatic Focus.   An automatic focus system, using an infrared light beam aimed at objects positioned before the center of the lens, keeps moving objects in focus. This automatically maintains a sharp image, even during zooms. The automatic focus can be switched off for manual focus control (generally to create special effects).

Automatic Iris.   To assure correct exposure (for proper picture brightness and contrast), the camcorder automatically responds to available light conditions and adjusts the aperture accordingly. A manual override is provided for unusual lighting conditions.

Microphone and Earphone.   Most camcorder microphones are front-mounted for increased sensitivity to audio from the subject being shot by the video camera. Some camcorders also include an accessory jack for an external microphone. Usually, camcorders include an earphone jack to monitor both record and playback audio.

High-Quality VHS (HQ).   Many late-model camcorders have circuits to enhance the picture quality by sharpening image definitions. This is the same as the HQ circuits discussed in Chap. 5 and allows complete compatibility with recordings made on other VHS devices.

Pause and Quick Review.   Many camcorders have pause and quick-review functions. In the record mode, the pause function is controlled by a push button. You press the button to start recording and press it again to stop. For a quick look at the last 3 to 4 seconds of the previously recorded scene (on the EVF), you press the review button (with the camcorder in the pause mode).

Search and Stop Action.   On camcorders with a search or stop function, you use the EVF to scan forward or reverse through material to find desired program segments. Or you can examine details easily missed during recording by stopping the action at any point on the tape. These special effects can also be used when viewing tapes on a monitor or TV.

Input-Output Connections.   Most camcorders feature considerable flexibility for use with external components. For example, *audio-video output connections* send playback signals to a VCR or monitor TV for dubbing or viewing, using cables. Likewise, the camcorder output can be applied to a conventional TV through an RF output adapter. *Audio-video input connections* permit recording from selected external sources, such as a VCR, tuner, and so on, using an optional adapter.

7.2   USER CONTROLS

This section describes basic user controls for a camcorder. We concentrate on the controls to establish a basis for troubleshooting. We do not include such things as operating procedures, accessories, and interconnections since these vary widely from unit to unit. Remember that you must study the operating control and indicators for any camcorder that you are troubleshooting. (Again, try to get the operation or user manual.) The following procedures are typical, but there are subtle differences you must consider.

For example, most camcorders have displays that show a low-battery condition and a tape count. In some camcorders, these displays are in the form of LED readouts, while in others both displays appear on the EFV screen. Further, the EFV screen of some (but not all) camcorders shows the recording time remaining and the date and time. Nothing is more frustrating than troubleshooting a failure symptom when the unit is supposed to work that way.

7.2.1   Typical Operating Controls and Indicators

The A/V out jack receives the accessory audio-video connector or cord, which allows video and audio to another device (typically a monitor TV). The audio-video output can also be connected to a standard TV through an RF adapter. (Some camcorders do not have a built-in RF modulator.)

The autofocus switch allows selection of manual focus or autofocus. The power indicator turns on to show that power has been applied. It also functions as a *dew indicator* (by flashing to indicate excessive moisture in the recorder portion of the camcorder). When the power indicator flashes, the camcorder will not operate (as is the case with most VCRs).

The earphone jack connects an earphone to the camcorder, the eject switch is pressed to remove or insert a cassette tape, and the electronic viewfinder, or EVF, displays the scene being observed by the camera.

The *autofocus window* receives an infrared signal that controls the *autofocus system* (Sec. 7.4). An infrared signal (transmitted through the camera lens) is reflected from the object located in front of the camera and is returned to the autofocus system through the autofocus window. The autofocus window is also called the *distance detection window* in some camcorders. With the autofocus switch in the manual position, the camera can be focused by viewing the picture displayed on the EVF while adjusting the *focus ring* for proper focus.

With the auto-manual iris switch in auto, the camera automatically adjusts the iris. The switch is set to manual when the iris is to be adjusted manually.

The *power zoom lens* directs the incoming light onto the camera tube or pickup. The picture size can be magnified 6 times with the zoom feature. For manual zoom, set the autofocus switch to manual and rotate the *zoom ring* in one direction for close-up (T, or telephoto) or in the other direction for wide-angle (W) pictures. For *macro close-ups,* set the autofocus switch to manual and rotate the zoom ring in the wide-angle direction, while pulling the macro switch located on the zoom lever.

The zoom feature is also motor-driven and controlled by the telephoto–wide-angle switches. When the telephoto (T) switch is pressed, the zoom ring moves in the telephoto direction, providing a close-up view of the subject. When the wide-angle (W) switch is pressed, the zoom ring moves in the wide-angle direction, increasing the area of the scene.

A digital *tape counter* indicates the relative position of the program on tape. Press the reset button to reset the counter display to 0000. When the memory switch is on during rewind, the tape moves to an indication of 0000 on the counter.

Press the power switch to apply or remove power. The *power connector* receives the adapter-charger or the battery to power the camcorder. When the standby-power switch is set to standby, power consumption is minimized.

Proper color balance is maintained automatically when the white balance switch is set to auto. For special lighting conditions, the white balance may be set manually by holding the white balance switch in auto set until the record indicator in the EVF stops flashing. The white balance then remains in the selected condition until the white balance switch is set to auto. Note that on some camcorders, the white balance switch can be set to a manual position and varied by a control to get a desired color condition.

The tracking control provides the familiar tracking function found on most VCRs (Chaps. 5 and 6). When playing prerecorded tape, or tapes recorded on other units, streaks may appear on the TV screen. The tracking control is adjusted slowly in either direction until the streaks disappear.

7.3 TEST EQUIPMENT, TOOLS, AND ROUTINE MAINTENANCE FOR CAMCORDERS

As a practical matter, the service procedures for the VCR section of a camcorder can be performed using the same test equipment and tools as for a VCR. Likewise, the routine maintenance, cleaning, and lubrication are the same as for a VCR. All of this is

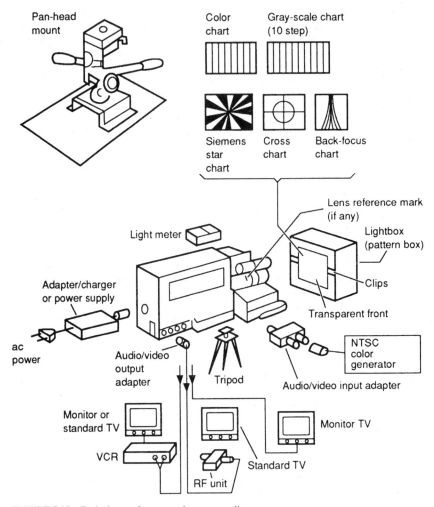

FIGURE 7.10   Typical setup for camcorder camera adjustments.

covered in Chap. 5. A possible exception is cleaning of the camera lens (which we discuss in Sec. 7.3.7). On the other hand, the camera section of a camcorder requires specialized test equipment, particularly for adjustment. Figure 7.10 illustrates a typical setup for camcorder camera adjustment. The following sections discuss the features of such equipment.

7.3.1   Tripod or Pan Head

Obviously, the camcorder must be held steady during camera adjustment. A tripod is your best bet, although some technicians prefer to mount the camcorder on a bench using a pan head. Either way, make certain that the tripod or mount matches the mounting hole on the camcorder bottom. Also check to see if the camcorder has a *lens reference mark*. For many adjustments, the camera must be mounted a precise distance from the object being photographed. The lens reference mark (if any) is usually located on top of the lens and is used as a reference point for measurement to the object. In most cases, the service literature simply specifies a distance from the front end of the lens.

7.3.2 Light Source and Light Meter

Many adjustments require that the object be photographed using a particular light source. Typically, the light source must have a color temperature of 3200 K. (A 3200 K quartz lamp is often recommended.) No matter what the source, use the *correct color temperature* when making any camera adjustments. If not, the colors will never be quite right.

Any light meter can be used. Preferably, the meter should read in lux rather than in footcandles since most service literature specifies lux. For example, a typical camera adjustment procedure specifies 100 lux reflected from an object (a gray-scale chart), 6 feet away from the lens reference mark. If your light meter reads only in footcandles, remember that 10 lux is approximately equal to 1 footcandle.

7.3.3 Lightbox or Pattern Box

A lightbox (also known as a pattern box) is essentially a light source (typically a 3200 K quartz lamp producing about 100 lux) within a box. The front of the box is transparent (Fig. 7.10) and has clips or slots to hold various adjustment charts.

7.3.4 Adjustment Charts

Although there is no standardization, a typical set of adjustment charts includes a *gray-scale chart,* an NTSC *color chart,* and *autofocus chart,* and a *backfocus chart.* The gray-scale chart provides for black and white adjustment, while the NTSC chart provides for color adjustments. The other charts shown in Fig. 7.10 (*Siemens star, backfocus, cross,* etc.) provide a fixed reference for both check and adjustment of the camera.

7.3.5 Monitor-TV

The monitor-TV sets designed specifically for VCRs, video-disc players, and video games provide the most practical means of monitoring a camcorder, both during and after adjustment and troubleshooting. However, if you are planning to go into camcorder service on a grand scale, you may want to consider a receiver-monitor such as used in studio or industrial video work (as discussed in Chap. 5). No matter what is used as the monitor, make certain that the colors are *properly adjusted.* If you adjust a camcorder to produce good colors on an improperly adjusted monitor, you are in trouble.

7.3.6 Vectorscope

The vectorscope described in Chap. 3 is often required for adjustment of some camcorders.

7.3.7 Camera Lens Cleaning

The camera lens requires particular care in cleaning. The lens surface must be clean (and free of moisture) for proper operation of a camcorder. Try not to touch the lens surface, even if your fingers are clean. Body oils can leave smudges on the lens. Keep the lens cap on (except when the cap must be off for service and/or adjustment). Dust can be removed from the lens with an air blower (designed for use on cameras). Dirt can be removed with camera-lens cleaners (cleaning papers).

7.4 CAMERA CIRCUITS

Figure 7.11 is a block diagram of the basic camera circuits. Although the camcorder shown in Fig. 7.11 uses a Saticon tube, the same circuits are found in camcorders with Newvicon tubes. Likewise, some of the circuits are the same for camcorders with MOS and CCD pickups. We discuss circuit differences between tube and MOS and CCD pickups in Sec. 7.4.6.

The camcorder in Fig. 7.11 has five circuit boards in the camera section. Most of these boards are located in the camcorder handle. (This arrangement is typical for many camcorders and video cameras.) The *deflection board* includes circuits that bias the pickup tube, as well as provide power for both the horizontal and vertical deflection of the tube beam (through the horizontal and vertical windings of the deflection yoke). The deflection circuits receive timing and sync signals from a master sync generator on the *regulator board.* Various types of power-supply regulators, *B+* switching circuits, and video switching circuits are also located on the regulator board.

The signal (or carrier) from the pickup tube is removed by the target contact (the ring around the front of the tube) and is applied to the *preamp board.* The target signal is then routed to the *process board.* The majority of the signal processing for the target signal is done on the process board. One of the main functions of the process board is to convert the target signal into an NTSC video signal. Note that the camcorder in Fig.

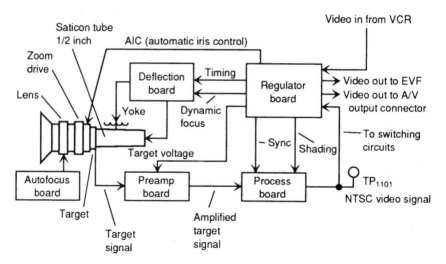

FIGURE 7.11   Basic camera circuits.

7.11 has a test point TP_{1101} to monitor the NTSC video signal. Virtually all camcorders have a similar test point, which is quite important in troubleshooting (and is comparable to the looker point at the output of a TV and VCR tuner).

The NTSC signal from the process board is passed back to the video switching circuits on the regulator board. This video is applied to the EVF and is also available at the audio-video output connector (and can be applied to a TV or VCR through an RF adapter).

The optics and lens assembly of the camcorder uses an IR autofocus system. Circuits on the autofocus board (usually mounted on the lens assembly) generate an IR signal that is transmitted through the lens to the object being photographed. The reflected IR signal is passed to sensors (photodiodes) that operate the autofocus motor.

When autofocus is selected, the motor drives the lens focus ring as necessary to produce correct focus. The lens can also be focused by a power zoom drive. As with most video cameras, there are two zoom buttons (on or near the handle grip) or a single rocker-type button with two positions. One position is for telephoto (T); the other is for wide angle (W).

The lens assembly also has a variable opening, or aperture, called the *iris* (or *iris assembly*). This iris is controlled by signals from the regulator board. The variable iris limits the amount of light that reaches the pickup-tube target surface. The use of a variable iris makes it possible to have a very sensitive target material but still protect the target from excessive light. When light is strong, the iris closes to limit light hitting the target. The opposite occurs when the light is weak.

7.4.1 Camera Optics and Lens Assembly

It is assumed that you have read Sec. 7.1 and that you are familiar with the basics of light, color temperature, and the striped filter used in the optics of camcorders with pickup tubes, as well as MOS and CCD. So we do not dwell on these subjects here. Also, *in most camcorders, the optics or lens must be replaced as a complete assembly* (exclusive of the iris, zoom motor, focus motor, lens hood, pickup-tube retainer or rear block, switches, and circuit boards). So we do not go into details of lens construction.

However, it may be of some practical help to understand the optics between the incoming light (from the scene being shot) and the pickup tube. So let us go through this before we get to camera circuit details. Figure 7.12*a* shows the sequence of light through the optics or lens assembly of the pickup tube.

Wavelength of Light. In addition to the color temperature discussed in Sec. 7.1, another important consideration for light is the wavelength. Light is made up of electromagnetic waves that have very short duration. As shown in Fig. 7.12*a*, electromagnetic waves with a length of about 0.4 to 0.5 μm are seen by the human eye as blue light. Light with wavelengths of about 0.5 to 0.6 μm appear as green light. Wavelengths of about 0.6 to 0.7 μm are seen as red light. White light is produced when all of these wavelengths are combined in the proper ratio.

Waves that are shorter than about 0.4 μm (ultraviolet) or longer than about 0.7 μm (infrared, IR) are generally not detected by the human eye. As a result, the pickup target must be sensitive only to light waves in the range from about 0.4 to 0.7 μm if the correct color video signal is to be generated.

Unlike the human eye, the target material for most pickup tubes is sensitive to IR light and produces an erroneous signal when subjected to IR radiations. To prevent this condition, all IR is filtered out by an *IR-cut filter* between the lens and target. Figure 7.12*a* shows a typical IR-cut filter response.

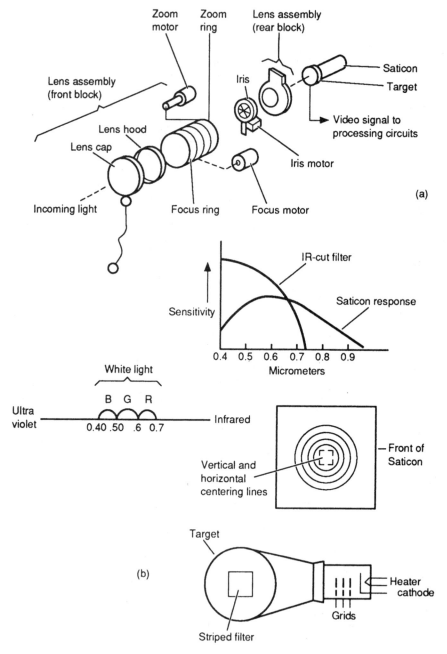

FIGURE 7.12 Camera optics and lens assembly.

Focus Ring.   The light reflected from the subject is passed to the lens assembly through the lens hood. The lens assembly has two adjustable rings. Adjusting the first (or focus) ring alters the position of the lenses so that the image being viewed is properly focused on the target surface of the pickup tube.

From a troubleshooting standpoint, remember that the focus ring has nothing to do with focus of the beam (on the other side of the target) within the tube. If the focus ring is improperly adjusted, the image is blurred. Also keep in mind that the focus ring can be adjusted manually or automatically, and the focus motor can be replaced as a separate component (on most camcorders).

Zoom Ring.   The second adjustable ring on the lens provides a variable zoom control of the lens assembly. Adjusting the zoom ring controls a 6-to-1 magnifying lens system. As a result, the image can be changed in perceptive distance by a ratio of 6 to 1. As with the focus ring, the zoom ring can be adjusted manually or automatically, and the zoom motor can be replaced as a separate component.

Iris and Lens.   Also included in the lens assembly is the iris that controls the amount of light passing through to the pickup target. In most camcorders, the iris is replaced as an assembly or block (including the iris motor) separate from the lens assembly. Generally, it is necessary to remove the rear block shown in Fig. 7.12*a* to replace the iris. As usual, *never* try to remove or replace any part of the lens assembly or pickup tube without consulting the service literature. Camera optics are all similar but *not identical.*

Video Signal.   After passing through the lenses, iris, and IR filter (often part of the rear block), incoming light information is converted into an electrical signal by the pickup tube. As discussed in Sec. 7.1, the tube is much like a conventional cathode-ray tube, except that the electron beam strikes a photosensitive material rather than a luminescent phosphor (as is the case with the cathode-ray tube).

As the beam is scanned across the target surface, a current is generated proportionally to the light falling on that portion of the target. The target current is supplied (via the target lead) to the preamplifier circuit where a video signal is generated.

Pickup Alignment.   As shown in Fig. 7.12*b*, the target surface of the pickup tube is a square area in the center of the tube. Note that it is the target surface that is sensitive to the incoming light that generates the current. If you are curious, you can see the target surface by removing the pickup tube and looking directly at the tube front.

In the case of the Saticon tube shown in Fig. 7.12*b*, you can see the four lines pointing in from both sides as well as from the top and bottom of the tube. These are the vertical and horizontal centering lines used during alignment to position the vertical and horizontal scan across the target surface. In most cases, it is necessary *to align the tube mechanically and then adjust electrically* for proper centering. Always check the service literature.

The author suggests that you not become too curious about the pickup tube or lens assembly. Unless required for service, do not pull the tube or disassemble the lens. If you must handle the tube, take great care to avoid touching or contaminating the front surface. All of these cautions also apply to MOS and CCD pickups.

Dust and Dirt.   With any type of pickup, fingerprints or small specks of dust can prevent the incoming light from reaching the target surface. This can create distortions and spots in the video. If there is dust on the front of the picture tube, wipe the dust with a very soft lint-free cloth (or use a brush or blower). Always handle the pickup with extreme care (tube, MOS, or CCD).

7.4.2 Camera Deflection Circuits

Figure 7.13 shows the deflection circuits for the camera section of a typical camcorder. Sync generator IC_{1106} produces a variety of crystal-controlled sync signals used by the deflection circuits (and the signal-processing circuits, Sec. 7.4.3). The three major signals produced by IC_{1106} are the wide horizontal drive (WHD), the vertical pulse (VP), and the clamp pulse 2 (CP_2). These signals ensure that all functions of the deflection and signal-processing circuits are synchronized as to time.

The WHD signal is applied to the flyback drive circuit which, in turn, drives the high-voltage circuit. Output voltages from the high-voltage circuit are used to bias the pickup tube grids. An additional output from the flyback drive is applied to the blanking amplifier.

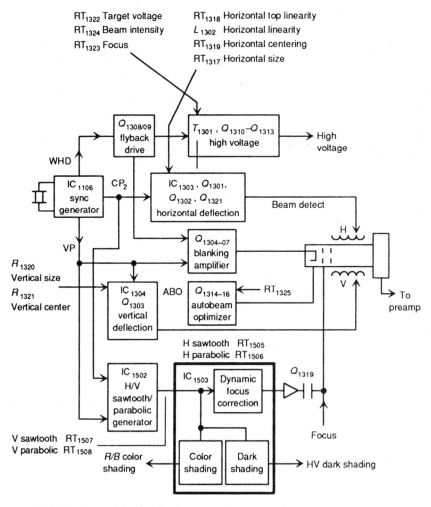

FIGURE 7.13   Camera deflection circuits.

The VP signal is applied to the blanking amplifier, to the vertical-deflection circuit IC_{1304}, and to the horizontal and vertical sawtooth and parabolic signal generator IC_{1502}. The blanking amplifier turns off the cathode emission of the pickup tube during the vertical and horizontal blanking periods of the scan.

The CP$_2$ signal is applied to IC_{1502} and to the horizontal deflection circuits.

Horizontal Deflection Circuits. When CP_2 is high (scanning interval), a sawtooth output from IC_{1303} flows to ground through the horizontal deflection yoke. This produces the usual beam scan, similar to the horizontal scan of a TV picture tube.

The horizontal controls are also similar to those of a TV picture tube. In this camcorder, both the *horizontal top-linearity* control RT_{1318} and *horizontal linearity* control L_{1302} affect the sawtooth current output from IC_{1303} and, thus, determine linearity of the horizontal sweep. *Horizontal-centering* control RT_{1319} sets the fixed reference voltage applied to IC_{1303} and, thus, sets the horizontal position of the beam on the target surface. The *horizontal size* control RT_{1317} controls the amount of direct current through the horizontal deflection yoke and, thus, determines the size (or width) of the horizontal scan.

Vertical Deflection Circuits. The VP pulse is applied to the vertical deflection circuits, which produce a sawtooth sweep at the output. The sweep is applied through the vertical deflection coil and produces vertical deflection similar to that of a TV picture tube.

The vertical controls are also similar to those of a TV picture tube. *Vertical size* control R_{1320} determines the charge time of a capacitor and thus determines the size of the vertical sweep. The *vertical center* control R_{1321} sets the fixed reference voltage applied to the vertical deflection yoke and thus sets the vertical position of the beam on the target surface.

Horizontal and Vertical Sawtooth and Parabolic Signal Generator. This circuit generates H and V sawtooth and parabolic correction signals, as well as a *window gate pulse* applied to the iris control in the signal-processing circuits (Sec. 7.4.3). The horizontal signals are generated by the CP_2 pulse, whereas the vertical signals are produced by the VP pulse. Both H and V parabolic signals, as well as the vertical sawtooth signal (but not the H sawtooth) produce the window gate pulse in IC_{1502}. There are no adjustments in these circuits.

Dynamic Focus Correction Circuit. Generally, the level of focus voltage that produces the best focus of the pickup-tube beam in the center part of the tube is different from that which produces the best focus at the edges of the pickup tube. As a result, the modulation depth for the center of the tube is different from the depth at the edges. (This problem does not usually occur in MOS or CCD pickups.) When a pickup tube is aimed at an evenly illuminated white object, the red and blue signals (modulated at 3.58 MHz) do not have a uniform level. As a result, *color-shading* appears (typically a *greenish color shading*).

The dynamic-focus circuits correct this condition by varying the dc focus voltage (also applied to grid 4 of the pickup tube) as required to offset any unevenness in modulation.

The dynamic focus circuits use the four signals from IC_{1502} (H and V sawtooth and H and V parabolic) to produce the correction signals. The four signals from IC_{1502} are applied to a differential amp in IC_{1503} through corresponding adjustment controls RT_{1505} through RT_{1508}. The signals overlap and produce a synthesized differential signal, which is amplified in IC_{1503} and combined with the dc focus voltage to produce an even depth of modulation across the scanning area.

Blanking and Beam-Detection Circuits.   These circuits generate blanking pulses applied to the pickup tube cathode, thus eliminating scanning-retrace lines. The WHD signal is used to produce the flyback signals for the pickup tube scan (and for the high-voltage circuits). The VP pulse is used to control blanking. The pickup tube is blanked when the flyback pulses are high. When both the VP and flyback pulses are low, the blanking function is removed and scanning occurs.

When the pickup-tube filaments are warming up or when power is low, the pickup-tube beam is absent or weak. This causes the iris to open. If you take pictures of high-luminance objects under these conditions, the pickup tube can be burned out. The beam-detection circuits prevent such a disaster.

During blanking or when there is very little cathode current (during warmup), the circuits generate a high that is applied to the iris. This high shuts the iris and prevents light from reaching the pickup tube surface. When the beam is on full, the circuits produce a low that keeps the iris open. Operation of the automatic iris control (AIC) circuits is discussed in Sec. 7.4.3.

Autobeam Optimizer (ABO) Circuits.   The ABO circuits control the beam of the pickup tube so that the beam matches the object light. Output from the ABO is mixed with the beam-control current in the high-voltage circuit and adds to (or subtracts from) the potential applied to grid 1 of the pickup tube.

Input to the ABO is an ABO signal from the signal-processing circuits (Sec. 7.4.3). Variations in light cause the ABO signal to vary. In turn, the ABO circuits vary the pickup tube beam to maintain the beam at an optimum level. The amount of control produced by the ABO circuit is set by RT_{1325}.

High-Voltage Power Supply.   This circuit provides voltages for the pickup tube grids and target. The circuit also provides a fixed dc potential (typically about 65 V) and a flyback pulse for the blanking and beam-detection circuit. Inputs to the high voltage include the WHD pulse, the ABO control signal, and the dynamic focus signal, as discussed. RT_{1322} sets the target voltage, RT_{1324} sets the pickup-tube beam intensity, and RT_{1323} sets the pickup-tube focus. These circuits are similar to those of a small picture tube or scope. Typically, the maximum voltage is 1400 V (at the final grid) with voltages of 700, 350, and 200 for other grids.

Camera Deflection-Circuit Troubleshooting.   Problems in the camera deflection circuits usually show up as picture problems in the EVF and/or monitor. For example, typical picture problems include no horizontal deflection, no vertical deflection, poor focus, erratic blanking, or no picture (indicating that the pickup is dead).

Before you condemn the deflection circuits, check operation on both a TV and the EVF. If the deflection appears normal on the TV, the problem is probably in the EVF circuits rather than the deflection circuits. If both the TV and EVF show the *same abnormality,* suspect the deflection circuits.

Next, check that all power sources are correct and that signals from the sync generator IC$_{1106}$ are present. For example, if there is no CP_2 signal, there can be no horizontal deflection. If there is no VP signal, there can be no vertical deflection, and so on.

If the picture is totally absent, check all of the voltages from the high-voltage power supply. Then try adjusting the high-voltage circuits. For example, if the target voltage is absent or abnormal, adjust RT_{1322}. If focus is poor, adjust RT_{1323}. If the grid 1 voltage is absent or abnormal, adjust RT_{1324}. If all of the voltages are good, try replacing the pickup tube.

If focus appears abnormal and cannot be corrected by adjustment of RT_{1323} on the high-voltage circuits, suspect the dynamic focus correction circuits. First make certain that the four drive signals from the H and V sawtooth and parabolic circuits are present and normal. Then try correcting focus problems by adjustment of RT_{1505} through RT_{1508}. There should be a measurable change in pickup-tube grid 4 voltage *as any of the* dynamic focus adjustments are made.

If there is no vertical deflection, check for drive signals to the vertical coil of the deflection yoke. If the drive signals are abnormal, try adjusting RT_{1320} (for vertical size) and RT_{1321} (for vertical centering).

If there is no horizontal deflection, check for drive signals to the horizontal coil of the deflection yoke. If the drive signals are abnormal, try adjusting RT_{1317} (for horizontal size), RT_{1319} (for horizontal centering), and L_{1302} and RT_{1318} (for horizontal linearity).

If you get "comet tails" or streaks in high-luminance areas or no picture in low-light areas, suspect the ABO circuits. Try adjusting R_{1325}. There should be a measurable change in pickup-tube grid 1 voltage as R_{1325} is adjusted.

If blanking appears to be abnormal, check for proper blanking pulses at the pickup-tube cathode. Remember that if there is a problem in the blanking circuits, the beam-detect circuit can hold the iris closed.

7.4.3   Camera Signal-Processing Circuits (with Pickup Tube)

Figure 7.14 shows the signal-processing circuits for the camera section of camcorders with pickup tubes. (MOS and CCD pickup signal-processing circuits are discussed in

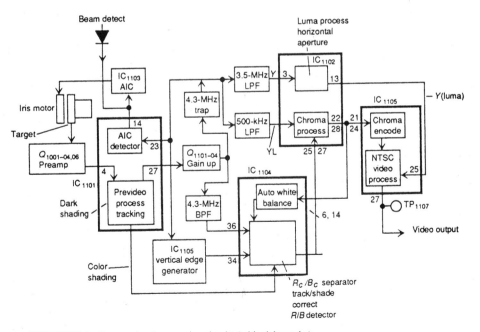

FIGURE 7.14   Camera signal-processing circuits (with pickup tube).

Sec. 7.4.6.) The signal from the target surface is amplified by a low-noise *preamp system.* The output signal from the discrete-component preamp is applied to the *prevideo processing and tracking generator* in IC_{1101}. A *dark-shading* correction signal from IC_{1503} (Fig. 7.13) is applied to the prevideo processor. The target signal from the prevideo processing circuit is applied to a discrete-component gain-up amplifier.

The output signal from the tracking generator circuit is applied to the tracking and shading correction circuit in IC_{1104}, along with the *color-shading* signals from IC_{1503} (Fig. 7.13). These correction signals are used to optimize gain of the red and blue process amplifier in IC_{1102} to compensate for target-surface output variations under different light levels.

The output signal from the gain-up amplifier is applied to both a 4.3-MHz trap and a 4.3-MHz BPF. The 4.3-MHz BPF passes the red and blue signals (R_c and R_b) to the input of IC_{1104}. The 4.3-MHz trap output is applied to the AIC detector in IC_{1101}, to the vertical-edge correction generator in IC_{1105}, and to the luminance-processing circuits in IC_{1102}.

The 4.3-MHz signal to the AIC detector represents the target signal level and thus varies with light. The AIC detector output is applied to the iris through the AIC amplifier in IC_{1103}. The iris is opened and closed as necessary to compensate for variations in light reaching the target (light increases close the iris, and vice versa). The beam-detect signals (Sec. 7.4.2) are also applied to IC_{1103} and keep the iris closed if there are problems in the blanking circuits.

The 4.3-MHz signal to IC_{1102} is applied through a 3.5-MHz LPF and a 500-kHz LPF to produce Y and YL signals, respectively. The signal component below 500-kHz is the YL-signal, which can be considered the *green element* of the picture. The signal component at 3.5-MHz and lower is referred to as the Y, or *luma, information.*

Within IC_{1102}, the red and blue chroma signals from IC_{1104} are combined with the YL signals to produce R-YL and B-YL. These output signals are passed to the *chroma encoder* in IC_{1105} and to the *automatic white-balance* circuits of IC_{1104}.

The Y signal is processed by compensation circuits in IC_{1102} and applied to the NTSC *video-process* amp in IC_{1105}. The chroma and luma signals, along with various sync signals, are combined in IC_{1105} to produce a standard NTSC video signal. In the camcorder in Fig. 7.14, the composite NTSC signal can be monitored at TP_{1107}.

The *vertical edge generator* produces a correction signal that is applied to the red and blue carrier amplifier in IC_{1104}. The correction signal shifts amplifier gain accordingly when a vertical edge is detected. This minimizes *vertical edge distortion* occurring on certain types of scenes (such as horizontal stripes, where there is a sharp, white-to-black or black-to-white vertical transition).

Cameral Signal-Processing Circuit Troubleshooting.   Problems in the camera signal-processing circuits usually show up as picture problems in the EVF and/or monitor. There is one major exception. Since the EVF is black and white, you can get a poor color picture on the TV (because of a problem in the color circuits) but a good picture on the EVF. On the other hand, a defect in the black and white circuits (the Y signal) usually produces problems in both color and black and white performance.

So, before you condemn the signal-processing circuits, check operation on both a TV and the EVF. If the picture appears normal on the TV, the problem is probably in the EVF circuits rather than in the signal-processing circuits. If both the TV and EVF show the *same abnormality,* suspect the signal-processing circuits.

Finally, before you rip into the signal-processing circuits, make certain that power distribution circuits and the deflection circuits (Sec. 7.4.2) are good.

Pay particular attention to the sync-generator (IC_{1106}) circuits (Fig. 7.13). Operation of the signal-processing circuits depends on pulses from the sync generator. The same is true of the sawtooth and parabolic signals from the H/V sawtooth and parabolic generator (IC_{1502}).

If there is no picture but the EVF deflection appears normal, check that the AIC and AGC (IC_{1101} and IC_{1103}, Fig. 7.14) circuits are good. It is possible that the AIC circuit has closed the iris aperture completely.

If you can see the aperture (or hear the iris motor, which is more likely), aim the camcorder at scenes with different light levels and check the response. If this is not practical, monitor the drive voltage to the iris motor and see if there are *voltage changes* with different light levels. Also check for proper *Y*- and *Y*L-signals at the input to IC_{1102}.

If the *Y* and *Y*L signals are normal (check the service literature for correct waveform and amplitude) and the iris motor is operating with changes in light, it is fair to assume that the AIC and AGC circuits are normal. If not, trace signals through the AIC and AGC circuits, starting at pin 14 of IC_{1101}.

Remember that there is a feedback loop in this circuit. For example, if the prevideo signal changes at pin 27 of IC_{1101}, this change is fed back to the AIC and AGC detector in IC_{1101} and produces a change in the detector output. In turn, the detector change varies the prevideo signal at pin 27 of IC_{1101}.

If the power supply, deflection circuits, and aperture appear normal but there is no picture, trace the video signal from the pickup-tube target to the chroma encoder and video processor (IC_{1105}). Pay particular attention to the following test points. (The testpoint values given here are typical. Check the service literature for correct values at corresponding points on the camcorder you are servicing.)

The video output of the preamp circuit is about 0.8 V_{p-p}, while the signal at the input (pin 4) of IC_{1101} is about 0.3 V_{p-p}. Typically, there is an adjustment control between these points.

The output of the prevideo circuits ($IC_{1101-27}$) is about 0.6 V_{p-p} and depends on the H and V sawtooth and parabolic signals from IC_{1502} and IC_{1503} which are usually adjustable.

The output of the AIC and AGC circuit can be checked at the iris motor or at the inputs to IC_{1102}. Note that it is in these circuits where the *Y*, *Y*L, and R_c and B_c signals are first formed. The *Y* signal is about 240 mV_{p-p}, while the *Y*L signal is about 200 mV_{p-p}. Both signals are adjustable. The *Y* output from the luma processing circuits can be monitored at pin 25 of IC_{1105} and is about 650 mV_{p-p}, depending on adjustment.

The combined red and blue carrier signals (R_c and B_c) can be monitored at pin 36 of IC_{1104}. Once the carrier signals are separated into R_c and B_c within IC_{1104}, the signals can be measured at pins 25 and 27 of IC_{1102}. Although both signals depend on several adjustments, the signals are about 200 mV_{p-p} at IC_{1102}.

The circuits that most affect the signals at $IC_{1102-25/27}$ are the vertical-edge correction, white-balance, and H and V sawtooth and parabolic signals. The vertical-edge correction signal can be monitored at $IC_{1104-34}$ and is adjustable. The white-balance signal is adjusted automatically in IC_{1104} (and manually on some camcorders).

The output of the chroma-encoder and video-processor circuits is (or should be) a standard NTSC video (1.0 V_{p-p}) signal (complete with chroma, luma, burst, blanking, and sync). The video output can be monitored at TP_{1107}.

If the signal at TP_{1107} appears to be normal, the entire camera signal-processing circuits can be considered to be good. As a result, some technicians prefer to check here first. Of course, if the output at TP_{1107} is not normal, you must trace back to the pickup tube, as we have discussed in this section.

7.4.4 Automatic Focus Circuit

Figure 7.15*a* shows the main components of an automatic focus (autofocus) system as well as the measurement principles. Figure 7.15*b* shows the autofocus circuits in block form. In some camcorders, the discrete components shown are combined in an IC.

In the camcorder in Fig. 7.15, the autofocus circuit is built into the zoom-lens assembly. The circuit uses infrared light (with a center frequency of 870 nm). IR light from LED D_6 is passed through a projecting lens to a full-reflection mirror, where the IR light is reflected back into a dichroic (two-color) mirror. The dichroic mirror passes visible light (from the scene back to the pickup) but reflects IR light.

The reflected IR light from the dichroic mirror is passed to an object in the scene through a complex lens and is reflected back from the object through a condensing lens to a sensor. The sensor (two IR-sensitive photodiodes, TL_1 and TL_2) produces a signal that adjusts the lens focus so that the reflected IR light is equal on both sensor photodiodes. When this occurs, the lens assembly is properly focused on the object. This same technique is generally called the *delta measurement principle* and is essentially the same as that used by a twin-lens reflex rangefinder to get the distance between the lens and object.

When autofocus is selected, microprocessor IC_2 produces a 9.1-kHz signal that is applied to IR LED D_6 through a drive circuit. The IR signal from D_6 is applied to and returned from the object through the lens assembly.

The reflected light is converted to a current by photodiodes A (far) and B (near). The outputs of photodiodes A and B are amplified and integrated in IC_1. The integrated far and near outputs from IC_1 are compared with a reference voltage from IC_1 to generate four comparison signals in IC_3. These signals are applied to IC_2 and determine which focus-motor drive direction is required so that the output from photodiode A equals the output of photodiode B. When the outputs are equal, the output circuit turns off, and the focus motor stops.

RT_1 sets the gain of IC_1, while RT_2 and RT_3 adjust the offset or balance between the A and B channels. RT_5 sets the gain of the focus-motor drive.

Terminals SL-A and SL-B provide a means to short resistors in or out of the IC_2 oscillator circuit. This sets the width or period of the sync pulses (typically 118 ± 15 μs). The sync signal is applied to both the IR LED D_6 (through the drive circuit) and to the autofocus-signal amplifier IC_1.

Terminals SL-C and SL-D provide a means to short resistors in or out of the IR LED D_6 drive circuit. This sets the amount of drive current to D_6. When the lens moves to the maximum far direction (infinity), the far-end switch S_1 is activated mechanically. This stops the focus motor, preventing damage to the lens mechanism.

Troubleshooting for the autofocus system is discussed in Sec. 7.11 and is not duplicated here.

7.4.5 Electronic Viewfinder Circuits

Figure 7.16 shows typical EVF circuits. The EVF displays video signals from the camera, as well as the video signal played back by the VCR section. The EVF is essentially a small-screen black and white monitor. (Compare the circuits of Fig. 7.16 to those shown in Chap. 2.) The composite video input signal is received from the camera or the VCR. Most of the EVF circuits are contained in IC_{1801}.

The horizontal and vertical sync signals are separated from the composite video signal and applied to the horizontal (H) and vertical (V) deflection circuits. In turn, the H and V deflection circuits provide signals to the horizontal and vertical yokes of

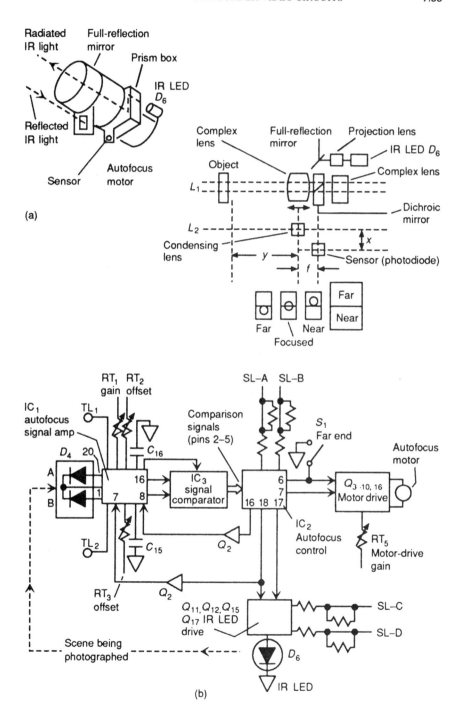

FIGURE 7.15 Automatic focus circuit (camera lens).

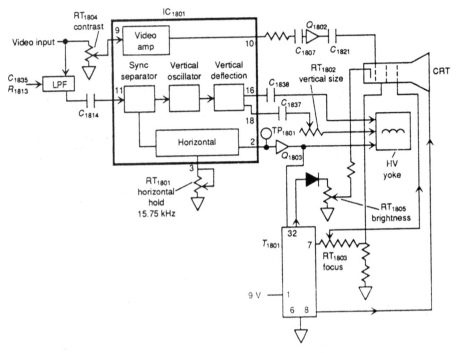

FIGURE 7.16   Electronic viewfinder circuits.

the EVF tube (a small black and white CRT). The horizontal deflection circuit also generates the input signal to the high-voltage power-supply circuits.

Horizontal hold control RT_{1801} sets the frequency of the horizontal oscillator in IC_{1801} (typically at 15.75 kHz).

Vertical size control RT_{1802} sets the amplitude of the vertical drive signal from IC_{1801} to the vertical deflection yoke.

Focus control RT_{1803} sets the focus voltage from IC_{1801} to the focus grid of the EVF tube.

Contrast control RT_{1804} sets amplitude of the video signal to IC_{1801} and, thus, determines the amount of contrast on the EVF tube.

Brightness control RT_{1805} sets bias voltage from IC_{1801} to the cathode and filament of the EVF tube and, thus, determines the amount of brightness on the EVF screen.

Troubleshooting for the EVF is discussed in Sec. 7.11 and is not duplicated here.

7.4.6   Camera Signal-Processing Circuits (with MOS Pickup)

Figure 7.17 shows the signal-processing circuits for the camera section of camcorders with MOS solid-state pickup (or *color image sensors* as they are called in some literature). Compare these circuits with those in Fig. 7.13 (for camcorders with pickup tubes) and Fig. 7.8 (for CCD color image sensors).

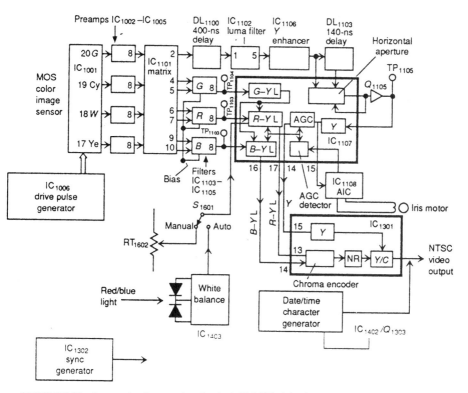

FIGURE 7.17   Camera signal-processing circuits (with MOS pickup).

As shown in Fig. 7.17, the MOS pickup IC_{1001} develops four color signals—cyan (Cy), green (G), yellow (Ye) and white (W)—when driven by signals from drive-pulse generator IC_{1006}. The four signals are applied to preamplifier ICs IC_{1002} through IC_{1005}.

The preamplifiers amplify the color signals to form prevideo signals. In turn, the prevideo signals are applied to matrix IC_{1101}.

IC_{1101} converts the four color signals (G, Cy, Ye, and W) into the luma signal (Y) and chroma signals (R, B, and G). The luma and chroma signals are applied to corresponding filters IC_{1102} through IC_{1105}.

Delay line DL_{1101} delays the luma signal by 400 ns to match the phase with the chroma signals. DL_{1101} also eliminates the 5.43-MHz horizontal shift-register clock pulse. Luma filter IC_{1102} is applied to Y-enhancer IC_{1106}, where the vertical edges of the picture are emphasized.

The luma and chroma processing circuits in IC_{1107} correct the luma, or Y signal, and process the chroma signals (R, B, G, YL) to generate two color-difference signals R-YL and B-YL. The three signals (Y, B-YL, and R-YL) are applied to NTSC encoder IC_{1301}.

IC_{1301} combines the chroma and luma signals in a Y/C mixer to produce a standard NTSC signal. This signal is combined with that of the date and time character generator.

The date and time character generator IC_{1402} and Q_{1303} generates date and time character signals (to be displayed on the EVF and recorded on tape). The date and time signals are mixed with the NTSC video from IC_{1301}.

Sync generator IC_{1302} generates sync pulses and scan pulses for producing the signal and for synchronization with the MOS drive-pulse generator IC_{1006}.

The AIC circuit IC_{1108} controls the iris opening (as determined by the level of the Y signal in IC_{1107}).

The automatic white-balance control IC_{1403} detects the color temperature (by sensing red and blue light) and adjusts the red and blue gains (in IC_{1107}) accordingly to maintain the correct white balance.

MOS-Sensor Signal-Processing Circuit Troubleshooting. The basic troubleshooting approach described in Sec. 7.4.3 also applies to the signal-processing circuits of camcorders with MOS sensors (and CCD sensors). Of course, there are obvious differences. Compare Figs. 7.14 and 7.17. When tracing the video signal from the MOS sensor to the chroma encoder (Fig. 7.17), pay particular attention to the following test points. Check the service literature for correct values at corresponding points on the camcorder you are servicing.

The output of the MOS sensor can be monitored at pins 17 through 20 of IC_{1001}. Typically, these outputs are in the 40- to 50-mV$_{p-p}$ range. The *outputs should increase* when the camera is aimed at a color chart. (This generally applies to all video-signal levels in the path from the MOS sensor to the NTSC encoder.)

If there is no output from IC_{1001}, check that IC_{1001} is receiving drive signals from IC_{1006}. The drive signals should be about 5 V$_{p-p}$.

The output of the video preamps can be monitored at pin 8 of IC_{1002} through IC_{1005}. These outputs to IC_{1101} should be about 4 V$_{p-p}$.

The outputs of the matrix can be monitored at pins 2, 4, 5, 6, 7, 9, and 10 of IC_{1101}. The Y-signal output at pin 2 should be about 1.35 V$_{p-p}$. This signal drops through DL_{1101} to about 1.3 V$_{p-p}$ at pin 1 of IC_{1102}. The color-signal outputs from IC_{1101} should be (approximately) as follows: $-G$ pin 4 = 0.8 V, $+G$ pin 5 = 4 V, $-R$ pin 6 = 3.6 V, $+R$ pin 7 = 0.7 V, $-B$ pin 9 = 2.2 V, and $+B$ pin 10 = 2.8 V (all voltages are peak to peak).

The outputs of the filters can be monitored at pin 5 of IC_{1102} and pin 8 of IC_{1103} through IC_{1105} (or at TP_{1102} through TP_{1104}). The Y signal at pin 5 of IC_{1102} should be about 3.9 V$_{p-p}$. The color signal at pin 8 of IC_{1103} through IC_{1105} should be about 1.2 or 1.3 V$_{p-p}$.

The color-signal outputs from processor IC_{1107} should be about 0.65 V$_{p-p}$ at pin 16 and 0.15 V$_{p-p}$ at pin 17, while the Y-signal output at pin 14 should be about 0.9 V$_{p-p}$. The AIC control output at pin 15 of IC_{1107} should be about 0.44 V$_{p-p}$.

If there are good signals at pins 13 through 15 of IC_{1301} but you do not get a good NTSC signal (video output), suspect IC_{1301}.

If there is good video output but no date and time display, suspect IC_{1402} and Q_{1303}.

If the iris appears to be inoperative, suspect the iris motor and IC_{1108}. (You may not be able to see the aperture, but you should hear the iris motor.)

If there appears to be a problem with white balance, here is a simple trick that applies to most camcorders. Set S_{1601} to manual and check to see if good white balance can be obtained manually by adjusting RT_{1602}. Then set S_{1601} to auto and check automatic operation of the white-balance circuits. If you get good white-balance manually with RT_{1602} but not automatically, suspect IC_{1403}. If you do not get good white balance manually or automatically, suspect IC_{1107} (but first check the signals at TP_{1102} through TP_{1104}).

Remember that all the circuits shown in Fig. 7.17 depend on pulses from sync generator IC_{1302}. If any one pulse is absent or abnormal, one (or more) circuits may be inoperative. In the camcorder in Fig. 7.17, IC_{1302} is crystal-controlled (14.3 MHz) and the oscillator frequency is adjustable.

7.5   TAPE TRANSPORT AND SERVO SYSTEMS

Figure 7.18 shows the arrangement of mechanical parts (from the top side of the tape transport) for a typical VHS camcorder. Figure 7.19 shows a typical VHS camcorder servo system. Compare Fig. 7.19 with Figs. 5.4, 5.16, and 5.18. As discussed in Sec. 7.1, camcorder tape-transport mechanisms are smaller than those of VCRs but contain all the same elements. The cylinder and capstan servo systems are also very similar to those of VCRs.

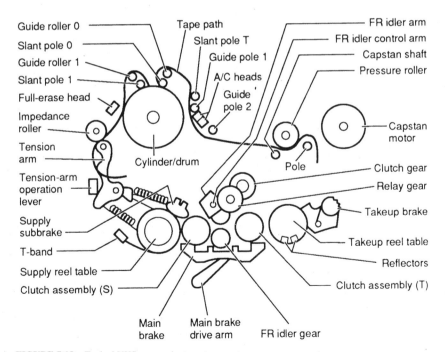

FIGURE 7.18   Typical VHS camcorder tape transport.

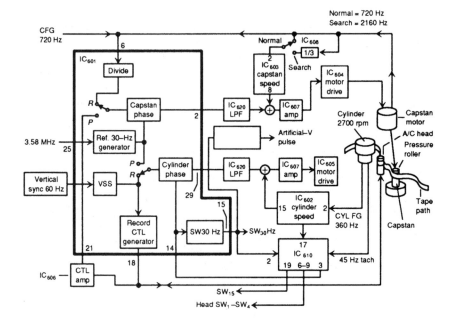

Motor	System	Mode	Reference signal	Comparison signal
Cylinder	Phase	Record Play	1/2–V sync Ref. 30 Hz	30–Hz PG tach pulse
	Speed	Shared	Cylinder FG (360 Hz)	
Capstan	Phase	Record Play	Ref. 30 Hz	Capstan FG (720 Hz) CTL pulse (30 Hz)
	Speed	Shared	Capstan FG (720 Hz)	

FIGURE 7.19   Typical VHS camcorder servo system.

7.5.1 Camcorder Servo-System Basics

If you have read Chap. 5 (and you should), you will recognize most of the circuit elements, with the possible exception of IC_{610}, which produces the pulses to accommodate the four-head cylinder that rotates at 2700 rpm. When the cylinder rotates at 2700 rpm (instead of the typical 1800 rpm for VHS), a 45-Hz cylinder-tach pulse is developed (instead of the usual 30-Hz pulse). IC_{610} converts the 45-Hz pulse from the cylinder into a 30-Hz pulse that is applied to IC_{601}. (The circuits in servo IC_{601} are almost identical to those found in VCR servo ICs.)

The 30-Hz pulse (tach pulse) from IC_{610} is applied to IC_{601} and is returned to IC_{610} as the 30-Hz switching pulse (SW_{30}), together with a 360-Hz cylinder FG (CFG) pulse. These two signals are used to generate the proper head-switching signals for the four video heads (Sec. 7.6).

The servo system shown in Fig. 7.19 also includes a lock-search circuit IC_{608} during search operation. Lock search is achieved by dividing down both the capstan FG and the control-track pulse (CTL pulse) by a factor of 3 (from 2160 to 720 Hz) in the search mode (either forward or reverse). This maintains normal operation of the capstan servo system during search operation.

Camcorder Servo Troubleshooting.   The troubleshooting notes for the VCR servo (both cylinder and capstan) described in Sec. 5.3 also apply to VHS camcorders. Specific camcorder trouble symptoms caused by servo problems are covered in Sec. 7.11.

7.6   VIDEO SIGNAL-PROCESSING CIRCUITS

The video signal-processing circuits of a VHS camcorder are very similar to those of a VHS VCR, with two major exceptions. First, the video for a camcorder is usually taken from the camera section (Sec. 7.4) rather than from a tuner (as is the case with a VCR). On most camcorders, it is possible to receive video from other sources (character generator, VCR, etc.). The video from the camcorder is applied to the EVF as well as to the TV. Because of these factors, special video in- and out-selection circuits are required for a camcorder. Second, because of the four-head video-recording configuration, special head-switching circuits are required for VHS camcorders.

Remember that the luma and chroma circuits for VHS camcorders are contained in three or four ICs, as is the case with most present-day VCRs. If all else fails, you can replace the few ICs, one at a time, until the problem is solved. The one major exception to this applies to the adjustment of controls in the video circuits. The adjustment controls are found outside the ICs. However, when you go through the adjustment procedures recommended in the service literature, you simultaneously localize faults in the adjustment-control circuits.

Also remember that each camcorder has a unique set of circuits which you must check out during troubleshooting (using the service literature). However, to give you a head start in video-circuit troubleshooting, we conclude each of the following sections with troubleshooting tips that apply to virtually all camcorders.

7.6.1   Luma and Chroma Record Process

Figure 7.20 shows the combined luma and chroma signal-processing functions in record for a typical VHS camcorder. Video from the video in- and out-selection circuits is applied to the input of both luma-record IC_{203} and chroma-record IC_{204}. The luma portion of the composite video signal is converted to an FM carrier at a modulation deviation of 3.4 to 4.4 MHz. The FM luma carrier is applied to the luma input of IC_{202}. The chroma portion of the composite video signal is down-converted from 3.58 MHz to 629 kHz by IC_{204}. The 629-kHz chroma signal is applied to the chroma input of IC_{202}, summed with the FM luma signal, and then routed to the video heads through IC_{201}.

As discussed, VHS camcorders use four video heads with 270° tape wrap. Since each head is positioned 90° apart, more than one head is in contact with the tape at any given time. To prevent overwriting adjacent tracks during record, the record current to the video heads must be switched at precise intervals.

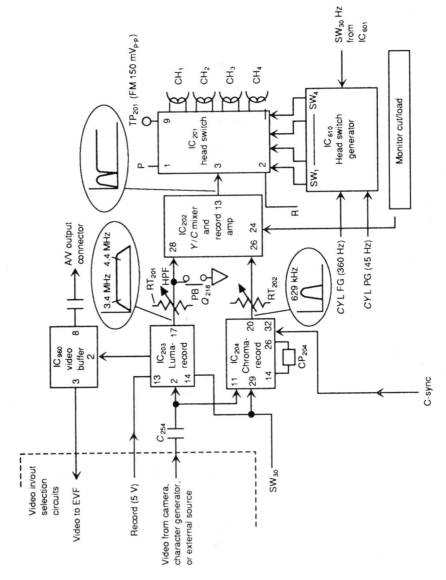

FIGURE 7.20 VHS camcorder record functions.

The signals for head switching are generated in IC_{610}, while the actual head-switching process takes place in IC_{201}. As discussed in Sec. 7.5.1, IC_{610} uses the SW_{30}, the 360-Hz cylinder FG, and the 45-Hz cylinder PG signals to produce the synchronized head-switching signals SW_1 through SW_4.

Luma and Chroma Record Troubleshooting. As in the case of a VCR, the basic approach for troubleshooting of the luma and chroma circuits in a camcorder is to trace signals through the luma and chroma ICs. For the record process (Fig. 7.20), trace from the video in- and out-selection circuits to the video heads, using the service data to locate inputs and outputs. When tracing, make certain to check any components outside the ICs. For example, IC_{204} has three external capacitors and an external bandpass filter, all of which are required to complete the chroma signal path within IC_{204}.

In the circuit in Fig. 7.20, start by checking for a composite video signal at pin 2 of IC_{203} and pin 11 of IC_{204}. If it is missing, suspect the video in- and out-selection circuits.

If there is good video at the input to the luma and chroma ICs, check for video at the corresponding outputs.

Trace the 629-kHz chroma signal from pin 20 of IC_{204} to pin 26 of IC_{202}, through chroma-record control RT_{202}. If necessary, adjust RT_{202} as described in the service data.

Trace the 3.4- to 4.4-MHz luma signal from pin 17 of IC_{203} to pin 28 of IC_{202}, through the luma-record control RT_{201}. If necessary, adjust RT_{201} as described in the service data.

If there is a good composite video signal at pin 2 of IC_{203} but the output at pin 17 of IC_{203} is absent or abnormal, suspect IC_{203} or the associated external components.

Also check that *all the signals and voltages* are available to IC_{203} *for record*. For example, there must be SW_{30} signals at pin 14 to operate the FM modulator in IC_{203}. Likewise, there must be 5 V at pin 13 of IC_{203} to turn on the record functions. On the other hand, pin 16 of IC_{203} must be low. If not, IC_{203} goes into playback mode. (This signal and voltage data must be obtained from the service schematic.)

If there is a good output at pin 17 of IC_{203} but not at pin 28 of IC_{202}, check that Q_{218} is off. Q_{218} is turned on when playback is selected. This bypasses record signals to ground during playback.

If there is a good composite video signal at pin 11 of IC_{204} but the output at pin 20 of IC_{204} is absent or abnormal, suspect IC_{204} or the external components.

Also check that *all the signals and voltages* are available to IC_{204} for record. For example, there must be SW_{30} signals at pin 29 to operate the phase shifter in IC_{204}. Likewise, there must be C-sync signals at pin 32 to operate the burst gate. Pin 12 of IC_{204} must be low. If not, IC_{204} goes into playback. Again, the service schematic must be consulted for signal and voltage data.

If both the luma and chroma signals are present at the input of IC_{202} but the output at pin 13 of IC_{202} is absent or abnormal, suspect IC_{202} or the external components.

Check the status at pin 24 of IC_{202}. This pin goes high during load and momentarily during an edit function (known as *phase match*). If pin 24 is high, the record output of the *Y/C* mixer in IC_{202} is shorted to ground and the record signal does not pass to the video heads. Likewise, the current-up signal at pin 17 of IC_{202} should be on only for about 1 s, following a record-pause-record operating sequence (which is where phase match is used to edit out any blanks in tape).

If the output from pin 13 of IC_{202} is good and appears at pin 3 of IC_{201} but the signal is not being recorded on one or more of the video heads, suspect IC_{201}, the video heads, or the external components.

Check for switching signals SW_1 through SW_4 at corresponding pins of IC_{201}. If the switching signals are absent, suspect IC_{610}. Also check that IC_{201-1} is low and IC_{202-2} is high. If not, IC_{201} may be in playback.

7.6.2 Luma and Chroma Playback Process

Figure 7.21 shows the combined luma and chroma signal-processing functions in playback for a typical VHS camcorder. Although the basic luma and chroma functions for playback in a camcorder are essentially the same as in a VCR, the video head-switching operation is unique to a camcorder. The signals from the four video heads must be selected in the correct sequence, and at exactly the correct time, to combine the signals properly.

During playback, IC_{201} grounds the return end (low end) of the corresponding video head during the period when the head is picking up (reading) the video signal on tape. Also, when one head is turned on (or active), the other three heads are turned off (shorted out). This minimizes any interference that could be developed as a result of the three inactive heads picking up signals on tape.

The output signals from the video heads are applied to head preamplifier IC_{202}. The SW_{30} and SW_{15} signals are used by IC_{202} to select the correct video-head FM carrier and to produce a continuous FM-carrier signal.

The FM carrier output from IC_{202} is filtered, and the luma portion is passed to the input of the luma-playback circuits in IC_{203}. The chroma-playback signal, filtered out of the FM carrier signal, is applied to IC_{204}. The 629-kHz chroma signal is up-converted to the original 3.58-MHz and coupled back into IC_{203}. The chroma signal is mixed with the demodulated luma FM signal in IC_{203} to form a composite video signal at pin 11 of IC_{203}.

The composite video signal is applied to buffer IC_{960} where the two video signals are developed. One signal is applied to the video in- and out-selection circuits, while the other signal is applied to the audio-video output connector. The signal at pin 3 of IC_{960} is also applied to the EVF.

Luma and Chroma Playback Troubleshooting.   As with record, the troubleshooting approach for luma and chroma playback is to trace signals. For the playback process (Fig. 7.21), trace from the video heads to the output of IC_{960} (or directly to the EVF and audio-video output connector).

Again, when tracing signals through the ICs, make certain to check any external components. For example (as shown on the service schematic), IC_{202} has a peaking filter, a chroma trap, a 1H-delay line, a phase comparator, and a buffer, all of which are required to complete the luma and chroma signal path within IC_{202}.

Start by checking the obvious functions. For example, if playback is good on the monitor TV but not on the EVF (or vice versa), check the outputs at pins 3 and 8 of IC_{960}.

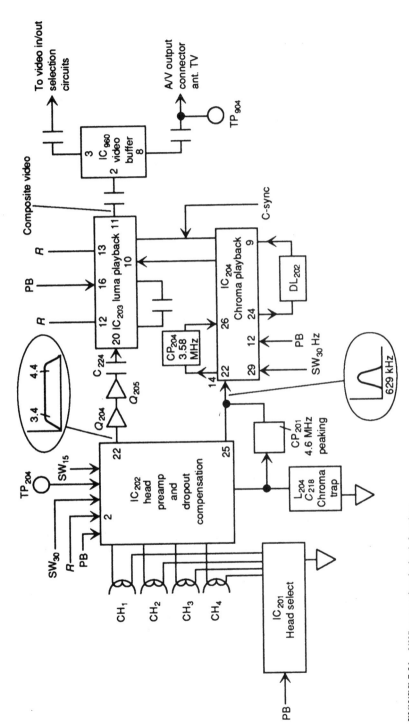

FIGURE 7.21 VHS camcorder playback functions.

7.43

If both outputs from IC_{960} are absent or abnormal, check for a composite signal at pin 11 of IC_{203} and pin 2 of IC_{960}.

If there is no composite video at pin 11 of IC_{203}, check for proper inputs at pins 20 and 10 of IC_{203}. If they are missing, suspect IC_{202} and/or IC_{204}. If they are present, suspect IC_{203}.

Before you pull IC_{203} or IC_{204} (or any other IC), check that all the signals and voltages are available for playback. For example, $IC_{203\text{-}16}$ and $IC_{204\text{-}12}$ must be high for playback, as shown in the service schematics.

It is usually very difficult to check the outputs of individual video heads on most camcorders (as it is on many VCRs). That is why test point TP_{204} is provided. TP_{204} monitors the continuous FM signal, after some amplification and reconstruction by IC_{204}. In most camcorders, it is practical to monitor the 629-kHz chroma output at $IC_{202\text{-}25}$ and the luma output at $IC_{202\text{-}22}$.

7.7 SYSTEM CONTROL

It is difficult to generalize about system-control troubleshooting problems. In camcorders, as in VCRs, system control is performed by a microprocessor. In the simplest of terms, you press buttons to initiate a given operating mode, and the microprocessor produces the necessary control signals (to operate relays, motors, etc.). Each camcorder has unique system-control functions. *You must learn these functions.* Likewise, system control for all camcorders has certain automatic-stop functions, such as end-of-tape stop, condensation detector (dew sensor), reel-rotation detector, and so on, which must be accounted for in service.

From a practical troubleshooting standpoint, all of the notes concerning VCR system control (Sec. 5.3) apply to camcorders. Compare these notes to the system-control circuits you are servicing.

7.8 ELECTRICAL AND MECHANICAL ADJUSTMENTS

We do not describe adjustments in this chapter for several reasons. First, the adjustment procedures are usually very complex, consuming considerable space (more than available here). Of much greater importance, the adjustments can apply to only one specific camcorder. There are virtually no "universal" or "typical" adjustments for camcorders (as there are for VCRs, covered in Chap. 5). However, in Sec. 7.11, we do describe troubles that are associated with improper adjustment. If you experience these troubles during service, look for corresponding adjustments on the camcorder you are troubleshooting.

7.9 OPERATIONAL CHECKLIST

Before we get into detailed troubleshooting, let us review some simple, obvious steps to be performed first. The following checklist describes common symptoms and possible causes for some basic camcorder troubles. Make these checks before you tear into the camcorder with soldering tool and hacksaw.

Camcorder cannot be turned on. No power: First, make sure the power button is on and that the standby button is not on. If you are using a battery, make sure the battery is charged. If you are using an adapter-charger, make sure that it is properly plugged in. Check the dew indicator. The camcorder should not operate if the dew indicator is on. Make sure that all cables (if any) are connected correctly and firmly. Finally, if power seems to be on but you get no picture on the EVF, make sure the lens cap is off.

Cassette cannot be inserted: Check that the power button is on. Insert the cassette with the window side facing out and the safety tab facing up (in most camcorders).

Cassette cannot be ejected: Check that the power button is on. Then press the eject button to open the cassette compartment.

Camcorder cannot be operated in any mode. Power good: Check the dew indicator. The camcorder should not operate if the dew indicator is on.

Camcorder cannot be operated in record: Make sure that the safety tab on the back of the cassette is in place. Make sure the battery and dew indicators in the EVF are not on.

Focus is not sharp: Make sure the lens is properly focused. First, set the focus switch to auto. Then try focusing the lens manually. Also make sure the lens surface is not dirty or dusty. Make certain that there are no obstructions over the focus window (Fig. 7.9). Note that if the focus is good on a monitor or TV but not on the EVF, you have problems in the EVF focus circuits (Fig. 7.16), which are separate from the camera focus circuits.

Color balance is not proper: Check to see if the condition can be corrected by adjustment of white balance (either automatic or manual).

Color picture is not satisfactory: Make certain that the monitor or TV color circuits are properly adjusted. If the camcorder output is being fed to the TV through an RF adapter (to the TV antenna terminals instead of the video-in jack), make certain that the TV fine tuning (if any) is properly adjusted.

Playback picture is noisy or contains streaks: Try correcting the condition with the tracking control.

Playback picture shows blurred action or indicates motion: If this condition occurs when the subject is moving rapidly (or when the camcorder is being panned rapidly), this may be normal (even for camcorders with MOS or CCD pickups, which are supposed to be less prone to blurred action).

Top of playback picture on TV waves back and forth. EVF display is good: This condition indicates a possible compatibility problem between the camcorder and TV. Camcorder signals are not as stable as off-the-air TV signals. This symptom is usually more noticeable when a prerecorded tape, recorded on another camcorder or VCR, is played back. In very extreme cases, it is necessary to alter the TV circuits (Sec. 5.5.3). Before going to such drastic measures, try correcting the problem by adjustment of the horizontal hold on the TV.

7.10 CAMCORDER TROUBLESHOOTING AND REPAIR NOTES

The basic troubleshooting procedures in Sec. 5.5 (for VCRs) also apply to camcorders. These procedures should be followed before going to the specific trouble symptoms discussed in Sec. 7.11.

7.11 CAMCORDER TROUBLE SYMPTOMS AND THEIR SOLUTIONS

The following two sections describe troubleshooting approaches, based on trouble symptoms. The first section covers symptoms that can be caused specifically by improper adjustment. The second is devoted to step-by-step troubleshooting for symptoms that can have many causes.

7.11.1 Trouble Symptoms Related to Adjustment

The following notes describe trouble symptoms that can be caused by improper adjustment. Remember that the same symptoms can also be caused by circuit failure.

Picture displaced in the line following head switching: This problem, often called skew error or skew, is often the result of improper *back-tension adjustment.*

Tape stretched, broken, loose in cassette, or spilled from cassette: These problems can usually be corrected by *cleaning the brakes.* However, if the symptoms remain after cleaning, it may be necessary to replace the brakes.

Tape creased or frilled at the edges: These problems can usually be corrected by *tape-guide* or other *tape-path* adjustments.

No color, no color sync, or improper tint: Assuming a known-good monitor-TV (properly fine tuned if necessary), these problems are usually the result of an *improperly adjusted subcarrier* oscillator but can also be caused by *chroma carrier-level* adjustment.

No picture on EVF or monitor-TV during record: This can be caused by a missing or improperly adjusted target voltage.

Image appears to be dark or light under normal lighting conditions: This condition can be caused by *improper adjustment of the AIC level.*

Picture heavily shaded or washed out: This condition can be caused by *improper adjustment of the beam current* (in a pickup tube).

Picture streaked in the horizontal direction: This condition can be caused by *improper adjustment of the streaking* control (in the prevideo circuits).

Picture does not become totally dark, or there is color noise, when camera is aimed at a dark area: This condition is usually the result of improperly adjusted *dark-shading, dark-offset,* or *chroma setup.*

Focus does not track throughout the entire zoom range: This problem is usually the result of improper *backfocus adjustment* (a mechanical adjustment found on some camcorders).

Distorted images or space at the top or bottom of the picture: This problem is usually the result of improper *vertical size* adjustment.

Distorted images together with a shift in color balance: This problem is usually the result of improper *horizontal size and linearity* adjustment.

Space on either side of the picture: This problem is usually the result of improper *horizontal and/or vertical center* adjustment.

Colored edges in the line following a dark-to-light or light-to-dark transition: This problem is usually the result of *improper vertical edge correction* adjustments.

Uneven red or blue colors: Look for improper *R*-signal or *B*-signal *shading* adjustments.

Red and blue colors change with changes in light levels: Look for improper *tracking* adjustments.

EVF picture tilted or not centered: Look for improper *EVF-tilt* and *centering-magnet* adjustments.

EVF picture out of horizontal sync, distorted, out of focus, or lacks contrast or brightness: Look for improper adjustment of controls shown in Fig. 7.16.

No autofocus. Manual focus is good: Look for improper *autofocus* adjustments.

Horizontal instability and/or picture noise: Look for improper *cylinder motor-speed* adjustment.

Vertical instability and/or picture noise: Look for improper *capstan motor-speed* adjustment.

Horizontal and vertical instability: Look for improper *30-Hz reference-frequency* adjustment.

Vertical jitter (head-switching noise):   Look for improper *PG-shifter* adjustment.

Noise in picture, but picture stable: Look for improper *tracking preset* adjustment, but this can also be caused by user *tracking* control.

Color degraded or diamond-shaped beats in picture: Look for improper *record chroma-level* adjustment.

Picture "overloaded" or "washed out":   Look for improper *record luma-level* adjustment.

7.11.2   Typical Camcorder Trouble Symptoms

The remainder of this section is devoted to step-by-step troubleshooting procedures for typical camcorders. The troubleshooting approach here is based on trouble symptoms (the most common troubles reported to manufacturing service personnel). These symptoms can apply to any camcorder but are related specifically to the camcorders described in this chapter.

   After selecting the symptom (or symptoms) which match(es) those of the camcorder being serviced, follow the steps in the troubleshooting procedure. The procedures help isolate the problem to a defective IC or component shown in the diagrams referenced from the procedures. If the procedures recommend adjustment as a troubleshooting step, consult the service data.

No Picture on EVF, or Monitor, with Power on in Camera Mode.   Figures 7.20 and 7.21 show the circuits involved. This symptom assumes that the camera mode is selected, there is a raster on both the EVF and monitor, but there is no picture on one or more display.

If there is a picture on the monitor but not on the EVF, check for video at pin 3 of IC_{960}. If it is absent, suspect IC_{960}. If it is present, suspect the EVF circuits (Fig. 7.16).

If there is a picture on the EVF but not on the monitor, check for video at IC_{960-8}. If it is absent, suspect IC_{960}.

If there is no picture on either display, apply a color-bar signal (1 V_{p-p}) to C_{254} or IC_{203-2}. If you get a good color-bar picture on the monitor and EVF, suspect the camera circuits (Figs. 7.14 and 7.17).

Next, check for video (1 V_{p-p}) at pin 11 of IC_{203}. If present, check for video at pins 3 and 8 of IC_{960}. If it is absent at both pins, suspect IC_{960}.

If video is absent at IC_{203-11}, check that pin 13 of IC_{203} is at 5 V (high). If it is not, check the record line. Also check that pin 16 of IC_{203} is not high. If IC_{203-16} is high, check the playback line.

No Record Video. Figures 7.20 and 7.21 show the circuits involved. This symptom assumes that the camera mode is selected, the camcorder is in record, the circuits check out as described in the previous "no picture" symptom, but there is no video recorded on tape.

Start by checking for camera video of about 4 V_{p-p} at IC_{202-13}. If it is present, check for an FM signal of about 150 mV_{p-p} at TP_{201}. If it is present, suspect the cassette tape, video heads, or IC_{201}. Also check that pin 2 of IC_{202} is at 5 V (high). If it is not, check the record line.

Next check for a luma FM signal of about 350 mV_{p-p} at IC_{202-28}. If it is present, check the monitor cut and load line at IC_{202-24}. This line should be low during record. If it is not, suspect the system-control circuits.

If the luma FM signal is absent at IC_{202-28}, check for a luma FM signal of about 650 mV_{p-p} at IC_{203-17}. If it is present at IC_{203-17} but absent at IC_{202-28}, suspect RT_{201} or the HPF between IC_{202} and IC_{203}. Also check that Q_{218} has not been turned on by a playback signal. If the luma FM signal is absent at IC_{203-17}, suspect IC_{203}.

No Playback Video. Figures 7.20 and 7.21 show the circuits involved. This symptom assumes that playback mode is selected, the circuits check out as described in the previous "no picture" symptom, but there is no video played back from a known-good tape.

Start by checking for video of about 640 mV_{p-p} at pin 22 of IC_{202}. If it is absent, suspect IC_{202}, IC_{201}, or the video heads. However, before pulling any of these components, check for SW_{15}, SW_{30}, and playback signals to IC_{202} and IC_{201}. If any signal or voltage is absent, check the corresponding line.

If there is video at IC_{202-22} but no playback, suspect IC_{203}. Before pulling IC_{203}, check that IC_{203-12} is low and IC_{203-16} is high. If it is not, suspect system control. Also check Q_{204}, Q_{205}, C_{224}, and all of the components connected to IC_{203}.

No Record Chroma. Figures 7.20 and 7.21 show the circuits involved. This symptom assumes that there is a picture (but no color) recorded on tape.

Start by checking for chroma signals at IC_{202-26}. If they are present but no color is being recorded, suspect IC_{202}.

Next check for about 65 mV$_{p-p}$ at IC$_{204-20}$. If the signal is present but not at 65 to 80 mV, suspect adjustment of RT$_{202}$.

If there is no signal at IC$_{204-20}$, check for about 350 mV$_{p-p}$ at IC$_{204-14}$ and about 260 mV$_{p-p}$ at IC$_{204-26}$. If the signal is present at pin 14 but not at pin 26, suspect CP$_{204}$. If the signal is absent at pin 14, suspect IC$_{204}$. Also check that pin 12 of IC$_{204}$ is not high (at 0 V). If IC$_{204-12}$ is high, suspect the playback line.

No Playback Chroma. Figures 7.20 and 7.21 show the circuits involved. This symptom assumes that there is a picture but no color played back from a known-good tape.

Start by checking for chroma of about 130 mV$_{p-p}$ at IC$_{203-10}$. If it is present, suspect IC$_{203}$.

Next check for chroma of about 600 mV$_{p-p}$ at IC$_{204-14}$. If it is absent, check that IC$_{204-12}$ is at 5 V (high). If it is so, suspect IC$_{204}$. If it is absent, check the playback line.

If there is chroma at IC$_{204-14}$, check for chroma of about 540 mV$_{p-p}$ at pin 26, 1 V$_{p-p}$ at pin 24, and 200 mV$_{p-p}$ at pin 9.

If the signal is absent at pin 26, suspect CP$_{204}$. If the signal is absent at pin 24, suspect IC$_{204}$. If the signal is absent at pin 9, suspect DL$_{202}$.

Video Level Too High or Low: Figure 7.14 shows the circuits involved. This symptom assumes that the picture is either "overloaded" or "washed out," with normal lighting. A malfunction in either the AIC or AGC circuits can cause such a condition.

Start by checking for video of about 640 mV$_{p-p}$ at IC$_{1101-23}$. If the video is good at pin 23 of IC$_{1101}$, suspect IC$_{1101}$.

Next rotate the AIC adjustment, and check that the voltage at IC$_{1101-14}$ changes and that the drive voltage to the iris motor varies from about 0 to 3.5 V. If the iris motor voltage does not vary but the voltage at pin 14 of IC$_{1101}$ changes, suspect IC$_{1103}$. If neither voltage changes, suspect IC$_{1101}$.

No Picture. Figures 7.13 and 7.14 show the circuits involved. This symptom assumes that the camcorder has been checked as described in the previous "no picture" symptom but there is no picture, and the camera mode is selected.

Start by checking out the deflection and signal-processing circuits as described in the troubleshooting portion of Secs. 7.4.2 and 7.4.3, respectively.

If both the deflection and signal-processing circuits appear to be good, check that noise appears (on the EVF and/or monitor) when the pickup-tube target lead is touched. If it does not, suspect Q_{1001} through Q_{1006}. If noise appears, check that the iris is open and that the AIC circuits are operating normally as described in the previous "video level too high or low" symptom.

If the iris is open, all the voltages to the pickup tube are good, and noise appears when the target lead is touched, but there is no picture, suspect the pickup tube.

No Autofocus. Figure 7.15 shows the circuits involved. This symptom assumes that the camcorder can be focused manually, but there is no automatic focus when the focus switch is set to auto.

First make certain that there is power to all the autofocus circuits. Also check for 3 V at $IC_{1\text{-}16}$. If it is absent or abnormal, suspect IC_1.

Check for signals at TL_1 and TL_2. If they are absent, check for signals at pins 1 and 20 of IC_1. If absent, check for a drive signal to IR LED D_6. If it is absent, suspect Q_{11}, Q_{12}, Q_{15}, Q_{17}, and IC_2. If there is drive to D_6 but no signals at pins 1 and 20 of IC_1, suspect D_4 to D_6. (Also make sure the lens cap is off.)

Note that the drive signal from IC_2 and D_6 is controlled by the IC_2 oscillator which, in turn, is controlled by SL-A and SL-B. The oscillator frequency should be about 833 kHz and produce sync signals of about 100-μs duration at pin 18 of IC_2.

If there are signals at TL_1 and TL_2, check for signals at pins 2 through 5 of IC_2. If they are absent, suspect IC_3. If they are present, check for signals (far and near signals) at pins 6 and 7 of IC_2. If they are absent, suspect IC_2.

If the far and near signals are correct but the focus motor does not respond properly, suspect Q_3 through Q_{10} and Q_{16}. Also check adjustment of RT_5, and determine that far-switch S_1 is properly actuated.

Before pulling IC_1, make certain that the clear and sync signals (at pins 7 and 8 of IC_1, respectively) are present. If they are not, suspect Q_1 and Q_2. Check for inverted clear and sync signals at pins 16 and 18 of IC_2. Also check adjustment of RT_1 (gain), RT_2 (B-channel offset), and RT_3 (A-channel offset). Finally, if the TL_1 and TL_2 signals are absent or abnormal but all signals to IC_1 are correct and all IC_1 adjustments have been made, check C_{15} and C_{16}.

No Color.   Figure 7.14 shows the circuits involved. This symptom assumes that there is a picture, but no color, on a known-good monitor TV and that the circuits have been checked as described in the previous "no picture" symptom.

Start by checking out the signal-processing circuits as described in the troubleshooting portion of Sec. 7.4.3.

Then perform all of the adjustments specified for the circuits involved.

Incorrect Color Shading.   Figures 7.13 and 7.14 show the circuits involved. This symptom assumes that there is a color picture but that color shading is incorrect.

Start by checking out the deflection and signal-processing circuits as described in the troubleshooting portion of Secs. 7.4.2 and 7.4.3, respectively. Then perform all of the adjustments specified for the circuits involved.

If all the circuits and adjustments appear to be good but color shading is incorrect, try adjustment of the pickup-tube focus. Then check the pickup tube and deflection yoke. Improper focus and/or deflection problems can cause incorrect color shading.

Incorrect White Balance (Incorrect Color Balance).   Figure 7.17 shows the circuits involved. This symptom assumes that color balance is incorrect, and it is suspected that white-balance circuits may be the cause. (If color balance is good, leave the white balance alone.)

If there appears to be a problem with white balance, set S_{1601} to manual and check to see if good white balance can be obtained manually by adjustment of RT_{1602}. Then set S_{1601} to auto, and check automatic operation of the white-balance circuits.

If you get good white balance manually with RT_{1602} but not automatically, suspect IC_{1403}. If you do not get white balance manually or automatically, suspect IC_{1107}.

No EVF Raster. Figure 7.16 shows the circuits involved.

Start by checking for 9 V at pin 1 of T_{1801}. If it is missing, check the 9-V line. Next, check for voltages from T_{1801} to the EVF CRT as shown in Fig. 7.16. If all the voltages are absent, check for a flyback pulse from Q_{1803}. If one or more of the voltages is absent, check the corresponding circuit. For example, if there is no 2.6-kV power to the CRT, with a good flyback pulse, suspect T_{1801}.

If all voltages from T_{1801} to the CRT are good but there is no raster (or dot or trace), suspect the CRT.

If there is a horizontal trace but no vertical deflection, check for vertical signals from IC_{1801} to the CRT deflection yoke. The vertical drive signal at $IC_{1801-16}$ is about 7 V. If it is absent, suspect IC_{1801}. If it is present but there is no vertical deflection, suspect C_{1838}, C_{1837}, RT_{1802}, and the deflection yoke.

Note that if there is no horizontal drive signal at IC_{1801-2}, there will be no flyback pulse at T_{1801} (and no voltages to the CRT).

No EVF Picture. Figure 7.16 shows the circuits involved. This symptom assumes that there is an EVF raster and there is video on the monitor-TV but no video on the EVF.

Start by checking for a video signal of about 1.5 V_{p-p} at $IC_{1801-10}$, 300 mV_{p-p} at IC_{1801-9}, and 900 mV_{p-p} at $IC_{1801-11}$. If video is present at IC_{1801-9} but absent at $IC_{1801-10}$, suspect IC_{1801}. If video is present at IC_{1801-9} but absent at $IC_{1801-11}$, suspect C_{1835}, R_{1813}, and C_{1814}. If video is present at $IC_{1801-11}$ but absent at IC_{1801-9}, suspect RT_{1804}.

If video is absent at all three points, trace the video back to the EVF input connector (Fig. 7.20).

Finally, check for video of about 20 V_{p-p} at the CRT grid. If it is present but there is no video on the EVF screen, suspect the CRT. If there is no video at the CRT grid but video at $IC_{1801-10}$, suspect C_{1807}, Q_{1802}, and C_{1821}.

Servo Problems. The basic servo systems for both camcorders and VCRs are substantially the same. Compare Fig. 7.19 with Figs. 5.4 and 5.16 through 5.20. All of the troubleshooting information in Secs. 5.3.9 and 5.3.10 applies to camcorders. Of course, there are differences. (The camcorder cylinder rotates at 2700 rpm, to produce the 45-Hz tach pulse, and there is usually no VCR function similar to that performed by IC_{610} in our camcorder.) For these reasons, we list the following trouble symptoms that apply to camcorder servo circuits such as those as shown in Fig. 7.19.

If there are horizontal stripes on the display (both on the EVF and monitor), check adjustment of the *cylinder-speed control* as described in the service data. If this does not cure the problem, check all of the pulses to and from IC_{602}. Pay particular attention to the output from IC_{602} to the motor-driver IC_{605} through IC_{607}. If this signal is good but there are horizontal stripes, suspect IC_{607} and/or IC_{605}. If the signal at IC_{602} is not good, with all other inputs good, suspect IC_{602}.

If the cylinder does not rotate, check that the camcorder is not in stop, rewind, or fast forward (or any mode where the cylinder is not supposed to rotate). With the camcorder in either playback or record, check all voltages and signals at IC_{605}. Pay particular attention to the drive voltage for the cylinder motor. If the drive voltage is good, suspect the motor. If the drive voltage is absent or abnormal, suspect IC_{605}. If the inputs to IC_{605} are absent or abnormal, suspect IC_{607}, IC_{602}, and IC_{601}, in that order.

If the capstan does not rotate, check that the camcorder is not in any mode where the capstan is not supposed to rotate. With the camcorder in either playback or record, check all voltages and signals at IC_{604}. Pay particular attention to the drive voltage for the capstan motor. If the drive voltage is good, suspect the motor. If the drive voltage is absent or abnormal, suspect IC_{604}. If the inputs to IC_{604} are absent or abnormal, suspect IC_{607}, IC_{603}, and IC_{601}, in that order.

If the picture swings horizontally or an alternate noise and clean picture appears in playback, play back a known-good tape recorded on another camcorder or VCR. If the symptoms are removed, it is possible that the sync signal is not being recorded on tape during the camera (or record) mode.

Check the sync signals at IC_{601}. The sync signal is taken from IC_{601-18} and applied to the control (A/C) head during record. The recorded sync signal is taken from the control (A/C) head and applied to IC_{601-21} (through IC_{606}) during playback. These functions are determined by commands from system control, so make sure that the correct commands are given to IC_{601} during both record and playback.

If the symptoms are the same when a known-good tape is played back, check for the cylinder-phase signals to IC_{605} through IC_{620} and IC_{607} from IC_{601-29}. Then check for SW_{30} signals from IC_{601} to IC_{610}. If any of the pulses are absent from IC_{601}, make certain that all record commands from system control are present. (For example, pin 7 of IC_{601} must be high during record and low during playback, in our camcorder.)

If the SW_{30} signal is good, check that IC_{610} also receives both 45-Hz tach pulses from the cylinder motor and the 360-Hz signals from IC_{602}. (However, the 360-Hz pulses are probably good if the cylinder motor is operating at the correct speed.)

Finally, it is possible that the PG-shifter adjustments applied to IC_{601} are not correct. However, this will usually show up when the servo system is adjusted as described in the service data.

If the picture swings vertically or an alternate noise and clean picture appears in playback, play back a known-good tape recorded on another camcorder or VCR. If the symptoms are removed, switch to record and check for the 720-kHz signals at IC_{601-6}. If they are absent, suspect the capstan motor. If the signals are good at IC_{601-6}, check for 437-Hz pulses at IC_{601-2}. If they are present, try correcting the problem by adjustment of the tracking control and any tracking-preset controls. If the 437-Hz pulses are absent at IC_{601-2}, check all input pulses to IC_{601}. Pay particular attention to the control pulse at IC_{601-21}. If they are absent or abnormal, suspect the control head and/or IC_{606}.

If there is considerable noise on the screen in playback (on both the EVF and monitor), start by checking the adjustment of the *capstan-speed control* as described in the service literature. If this does not cure the problem, check the pulses at IC_{603-8}. If they are absent or abnormal, check for capstan FG pulses at IC_{603-2}. If the capstan pulses are absent or abnormal, suspect the capstan motor. If the pulses are good at IC_{603-2} and IC_{603-8}, check all voltages and signals at IC_{603}.

If an alternate noise and clean picture appears (only when a prerecorded tape is played back), check that IC_{610} receives the correct playback and record commands from system control. If it does not, check the system-control microprocessor. Next check for SW_{15} (at pin 19 of IC_{610}) and SW_{30} (at pin 2 of IC_{610}). If the pulses are present at pin 2 but not at pin 19, suspect IC_{610}.

If the pulses are absent at IC_{610-2}, check for pulses at IC_{601-14} and IC_{610-3}. Also check for head-switching pulses at pins 6 through 9 of IC_{610}. If any of these pulses are absent, suspect IC_{601} and/or IC_{610}. If all of the pulses appear to be good, suspect IC_{201}, IC_{202}, and the video heads (Figs. 7.20 and 7.21).

If an alternate noise and clean picture appears (only when a self-recorded tape is played back but prerecorded tapes are good), do not start by saying that this cannot happen. It did. The problem is in the record circuits, since the camcorder plays back a known-good tape. The video record circuits (Figs. 7.20 and 7.21) depend on signals from the servo (30-, 45-, and 360-Hz pulses).

For example, it is possible that IC_{610} is not getting correct servo pulses. (The pulses may be totally absent or not properly timed.) IC_{610} uses the 45- and 360-Hz pulses *directly during record* but not during playback. If either or both of these pulses are abnormal, prerecorded playback can be good, but the self-recorded tape does not get the proper head-switching signals.

7.12   TYPICAL CCD-CAMERA CIRCUITS

This section describes CCD-camera circuits found in video cameras and camcorders (both VHS and 8 mm). The basic principles of the CCD color image sensor are described in Sec. 7.1.6. The circuits described here are found in an 8-mm camcorder but also apply to video cameras and VHS camcorders in general.

7.12.1   Overall Camera Functions

Figure 7.22 shows the overall functions of the CCD-camera circuits. The circuits that apply to the tape transport and servo are discussed in Chap. 8. The CCD camera can be divided into eight main circuits: the CCD, timer, sync generator, CCD drive circuits, process, white balance and iris control, matrix, and encoder. In this camcorder, all of these circuits are ICs, with the usual support components for adjustment and interconnection.

7.12.2   Timing Generator

Timing generator IC_{707} generates all of the timing pulses required for operation of the camera circuits. The timing pulses on the left side of IC_{707} are used to operate the CCD imager. XSG_1 and XSG_2 develop a 60-Hz pulse. The combination of this 60-Hz pulse and the 180° out-of-phase 15,750-Hz pulses from XV_1 and XV_3 produce the V_1 and V_3 pulses, which shift charges out of the sensors, as described in Sec. 7.1.6. The XV_2 and XV_4 pulses are also out of phase at a frequency of 15,750 Hz and are used to produce the V_2 and V_4 pulses. (The V_1 and V_3 pulses shift the charges out of the CCD sensors to the vertical registers and down the registers in sequence. The V_2 and V_4 pulses shift the charges down the registers.)

The XH_1 and XH_2 pulses are also 180° out of phase with each other but are at a frequency of 9.55 MHz. XH_1 and XH_2 are processed to develop the H_1 and H_2 pulses which, in turn, are used to shift, or clock, the charges from the H_1 and H_2 registers at the bottom of the CCD to the output. The output circuit of the CCD is essentially a sample and hold circuit. The sample and hold is operated by the PG pulse (developed from the XH_1 pulse).

The remaining timing pulses are as follows: the vertical area available (VAA) pulse is used together with the CLP_1 clamp to hold the red, green, and blue signals prior to multiplexing and insertion of the horizontal blanking. The sample and hold data (SHD) extracts the CCD pixel data from the output signal. The sample and hold precharge (SHP) extracts the CCD precharge level from the output signal. SP_1 and SP_2 (separation) separate green (*G*) picture information from the red and blue (*R-B*) picture information. ID separates the red and blue information and is also used in the color-multiplexing circuits of the matrix IC_{702}.

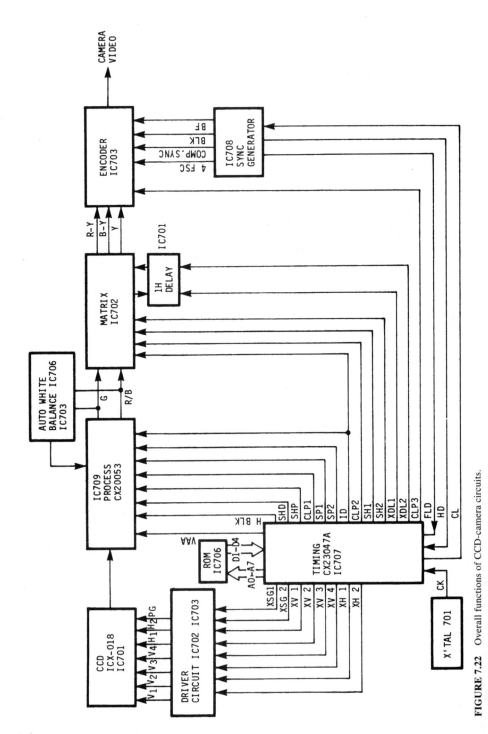

FIGURE 7.22 Overall functions of CCD-camera circuits.

CLP_2 and CLP_3 clamp the G and R-B signals. SH_1 and SH_2 (sample and hold) operate the sample and hold circuits in matrix IC_{702} to develop the Y signal. XDL_1 and XDL_2 (delay) drive a clock in the 1H delay IC_{701} (to matrix the G and R-B signals). The horizontal drive (HD) and field drive (FLD) pulses from sync generator IC_{708} inform IC_{707} of line and field timing.

Timing generator IC_{707} is operated by a 28.6-MHz clock. ROM IC_{706} is programmed to correct flaws created in the CCD during manufacture. In this camera, each CCD is provided with a custom ROM to compensate for flaws at various spots on the particular CCD, as discussed in Sec. 7.12.8.

7.12.3 V_1 and V_2 Drive Circuits

Figure 7.23 shows the circuits used to generate the V_1 and V_3 drive pulses. Figure 7.24 shows typical scope displays for V_1 and V_3. As discussed, these drive pulses shift the charges of the odd- and even-field sensors, respectively, to the vertical registers. V_1 and V_3 are also used to shift the charges down the vertical registers. IC_{702} and Q_{718} combine to couple timing pulses from IC_{707} and to develop the required drive-voltage level.

The XV_1 and XV_3 pulses developed by IC_{707} are applied to pins 1 and 3 of IC_{702}. When XV_1 is pulsed low, the NAND-gate output at pin 7 rises to the level of the Vcc_2 voltage at pin 5. (Because of the voltage drop across circuit components within IC_{702}, between pins 5 and 7, the actual voltage at pin 7 is about 7 V_{p-p}.) The pulsed voltage at pin 7 is coupled across C_{718} and clamped at -5 V by D_{705} and R_{723}. As a result, the pulse rises to $+2$ V and falls to -5 V, as shown in Fig. 7.24.

The XSG_1 and XSG_2 pulses from IC_{707} are applied to pin 5 of IC_{702} through AND gate D_{711} and Q_{718}. These pulses are normally high, except during the time of vertical blanking when one of the pulses goes low. This creates a low at the output of D_{711} at the vertical-blanking rate of 60 Hz. The low turns Q_{718} on and applies 12 V to pin 5 of IC_{702} through C_{714}. The other side of C_{714} is at Vcc_2 (about 9 V), creating a total voltage at pin 5 of about 20 V. With XV_1 low, pin 7 of IC_{702} (normally at about 7 V) rises to about 18 V_{p-p} (or the 20-V Vcc, less the voltage drop across components between pins 5 and 7). The 18-V signal is then coupled across C_{718} and clamped by D_{705} and R_{723}. The resulting signal (called V_1) is used to transfer the charges on the odd-field sensors to the V_1 vertical register. The V_3 signal is developed in essentially the same way as V_1 and is used for the even-field sensors.

7.12.4 V_2 and V_4 Drive Circuits

Figure 7.25 shows the circuits used to generate the V_2 and V_4 drive pulses. As discussed, these drive pulses shift the charges of the CCD sensors down the vertical registers.

The XV_2 and XV_4 pulses developed by IC_{707} are applied to pins 1 and 3 of IC_{703}. When XV_2 is pulsed low, the NAND-gate output at pin 7 rises to the level of the Vcc_2 voltage at pin 5. (Because of the voltage drop within IC_{703}, the voltage at pin 7 is about 7 V, instead of the 9-V Vcc_2.) The pulsed voltage at pin 7 is coupled and clamped, producing a V_2 pulse that rises to $+2$ V and falls to -5 V. The V_4 signal is developed in essentially the same way as V_2. However, note that the V_2 and V_4 pulses do not have a 12-V pulse superimposed on them (as do the V_1 and V_3 pulses) because the V_2 and V_4 pulses are not used to transfer charges from the sensors to the vertical registers (Sec. 7.12.3).

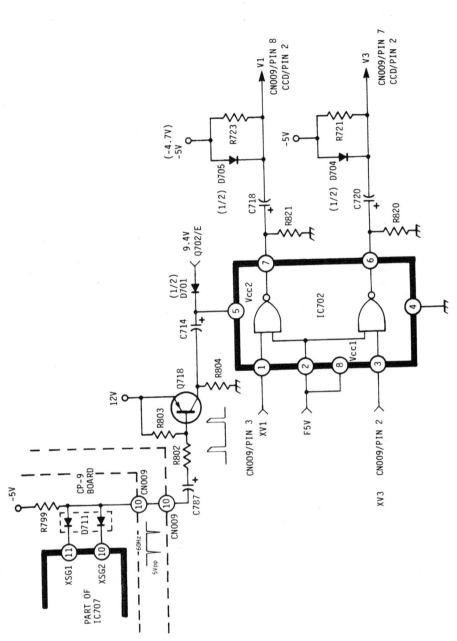

FIGURE 7.23 V_1 and V_3 drive circuits.

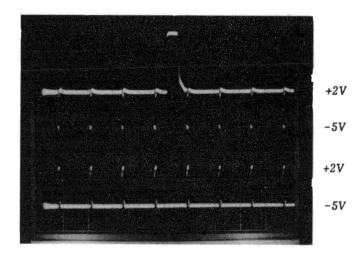

Top: V1 IC701/Pin 5 5V/d 50μsec/d
Bottom: V3 IC701/Pin 2 5V/d 50μsec/d

FIGURE 7.24 Typical V_1 and V_3 scope displays.

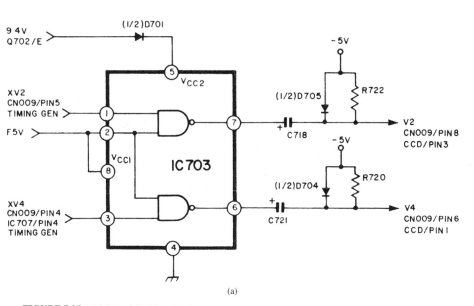

(a)

FIGURE 7.25 (a) V_2 and V_4 drive circuits.

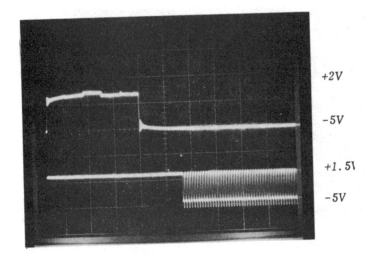

+2V

-5V

+1.5\

-5V

Top: V4 IC701/Pin 1 5V/d 1μsec/d
Bottom: H1 IC701/Pin 18 5V/d 11μsec/d

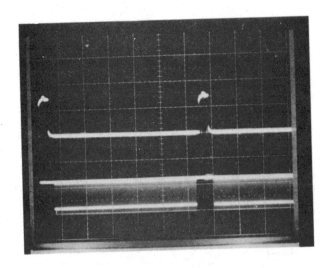

Top: V4 IC701/Pin 1 5V/d 10μsec/d
Bottom: H1 IC701/Pin 18 5V/d 10μsec/d

FIGURE 7.25 (*b*) Scope displays.

7.12.5 H_1 and H_2 Drive Circuits

Figure 7.26 shows the circuits used to generate the H_1 and H_2 drive pulses. As discussed, these drive pulses shift the charges from the CCD imager to the output circuit. During horizontal blanking, the V_4 and H_1 registers are pulsed to about 2 and 1.5 V,

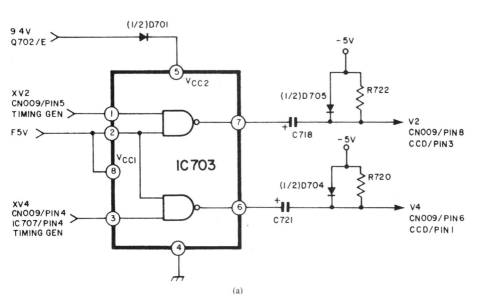

(a)

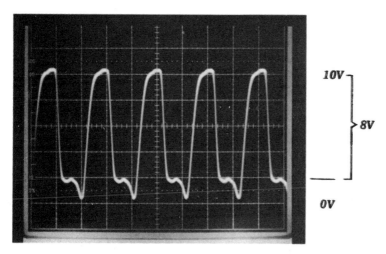

PG Pulses: IC701/Pin 15   2V/d   0.05μsec/d

FIGURE 7.26   (a) H_1 and H_2 drive circuits and (b) scope displays.

respectively. About 1.6 μs before the end of the horizontal blanking, V_4 goes to -5 V, but H_1 remains at 1.5 V. The charges in the V_4 register are then transferred to the H_1 register and pulsed by H_1 and H_2 to the output circuit.

The XH_1 and XH_2 pulses are generated by IC_{707} (at a frequency of 9.55 MHz) and applied to IC_{705}. XH_1, a 4.5-V_{p-p} pulse, is applied to pin 2 of IC_{705} and amplified to about 6 V. The amplified and inverted pulse at pin 5 is clamped to about 1.3 or 1.5 V by D_{707} and R_{737}. The resulting H_1 signal rises to 1.5 V and falls to about -4 V. The H_2 pulse is developed in essentially the same way as H_1.

7.12.6 PG Drive Circuit

Figure 7.27 shows the circuits used to generate the PG pulses. As discussed, these pulses are developed from the XH_1 pulse and are used to control the output circuit of the CCD (essentially a sample and hold circuit).

When the XH_1 pulse rises to 5 V, Q_{714} is turned on, as is Q_{717}. The output at the collectors of Q_{717} and Q_{716} goes low and is clamped at about 2 V by D_{706} and R_{735}. When XH_1 goes low, Q_{714} turns off, and Q_{715} turns on. This turns Q_{717} off and Q_{716} on. The Q_{716} and Q_{717} collectors go high (about 8 V) and are clamped at about 10 V.

7.12.7 Sync Generator

Figure 7.28 shows the internal circuits of sync generator IC_{708}, which produces all necessary NTSC timing pulses required by the CCD processing circuits. (The timing pulses are required because the CCD output is not directly NTSC compatible.) A 14.3-MHz timing signal, originating in timing generator IC_{707}, is applied to pin 26 of IC_{708} and is used as the timing source for all pulses developed in IC_{708}.

The timing pulses produced by IC_{708} are as follows:

Pin 3: BF (Burst Flag) to the encoder

Pin 4: Comp Sync (Composite Sync) to the encoder

Pin 5: FLD (Field Drive) to the timing generator IC_{707} for field identification

Pin 6: BLK (Blanking) to the encoder

Pin 8: HD (Horizontal Drive) to the automatic white-balance (AWB) circuit, and back to IC_{707} for line identification

Pin 12: VD (Vertical Drive) to the AWB circuits

Pin 25: 4 FSC (Frequency of Color Subcarrier) to the encoder

7.12.8 CCD Output Demodulation Circuits

Figure 7.29 shows the CCD output demodulation circuits. As discussed, the CCD signal is not directly compatible with NTSC standards and must be converted. The first step in conversion to NTSC is demodulation of the CCD output.

The output of the CCD is a 9.55-MHz signal containing the picture information. This signal is coupled from pin 11 of IC_{701} to pin 47 of IC_{709} (the processing IC) by Q_{704}. IC_{709} contains a sample and hold circuit used to separate picture information from the 9.55-MHz carrier.

The SHP and SHD signals from IC_{707} are applied to pins 44 and 45 of IC_{709}, respectively. The SHD and SHP signals are 180° out of phase, at a frequency of 9.55

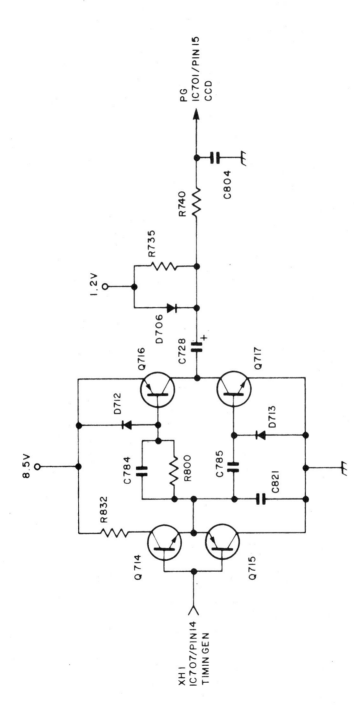

FIGURE 7.27 PG drive circuits.

7.61

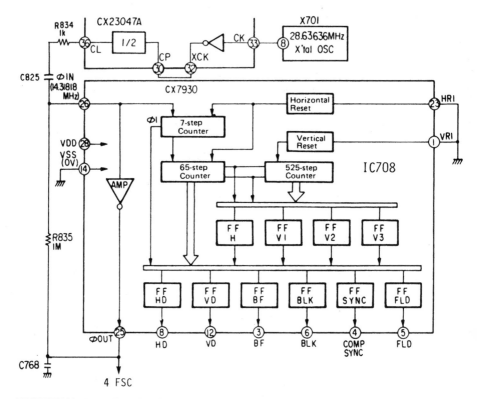

FIGURE 7.28   Internal circuits of sync generator.

MHz. The SHD pulse goes high when there is picture information in the CCD output signal. With SHD high, SW_1 in IC_{709} closes and C_{749} charges. When SHD goes low, SW_1 opens and C_{749} maintains the charge until SW_1 closes again. C_{749} then charges to a new level. The continuous charge variations of C_{749} (because of output variations from the CCD), cause the voltage across R_{716} to vary. This varying voltage represents the CCD output signal (picture information), but without the 9.55-MHz carrier signal.

During manufacture of this particular CCD, some pixels may become defective. If these flaws are not corrected, there will be black or white spots in the reproduced picture. (To control and correct this problem, only CCDs with less than 12 defective pixels are used.)

Because of the way in which the pixels are arranged on the CCD and because of the method used to remove the charges, the exact time of charge removal can be computed. Each CCD with defective pixels has a ROM programmed with this information. The ROM causes the timing generator to delete the SHD pulse when the charge of a defective pixel would normally be converted. The signal from the previous pixel is used in place of that from the defective pixel because C_{749} is not allowed to change its charge-voltage level. In the chroma-separation circuits (Sec. 7.12.10), the SP_1 and SP_2 pulses used to control the sample and hold circuits are also controlled by the ROM to correct pixel flaw problems that could affect color information.

SW_2 and a comparator at pins 3 and 7 of IC_{709} are used to remove glitches (or

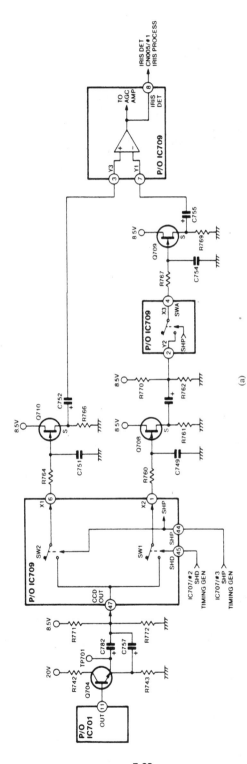

FIGURE 7.29 (a) CCD output demodulation circuits.

(a)

7.63

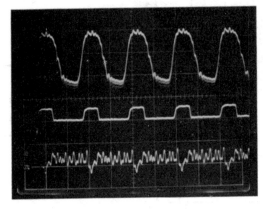

CCD Output TP701
0.5V/d 50nsec/d

SHP Pulse
10V/d 50nsec/d

Precharge Output
Q710/d 0.2V/d 50nsec/d

FIGURE 7.29 (*b*) Scope display.

switching noise) caused by the rapid switching of SW_1 and the sudden changing of the charge levels on C_{749}. When SHP is high, SW_2 is closed during the time that the CCD output does not have pixel information. This time is called the *precharge time*. SW_2, C_{751}, and Q_{710} combine to extract the voltage levels established from the CCD during precharge time. The resulting precharge voltage level found across R_{766} is coupled to pin 3 of IC_{709} (the noninverting input of the comparator).

Glitches across R_{761} are removed by another sample and hold circuit at pins 2 and 4 of IC_{709}, operated by the SHP signals. The signal at pin 2 of IC_{709} is sampled and held by C_{754}. This signal appears across R_{796}, with any glitches referenced to SHP, and is applied to the inverting input of the comparator. The two signals are compared, with one signal canceling the other. The resulting output signal is void of glitches and displays only the pixel information. The comparator output is applied to the AGC amplifier within IC_{709} (Sec. 7.12.9) and to the iris-detect circuits within IC_{706} (Secs. 7.12.14 through 7.12.16).

7.12.9 AGC and Chroma Separation Circuits

Figure 7.30 shows the AGC and chroma-separation circuits. The signal leaving the comparator in IC_{709} does not maintain a constant strength, and must be processed by an AGC circuit to properly reproduce the color and brightness information. The demodulated CCD signal is applied to a gain-controlled amplifier (GCA) within IC_{709}. Maximum gain of this amplifier is adjusted manually by RV_{707}. Automatic or variable gain is controlled by the voltage at pin 10 of IC_{709} from pin 16 of IC_{706}, as discussed in Secs. 7.12.17 through 7.12.20.

7.12.10 Chroma Separation Circuits

Figure 7.31 shows the chroma-separation circuits. The pixels in the CCD are arranged in the following manner for field one: Line 1 = green-red, green-red, etc.; line 3 = blue-green, blue-green, etc.; line 5 = green-red, green-red, etc.; line 7 = blue-green, green-blue. This arrangement continues throughout the CCD so that, by properly tim-

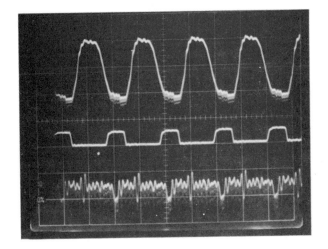

Top: CCD Output TP701 0.5V/d 50nsec/d
Center: SHD Pulse IC709/Pin 45 10V/d 50nsec/d
Bottom: Q708/S 0.2V/d 50nsec/d

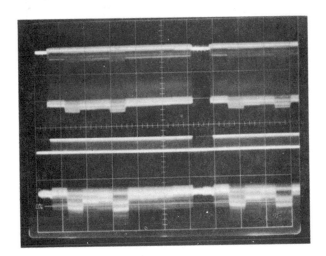

Top: CCD Output TP701 0.5V/d 10μsec/d
Center: SHD Pulse IC709/Pin 45 10V/d 10μsec/d
Bottom: Q708/S 0.2V/d 10μsec/d

FIGURE 7.29 (c) Scope displays.

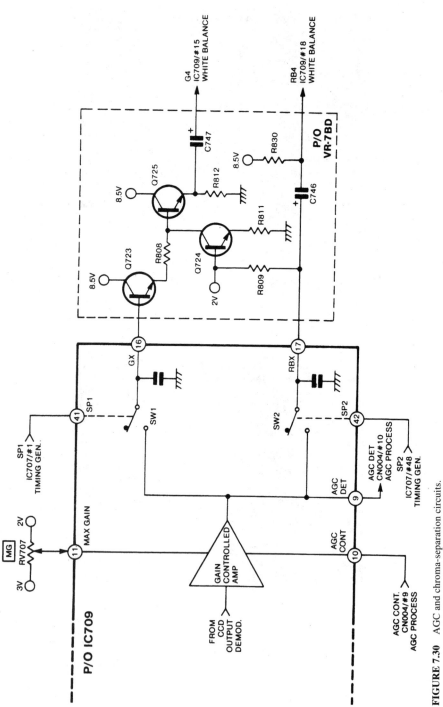

FIGURE 7.30 AGC and chroma-separation circuits.

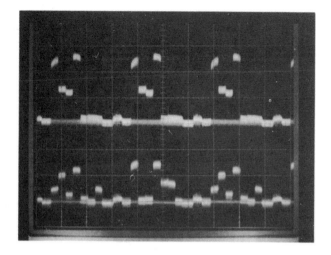

Top: GX IC709/Pin 16 0.1V/d 20μsec/d

Bottom: RBX IC709/Pin 17 0.1V/d 20μsec/d

FIGURE 7.31   Typical chroma-separation scope displays.

ing the output signals, the colors can be separated. Two 4.77-MHz timing pulses, SP_1 and SP_2, from IC_{707} are applied to pins 41 and 42 of IC_{709}, respectively. These timing pulses are used to separate the output into G and R-B signals, shown in Fig. 7.31.

A high SP_1 closes SW_1 when picture information from a green pixel is present. A high SP_2 closes SW_2 when a red or blue pixel is present. In effect, SW_1, SP_1, and the related capacitor form a sample and hold circuit for green, with SW_2, SP_2, and the related capacitor forming a sample and hold for red-blue. The green picture informa-tion (now called GY) is buffered by Q_{723} and Q_{725} and returned to pin 15 of IC_{709} for further processing. Q_{724} acts as a constant load on the output of Q_{723}. The red-blue information (now called RBY) is applied to pin 18 of IC_{709} through C_{746}, without buffering.

7.12.11   White-Balance Gain-Control Circuits

Figure 7.32 shows the white-balance gain-control circuits. White balance is necessary to compensate for color sensitivity of the camera and for changes in color temperature of the light that illuminates the scene. This is done by adjusting the amplitude of the RY and BY signals. The separated GY signal is amplified by a fixed amount and applied to the green clamp (Sec. 7.12.12) through two GCAs. The RBY signal is split into two paths and applied to red and blue clamps. Both RBY paths have a manual gain adjustment (RV_{705} and RV_{706}) as well as an automatic white-balance (AWB) adjust-ment voltage (Secs. 7.12.21 through 7.12.27).

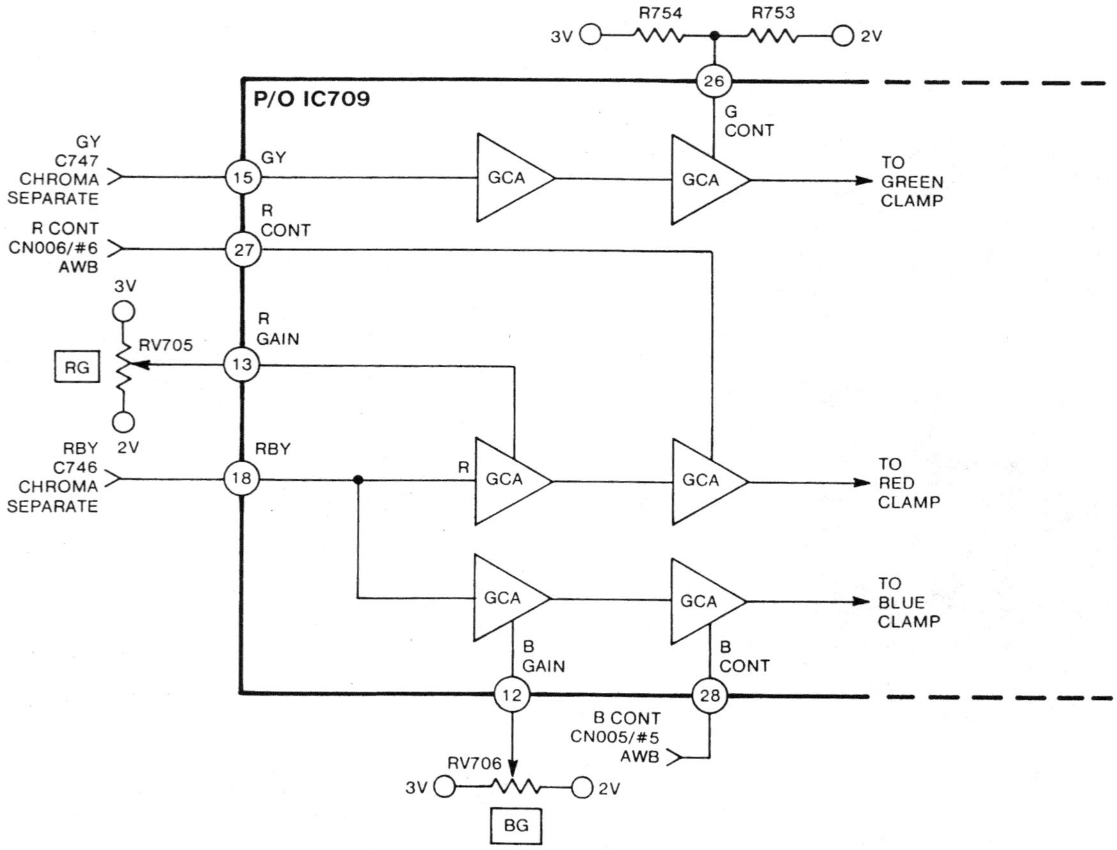

FIGURE 7.32   White-balance gain-control circuits.

7.12.12   Clamp Circuits

Figure 7.33 shows the clamp circuits. Up to this point, the processed signals are coupled from one circuit to another by capacitors. This removes the black reference from the signal and, before the blanking pulse can be inserted, the black reference must be restored by means of clamps, one for each color. This is done by the clamp circuits of Fig. 7.33.

The outputs of the red, green, and blue white-balance GCAs are fed to the noninverting input of individual amplifiers in IC_{709}. Each GCA output is also switched to the inverting input through SW_1, SW_2, and SW_3, which are controlled by CLP_1 at the horizontal frequency. VAA is a 60-Hz pulse which goes low during the vertical-blanking time and is used to switch C_{740}, C_{741}, and C_{742} out of the clamp circuit during vertical blanking.

When VAA is high (during the time of a field), Q_{705}, Q_{706}, and Q_{707} are all on, thus connecting C_{741}, C_{740}, and C_{742} to ground. At the end of the time for one line, CLP_1 goes high and closes SW_1, SW_2, and SW_3. C_{740} through C_{742} charge to the no-signal voltage level. When CLP_1 goes low, SW_1 and SW_3 open and the charge on C_{740} and C_{741} becomes the reference (clamp) voltage level at the inverting input of the amplifiers. Picture information of the succeeding line (lasting about 53.6 μs) follows and is applied to the noninverting input. The amplified output signal is referenced to the clamp voltage level maintained by C_{740} and C_{742}.

The clamping process continues for the time of one field until VAA goes low during the vertical-blanking time. A low VAA switches Q_{705} through Q_{707} off, removing C_{740} through C_{742} from the circuit. During the vertical-blanking time, CLP_1 pulses are not generated, so SW_1 through SW_3 are open, and the inverting input is floating. Because the noninverting input is at the vertical-blanking level, the amplifier outputs go to zero and remain zero until VAA goes high and CLP_1 pulse are again generated to restart the clamping process (on each line of the succeeding field). The blue and red offset controls RV_{703} and RV_{704} adjust the blue and red outputs with reference to the green output.

The clamp outputs are fed to the blanking and AWB circuits. The red and blue clamp signals are alternately selected by SW_4, which is operated by a 7.5-kHz ID pulse. ID is high for the blue-green line, during which time SW_4 is in the blue clamp-output position. Blue picture information is applied to the $R\text{-}B$ blanking circuit in IC_{709} and to the AWB circuit through pin 33. The red clamp output is applied to $R\text{-}B$ blanking and the AWB circuits when ID is low. The green clamp output is not switched at this time because green is separated from red and blue in the AGC and chroma-separation circuits of Fig. 7.30, as described in Sec. 7.12.10.

7.12.13   Blanking, Gamma Correction, White Clip, and Pedestal Circuits

Figure 7.34 shows the blanking, gamma-correction, white-clip, and pedestal circuits.

The blanking function is required to blank the outputs at the end of each line so that a blanking pulse (with pedestal) can be added. The outputs of the green clamp and the $R\text{-}B$ switching circuit are amplified and inverted by individual amplifiers in IC_{709}. An H BLK pulse (at the horizontal frequency) is applied (at pin 38) to the G and $R\text{-}B$ outputs after inversion and amplification. The H BLK pulse occurs at the end of each line, removing any information that might be present.

The gamma-correction circuit corrects the nonlinear reproduction of light levels between a CCD camera and the CRT of a television set. Typically, the CCD camera has a gamma of 1, which means that the output signal is directly proportional to the

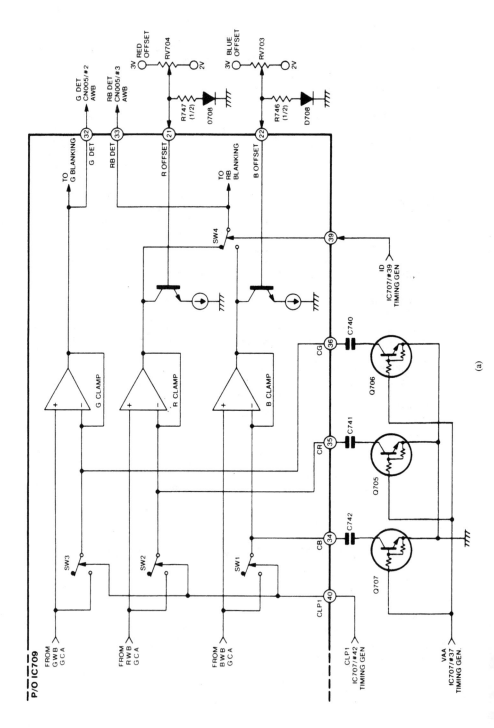

(a)

7.70

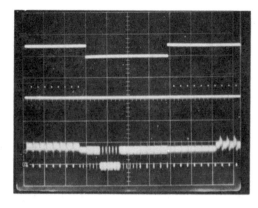

1

VAA IC707/Pin 37
10V/d 0.2msec/d

CLP1 IC709/Pin 40
10V/d 0.2msec/d

Camera Video Output
TP711 0.2V/d 0.2msec/d

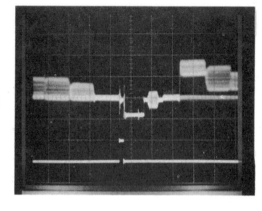

2

Camera Video Output
TP711 0.2V/d 5μsec/d

CLP1 IC709/Pin 40
5V/d 5μsec/d

FIGURE 7.33 (*b*) Scope displays.

light levels of the scene being photographed. A television CRT has a gamma of 2.2, which means that the blacks and low-level grays are suppressed, with the whites and highlights boosted. To produce proper reproduction of light levels by the entire system (camera and TV), the CCD gamma is adjusted electronically to a level of 0.455, which results in a product of 1.0 when multiplied by 2.2 ($2.2 \times 0.455 = 1$).

The outputs of the G and R-B amplifiers are controlled by diodes D_1 through D_6. In turn, the diodes are controlled by bias voltages from adjustment controls RV_{701}, RV_{702}, and RV_{708}. The bias voltages determine the level at which the diodes conduct and thus set the level of the G and R-B amplifier outputs. A white (bright) signal produces a higher voltage at the G and R-B outputs, thus increasing conduction in the diodes. In turn, this lowers the output signal level. A darker signal produces lower voltages at the amplifier outputs. This reduces conduction in the diodes and increases output level.

In the case of the gamma-correction control RV_{702}, the D_1 and D_2 diode bias is set so that the G and R-B output levels are at 0.455, as discussed. Diodes D_1 and D_2 are also controlled by a voltage from the iris-control circuits (Secs. 7.12.14 through 7.12.16). This voltage controls G and R-B outputs for backlight and high-light conditions. When a

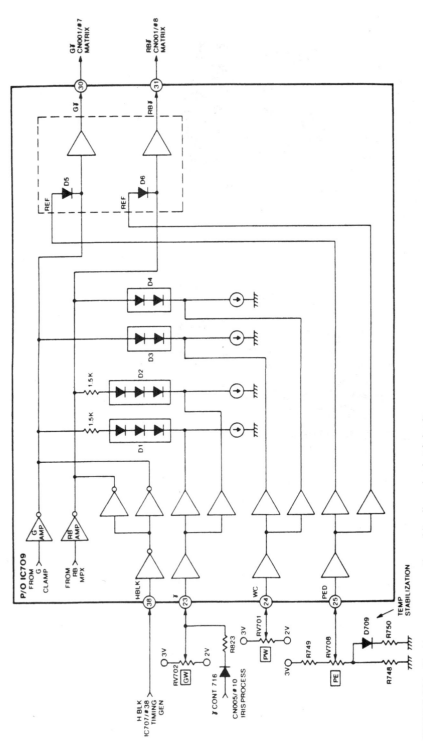

FIGURE 7.34 Blanking, gamma-correction, white-clip, and pedestal circuits.

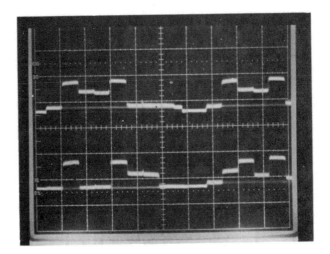

Top: GY TP701 IC709/Pin 30   0.2V/d   10µsec/d

Bottom: RBY TP702 IC709/Pin 31   0.2V/d   10µsec/d

FIGURE 7.35   Blanking, gamma-correction, white-clip, and pedestal scope displays.

scene with backlight is being photographed, 5 V is applied across D_{716} and amplified. This reverse biases D_1 through D_2 and allows maximum white (brightness) output from G and R-B amplifiers. When the scene is under high-light conditions, D_{716} is grounded and D_1 through D_2 conduct heavily to reduce the bright output levels.

The white-clip circuit ensures that the white peaks of the final output stay within the range (set by the industry) to prevent overmodulation. White-balance control RV_{702} is set so that D_3 and D_4 conduct only during excessive white peaks (typically above 100 IEEE units, as discussed in Sec. 3.3.4). This in no way reduces picture quality because the detail found during white peaks is not important.

The pedestal-adjust circuit sets the pedestal level of the H BLK pulse which is added to the outputs of the G and R-B amplifiers. The G and R-B signals are clamped at the black (no-signal) level and do not turn on the pedestal circuits (represented by D_5 and D_6). However, when the H BLK pulse goes low, D_5 and D_6 conduct and clamp the signal to the reference (REF) level set by RV_{708}. (Figure 7.35 illustrates the pedestal and blanking signals.)

In this particular CCD, the REF voltage is adjusted to produce an output signal with a pedestal at 25 mV. Adjusting the size of the pedestal is critical because it establishes the reference level of the final camera output signal. If adjusted too low, the picture will lack contrast. If too high, the picture will be too dark.

7.12.14 Iris-Processing Circuits

Figure 7.36 shows the first of the iris-control circuits. These circuits control operation of the iris (part of the CCD lens system). The amount of light that can pass through the camera lens is controlled by the aperture. The size of the aperture is variable, allowing

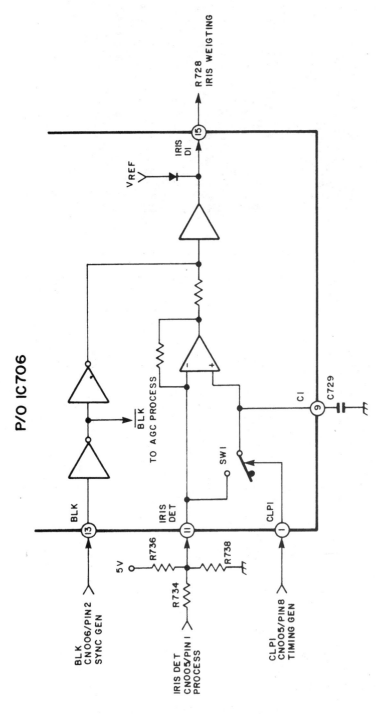

FIGURE 7.36 Iris-control circuits.

more or less light to enter the lens system. Control of aperture size is done electronically by circuits which monitor the voltage resulting from the light input. The circuits then develop a drive voltage used to operate a motor. In turn, the motor moves the iris mechanism to open and close the aperture.

As discussed in Sec. 7.12.8, the demodulated output of the CCD is applied to the AGC circuits and to the iris-detection circuits at pin 8 of IC_{709} (Fig. 7.29). The signal leaving pin 8 is made up of G and R-B signals. Because of the capacitive coupling between the CCD and the input at pin 47 of IC_{709}, the iris-detection signal at pin 8 does not have a black (no-signal) reference. The circuit of Fig. 7.36 provides this reference.

The CLP_1 pulse at pin 1 of IC_{706} goes high at the end of the line (when there is no signal information) and closes SW_1. With SW_1 closed, C_{729} at the noninverting input of the amplifier charges to the black voltage level. When CLP_1 goes low, SW_1 opens and the black voltage is stored by C_{729}. This voltage becomes the reference to the amplifier for the next line of G and R-B information at the inverting input. The signal at the output of the amplifier is thus clamped (or referenced) to the black level maintained by C_{729}.

The BLK pulse from IC_{708} (Fig. 7.22) enters IC_{706} at pin 13 and is inverted twice before blanking the amplifier output. The signal is then coupled through another amplifier, and the pedestal of the blanking is set at a fixed level by the VREF voltage. The clamped signal (with the pedestal at about 400 mV) exits IC_{706} at pin 15 and is applied to the iris-weighting circuit.

7.12.15 Iris-Weighting and Detect Circuits

Figure 7.37 shows the iris-weighting and detect circuits. These circuits inhibit the effect of the brightness for the first 25 percent of each field. Typically, the brightest section of a scene is in the top 25 percent of the field. If the iris is controlled without considering this effect, the picture will have improper contrast.

The signal from the iris-process circuit (Fig. 7.36) is applied to the iris-motor drive (Sec. 7.12.6) through Q_{715} and Q_{716}. The output from Q_{716} is a dc voltage which represents the overall brightness of the scene. The dc voltage is adjusted by Q_{709} and Q_{714} to account for the effect of the increased brightness at the top 25 percent of the picture.

At the end of a field, a high 60-Hz VD pulse turns Q_{709} on. C_{717} discharges and turns Q_{714} on. With Q_{714} on, Q_{715} turns off, producing no output to Q_{716}. When VD goes low, Q_{709} is turned off, and C_{717} begins to charge through R_{718}, turning Q_{714} off gradually. At this time, the first 25 percent of the field is applied to Q_{715}. Q_{714}, which is still conducting (but not at maximum) attenuates the input signal for 25 percent of the field until C_{717} fully charges and Q_{714} is off. (C_{717} charges in about 3 ms.) Q_{714} remains off until the end of the field, when VD starts the process again. With no signal passing when VD is high and a reduced signal when C_{717} is charging, the average dc produced by the top of the scene (to control the iris) is decreased. As a result, the bottom 75 percent of the scene has greater effect on the iris.

7.12.16 Iris-Motor Drive Circuits

Figure 7.38 shows the iris-motor drive circuits. The average dc voltage developed by the iris-weighting circuit (Sec. 7.12.15) is applied to the inverting input of an op amp at pin 30 of IC_{706}. The op-amp output is used to drive the iris motor. The reference voltage on the noninverting input of the op amp is set by iris adjust RV_{710} and a user aperture switch connected through R_{780}.

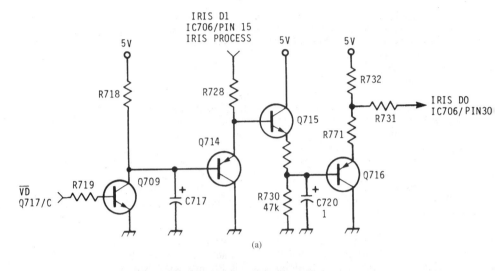

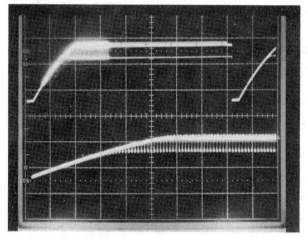

Top: Q715/B 1V/d 2msec/d

Bottom: Q715/B 1V/d 0.5msec/d

FIGURE 7.37 (*a*) Iris-weighting and detect circuits and (*b*) scope displays.

The user can alter the reference voltage by means of the aperture switch. When the switch is in the Backlight position, pin 37 is connected to 5 V. This increases the reference voltage at pin 31 and causes the aperture to open wider (the iris motor is driven in a direction to open the aperture by the voltage at pin 33). In the High-Light position, pin 37 is grounded and the iris motor is driven to close the aperture more and reduce the light.

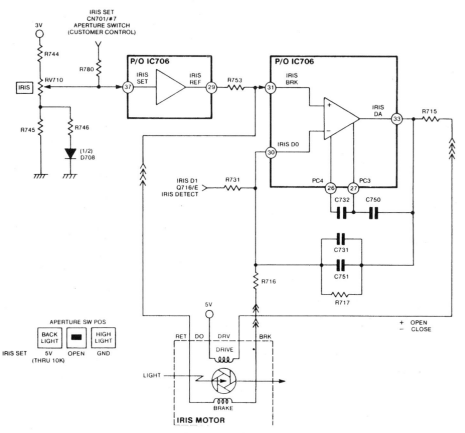

FIGURE 7.38   Iris-motor drive circuits.

During the iris setup procedure, RV_{710} is adjusted so that the iris is open enough to allow light from a standard test pattern to enter the camera and produce an iris-detect voltage of 5 V at the output of the iris-motor drive amplifier. The other end of the iris-motor drive winding is connected to 5 V. As a result, the iris motor remains in the setup position (5 at both the *DO* and DRV terminals of the iris motor).

During operation where lighting conditions change, the iris-detect voltage varies. In turn, this changes drive to the iris motor, opening and closing the iris as necessary to balance the voltage across the iris-motor drive winding. The motor stops when the voltage at pin 33 of IC_{706} is 5 V. At pin 33 an output voltage above 5 V opens the iris. A voltage below 5 V closes the iris.

A brake coil, wound close to the motor-drive winding, is used to ensure smooth movement of the iris. One end of the brake coil is connected to the reference voltage at pin 31 of IC_{706}, with the other end connected to pin 30 (the iris-detect input). As the iris-detect voltage increases, the iris-motor drive output decreases, and the iris motor attempts to turn (at a high speed) to open the iris. This induces a voltage in the brake winding, making pin 30 of IC_{706} negative with respect to pin 31 to slow the motor speed, even though the motor continues to open the iris. When the iris-detect voltage decreases, the same condition occurs but in the opposite polarity (the iris closes, but not at a high speed, so the closure is smooth). Capacitors C_{732} and C_{750} provide feedback to the iris-motor op amp and prevent oscillation when the iris is near the balance point.

7.12.17 AGC Process

Figure 7.39 shows the first of the AGC circuits. The purpose of these circuits is to maintain the CCD output constant (within limits) in spite of changes at the input. As discussed in Sec. 7.12.8, the demodulated output of the CCD is applied to the AGC circuits and to the iris-detection circuits at pin 8 of IC_{709} (Fig. 7.29). As shown in Fig. 7.30, the CCD output is amplified by a GCA in IC_{709} and appears at pin 9. Gain control RV_{707} at pin 11 sets the gain of the GCA and thus the voltage at pin 9. The GCA is also controlled by the AGC circuit through pin 10 in response to fluctuations in camera signal strength.

The signal at pin 9 of IC_{709} is applied to the inverting input of an op amp within IC_{706} through R_{735} and pin 12 of IC_{706} (Fig. 7.39), where a black (no-signal) reference is added. This operation is similar to that for adding the reference to the iris-processing circuits described in Sec. 7.12.14, except that the output at pin 14 of IC_{706} is clamped (or referenced) to the black level maintained by C_{730}. The clamped signal (with a pedestal at about 400 mV) exits IC_{706} at pin 14 and is applied to the AGC-weighting and detect circuit.

7.12.18 AGC-Weighting and Detect Circuits

Figure 7.40 shows the AGC-weighting and detect circuits. These circuits inhibit the effect of brightness on AGC for the first 25 percent of each field, as described for the iris circuits in Sec. 7.12.15. The signal developed by the circuits of Fig. 7.40 is an average dc voltage, corrected for the first 25 percent of the scene and applied to the AGC process-2 circuits through R_{726}.

7.12.19 AGC Process-2 Circuits

Figure 7.41 shows the AGC process-2 circuits. The average dc voltage developed by the AGC-weighting circuit (Sec. 7.12.18) is applied to the inverting input of an op amp at pin 18 of IC_{706}. The op-amp output is used to control gain of the GCA in IC_{709} at pin 10 of IC_{907} (Fig. 7.30). The reference voltage to the noninverting input of the op amp in IC_{706} is developed by AGC control RV_{709}, through an amplifier in IC_{706} and Q_{718}.

As the light input to the camera varies, the average AGC-detect voltage applied to the inverting input changes. A voltage lower than the reference voltage at pin 20 increases the output control voltage and thus increases gain of the AGC amplifier. A voltage higher than the reference voltage decreases the output-control voltage and thus reduces AGC gain.

The user can alter the reference voltage with the Aperture switch. When the switch is in the Backlight position, 5 V is applied to the cathode of D_{715}. This reverse biases D_{715}, and there is no effect on the reference voltage. When the switch is in the High-Light position, the cathode of D_{715} is grounded. This lowers the reference voltage and causes the AGC control voltage to decrease the AGC gain.

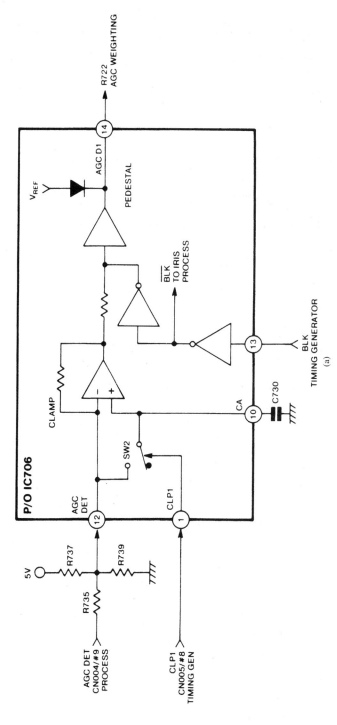

FIGURE 7.39 (a) AGC-process circuits.

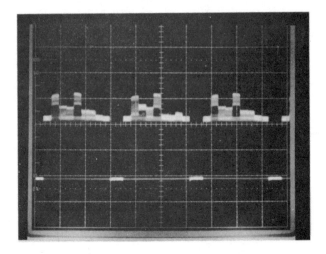

$$IC706/Pin\ 14\quad 0.2V/d\quad 20\mu sec/d$$

FIGURE 7.39   (*b*) Scope displays.

The output of the op amp at pin 16 of IC_{706} is applied to the GCA within IC_{709} (Fig. 7.30) through Q_{730}. The position of the Aperture switch also affects operation of Q_{730}. For a normally lighted scene, and a scene with high light, the base of Q_{730} is grounded by the Aperture switch. This permits the output of the op amp to control AGC gain in the normal manner as described. With the Aperture switch set to Backlight, 5 V is applied to Q_{730}, preventing the AGC signals from passing. The AGC gain is then set to a fixed value by RV_{711}. The AGC control voltage at pin 16 of IC_{706} is also applied to low-light alarm (LLL) circuits within IC_{706}.

7.12.20   Low-Light Alarm Circuits

Figure 7.42 shows the low-light alarm circuits. These circuits inform the user that a low-light condition exists by means of an LED D_{001} in the electronic viewfinder (EVF). The AGC control voltage at pin 16 of IC_{706} is applied to the noninverting input of a comparator within IC_{706}. The inverting input is connected to low-light sensor adjust RV_{708}. Under low-light conditions, the comparator produces a high output which is inverted to a low at pin 34. This completes the circuit through Q_{732}, turns Q_{708} on, connects the anode of D_{001} to 5 V, and turns D_{001} on to indicate the low-light condition.

During a low-light condition, the AWB control circuits are turned off. This is done when the cathode of D_{710} is connected to a low through Q_{732}.

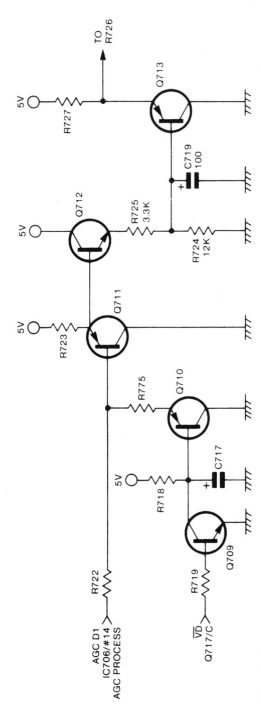

FIGURE 7.40 AGC-weighting and detect circuits.

7.81

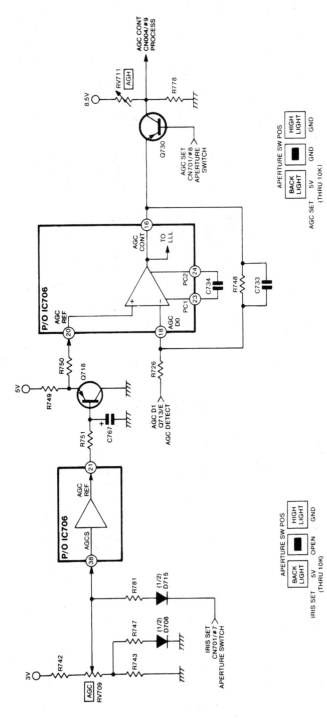

FIGURE 7.41 AGC process-2 circuits.

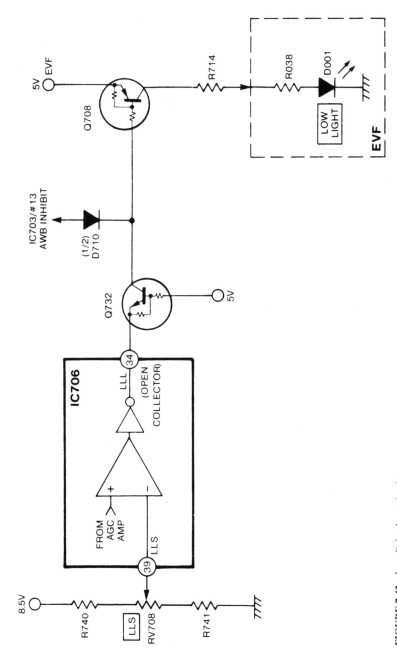

FIGURE 7.42 Low-light alarm circuits.

7.12.21 Automatic White Balance Circuits

Figure 7.43 shows the overall functions of the automatic white-balance (AWB) circuits for this particular CCD camera. The G and R-B signals from the process circuits (Secs. 7.12.8 through 7.12.13) are combined in IC_{706} to produce the R-G and B-G outputs. During white-balance setup (typically using a white light box with the test pattern removed), the values of R, G, and B are made equal for a white scene. In this way, the values of R-G and B-G indicate if white balance is correct. If the values are low, the R or B is excessive. G is excessive if the values are high.

Most of the AWB control is performed by IC_{703}, which varies the gain of the red and blue amplifiers through the RA and BA outputs. These outputs are buffered and applied to IC_{709}. This feedback keeps R-G and B-G at the correct value to produce proper white balance under varying light conditions.

7.12.22 AWB Clamp and Offset Circuits

Figure 7.44 shows AWB clamp and offset circuits. The G and R-B DET signals from the process circuits are applied to IC_{706} at pins 6 and 8. The signals are clamped by CLP_1, which establishes the no-signal reference level on capacitors C_{726} and C_{728}. When the AWB switch is pressed, the AWB control IC_{703} (Sec. 7.12.24) automatically tests the white balance and develops a signal state on pins 21, 22, and 23. The states are: BNG (Blue No Good), OK, and RNG (Red No Good). These three signals are connected to the anodes of D_{716} and D_{717}. The cathodes of D_{716} and D_{717} are connected to the base of Q_{734}.

During AWB (and after AWB if white balance is correct) all three signals are low and Q_{734} is off. The green amplifier is thus referenced to the no-signal level on C_{726}. If white balance is not correct, one of the color-signal lines (the one which needs adjusting) and the OK line go high. Q_{734} turns on and applies 5 V to pin 3 of IC_{706}. This increases the reference of the green amplifier after AWB has been attempted. During the time that OK is high, the white-balance LED control circuit is activated, causing an AWB LED to blink in the EVF, as described in Sec. 7.12.26.

The outputs of the R and B clamp circuits are adjusted by RV_{707} and RV_{708} to offset any dc-level differences between red or blue and the green output (which is not adjustable). After a stage of amplification, the RGB clamp outputs are applied to the color-separation circuits.

7.12.23 R-B Separation, Pedestal, R-G and B-G Circuits

Figure 7.45 shows the circuits used to form the R-G and B-G signals used by the AWB control circuit (Sec. 7.12.24). The signals at the outputs of the R- and B-clamp circuits have red and blue color information. The circuits of Fig. 7.45 separate red from blue and add a blanking pulse at the end of each line.

R-B signals from the R clamp and offset circuits are applied to the input of amplifier 2. The 7.5-kHz ID_2 pulse is high during a G-B line and low during a G-R line. When BLK and ID_2 are low, Q_2 is off. At this time, the red information in the R-B signal is applied to the R-G circuits (at pin 35) through amplifier 2. When ID_2 is inverted, Q_1 is turned on through OR gate 1. This prevents the red portion of the R-B signal from passing to the B-G output at pin 36.

When BLK goes high at the end of a line, Q_2 turns on and pulls the input to amplifier 2 low. BLK goes low again, but ID_2 goes high, keeping Q_2 on to shunt the signal

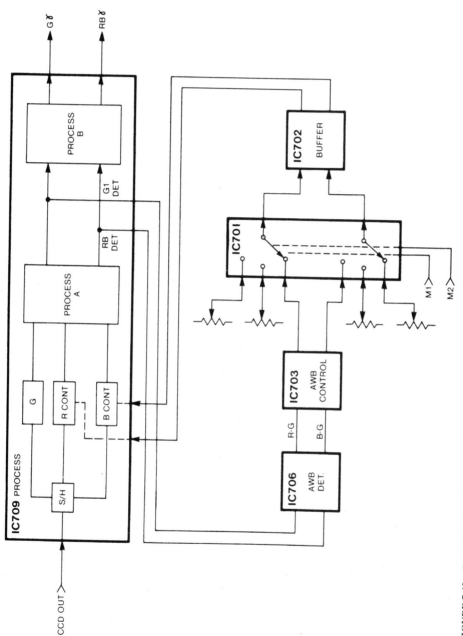

FIGURE 7.43 Overall functions of automatic white balance (AWB) circuits.

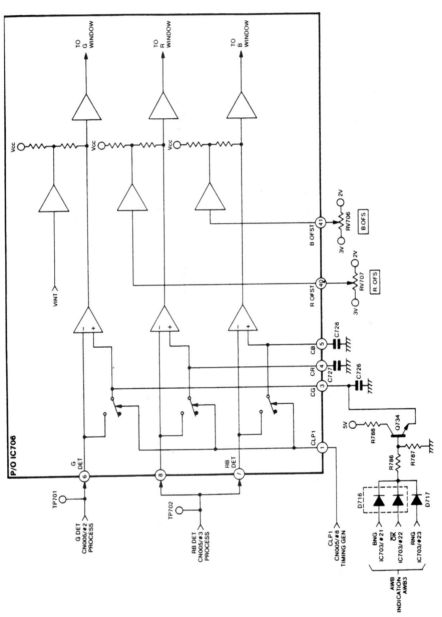

FIGURE 7.44 AWB clamp and offset circuits.

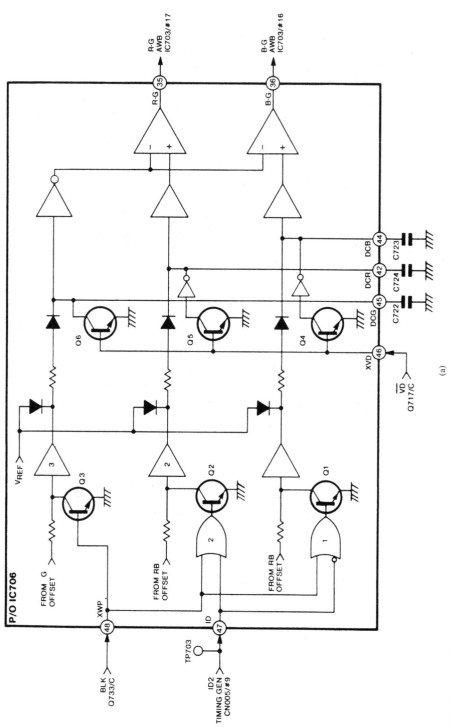

FIGURE 7.45 (*a*) *R-G* and *B-G* signal circuits.

7.87

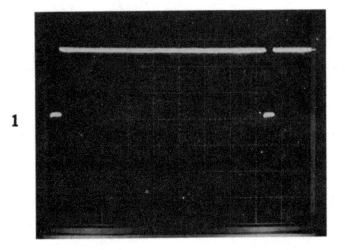

IC706/Pin 35   R-G Normal Operation

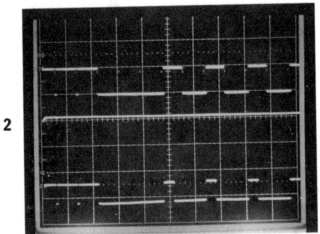

Top: IC706/Pin 35 R-G

Center: IC703/Pin 13 WRT

Bottom: IC706/Pin 36 B-G

FIGURE 7.45   (*b*) and (*c*) Scope displays.

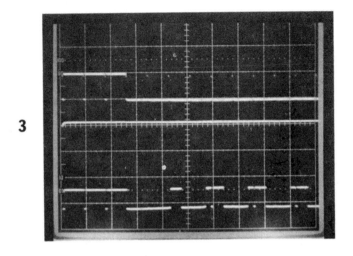

Top: IC706/Pin 35 R-G

Center: IC703/Pin 13 WRT

Bottom: IC706/Pin 36 B-G

FIGURE 7.45 (*d*) Scope display.

from the *R*-clamp and offset circuits. When ID_2 is high and BLK is low, Q_1 is off. At this time, the blue information in the *R-B* signal is applied to the *B-G* circuits (at pin 36) through amplifier 1. When BLK goes high again at the end of a line, Q_1 turns on and shunts the amplifier-1 input to ground. The signal from the *G*-clamp and offset circuit is not switched and is applied directly to amplifier 3, but with blanking (BLK) added through Q_3.

The pedestal level of the blanking signal is set by three diodes representing the pedestal circuit at the three amplifier outputs. Each field of color-signals (from each pedestal circuit) is rectified and then filtered by C_{722} and C_{724}. The dc voltages on each capacitor represent the color level of red, green, and blue during each field. At the end of the field, the VD pulse goes high and turns Q_4 and Q_6 on. These transistors quickly discharge C_{722} and C_{724}, preparing the capacitors for the next color field.

The green dc voltage is amplified and applied to the inverting input of the *R-G* and *B-G* comparators. The red and blue dc voltages are applied to the noninverting inputs of the comparators. If the red dc voltage is higher than the green, the output of the comparator will be low, and vice versa. The same is true for the *B-G* comparator. (Note that *R*, *G*, and *B* were previously inverted in the AWB clamp and offset circuit, Fig. 7.44). Because of the bias within IC_{706}, VD turns Q_4 and Q_6 on at the end of a field. This causes *R-G* and *B-G* (at pins 35 and 26) to go low.

7.12.24 AWB Control Circuits

Figure 7.46 shows the AWB control circuits. These digital circuits within IC_{703} are actuated when the white-balance switch S_{751} is set to the AWB position (Sec. 7.12.25).

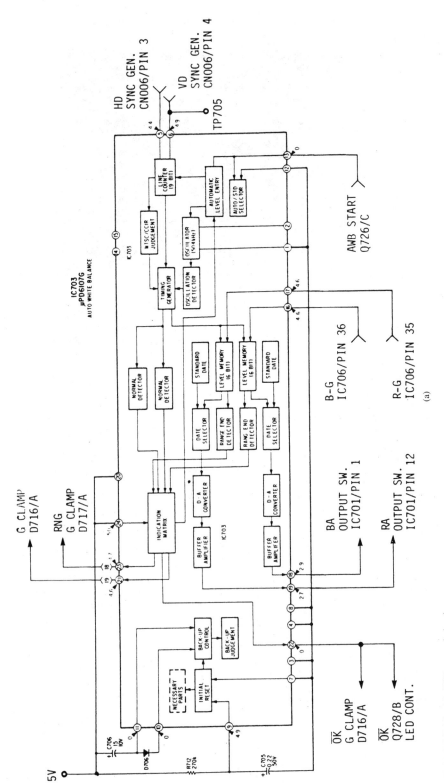

FIGURE 7.46 (*a*) AWB control circuits.

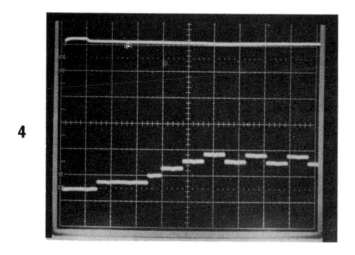

Top: IC703/Pin 13 WRT 2V/d 20msec/d
Bottom: IC703/Pin 19 RA 0.1V/d 20msec/d

FIGURE 7.46   (b) Scope displays.

This turns the AWB-initiate circuits (Sec. 7.12.27) on, placing 5 V at pin 13 of IC_{703} for 250 ms (about 15 fields). Operation of the AWB control circuits in Fig. 7.46 then depends on lighting conditions that exist when AWB is selected.

If the unit has been off for a long period of time, C_{706} at pin 11 of IC_{703} is discharged, and the standard white-balance reference is selected by the data selector in IC_{703} and applied to the D/A converter. This produces the BA and RA outputs at pins 18 and 19, representing an approximate white balance. Level-memory circuits within IC_{703} monitor the B-G and R-G inputs at pins 16 and 17. Using VD and HD pulses from sync generator IC_{708}, the level memory is stepped up one level, either in the positive or negative direction, depending on the R-G and B-G inputs. The data selector chooses this output, in place of the standard memory, and uses the output to change the RA and BA voltages, bringing the white balance closer to the correct value.

After one field, the level memory is again adjusted in the desired direction until the approximate white balance is obtained. When the correct white-balance level is passed, the levels of the B-G and R-G signals change, and the level memory changes direction (if the memory was counting up, it starts to count down). This causes red-attenuate (RA) and blue-attenuate (BA) outputs to change direction, as the correct white-balance position is passed. Typically, the B-G and R-G outputs toggle around the correct white-balance point.

The circuits of IC_{703} operate in a different way when the unit is brought in from the outdoor sunlight to be operated indoors with incandescent light. When AWB is activated under this condition, the output of the R-G comparator will go low because of excessive red (typical of incandescent light). The resulting output at pin 19 of IC_{703} is a voltage that increases in small steps during each field. The voltage adjusts the gain of the white-balance amplifier in the process circuits. Each increment reduces gain, the effect of which is reflected in the R-G comparator output voltage. At the end of

250 ms (the AWB setup time), the changes stop, and a steady dc voltage maintains the gain of the red amplifier. (Note that this same process occurs in the B-G loop but with the output at pin 18, if the unit is operated where the dominant light source is blue.)

Operation of the AWB circuits is synchronized to the line and field rates. During the 250-ms period that the AWB function is active, approximately 15 fields are sampled. The line and field count is made by the inputs at pins 6 and 5, respectively, of IC_{703}, using the HD and VD pulses.

The RNG, BNG, and OK signals used by the AWB clamp circuit (discussed in Sec. 7.12.22) are generated by the indicator matrix in IC_{703}. These signals (at pins 21, 22, and 23) change when the level-memory circuit reaches the end of its range (when the AWB cannot compensate for a color imbalance). The change indicates that AWB has not been reached. The OK line at pin 22 is also used to control the white-balance LED in the EVF (Sec. 7.12.26). (The LED turns on and continues to flash when AWB cannot be obtained.)

The circuits within IC_{703} are reset by R_{712} and C_{705} at pin 9 when the camera is first switched on. To prevent reset (which removes any AWB setting) when power has been off for only a short time, backup circuits in IC_{703} charge C_{706} to 5 V. If there is a momentary power interruption, C_{706} discharges and prevents reset.

7.12.25   AWB Switching

Figure 7.47 shows the AWB output-switching functions, which are controlled by the White Balance switch S_{751} of Fig. 7.48. Note that this switch has a button and a three-position selector. There are three modes of operation, Auto, Incandescent (3200 K), and Sunlight (5800 K), which are selected by S_{751}.

To select AWB, S_{751} must be in the M_3 (or Auto) position when the AWB button is pressed. This ensures that positions M_1 and M_2 are low (no 5-V power). With both M_1 and M_2 low, and the AWB button pressed, IC_{701} (Fig. 7.47) switches pins 3 and 13 to pins 1 and 12, which carry the red-attenuate (RA) and blue-attenuate (BA) control signals from pins 18 and 19 of IC_{703} (Fig. 7.46) for automatic white-balance operation.

When S_{751} is in the M_1 or M_2 position, IC_{701} connects pins 3 and 13 to controls (RV_{701} and RV_{704}) which are adjusted to give an approximate white balance for conditions of extreme sunlight or incandescent light. (These are conditions beyond the capability of the AWB circuits.) The outputs of IC_{701} at pins 3 and 13 are buffered by circuits in IC_{702} and coupled to the white-balance GCAs in the process circuits of IC_{709} (Fig. 7.32).

7.12.26   White-Balance LED Control Circuits

Figure 7.49 shows the white-balance LED control circuits. The white-balance LED D_{009} in the EVF remains on when the White Balance switch S_{751} (Fig. 7.48) is in either the Sunlight or Incandescent positions. With either M_1 or M_2 high (5 V), both Q_{723} and Q_{709} are turned on by the 5 V at the anodes of D_{702}. LED D_{009} turns on when Q_{707} is on. D_{009} is also pulsed on when the AWB button of S_{751} (Fig. 7.48) is pressed, as described in Sec. 7.12.27.

When automatic white balance cannot be obtained, the OK signal from pin 22 of IC_{703} (Fig. 7.46) is high. In addition to making adjustments to the green AWB clamp circuit (Fig. 7.44), a high OK signal turns Q_{728} on. In turn, this turns Q_{729} on and applies power to the multivibrator Q_{721} and Q_{722}. The multivibrator oscillates at about

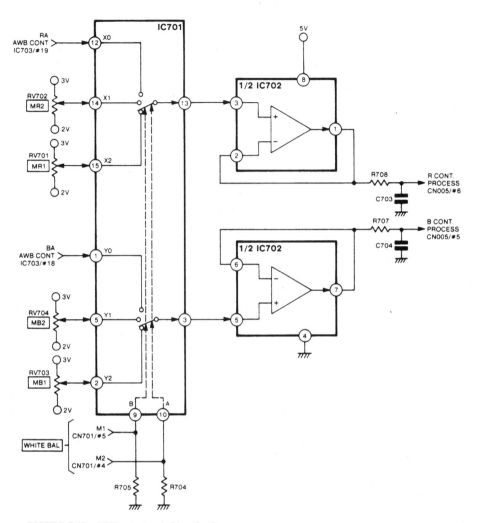

FIGURE 7.47   AWB output-switching circuits.

6 Hz and pulses D_{009} on at that rate, through D_{709}, Q_{723}, and Q_{707}. D_{009} continues to pulse or flash at the 6-Hz rate until the OK signal returns low, indicating that white balance is obtained.

7.12.27   AWB Initiate Circuits

Figure 7.50 shows the AWB initiate circuit. The main purpose of this circuit is to develop the 250-ms AWB-start pulse used by the AWB control IC_{703} as described in Sec. 7.12.24.

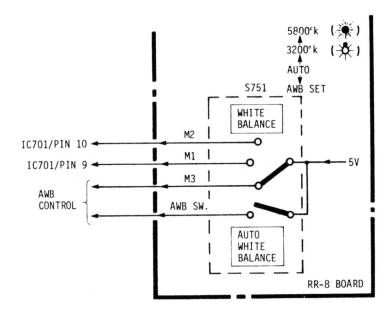

M1	M2	Auto	Incandescent	Sunlight
0	0	1	—	—
0	1	—	—	1
1	0	—	1	—
1	1	X	X	X

FIGURE 7.48   White-balance switch circuits.

When White Balance switch S_{751} (Fig. 7.48) is in Auto, 5 V is applied to the M_3 line. This voltage is used to power the monostable (one-shot) multivibrator (MMV) and control Q_{724} and Q_{726}. Normally, the collector of Q_{726} is low. When S_{751} is pressed, pin 2 of CN_{701} goes high to turn Q_{725} on and Q_{726} off. The collector of Q_{726} goes high for approximately 250 ms and initiates the AWB control function.

When S_{751} is in Auto (M_3), C_{744} charges to 5 V through D_{711}. When S_{751} is moved away from M_3, the MMV attempts to switch states (or toggle). If this is allowed, the AWB circuit is again activated and produces a start pulse. Toggling is prevented by Q_{724}, which turns on when S_{751} is moved from M_3. With Q_{724} on, the 5-V charge on C_{744} is applied to the base of Q_{726} (through Q_{724}), keeping the collector of Q_{726} low.

7.12.28   *RGB* Delay-1 Circuits

Figure 7.51 shows the first of the *RGB*-delay circuits. These circuits produce the 1H and 2H delayed *G* and *R-B* signals that are combined in the matrix (Sec. 7.12.33) where the color and *Y* (luminance) signals are generated.

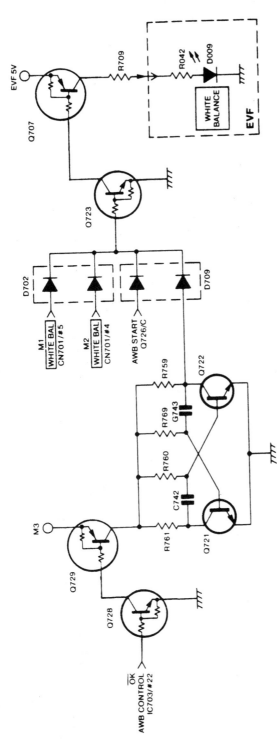

FIGURE 7.49 White-balance LED control circuits.

7.95

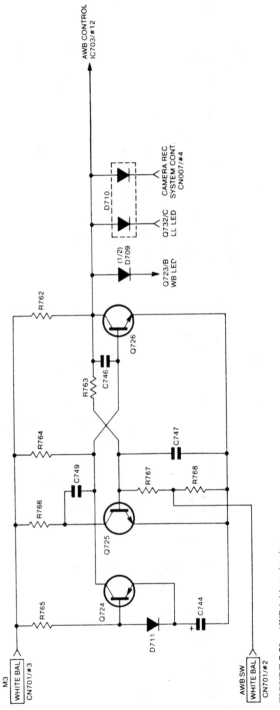

FIGURE 7.50 AWB initiate circuits.

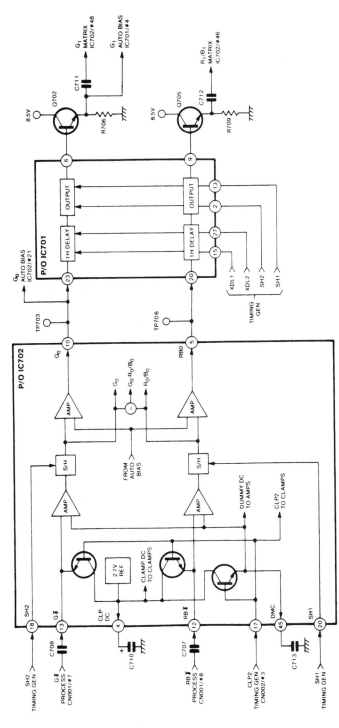

FIGURE 7.51 *RGB* delay-1 circuits.

7.97

The G and R-B signals from the gamma-correction circuit (Fig. 7.34) are applied to pins 12 and 13 of IC_{702} and are clamped to 2.7 V by CLP_2. The G and R-B signals are then amplified by two differential amplifiers (that also use the same 2.7 V as a reference voltage). The outputs of the differential amplifiers are sampled by SH_1 and SH_2 and produce the G_0 and R_0-B_0 outputs. These outputs are also subtracted within IC_{702} to produce a third output, the G_0 and R_0-B_0 component. The three outputs are combined in the color matrix as described in Sec. 7.12.33.

G_0 and R_0-B_0 are further amplified by a second pair of amplifiers which are referenced to a dc voltage from the autobias circuits (Sec. 7.12.31). The autobias circuit produces a reference voltage from the G_0 signal going to the first delay line. This reference voltage is used throughout IC_{702} to ensure proper clamping voltage for the signals going through the various delay lines.

The G_0 and R_0-B_0 signals are applied to 1H delay lines in IC_{701} and are clocked through the delay lines by 9.77-MHz pulses XDL_1 and XDL_2. The outputs of the delay lines are clamped by SH_1 and SH_2 and buffered by Q_{702} and Q_{705} to produce the G_1 and R_1-B_1 signals. (The designation G_1 designates G delayed by the time of one horizontal line.)

7.12.29 *RGB* Delay-2 Circuits

Figure 7.52 shows the second of the *RGB*-delay circuits. Here, the G_1 and R_1-B_1 signals produced by the 1H delay lines are further processed to ensure accurate color reproduction and drive a second set of delay lines to produce the 2H delayed signals necessary to produce the color-difference signals at the matrix.

The G_1 and R_1-B_1 signals are applied to pins 46 and 48 of IC_{702} and are clamped to 2.7 V by CLP_2. The signals are then amplified by differential amplifiers as in the case of the *RGB* delay-1 circuits. However, the gain of the *RGB* delay-2 circuits is made adjustable (by RV_{701} and RV_{702}) to compensate for differences in signal loss as the signals pass through the 1H delay lines. (If these adjustments are improperly set, multiple color spots appear for each color on a vectorscope display, Sec. 3.10.)

The G_1 and R_1-B_1 signals are combined to produce a G_1 and R_1-B_1 signal, and the three outputs are used by the matrix in IC_{702}. The G_1 and R_1-B_1 signals are further amplified (by amplifiers using the same autobias voltage as in the delay-1 circuits) to produce amplified G_1 and R_1-B_1 outputs at pins 6 and 9 of IC_{702}. These outputs drive a second pair of 1H delay lines in IC_{701} and produce the G_2 and R_2-B_2 outputs, which are buffered by Q_{703} and Q_{704}.

7.12.30 *RGB* Delay-3 Circuits

Figure 7.53 shows the third of the *RGB*-delay circuits. Here, the G_2 and R_2-B_2 signals reenter IC_{702} at pins 42 and 44, are clamped by CLP_2, and are amplified by differential amplifiers, as in the case of the previous signals. Again, the gain of the signals is made adjustable (by RV_{703} and RV_{704}) to compensate for the different gains in the delay lines and to ensure color purity in the camera output.

The G_2 and R_2-B_2 signals are combined to produce a G_2 and R_2-B_2 signal, and the three outputs are used in the matrix. The G_2 and R_2-B_2 signals can be monitored at test points TP_{705} and TP_{708}, respectively.

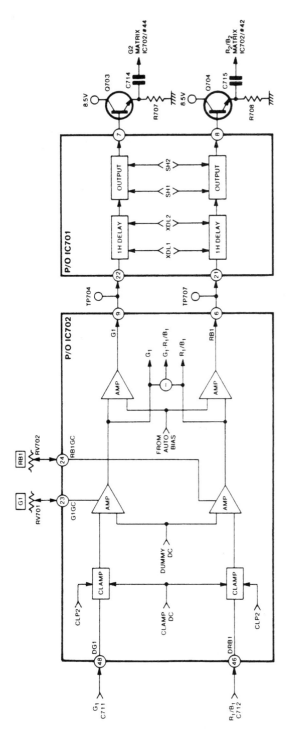

FIGURE 7.52 *RGB* delay-2 circuits.

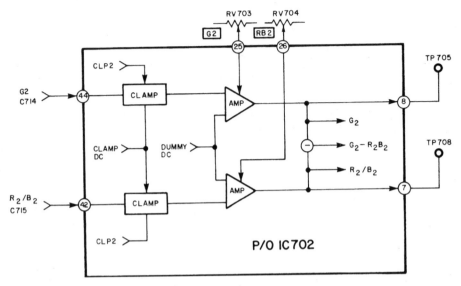

FIGURE 7.53   *RGB* delay-3 circuits.

7.12.31   Autobias Circuits

Figure 7.54 shows both autobias circuits. These circuits are located in the delay-line IC so that both the autobias and delay-line circuits will be subjected to the same environmental conditions.

The G_0 output from pin 10 of IC_{702} (Fig. 7.51) is applied to a comparator within IC_{702}. The reference for the comparator is the dc voltage at pin 22. This voltage, developed by the autobias circuit within IC_{701}, is sampled by CLP_2 and held by C_{708}. The output from this sample and hold circuit is applied to all RGB delay amplifiers in IC_{702}, providing a bias for the amplifiers which drive the delay line. A common bias ensures compatibility among all of the circuits in IC_{702}.

The second autobias circuit is contained within IC_{701}. The G_1 output from the delay line is clamped by CLP_2, and the dc voltage from the clamp circuit is fed back to the autobias circuits through pin 14 of IC_{701}. This controls the bias level on the output amplifiers for the delay lines, ensuring compatibility at the outputs.

7.12.32   Delay-Line Power Supply

Figure 7.55 shows the 9-V power supply for the RGB delay lines in IC_{701}. With 5 V at the anode of current-shunt device IC_{704}, the reference input is connected to monitor the delay-line supply at pin 5 of IC_{701}. Under ideal conditions, pin 5 should be at 9 V, with the reference input of IC_{704} at 7.5 V. If the voltage at pin 5 increases, the reference voltage increases, causing IC_{704} to conduct more, decreasing the voltage at the cathode (K) of IC_{704}. This decreases the base voltage of Q_{701}, causing Q_{701} to conduct less and thus reduce the voltage at pin 5 of IC_{701}.

Transistors Q_{720} and Q_{721} provide protection to remove the supply voltage from the delay lines, if the voltage drops below a specified value. If the voltage at R_{764} and R_{765}

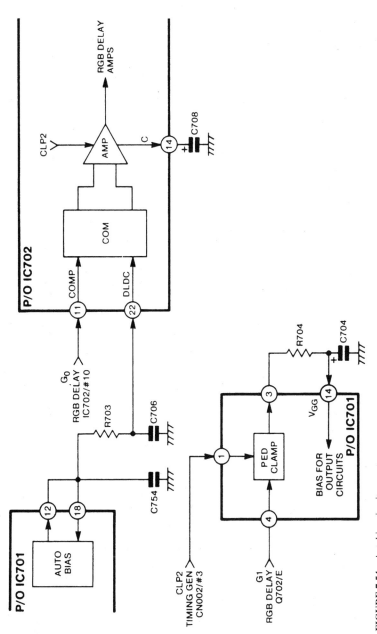

FIGURE 7.54 Autobias circuits.

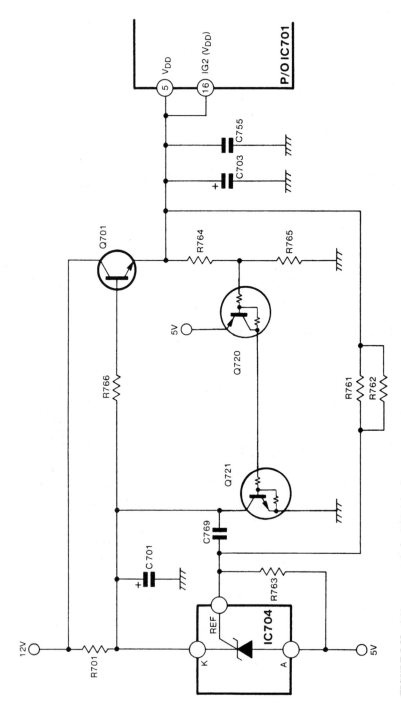

FIGURE 7.55 9-V power supply for *RGB* delay lines.

7.102

decreases, the base voltage of Q_{720} also decreases. If the decrease is large, Q_{720} turns on, as does Q_{721}. This removes the base bias from Q_{701}, turning Q_{701} off and removing all power from IC_{701}.

7.12.33 Color Matrix

Figure 7.56 shows how the various *RGB* components are combined to produce color signals. The ID_1 pulses from timing generator IC_{707} are applied to a multiplex switch in IC_{702} through pin 16. These pulses are low during a red line and high during a blue line. With ID_1 low, a red line is being clocked out of the CCD, but a blue line appears at the multiplex switch in IC_{702}. In the position shown, the switch selects G_1 and R_1-B_1 for the *G-B* term (without interpolation) and the interpolated value for the *G-R* term. With ID_1 high, a blue line is being clocked out of the CCD, but the red line is entering the multiplex switch. Under these conditions, the switch is in the opposite position to that shown, and the noninterpolated *G-R* and the interpolated *G-B* are selected.

The signals from the multiplex switch are applied to the matrix where the signals are combined to produce the *R-Y* and *B-Y* signals. In this particular camera, the matrix is not blanked, so pin 15 is tied to 5 V. The matrix outputs are also used to generate the luminance signal (Sec. 7.12.34).

The *R-Y* and *B-Y* signals are processed through offset amplifiers (although the offset function is not used in this camera) and through GCAs. Gain of the GCAs is set by RV_{706} and RV_{707}. The *R-Y* and *B-Y* outputs at pins 39 and 40 are applied to the chroma and luma encoders through low-pass filters.

7.12.34 Luminance Matrix Circuits

Figure 7.57 shows the luminance matrix circuits. These circuits are also located in IC_{702} and follow the multiplex switch (Fig. 7.56). Both *YL* and *YH* signals are produced by the luminance circuits. The G_1, *G-B*, and *G-R* signals from the chroma matrix are applied to the *YL* matrix to produce the *YL* signal. The *YH* matrix combines G_0 and G_1 with the R_0-B_0 and R_1-B_1 signals in another matrix. The outputs are further combined to produce the *YH* signal.

YH is subtracted from *YL* and appears at pin 36. This output contains the color error present in the *YH* signal and is combined with the *YH* signal (after further filtering) to produce a color-corrected high-resolution *Y* signal.

A second *YH* signal is produced by combining the *G* and *R-B* outputs through two sample and hold circuits. SH_1 and SH_2 sample the components of the *YH* signal, which are then combined and appear at pin 37 as *YH*. This signal contains the maximum of high-frequency detail in the picture but is still not a true NTSC luma signal. The output at pin 37 is combined with the *YL* and *YH* signal at pin 36 after filtering.

7.12.35 Luminance Signal Bandpass Circuits

Figure 7.58 shows the low-pass filtering between the matrix IC_{702} and encoder IC_{703}, as well as how the outputs from the matrix circuit are combined to form the composite NTSC signal. The four outputs from the matrix (*B-Y, R-Y, YL-YH,* and *YH*) contain all of the information necessary for the composite video output. The luma signals (Fig. 7.57) contain the color-correction and high-detail information. The chroma signals (Fig. 7.56) are used to modulate the 3.58-MHz NTSC color subcarrier. When these

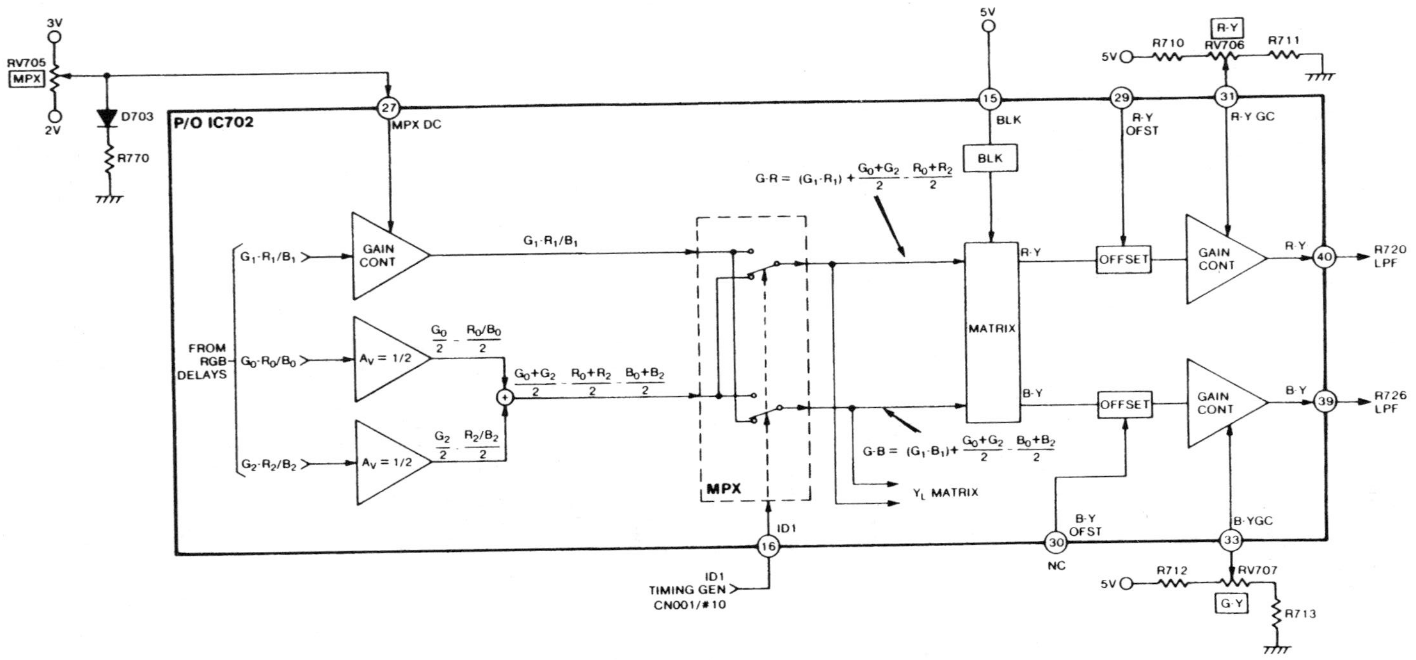

FIGURE 7.56 Color matrix circuits.

$$G\,R = (G_1 \cdot R_1) + \frac{G_0 + G_2}{2} - \frac{R_0 + R_2}{2}$$

$$G\,B = (G_1 \cdot B_1) + \frac{G_0 + G_2}{2} - \frac{B_0 + B_2}{2}$$

$$\frac{G_0 + G_2}{2} \quad \frac{R_0 + R_2}{2} \quad \frac{B_0 + B_2}{2}$$

$$\frac{G_0}{2} - \frac{R_0/B_0}{2}$$

$$\frac{G_2}{2} - \frac{R_2/B_2}{2}$$

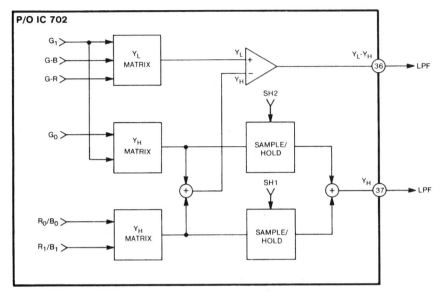

FIGURE 7.57   Luminance matrix circuits.

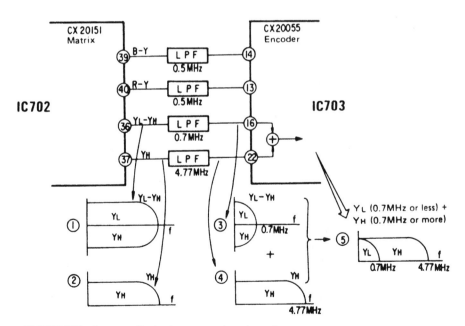

FIGURE 7.58   Low-pass filtering between matrix and encoder.

signals (luma and chroma) are combined, the composite NTSC video output is formed. (The encoder circuit within IC_{703} combines the information from the matrix circuit and produces the NTSC output, as described in Secs. 7.12.37 and 7.12.38.)

7.12.36 Aperture-Correction and Luminance-Mixing Circuits

Figure 7.59 shows the aperture-correction and luminance-mixing circuits. Upon entering IC_{703}, YH and YL-YH are clamped by CLP_3 to 3.6 V. The clamp outputs are then applied to two differential amplifiers, with a reference voltage consisting of the dummy dc voltage at pin 15 (from the same sources as in the matrix IC_{702}, Figs. 7.51 through 7.53).

The YH signal, containing the high-frequency detail, is applied to an aperture-correction circuit consisting of DL_{701}, an internal DTL (detail) amplifier, and another internal amplifier to enhance picture detail. YH exits at pin 18 and passes through 0.18-μs delay line DL_{701} before reentering at pin 20. The delayed YH is added to the YL-YH signal at the outputs of two internal amplifiers to produce the final luminance signal.

The undelayed YH signal is also applied to the DTL amplifier together with the YH signal which has been delayed 0.18 ms. The DLT amplifier compares these two signals, producing the detail-enhancement signals which exit the DTL amplifier. The amplitudes of these signals are controlled by the voltage at pin 21 (as set by RV_{710}).

After leaving the DTL amplifier, the signals are applied to both inputs of an internal amplifier. One input is filtered by the capacitor at pin 12 to remove the detail information but provides a low-frequency reference level for the video signal. The output of the second amplifier consists of the enhancement pulses, which are peak-limited by the diode network at pin 23. The dc voltage at pin 23 is set by RV_{711}. These aperture-correction pulses are added to the luminance signal (which is corrected for low-frequency color balance). The luminance signal is applied through a GCA within IC_{703} to the luminance-encoding circuits.

7.12.37 Luminance Encoding Circuits

Figure 7.60 shows the luminance-encoding circuits. The luminance signal from the aperture-correction circuit is fully processed to contain the correct luma information for color balance and detail (which has also been enhanced). The resultant signal is brought up to NTSC standards by adjusting the pedestal, limiting peak level, and adding sync, as shown in Fig. 7.60.

The Y signal from the aperture-correction circuits is applied to a GCA, which is set to a fixed gain by a dc voltage at pin 29 of IC_{703}. The amplified Y signal is blanked by the BLK signal at pin 5, and the pedestal level is set by RV_{716}. The white-clip circuit, controlled by RV_{717}, ensures that the luminance signal does not exceed the maximum value. The Y signal (with blanking, pedestal level, and white clip properly set) is taken from IC_{703} at pin 36 and returned to pin 37 through C_{744}.

The Y signal takes two paths from pin 37. The first path is to a mixer where sync is added. Composite sync from the sync generator enters at pin 3 and is amplified to a fixed level as determined by R_{741} and R_{742}. With the sync added, the Y signal is applied through C_{743} and the muting switch (always closed) and exits at pin 47 to ultimately appear in the viewfinder (EVF).

The second path for the Y signal is to a fader amplifier. The gain of the fader amplifier is controlled by the voltage at pin 33. This same voltage is used to control

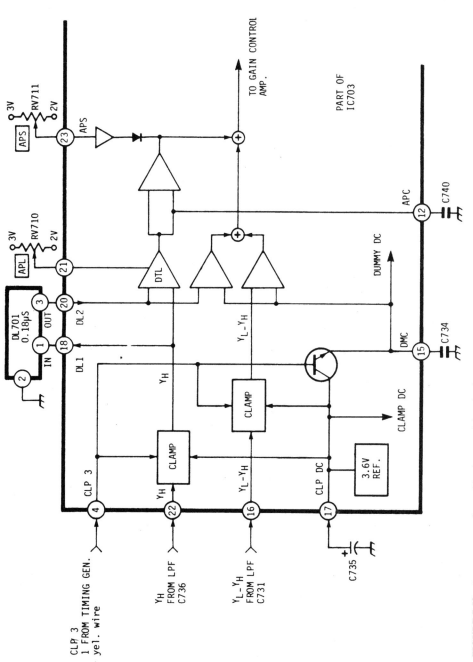

FIGURE 7.59 Aperture-correction and luminance-mixing circuits.

7.107

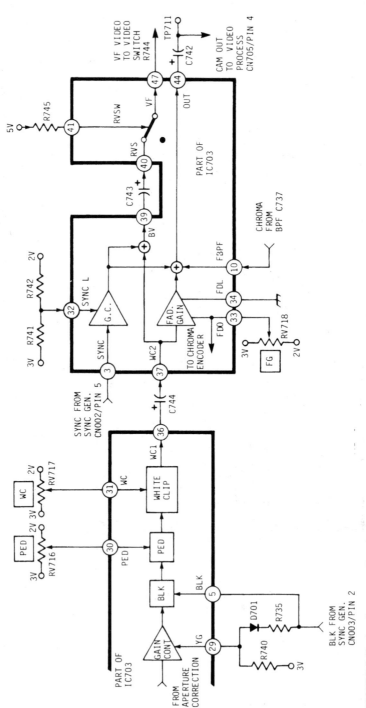

FIGURE 7.60 Luminance-encoding circuits.

the chroma-signal level as described in Sec. 7.12.38. The Y output of the fader circuit is mixed with sync, burst, and chroma before being applied to the video processor and camera output through pin 44 and C_{742}. Sync and burst are added to Y after the fader to prevent loss of servo control when the 8-mm tape is played. If sync and burst were added before the fader, the level could be incorrect, and the servo might not be locked to the sync. (Operation of 8-mm servos is discussed in Sec. 8.6.)

7.12.38   Chroma Encoding

Figure 7.61 shows the chroma-encoding circuits, most of which are in IC_{703} along with the luminance circuits. Color signals R-Y and B-Y from the matrix (Fig. 7.58) are applied to clamp circuits through pins 13 and 14. R-Y and B-Y are clamped by CLP_3 and applied to two amplifiers which use the dummy dc voltage as the reference. The output from these amplifiers is applied to chroma modulators where the 3.58-MHz chroma carriers (which differ by 90°) are modulated.

The color subcarrier is derived from a 4 FSC signal which enters at pin 8. This signal is divided by 4 to produce the 3.58-MHz chroma subcarrier and is phase-shifted to produce the two carriers (which differ from each other by 90°). The phase shift is fixed for normal operation. However, the shift can be varied by changing the voltages at pins 1 and 24 (to accommodate non-NTSC formats).

The color subcarriers are applied to the appropriate chroma modulators where the R-Y and B-Y signals control the output of the 3.58-MHz signals. Voltages at pins 25 and 26 control the modulator balance so that when no color signals are present (during a black scene), the outputs from the balanced modulators are zero. Modulator outputs are mixed and applied to a fader circuit (controlled by the same voltage which controls the Y-fader gain, Sec. 7.12.37). After the chroma fader, burst is added to the chroma signal.

The burst is produced by taking two 3.58-MHz subcarrier outputs (of different phases) from the phase-shift circuit and applying the outputs to balanced modulators. The gain of one modulator is controlled by the voltage at pin 28, as set by Hue control R_{715}. The burst is applied to a GCA, which produces a constant-amplitude burst, as set by B-G control R_{714}, and is switched by a burst flag at pin 2. The resultant burst signal is added to the chroma signal and is combined with the luminance signal after passing through pin 11, bandpass filter FL_{702}, C_{737}, and pin 10 (Fig. 7.60).

7.12.39   VF Video Switch Circuits

Figure 7.62 shows the video-switch circuits. The viewfinders in many cameras serve two purposes: to monitor the camera output and to monitor the output of the camcorder/VCR playback. In our camera, switching for these two modes is controlled by the VF video switch.

The camera output (VF Video) from pin 47 of IC_{703} (Fig. 7.60) is applied to Q_{712} through C_{745}. In the record mode, CAM REC is high, indicating that the camera is recording. Q_{717} is turned on, and Q_{716} is off, allowing VF video to appear on the base of Q_{712}. This video signal is buffered by Q_{712} and Q_{714} and appears in the viewfinder (EVF).

During playback, CAM REC is low. This turns Q_{717} off and Q_{716} on, muting VF Video from the camera. The low also turns Q_{715} off, allowing the return (or playback) video from the playback heads to be applied at Q_{713}. The playback video signal is buffered by Q_{713} and is applied to the viewfinder through Q_{714}.

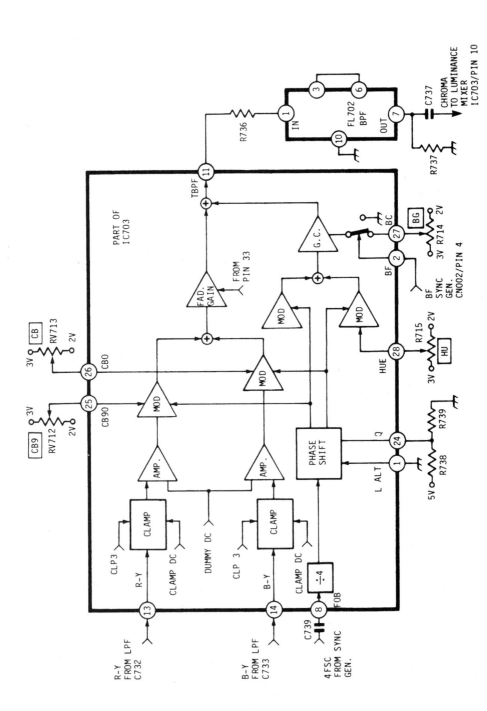

7.110

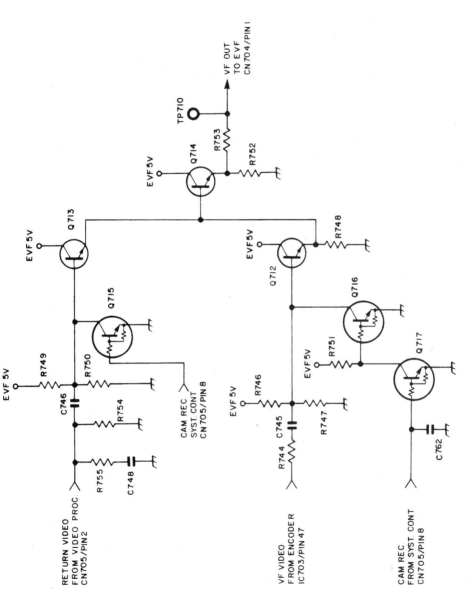

FIGURE 7.62 VF video-switch circuits.

7.111

CHAPTER 8
8-MM VIDEO FORMAT

This chapter describes the differences between the 8-mm format and the VHS camcorder formats discussed in Chap. 7. The 8-mm format uses a tape that is 8 mm wide. This results in a very small cassette, as well as a small drum required to play the tape, and a small mechanism to house the transport and drum assembly, making the 8-mm format ideal for camcorders.

At one time, 8-mm VCRs (that can play the 8-mm cassette directly) were under design. But 8-mm VCRs have been abandoned in favor of the popular VHS format. However, 8-mm camcorders are in wide use.

8.1 THE 8-MM CASSETTE

Figure 8.1*a* shows a comparison of the 8-mm cassette with other cassettes. Figure 8.1*b* shows details of the 8-mm cassette. As shown, the 8-mm cassette is about the same size as a conventional audio cassette. In addition to size, the 8-mm cassette has many features that distinguish it from either VHS or Beta.

8.1.1 8-mm Cassette Features

The following is a summary of the features shown in Fig. 8.1.

Record-Protect Switch.   Unlike the record-protect break-off tabs found on VHS and Beta cassettes, the 8-mm cassette has a record-protect, or record-proof, switch. The 8-mm cassette is ready to record when the switch is set to record (right) and cannot record when the switch is in the record-inhibit position (left). When in the inhibit mode, the window in the back of the cassette is red, telling the user that the cassette is incapable of recording.

Standard Designations.   The cassette designations (usually four or five characters, such as P6-90) describe the cassette characteristics. As shown in Fig. 8.1*b,* the first letter of the designation stands for either metal-powder tape (P) or metal-evaporated tape (E). The second number (6) in the example stands for NTSC system using 60 fields. If the 6 is replaced by a 5, the cassette is recorded in the PAL system using 50 fields. The last two numbers in the standard designation (90) represents the recording time in minutes (90 minutes in our example). There is no physical difference between an NTSC and a PAL tape. However, because of the differences in the number of fields, the recording time is different for the two systems.

Cassette type	Tape width (mm)	W-H-D(mm)	Volume ratio
8mm 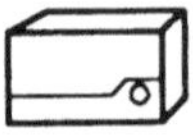	8	95 x 62.5 x 15	1
VHS-C 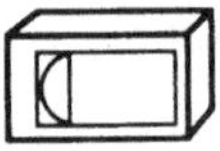	12.65	92 x 59 x 23	1.4
Beta 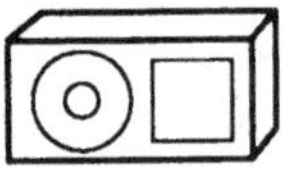	12.65	156 x 96 x 25	4.2
VHS 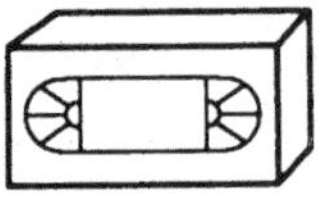	12.6	188 x 104 x 25	5.5
Audio 	3.8	102 x 63 x 12	0.87

(a)

FIGURE 8.1   The 8-mm cassette.

Mixed-Tape Recordings.   Note that a blank P6-90 tape can be used in a PAL machine. Of course, NTSC recordings cannot be played on a PAL machine and vice versa. However, the same blank tape can be used in either machine. When recording on a tape in a different format, the record time is not accurate. For example, if you record P6-90 tape in PAL, the PAL machine plays faster and records the full tape in less time.

Double Door with Lid Lock.   The 8-mm tape is protected by a double door, both on the outside and inside of the tape. This door can be released by pulling the lid lock on the right side of the cassette and folding the door up to expose the tape.

Reel Lock.   There is a reel lock at the bottom center of the tape. When the tape is loaded into the transport mechanism, a pin fits above the reel-lock tab. When the tape is threading, the reel lock is pulled back, releasing the lock mechanism (which can be observed through the window on the cassette top). The reels are free to turn when the reel lock is pulled back.

Alignment Guides or Holes.   The 8-mm tape uses two alignment guides or holes. These guides hold the cassette in position and greatly reduce mechanical alignment problems in the transport mechanism.

Tape-Sensor System.   A large hole in the middle of the cassette is for a tape-sensor lamp. This lamp is an LED (on the mechanism) which shines through the sides of the

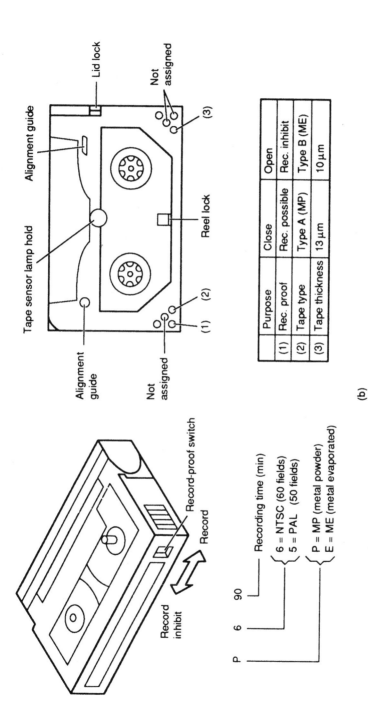

Lid lock

Alignment guide

Tape sensor lamp hold

Alignment guide

Not assigned

Reel lock

Not assigned

(1)   (2)

(3)

	Purpose	Close	Open
(1)	Rec. proof	Rec. possible	Rec. inhibit
(2)	Tape type	Type A (MP)	Type B (ME)
(3)	Tape thickness	13 μm	10 μm

(b)

Record-proof switch

Record

Record inhibit

P   6   90

Recording time (min)

$\left\{\begin{array}{l} 6 = \text{NTSC (60 fields)} \\ 5 = \text{PAL (50 fields)} \end{array}\right.$

$\left\{\begin{array}{l} P = \text{MP (metal powder)} \\ E = \text{ME (metal evaporated)} \end{array}\right.$

FIGURE 8.1 (*Continued*) The 8-mm cassette.

8.3

cassette. The tape is constructed with a short transparent leader at both ends (similar to VHS). When the tape reaches either end, light from the lamp shines through the leader and out the side of the cassette and is sensed by a light sensor on the side of the transport mechanism. One advantage to this type of tape sensor is that a cassette-in switch (such as found on VHS and Beta) is not necessary. When no cassette is in the transport, light passes from the lamp and is detected by sensors on *both sides* of the transport. When a cassette is installed in the transport, one path is always blocked. (Usually both paths are blocked, unless the tape is at either end.) When at least one path is blocked, the transport knows that a cassette is in place and proceeds with the loading and threading sequence (through operation of the system-control microprocessor).

Sensor Holes.   In addition to the alignment holes at the top of the cassette, there are automatic-sensor holes on the lower left and right corners of the cassette.

The record-proof hole is red (about $\frac{1}{16}$ in below the surface) when the record-proof switch is in the record position. This is the closed position. When the record-proof switch is in the inhibit position, the red is moved to the side, and the record-proof hole is about $\frac{1}{2}$ in deep (the thickness of the cassette).

The MP and ME hole is used to sense metal-powder or metal-evaporated tape (closed for powder, open for evaporated). This is used to change the equalization characteristics for optimum results with both types of tape.

The tape thickness hole is used to sense tape thickness (closed for 13 µm; open for 10 µm).

The three remaining holes are not generally used at this time but will be used in the future (depending on the fate of the 8-mm system).

8.2 8-MM TAPE FORMAT AND HEADS

Figure 8.2a shows the basic 8-mm tape format. Figure 8.2b shows the relationship of the tape and heads. Compare this to the VHS and Beta formats and head configurations described in Chaps. 5 and 6.

8.2.1 Format

As shown in Fig. 8.2a, the 8-mm format uses a helical-scan system similar to that in VHS and Beta. As discussed in Chaps. 5 and 6, helical scan produces the high video-head speed essential for recording the high-frequency video information (and hi-fi audio). However, the 8-mm helical-scan system has many additional features not found in VHS and Beta. First, the 8-mm format divides the tape into *four areas of information,* two of which are scanned by rotating video heads (to produce the slanted tracks) and two of which are recorded by fixed heads.

Video and FM Audio and Tracking.   The largest area of information contains the video (both luma and chroma), the FM audio (mono) signal, and a tracking signal (used in 8 mm to maintain tape tracking, instead of the control-head or CTL signal found in VHS and Beta). The large area of information is recorded by two video heads, spaced

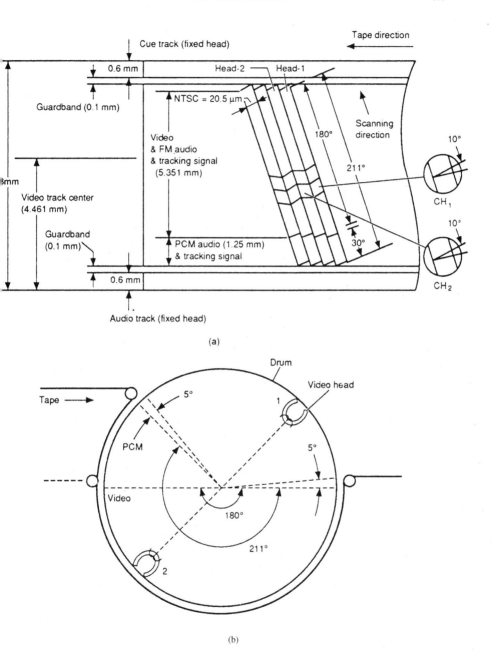

(a)

(b)

FIGURE 8.2 The 8-mm tape-format and video-head configuration.

180° apart, with ±10° azimuth (instead of the 6° and 7° azimuths found in VHS and Beta). The width of an 8-mm slant track is 20.5 μm (for the NTSC format).

PCM Audio.   The second area of information extends 1.25 mm below the video track. This area is for pulse code modulation (PCM) audio and a PCM tracking signal. The PCM area is not required for basic 8-mm operation but is reserved for possible stereo operation. A tape recorded with PCM audio also has the FM audio (mixed with video and tracking on the large area of information) to make all 8-mm tapes compatible.

Scanning and Tape Wrap.   Tape scan for the two areas of information is 221° (instead of the typical 180°). In 8 mm, the tape is wrapped about 40° more to provide a 1° guard band, 30° of additional tape contact, and 5° of tape contact at the entrance and exit. (It is during the additional 30° of tape contact that the entire audio signal can be recorded as PCM information, in addition to video and FM audio and tracking information.)

Cue Track.   The third area of information is an optional cue track, 0.6 mm wide at the top of the tape. This cue track is separated from the video tracks by an 0.1-mm guard band and is intended for recording of edit data. Like PCM, the cue track cannot be recorded without also recording the full video and FM audio and tracking, to maintain compatibility on 8-mm machines.

Audio Track.   The fourth area of information is an optional audio track at the bottom of the tape. Like the cue track, the audio track is 0.6 mm wide and is separated by a 0.1-mm guard band. Again, the audio track cannot be recorded without also recording the full video and FM audio and tracking, to maintain compatibility on 8-mm machines.

8.2.2   Relationship of Head and Tape

As shown in Fig. 8.2*b,* the drum or cylinder used in 8 mm is much smaller than that found in VHS or Beta (40 versus 74 mm). Because of the smaller drum size, 8-mm head speed is reduced to 3.75 m/s (compared to 6.97 m/s for Beta II). Normally, this could result in poor high-frequency response. However, the improved head design in 8 mm provides a frequency response similar to that found in Super Beta and S-VHS and hi-fi.

To get a long playing time (90 to 120 minutes) in 8 mm, the tape speed is decreased to 14.345 mm/s (or 0.56 in/s) compared to 20 mm/s (or 0.75 in/s) for Beta II. This relatively low tape speed does not affect speed of the video head but does affect the track pitch (width of the video tracks). In 8 mm, the track pitch is reduced to 20.5 μm, compared to 29.2 μm found in Beta II.

In either VHS or Beta, the RF-switching pulse is used to switch between the video heads and thus produce a continuous RF envelope (Fig. 5.10). In 8 mm, the RF-switching pulse is still used to select video information from head 1 and head 2. However, with the additional wrap for the PCM audio, the 8-mm RF-switching pulse occurs *between* the PCM and video and FM audio and tracking information, in the same relative position as it occurs in VHS and Beta (at the 180° position). If PCM is used, the PCM information is pulled from the tape by a different pulse.

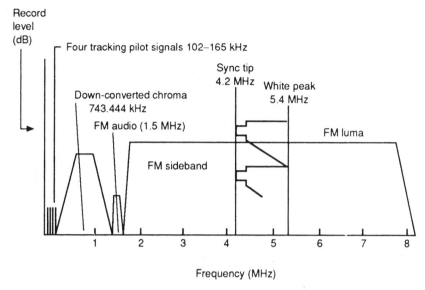

FIGURE 8.3   NTSC spectrum for the 8-mm format.

8.3   8-MM NTSC SPECTRUM

Figure 8.3 shows the NTSC signal spectrum for the 8-mm format. Compare this to the
VHS and Beta recording spectrums. There are four separate signals recorded in the 8-
mm format: (1) the FM luma, (2) the down-converted chroma, (3) an FM audio carrier, and (4) a tracking-pilot signal. The frequencies for these four signals are chosen to
minimize interference and optimize recording.

8.3.1   FM Luma

Most of the spectrum recorded in 8 mm is for the FM luma signal. As in Super Beta
(Fig. 6.10), the luma signal is used to modulate an FM carrier with the sync tip at 4.2
MHz and the white peak at 5.4 MHz.

8.3.2   Down-Converted Chroma

The chroma is down-converted in a manner similar to that of VHS and Beta. However,
the frequency for 8 mm is 743.444 kHz (instead of 688 kHz for Beta and 629 kHz for
VHS). As usual, the down-conversion process separates the luma and chroma signals.

8.3.3   FM Audio

The FM audio is carried on a single 1.5-MHz carrier (instead of the four carriers found
on Super Beta and Beta Hifi). Because of the single carrier, the 8-mm format does not

provide for stereo operation (although the mono quality is about the same as for Super Beta and Beta Hifi). The single carrier is possible because of the $\pm 10°$ head azimuth.

8.3.4 Tracking-Pilot Signals

The four tracking-pilot signals (that extend from about 102 to 165 khz) are unique to 8 mm. These tracking signals are changed (in sequence) as the tracks of information are changed and are repeated after four consecutive fields. As discussed in Sec. 8.5, the pilot signals produce considerable crosstalk. However, this is desirable since the crosstalk is used to ensure that the video is tracking properly.

8.4   8-MM FLYING ERASE

As in the case of many professional video-editing machines (such as the $\frac{3}{4}$-in U-matic), the 8-mm format uses a flying-erase head. Figure 8.4a shows a comparison of flying-erase and full-erase formats. Figure 8.4b shows the relationship of the tape and heads in 8-mm flying erase.

8.4.1   Comparison of Full Erase and Flying Erase

Most VHS and Beta machines use a full-erase head to eliminate previously recorded information from the tape so that the video heads can write on a clean tape. Although full erase is simple, problems can be created, particularly in sophisticated machines where special edit techniques are used (for example, in assembly recording, where a new program is recorded so that the new tracks are aligned with old tracks to eliminate breaks in the picture).

As shown in Fig. 8.4a, with full erase, an overlap area is recorded by the video heads at the start of record. Likewise, a blank is produced after record stop. The overlap produces rainbow patterns (caused by mixing new and old information). Typically, the rainbow appears to move down from the top of the picture and disappears at the bottom.

The blanks produced after record stop not only cause gaps in the video but can affect the control or CTL pulses. The gaps result in video muting or noise, or both. VCRs and camcorders with full erase often have many elaborate circuits to prevent overlap and gaps in the special-edit modes. The flying-erase head eliminates the need for such circuits.

8.4.2   Flying-Erase Head

As the name implies, a flying-erase head is constantly rotating since the head is on the video cylinder with the two video heads. As shown in Fig. 8.4b, the flying-erase head erases *two video tracks directly after the old recording,* at record start. This eliminates overlap and prevents the rainbow pattern. The now blank tracks (behind the flying-erase head) are then filled in by the two video heads. This eliminates noise-producing gaps.

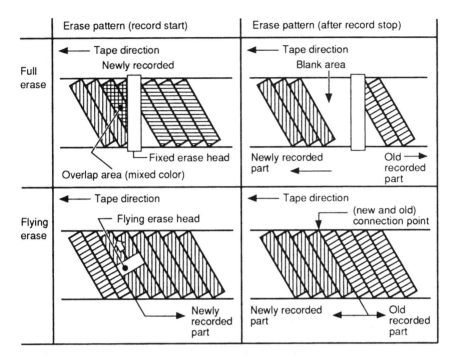

(a)

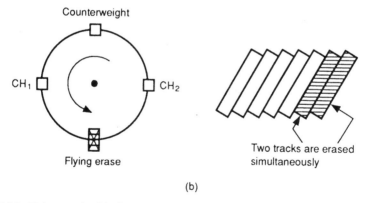

(b)

FIGURE 8.4 Flying-erase head for 8 mm.

When the recording is over (record stop), timing is such that the flying-erase head erases *only two tracks* and is turned off before the erased tracks are recorded upon. This eliminates the formation of blank areas, and a perfect edit exists between the new and old recordings.

As shown in Fig. 8.4*b*, the flying-erase head is located between head 1 and head 2.

The cylinder rotates counterclockwise so that the erase head precedes the head 1 track by $\frac{1}{4}$ field. The flying-erase head is twice the width of the video heads so that both tracks are erased simultaneously. The two tracks are filled in by the video heads before the flying-erase head contacts the tape again and begins to erase another two tracks. Both the timing of the flying-erase head and the switching of the video heads are determined by the system-control microprocessor. *As a troubleshooting tip, always check the timing and switching signals from system-control to the erase and video heads if you get a top-to-bottom moving rainbow pattern and/or excessive video noise.*

Note that a counterweight is positioned on the video cylinder opposite the flying-erase head. Also, because there is no control or CTL tracks required for 8 mm, the use of a flying-erase head eliminates the need for all fixed heads.

8.5   8-MM AUTOMATIC TRACK FINDING (ATF)

Figure 8.5a shows the principles of ATF, while Fig. 8.5b shows the relationship of the ATF pilots. Figure 8.6 shows typical ATF circuits. The 8-mm format uses ATF to identify when the video heads are properly positioned on the correct video tracks. The ATF system *constantly changes tape speed* (as necessary) to keep the heads and tape tracks properly aligned. This not only eliminates the need for a control CTL head but also provides continuous correction of video tracking. In turn, this eliminates the need for a user tracking control to ensure compatibility between tapes recorded on different machines. ATF also greatly simplifies tape-path alignment since ATF constantly corrects for tape position instead of correcting only once for each frame (as in VHS and Beta).

The four tracking-pilot signals discussed in Sec. 8.3.4 are recorded on tape and used to control the ATF system. As shown in Fig. 8.5, the pilot signals are repeated after four fields. However, since the RF-switching pulse occurs between the PCM, video, and FM audio and tracking areas, the *video and PCM are always offset by one track.*

In playback, when the video head scans the desired track (the track containing F_3, or pilot 3, in our example), the head also overlaps the adjacent (leading and trailing) tracks containing pilots. The frequencies for the pilot signals are relatively low (compared to luma and chroma) and crosstalk between adjacent pilots is not reduced by the $\pm 10°$ video-head azimuth. Also, the video-head width is 25 μm, compared to 20.5 μm for the track width. All of these factors guarantee a certain overlap of adjacent tracks.

When the camcorder is tracking properly and played back, the crosstalk between adjacent tracks is equal. The principle of equal crosstalk is used by the ATF servo to provide proper tracking. In effect, the ATF servo monitors the adjacent pilot signals and adjusts the tape position (in relation to the heads) so that crosstalk from the leading and trailing pilots is equal and constant.

The ATF system provides for switching the pilot signals and controlling the errors generated from the crosstalk to move the tape in the correct direction (and at the correct speed). The ATF servo (Sec. 8.5.2) controls the tape position for all four areas of information, even though control takes place only in the PCM and video section.

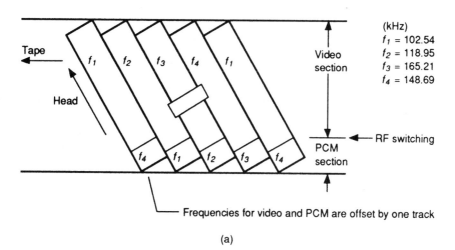

(a)

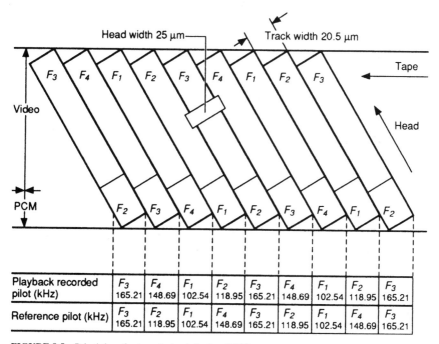

FIGURE 8.5 Principles of automatic track finding (ATF).

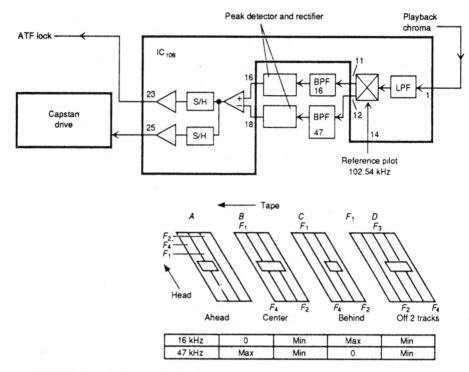

FIGURE 8.6   Typical 8-mm ATF circuits.

8.5.1 ATF Pilot Signals

The four ATF pilots (placed on tape during record) are chosen to provide corrections signals (of proper frequency) when mixed with a *reference-pilot signal* that occurs during playback. The reference-pilot signal is a function of the system-control micro-processor. System control knows which of the four fields the tape should cover at any given time and produces a reference pilot *for each field*. When the recorded pilot and reference pilot are mixed, an error signal is produced if tracking is not correct.

As shown in Fig. 8.5*b*, when the video head is tracking the F_3 pilot during play-back, the 165.21-kHz pilot (on tape) is mixed with a 165.21-kHz reference pilot, pro-ducing a zero-error or zero-crosstalk signal. When the pilot from the track containing the F_4 pilot (which trails the desired track) is mixed with the F_3 reference pilot, the mixed frequency is $165.21 - 148.69$, or about 16 kHz. This is the *trailing track fre-quency* to which the ATF servo responds.

When the crosstalk produced from the head overlapping the track containing the F_2 pilot (which leads the desired track) is mixed with the desired F_3 reference, the mixed frequency is $165 - 118.95$, or about 47 kHz. This is the *leading track frequency* to which the ATF servo responds.

By examining all combinations, it will be seen that the system produces zero error (tracking right on), a 16-kHz (trailing) signal, or a 47-kHz (leading) signal. These sig-nals are used by the ATF servo to move the tape (through action of the capstan) as necessary for proper tracking.

8.5.2 ATF Tracking Servo

As shown in Fig. 8.6, the playback chroma from the video heads is removed by an LPF in IC_{106}, leaving only the ATF pilot signals, which are mixed with the reference pilot. The resulting signals are applied to the capstan servo through corresponding BPFs (external to IC_{106}) and a comparator in IC_{106}. The leading and trailing track signals are detected and rectified by external circuits before being applied to the comparator.

Figure 8.6 shows the relationship of tracking signals to head position. Note that there are four possible conditions for the servo. In condition A, the tape is tracking ahead of the correct position (between F_1 and F_4). This produces a maximum 47-kHz difference signal (and a low or zero 16-kHz signal) and causes the tape to slow down (capstan motor-speed decreases) until tracking is correct.

In condition B, the tape is tracking properly. This produces minimum or zero 16- and 47-kHz signals, and there is no speed correction. In condition C, the tape is tracking behind the correct position (between F_1 and F_2). This produces a maximum 16-kHz signal, and a minimum or zero 47-kHz signal, causing the capstan and tape to speed up until tracking is correct.

In condition D, the tape is two tracks off (behind the correct position). This produces minimum or zero 16- and 47-kHz signals (as in condition B). However, as the head moves even slightly off center (say toward the F_2 track), crosstalk from the F_2 track increases. This produces the same condition as condition C, and the tape speeds up. Tape speed increases even further as track F_2 is passed and slows down only after the tape is tracking properly across F_1.

8.6 TYPICAL 8-MM SERVO AND ATF CIRCUITS

This section describes typical servo and ATF circuits found in 8-mm camcorders. Although the overall function is the same as for VHS (and Beta), these 8-mm circuits are different in detail, particularly in operation of the capstan servo and tape-counting functions. This is because there is no CTL track recorded on 8-mm tape. Instead, the playback signal is used to locate a position on tape and to change tape speed through operation of the capstan servo. As discussed, the 8-mm ATF system constantly corrects for tape position instead of correcting only once each frame (as in VHS or Beta).

8.6.1 Tape Counter Functions

Figure 8.7 shows the complete input-key matrix circuits. A portion of these circuits is used in the tape-counter function. The matrix is similar to that of many devices (camcorders, VCR, etc.) in that it provides control of the camcorder through operation of the system-control microprocessor. In simple terms, the microprocessor produces scan-out pulses which are returned to the microprocessor through switches in the matrix (after conversion by a binary-decimal decoder). When any of the switches are closed, the corresponding scan pulse is applied to the microprocessor at the scan-in terminals. This causes the microprocessor to issue a command to the camcorder.

The reel-sensor inputs (switches at the bottom of Fig. 8.7) are used in the tape-counter function. Reel-sensor S is located on the supply-reel table, with reel-sensors TA and TB located on the takeup reel table. These sensors are used in place of a real-time counter for 8-mm camcorders. A real-time counter is extremely difficult (if not

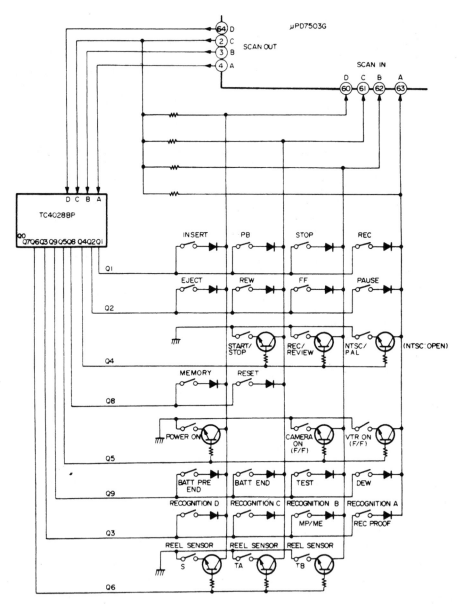

FIGURE 8.7   Input-key matrix circuits.

impossible) to implement in 8 mm because there is no CTL track. As a result, CTL pulses cannot be counted to provide an indication of location on tape at any given time (as can be done in VHS).

The reel-sensor switches are closed and opened by reel rotation. When the switches are closed, the scan pulses are returned to the microprocessor and provide an indication of tape count and position. Tape direction is indicated by the phase relationship between TA and TB switch closures.

TA and TB differ from each other in phase by 90°. In the Forward direction, TA leads TB. The opposite occurs in Reverse. The microprocessor converts this phase information to binary data. When TA is 1 before both TA and TB are 1, the tape is moving in a forward direction. When TB is 1 before TA and TB are both equal to 1, the tape is in reverse. Using this information, the microprocessor knows the direction of tape motion and, by counting pulses, can locate the exact position on tape at any given time.

8.6.2 Capstan Servo Circuits

Figure 8.8 shows the capstan-servo circuits. The capstan FG pulses are applied to a counter in IC_{502} through pin 23. Before reaching pin 23, the pulses pass through a programmable divider. In the Play mode, the divider is set to divide the FG frequency by 1. The capstan counter also receives pulses from a capstan reference generator in IC_{502}. The counter combines the two sets of pulses and produces both a speed output and a phase output. These outputs are converted by digital-analog circuits (DACs) in IC_{502} to produce a speed servo voltage at pin 15 and a phase servo voltage at pin 16. The outputs are dc voltages which vary with the speed and phase of the capstan motor (connected to the reel tables through a timing belt).

The speed and phase outputs are combined with the ATF Error voltage (Secs. 8.6.6 through 8.6.9) and the CAP COMPS voltage (Sec. 8.6.5) to produce the CAP Error output. During the Record mode, there is no ATF Error voltage, so capstan-motor speed (and thus tape speed) is controlled by the capstan speed and phase outputs alone. (During Record, the ATF Error voltage is held at a fixed 2.5 V.)

During Playback, the ATF Error voltage is used in place of the capstan phase output (which is held at a fixed 2 V). As a result, the ATF Error voltage and the capstan speed output maintain servo control of the capstan motor.

The capstan servo is also used during the Cue and Review modes (Sec. 8.6.3). However, the programmable divider is changed (by a command from the microprocessor) so that the FG signal is divided by 9 instead of by 1 for the Play mode. As a result, the circuits in IC_{502} increase the output voltage to restore the capstan FG to the proper frequency. This speeds up the tape for the Cue mode.

In Review mode, commands from the microprocessor change the programmable divider so that the FG signal is divided by 7, and tape speed is changed accordingly by the circuits in IC_{502}.

In Fast Forward and Rewind modes, the programmable divider is changed so that the FG signal is divided by 15. This increases tape speed by a factor of 15 over that in the Play or Record modes.

8.6.3 Capstan Free-Speed Compensation Circuits

Figure 8.9 shows the capstan free-speed compensation circuits. The CAP Error voltage (containing both speed and phase correction data) is applied to the noninverting input of an amplifier in IC_{502} through IC_{504}. The IC_{502} amplifier compares the error voltage (pin 23) to a capstan bias at pin 24. The bias is produced by a resistor network adjusted by RV_{503} and RV_{504} and buffered by the other half of IC_{504}. The capstan bias is adjusted so that the capstan motor (and reel tables) will turn at the correct speed when the ATF servo output is missing. This allows the ATF servo to correct (in either direc-

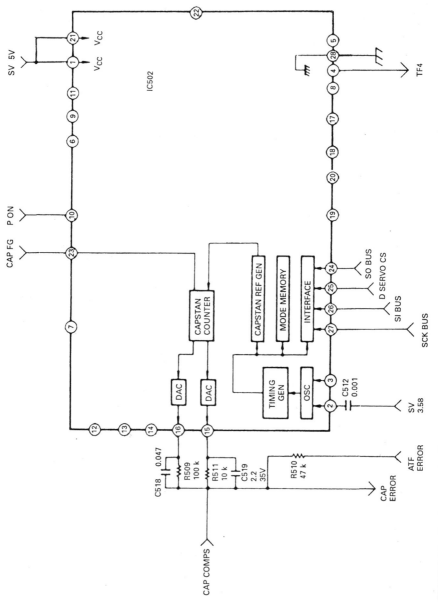

FIGURE 8.8 Capstan-servo circuits.

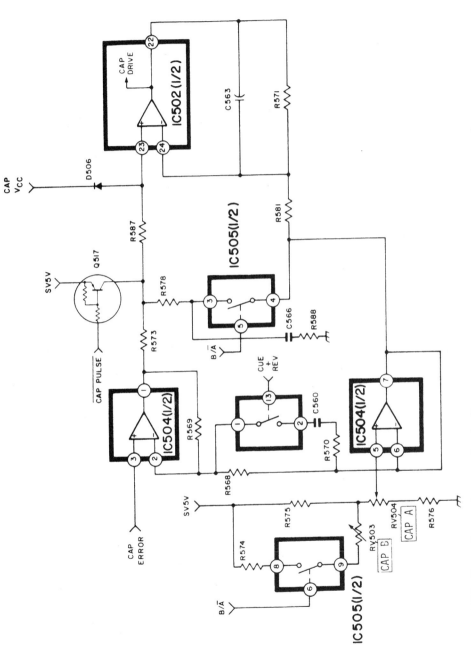

FIGURE 8.9 Capstan free-speed compensation circuits.

8.17

tion) for a relatively large range, thus assuring proper servo lockup. RV_{503} and RV_{504} are adjusted by disabling the ATF servo circuit and adjusting for correct capstan FG frequency.

The switches in IC_{505} are operated by system control to change the response time of the free-speed compensation circuits for different operating speeds. This is necessary because of the drastic change in tape speed during Rewind, Fast Forward, Cue, and Review.

To stop the capstan motor (and tape) quickly, a $\overline{CAP\ PULSE}$ from the microprocessor is applied through Q_{517}. This is done simultaneously with a command to reverse direction of the motor (Sec. 8.6.4). The combination of the two commands provides the same function as would a mechanical brake. Diode D_{506} prevents the CAP Error voltage, or the CAP PULSE voltage, from exceeding CAP VCC (the supply voltage for IC_{502}).

8.6.4 Capstan Motor Drive Circuits

Figure 8.10 shows the capstan-motor drive circuits. The CAP Error and CAP BIAS voltages are applied to comparator circuits in IC_{502} to produce a drive voltage for the capstan motor. The drive voltage is amplified and applied to a PWM modulator, which uses a sawtooth at pin 17 of IC_{502} to produce a PWM output at pin 16. The PWM output is applied to current amplifier Q_{502} after being amplified by Q_{503}. The current pulses are filtered, and the resultant dc voltage is applied to a current limiter (which removes bias from the capstan-servo voltage amplifier if the voltage at pin 15 exceeds CAP VCC).

The voltage at pin 15 is the drive voltage for the three-phase capstan motor and is used to operate three current-source circuits at pins 10, 11, and 12. The phase of the current through the motor windings is controlled by three Hall-effect devices mounted on the motor. The Hall-effect outputs are amplified and applied to a 120° logic block. This block determines which of the three current sources is on at any particular time.

The direction in which the capstan motor turns is determined by the logic block. In turn, the block is controlled by a CAP CW/CCW signal from the microprocessor (applied through pin 27 and Q_{513}). If pin 27 is high (above 0.7 V), the capstan motor turns counterclockwise, which is the normal direction for the motor. During Reverse or Review modes, pin 27 is low, and the capstan motor (and tape) moves in the clockwise direction. (The direction of the capstan motor is changed by controlling the phase relationship of the Hall-effect outputs which, in turn, control the current-source circuits.)

The capstan-motor drive circuit is turned on by a CAP ON command from the microprocessor. The motor drive circuits are on when pin 26 of IC_{502} is high. If the drive voltage at pin 15 of IC_{502} becomes excessive (above CAP VCC), Q_{514} turns on and applies a high to Q_{515}. This turns Q_{515} on, pulling the voltage at pin 26 low to turn the motor drive circuits off.

8.6.5 Error-Voltage Compensation

Figure 8.11 shows the error-voltage compensation circuits. Figures 8.12 and 8.13 show the compensation circuit truth tables. These circuits change the servo-error volt-

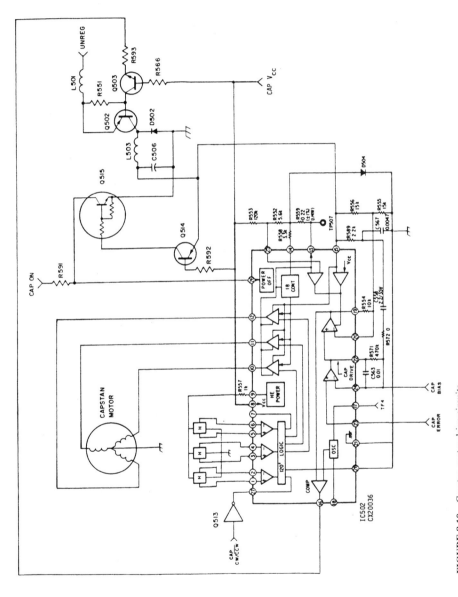

FIGURE 8.10 Capstan-motor drive circuits.

8.19

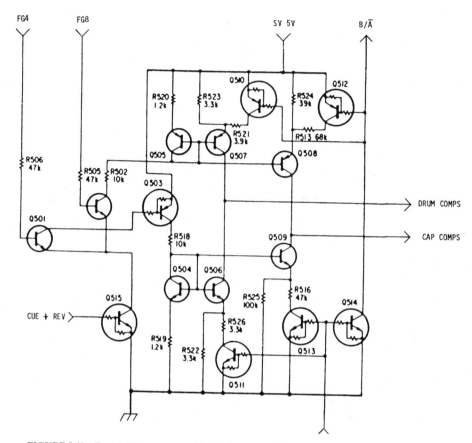

FIGURE 8.11   Error-voltage compensation circuits.

ages for the capstan (and drum) during Cue and Review modes. This is done so that the horizontal frequency does not change when Cue or Review modes are selected.

The circuits are turned on by commands (FG$_4$, FG$_8$, CUE + REV, and A/B) from the microprocessor. These commands operate transistors, turning them on or off (as shown in Figs. 8.12 and 8.13) to change capstan and drum speed as necessary. Note that the speed changes are very slight for both Cue and Review.

8.6.6   ATF Servo Circuits

Figure 8.14 shows the ATF servo circuits. Figures 8.15 and 8.16 show the timing diagram and pilot-signal frequencies, respectively. As discussed in Sec. 8.5, the ATF servo operates in Playback only by sensing four separate pilot signals recorded on tape during Record. The pilots have different frequencies (Fig. 8.16) for each of four consecutive fields. The frequencies are selected in such a way that the crosstalk can be used to indicate when the video head is tracking the correct track. Because the pilot signals are recorded along with video information, a separate track is not required for 8 mm.

MODE	LP		SP	
	CUE	REV	CUE	REV
SP/LP	L	L	H	H
CUE + REV	H	H	H	H
FG8	H	L	H	L
DRUM SYSTEM CORRECTION CURRENT	CURRENT SUPPLY BY Q507	CURRENT DRAIN BY Q506	CURRENT SUPPLY BY Q507	CURRENT DRAIN BY Q506
CAPSTAN SYSTEM CORRECTION CURRENT	CURRENT SUPPLY BY Q508	CURRENT DRAIN BY Q509	CURRENT SUPPLY BY Q508	CURRENT DRAIN BY Q509

FIGURE 8.12   Compensation-circuit truth table for LP and SP.

SIGNAL \ MODE	FWD	CUE	REV	FF/REW
CFG (Hz) (IC602⑥)	960	8903	6515	
CAPFG (Hz) (IC602⑫)	960	989	930	
FG2 (IC602⑪)	L	L	H	H
FG4 (IC602⑭)	L	L	H	H
FG8 (IC602②)	L	H	L	H
$\dfrac{CFG}{CAPFG}$	1/1	1/9	1/7	1/15

FIGURE 8.13   Compensation-circuit truth table for Cue and Review.

In Record, the ATF servo is not used to control the speed of the motor but is used to control the production of the pilot signals being recorded. The pilot signals are generated in IC_{105} by a 5.9-MHz oscillator and applied to various circuits through a programmable divider. The frequency of the signal at pin 7 of IC_{105} is determined by SEL 1 and SEL 2 commands from the microprocessor. Figure 8.16 shows the relationship among SEL 1 and SEL 2, the RF switching pulse, and the pilot frequency being recorded. Figure 8.15 shows the timing.

The output at pin 7 of IC_{105} is applied to mixer Q_{101} through low-pass filters Q_{112} and Q_{113} and REC ATF control RV_{103} (which sets the level of the ATF signal during Record). The low-pass filters convert the square-wave output from IC_{105} to a sine wave. In mixer Q_{101}, the Record ATF signal is combined with Record Chroma, Record AFM, and Record Y signals before being recorded on tape. The combined signal is applied through record amplifiers in IC_{101} to the video heads.

8.6.7   Playback ATF Servo Circuits

In the Playback mode, the ATF servo (Fig. 8.14) uses the four ATF pilots recorded on tape to produce the ATF Error voltage discussed in Sec. 8.6.2. In turn, the ATF Error is used to control capstan (and tape) speed. Figure 8.17 shows timing for the ATF servo circuits during Playback.

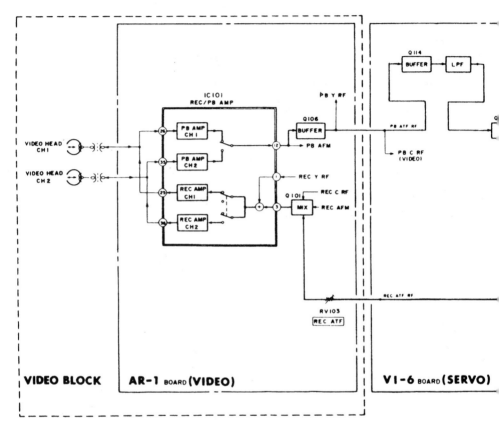

FIGURE 8.14   ATF servo circuits.

The ATF pilot signals are detected by the video heads (along with any video information) and amplified by the playback amplifiers in IC_{101}. At this point, the amplified signals include Y, Chroma, and Audio AFM information, as well as the pilots. The ATF pilots are separated from the Y and Chroma signals by buffer Q_{114} and a low-pass filter and are applied to circuits in IC_{106} through AGC amplifiers Q_{121} and Q_{123} and record-playback switch Q_{122}. The input at pin 1 of IC_{106} is muted by Q_{122} (in response to a command from the microprocessor) during Record. In Playback, the signal at pin 1 of IC_{106} is applied to a mixer through a low-pass filter.

In Playback, the pilot signal is combined in the mixer with a reference signal from IC_{105} through pin 14 of IC_{106}. The frequency of the reference signal is set by SEL 1 and SEL 2 (in response to commands from the microprocessor), which control the programmable divider in IC_{105}. The mixer outputs are applied to bandpass filters FL_{107} and FL_{108}. These filters permit only the crosstalk peaks of 16 and 47 kHz to pass. The

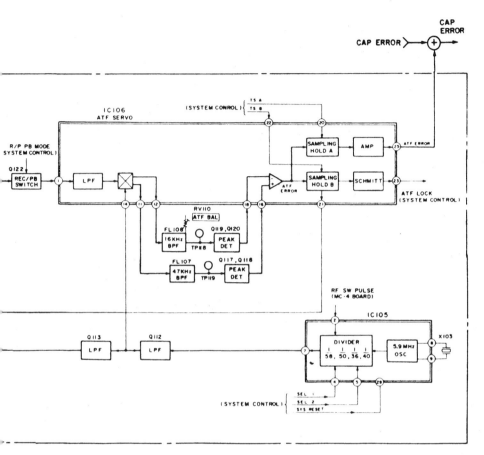

FIGURE 8.14   (*Contined*)ATF servo circuits.

16-kHz filter includes adjustment RV_{110} to compensate for different levels in the crosstalk outputs.

The outputs from the bandpass filters are peak-detected and applied to the inputs of the ATF error comparator, which compares the crosstalk for the leading and trailing tracks. If the crosstalk inputs are equal, the comparator produces a zero output. The comparator output is sampled by a TSA pulse (from the microprocessor) at pin 20 of IC_{106}. If the comparator output is zero (crosstalk inputs equal), the ATF Error signal is output at pin 25 of IC_{106} and is combined with the capstan-servo phase-error signal (Sec. 8.6.2) used during playback.

If the outputs from the bandpass filters are not equal, indicating that the correct track is not being followed, the output of the ATF error comparator produces an output that provides an ATF Error voltage to correct the position on tape. The ATF Error voltage is also sampled by a TSB pulse (from the microprocessor) to produce an ATF Lock output at pin 23 of IC_{106}. The ATF Lock output is high when the ATF servo is not

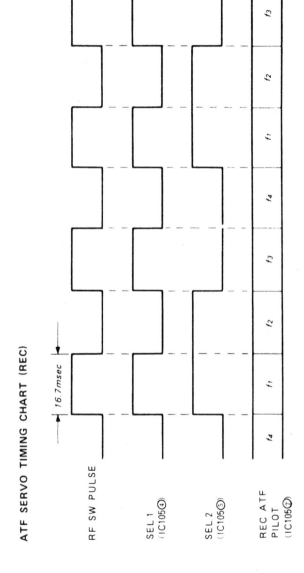

ATF SERVO TIMING CHART (REC)

RF SW PULSE

16.7 msec

SEL 1
(IC105④)

SEL 2
(IC105⑤)

REC ATF
PILOT
(IC105⑦)

f₄ f₁ f₂ f₃ f₄ f₁ f₂ f₃

ATF ERROR (IC106 ㉕) is held at approx. 2.5V dc during recording

FIGURE 8.15 ATF servo timing diagram.

8.24

ATF SERVO PILOT SIGNAL (VI-6 BOARD)

PILOT SIGNAL (IC105 ⑦)		CONTROL SIGNAL	
NAME	FREQUENCY	SEL 1 (IC105 4)	SEL 2 (IC105 5)
f_1	102.54kHz (378 58 f_H)	H	H
f_2	118.95kHz (378 50 f_H)	L	H
f_3	165.21kHz (378 36 f_H)	H	L
f_4	148.69kHz (378 40 f_H)	L	L

FIGURE 8.16 ATF servo pilot-signal frequencies.

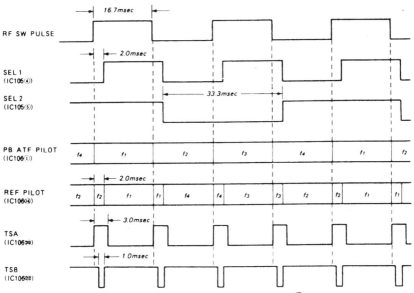

- The ATF ERROR voltage is detected and is output from Pin ㉕ of IC106 when the level of TSA is "L".
- ATF LOCK (phase lock) is detected and is output from Pin ㉓ of IC106 when the level of TSB is "L"

FIGURE 8.17 ATF servo playback timing diagram.

locked up and is used by the microprocessor to indicate correct or incorrect lockup of the capstan servo.

As shown in the timing diagram of Fig. 8.17, the reference pilot, or signal, at pin 14 of IC_{106} is changed for every field of information recorded. However, the change from one field to the next is delayed by 2 ms after the RF switching pulse transition. This relay results in an error in the ATF output for 2 ms at the beginning of each field. This is done so that the ATF servo produces two outputs: the ATF Error, where the correct reference pilot is used, and the ATF Lock, where the incorrect reference pilot is used. These outputs control the phase servo of the capstan motor and also provide an indication of correct tracking to the system-control microprocessor.

The ATF Error voltage is produced by sampling the output of the ATF servo during the time that the correct reference-pilot frequency is used. This is done with the TSA pulse. When low, the TSA pulse samples the output of the ATF servo. Because the TSA pulse is high for 3 ms after the RF switching pulse transition (Fig. 8.17), the incorrect ATF Error output is ignored. When TSA goes low (during the majority of the field), the ATF Error voltage is applied to the phase servo and constantly corrects the tape position through the entire track (including corrections for tape-path errors).

The incorrect portion of the ATF Error signal, which is produced when the reference pilot is not correct, is sampled by TSB. When the servo is locked up, the previous pilot combines with the pilot recorded on tape to produce a 16-kHz output. As a result, if the tape is correctly locked up at the transition of the RF switching pulse, a low ATF output is produced when sampled by TSB (pin 23 of IC_{106}, Fig. 8.14) is low. If the tape is not locked at the correct speed, pin 23 goes high, indicating to the system microprocessor that the tape speed is incorrect.

8.6.8 ATF Lock and Detection Characteristics

The ATF servo (Fig. 8.14) corrects the tape position to within four fields of video. If the ATF servo is tracking a tape two fields off, the ATF Error is still zero. Of course, the servo will eventually correct the problem, although it could take a relatively long time. For this reason, *rear-lock detection* is used to lock up the tape when the servo tracking is off (falling to the rear) by less than four fields. Real-lock timing, standard–long play discrimination, and overall lock-detection characteristics are shown in Figs. 8.18, 8.19, and 8.20, respectively.

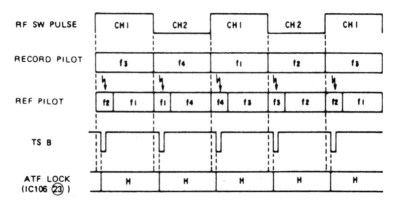

FIGURE 8.18   ATF rear-lock timing diagram.

LP RECORDING/SP PLAYBACK

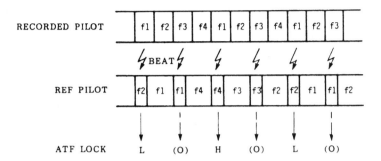

SP RECORDING/LP PLAYBACK

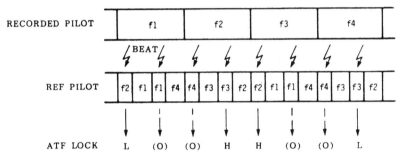

FIGURE 8.19   ATF standard play and long play discrimination.

When the tape is in the normal lock position, the ATF Error continually corrects to keep the tape locked. If a rear-lock condition occurs, the ATF signal pulls the tape from the normal lock, and the ATF lock signal (pin 23 of IC_{106} goes high to inform the system microprocessor that a rear lock condition exists (Fig. 8.18). The microprocessor responds by changing the SEL 2 output (pin 5, IC_{105}) 180°. This shifts the playback sequence of the reference pilots by exactly two video fields and permits the ATF servo to lock in the normal sequence.

The ATF lock signal at pin 23 of IC_{106} is also used by the microprocessor to detect speed changes. As shown in Fig. 8.19, if a long play (LP) recording is played back in standard play (SP), the ATF lock signal produces an output of low, zero, high, zero, low, zero, and so on. The microprocessor responds by changing the speed to LP. If an attempt is made to play back an SP recording in LP, the ATF output is low, zero, zero, high, high, zero, zero, low, and so on, causing the microprocessor to change the speed to SP. (Note that this particular camera is capable of playing in both SP and LP but records in SP only.)

8.6.9   Special ATF Operation

Operation of the ATF servo changes when Reverse is selected, such as during the normal Reverse and Review modes. Timing for normal Reverse and Review modes is

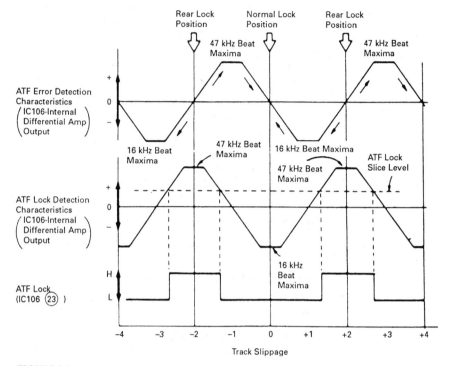

FIGURE 8.20   ATF lock and detection characteristics.

shown in Figs. 8.21 and 8.22, respectively. Operation is also different for the Cue mode.

In Reverse, the reference pilots are changed to correspond with the crossing of tracks. For example, in the normal Reverse mode of Fig. 8.21, the head traces across the F_2, F_1, and F_4 tracks. The next head traces across F_1, F_4, F_3, and so on. The system microprocessor recognizes this and produces the reference pilots accordingly (Sec. 8.6.6). Of course, the reference pilots are reversed from Forward operation (F_1 is used in place of F_3 and F_2 is used in place of F_4). This produces a Reverse ATF correction where the phase of the tape is constantly corrected, but in reverse direction.

In Review mode, several tracks are crossed by each pass of the head, but in the reverse direction, as shown in Fig. 8.22. Again, the microprocessor recognizes this and records the corresponding pilots. In Cue mode, several tracks are crossed by the heads (as in Review) but with the tape running in the opposite direction, producing the appropriate reference pilots, as shown in Fig. 8.23.

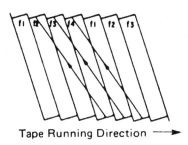

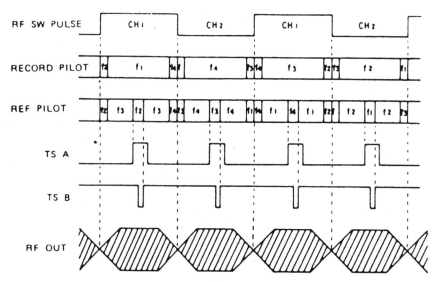

FIGURE 8.21 Reverse playback timing diagram.

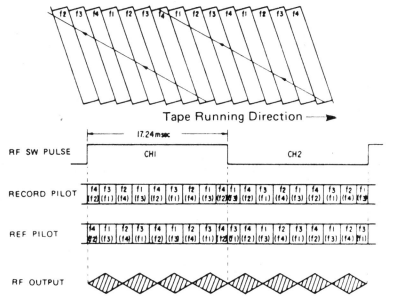

FIGURE 8.22 Review mode timing diagram.

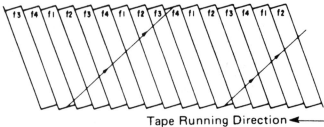

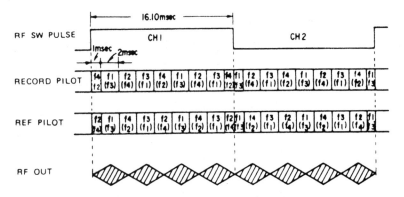

FIGURE 8.23 Cue mode timing diagram.

CHAPTER 9
VIDEO DISCS

This chapter is devoted to video discs and video-disc players. To understand the operation of such players, it is essential that you first understand the basics of laser-disc equipment. If you need a refresher on how such equipment works, read *Lenk's Laser Handbook* (McGraw-Hill, 1992). Here, we concentrate on information that the video technician can put to immediate use when troubleshooting video-disc players.

9.1 THE VIDEO-DISC SCENE

First, let us resolve the *disc* versus *disk* question. The author generally spells disk with a *k* rather than a *c*. There are those who feel disc should be used for consumer audio and video products and disk for magnetic data devices. Still others feel that disc should be used only with laser recording and playback, and disk for other video recording. As a practical matter, most audio and video player manufacturers have settled on disc, so we will do the same.

To further complicate the problem, various manufacturers produce laserdiscs, videodiscs, laservideo, Laservision, digital video discs, and so on; all of which are similar. That is, the discs have both video and audio information, are played by means of a laser beam, and can be shown on a TV or monitor. In this book, we call the devices on which any form of video discs are played *compact disc video* (CDV) players.

Of course, we recognize that there are *compact disc* (CD) players which reproduce audio only, and there are now *digital video disc* (DVD) players (to reproduce the newer form of laser disc). There are also compact disc interactive (CD-I) and CD-ROM devices that operate with lasers. (CD-ROMs are used with computers as a substitute for magnetic disk drives, both floppy and hard, and have few operating controls, beyond possible on-off switches, power LEDs, etc.)

For some time, there have been devices that will play both CDs and CDVs. Today, the DVD players will also accommodate CDVs. Likewise, there are DVD and DVD-ROM players. In this book, we concentrate on CDV players because they are still the most popular form of video-disc player and have all of the elements necessary to understand other forms of video-disc players.

9.1.1 Video-Disc History

One of the earliest video-disc systems, called TeD, used a pressure pickup (piezo-electric) and was introduced by Teledec Telefunken Decca of Germany in 1975. The TeD system was soon removed from the market. Another system that never quite got off the ground was the *transparent disc*, developed by Thomson-CFS.

The *reflective optical pickup* now used for CD, CDV, and DVD was developed by NV Philips in the Netherlands and by MCA. The system was called *video long play* at one time and was introduced in the United States in 1978.

A system called CED using *capacitive pickup* was developed by RCA Laboratories and was introduced in the United States in 1981. This was at a time when VCRs were first becoming popular. Because CED could not record (as can a VCR), the CED system soon disappeared. Still another system, known as video high density, or VHD, with a companion audio high density, or AHD, was developed by the Victor Company of Japan (JVC) and introduced in 1982. Both systems soon disappeared, primarily because of the VCR popularity.

The latest entries into the video-disc world are the DVD and DVD-ROM. DVD offers many advantages over CDV, mostly in the increased data capacity but also in reliability. The principal DVD features are discussed in Sec. 9.2.8.

9.2 INTRODUCTION TO CDV

The following paragraphs are for readers who are totally unfamiliar with the CDV system. As in the case of CDs, the CDV system uses discs on which information is recorded in the form of pits and flats. The pit and flat track format shown in Fig. 9.1 is the same for both CDs and CDVs and is similar to that for DVDs. Of course, much more information is required for CDVs and DVDs because both audio and video are involved.

Figure 9.2 shows the block diagram of a combination player for both CD and CDV discs. Figure 9.3 shows the frequency spectrum for CDV. As shown, the output of the player is both audio and video.

When the player is used for classic 8- or 12-in discs (Laservision discs), the composite video and left and right analog audio are output to a color monitor or TV. When the player is used with 3- or 5-in CDs, only the digital audio is output to a hi-fi stereo system. The player in Fig. 9.2 can also be used with 5-in discs (so-called Gold CD or CDV Single discs) that contain both digital and analog audio, as well as video.

9.2.1 CDV Frequency Spectrum and Encoding

As shown in Fig. 9.3*b*, the video signal is frequency modulated (FM) with a deviation of 1.7 MHz, from 7.6 MHz (TV sync tip) to 9.3 MHz (at TV white peak). The TV black level is located at 8.1 MHz.

The left analog audio channel (audio 1) is frequency modulated on a carrier at 2.3 MHz, with the right channel (audio2) using a carrier of 2.8 MHz. Both analog audio carriers have a maximum deviation of 100 kHz. The digital audio is pulse-width modulated (PWM) at frequencies below 1.7 MHz.

The signals are summed as shown in Fig. 9.3*b*. (Although four signals are involved, only two signals are shown.) The combined signal results in the PWM signal used to etch the pits onto the disc. We do not discuss disc encoding or decoding theory here because we are interested in practical troubleshooting for video-disc players. If you must have this information, read *Lenk's Laser Handbook* (McGraw-Hill, 1992). The author needs the money.

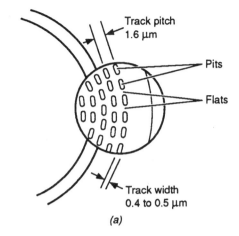

(a)

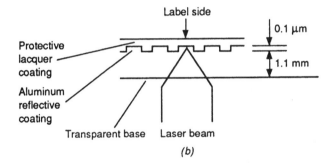

(b)

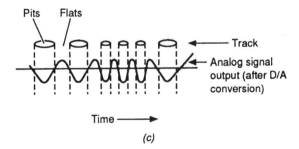

(c)

FIGURE 9.1   Magnified views of a CD showing tracks of pits and flats.

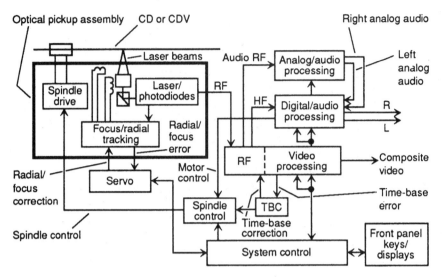

FIGURE 9.2   Block diagram of a combination player for both CD and CDV discs.

9.2.2 CDV Control Codes

Control codes are included with the composite video and audio information. These codes are placed on TV horizontal lines during the vertical-blanking interval (Chap. 2), as shown in Fig. 9.4. For example, line 17 of the first TV field contains a digital code which represents the picture (frame) number for a constant angular velocity (CAV) disc or the elapsed time for a constant linear velocity (CLV) disc (Sec. 9.2.3). Line 280 contains a code which represents the chapter. Lines 18 and 281 are duplicates of lines 17 and 280, respectively, which provide backup if the previous lines are lost because of dropouts. Other data may be included on other lines during the vertical-blanking interval, such as stop codes for interactive discs.

9.2.3 Disc Formats (CLV and CAV)

In addition to the variation in sizes, there are two different formats for placing video information on the disc: *extended play,* using CLV, and *standard play,* using CAV. Figure 9.5 shows the relationship among the disc sizes. Figure 9.6 shows the CLV and CAV formats (for a classic 12-in disc).

Standard Play CAV.   Each revolution of a standard-play CAV disc contains one frame of video information (Fig. 9.6a). One TV frame consists of two interlacing fields. Thus, the TV screen is scanned twice for each revolution of the CAV disc.
    Each field is separated by the vertical-blanking interval, which becomes longer as the interval extends to the other circumference of the disc. This format allows the disc to be played at the same speed of 1800 rpm, which is equivalent to the NTSC frame rate of 30 Hz (1800 ÷ 60 = 30) throughout the disc.
    The CAV format allows the disc to be interactive and provides for special playback effects such as still, slow, fast, and strobe. The disadvantage of CAV is that the video

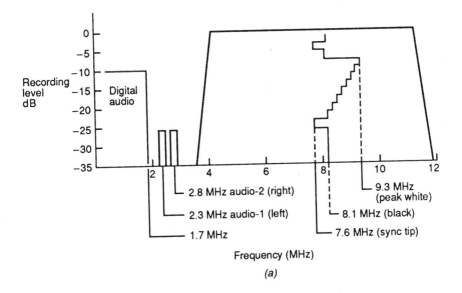

(a)

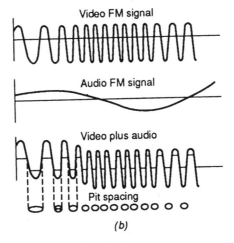

(b)

FIGURE 9.3 (*a*) Frequency spectrum; (*b*) encoding for CDV.

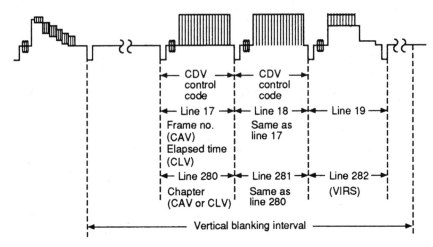

FIGURE 9.4 CDV control codes placed on TV horizontal line during the vertical blanking interval.

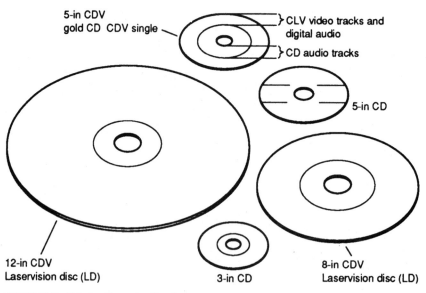

FIGURE 9.5 Relationship among disc sizes.

content is limited to 30 minutes per side on a 12-in disc (54,000 frames per side divided by 1800 rpm equal 30 minutes). The 8-in CAV disc contains about 14 minutes of video per side.

Extended Play CLV. CLV discs do not maintain the constant arrangement of vertical fields. Instead, the field is of equal length throughout the diameter of the disc, as

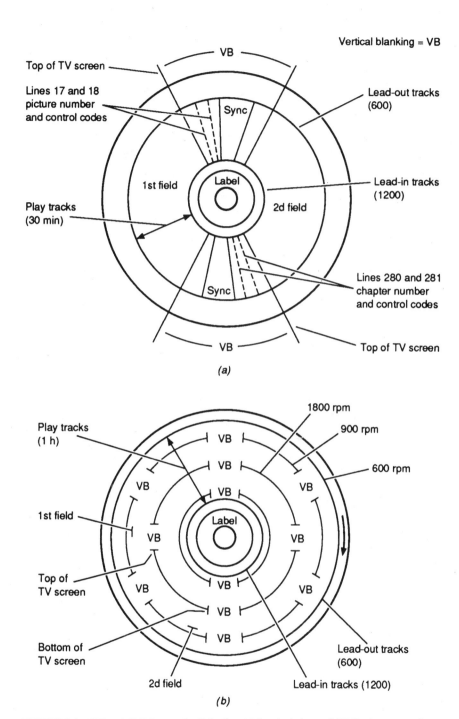

FIGURE 9.6 CLV and CAV formats for 12-in disc. (*a*) Standard play or CAV disc (motor speed constant, vertical field track length variable); (*b*) extended play or CLV disc (motor speed variable, vertical field track length constant).

shown in Fig. 9.6*b*. CLV allows the signal information to be packed more densely on the disc, which extends the play time of a 12-in disc to 1 hour per side (108,000 frames). The 8-in CLV disc contains about 20 minutes of video per side. In general, CLV also reduces the signal-to-noise (S/N) ratio and frequency response (because of the denser packing of signal information).

As a CLV disc plays, the rotation speed is reduced when the pickup laser beam tracks from the inside to the outside circumference. The velocity of the track passing the pickup beam is the same over the entire disc. Typically, CLV disc speed starts at about 1800 rpm when tracking closest to the center of the disc. Toward the outer diameter, disc speed changes to about 500 rpm.

9.2.4 Lead-In and Lead-Out Tracks

Both CAV and CLV discs have lead-in and lead-out tracks. Lead-in tracks are located prior to the start of program material and contain a *start code* used by the player to move the pickup beam to the start of the material. Lead-out tracks (typically a minimum of 600 tracks on both CAV and CLV) are located at the end of program material and contain an *end code* to identify the end of material. On most discs, there is also a *lead-out code* that instructs the pickup beam to return to the start of the disc.

9.2.5 Gold CD and CDV Single

As shown in Fig. 9.5, the Gold CD or CDV Single disc (5-in) contains video tracks in the CLV format, as well as standard CD audio tracks. The video tracks, with accompanying audio, are located toward the outer circumference of the disc. The CD audio tracks are located at the inner circumference. The outer tracks of the Gold CD are located about where the inner tracks begin on the 8- or 12-in disc. As a result, the Gold CD must spin faster than the larger discs. The speed of the video portion of the Gold CD is from 2700 to 1800 rpm. The CD audio-portion speed is from 500 to 300 rpm. Note that the audio portion of the Gold CD may also be played on any standard CD player.

9.2.6 Basic CDV Operation

Figure 9.2 shows the basic CDV functions. The following is a brief description of CDV operation. The boring details are covered in Sec. 9.6.

Laser Pickup. As in the case of CD, the CDV player reads prerecorded information with a noncontact optical pickup system using a low-power laser beam. The beam spot (with a diameter of 1 μm) strikes microscopic pits as the disc spins. This creates reflected light that varies in intensity according to the spacing and length of the pits. The laser beam is locked on the track of the pits by a servo-controlled system to keep the laser beam on track and in focus. Disc rotational speed is also controlled by the servo.

Chapters and Frames. Most CAV discs contain segments, or chapters, as well as numbered frames (pictures) which may be displayed on the monitor or TV. Usually, the number of the information being displayed appears on the TV screen (on-screen display, or OSD) and on the front panel of the player.

RF and Control-Code Processing. RF from the laser pickup is sent to the video-processing circuit where information read from the disc is converted to composite video. Control codes recorded on the disc are decoded and used by system control to operate the player and to provide display information to the front panel and OSD (for display on the monitor or TV screen).

Time-Base Correction (TBC). Video is also fed to the TBC circuit to provide timing information to the spindle-control circuit. In turn, the spindle-control circuit provides speed correction and start-up signals to the spindle (disc turntable) motor through the spindle-drive circuit. The TBC also provides time-base correction to the video signal before video is sent to the monitor or TV.

Video Processing. The video-processing circuits send audio RF to the analog audio-processing circuits for demodulation of the FM analog signal. The video-processing circuits also send high-frequency (HF) signals to the digital audio-processing circuits for decoding.

9.2.7 CDV Player Features

Disc Compatibility. Early laser-disc players are generally limited to 8- and 12-in discs. Most present-day CDV players can also play 3- and 5-in CDs, as well as the newer Gold CD and CDV Single discs.

Displays. Most CDV players provide for on-screen display of chapter and frame information on the monitor or TV. This same information is also displayed on the player front panel.

Special Effects. Most CDV players can provide multispeed, still, and step play.

Audio Features. Most CDV players have dual 16-bit digital-to-audio converters (DACs) and provide for CX *noise reduction.* The CX system serves to cut down noise by about 10 dB or more without compromising the frequency response. CX thus expands the dynamic audio range and enables reproduction of audio with a good S/N ratio.
    Note that when discs recorded with CX-encoded audio are played, it is necessary to press a CX button to activate the CX noise-reduction circuits. This sets the encoded audio signal to a normal level, but the noise is reduced. Also note that playing a normal (nonencoded) disc with the CX system on, or playing a CX-encoded disc without the CX system on, results in unnatural reproduction of the audio.
    Typical CDV players provide 4X oversampling of the audio and have stereo headphone jacks (with separate volume control). Many players provide for random play as well as for favorite-track selection (FTS) of the CD audio.

Miscellaneous Features. Most CDV players can be operated by remote, using an IR system (Sec. 4.3) and are capable of repeat play as well as chapter and track programming (15 or 20 selections). The more sophisticated (and expensive) players provide for S-VHS or S-Video outputs. Typically, a CDV player provides 425 lines of horizontal resolution.

Digital Special Effects. Typical effects available in the more sophisticated CDV players include still, step, multispeed, search, skip, picture freeze, strobe, memory, and jog (for continuous adjustment of playback speed).

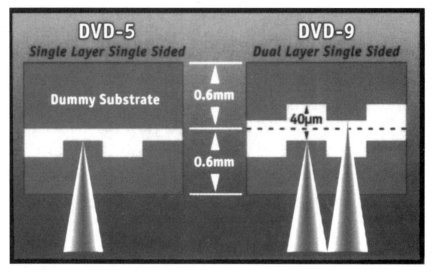

FIGURE 9.7 DVD disc.

9.2.8 Player Features

DVD players have several advantages over CDV players. The following is a brief summary.

CDs and CD-ROMs are typically 1.2 mm thick. The DVD disc uses two bonded 0.6-mm substrates, as shown in Fig. 9.7. This reduces the distance between the disc surface and the pits that hold information. The laser beam does not have to penetrate as much plastic, so the beam can be focused on a smaller area. Two different lens apertures can be included in the laser pickup, so information can be read from both DVD discs and standard CDs. In addition to holding more information, the two bonded sides strengthen the disc to prevent warping and make the disc more resilient to changes in temperature, humidity, and so on.

Decreased track-pitch pit length and pit width can be accommodated with an increased numerical aperture (NA) for the pickup lens and a shortened wavelength for the laser. The thin substrates assure wide tilt tolerance with small aberration of the laser spot, as shown in Fig. 9.8.

The specifications for a typical DVD are:

Diameter	120 mm/5 in
Thickness	1.2 mm (0.6-mm substrate $\times$ 2)
Minimum pit length	0.4 μm
Track pitch	0.74 μm
Sector layout	CLV
Modulation	8/16
Error correction	Reed Solumon product code
Wavelength of laser	650/635 nm
Numerical aperture of lens	0.6
Logical format (for DVD-ROM)	UDF/Micro UDF/SO-9660

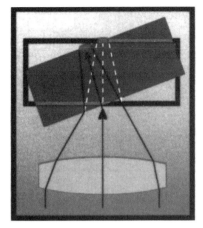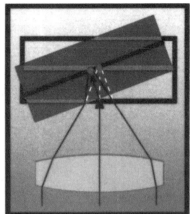

FIGURE 9.8 DVD disc tilt tolerance.

The storage capacity for typical DVDs are:

Single-sided discs: DVD5 (4.7 gigabyte/single layer); DVD9 (8.5 gigabyte/dual layer)

Double-sided discs: DVD10 (9.4 gigabyte = 4.7 gigabyte × 2); DVD18 (17 gigabyte = 8.5 gigabyte × 2)

Write once: DVD-R: 3.8 gigabyte per side

Overwrite: DVD-RAM: More than 2.6 gigabyte per side

A 4.7- or an 8.5-gigabyte DVD disc will hold enough information for a full-length movie or a typical computer application. On a 4.7-gigabyte disc, a laser reads the information that is pitted in the first layer of substrate. On an 8.5-gigabyte disc, the laser also reads through the first substrate to the information on the second layer of substrate. On a double-sided DVD disc the process is the same, except there are two discs bonded together to double the capacity.

9.3 CDV OPTICS

Figure 9.9 shows the optics for a typical CDV player. These optics are used to play both CD and CDV discs of all types (on this particular player).

9.3.1 Laser Beam

The laser diode emits a light beam which first passes through the grating to split the beam into one primary and two secondary beams. The center beam, the brightest of the three beams, is used to focus the beam of the reflective surface of the disc and to read the tracks on the disc. The secondary beams (one at each side of the main beam) are used strictly for radial tracking (and are known as the *radial beams*).

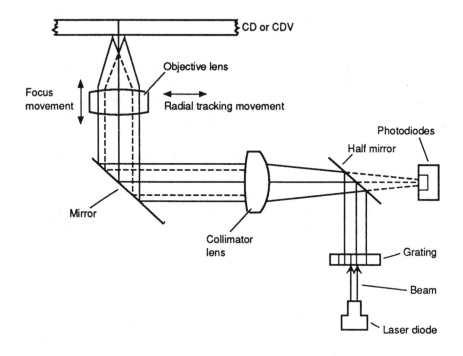

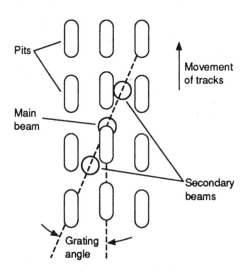

FIGURE 9.9   Optics for a typical CDV player.

Part of the light (some light passes through the mirror and is lost) from the three beams (light bundle) is reflected from the half mirror toward the collimator lens, which forms a column with the three beams parallel to each other. The light bundle exits the collimator lens toward the mirror and is directed up through the objective lens, which focuses the laser beam onto the reflective surface of the disc. The reflected light bundle returns through the objective lens, the mirror, and the collimator lens. About half of the reflected light bundle passes through the half mirror to strike the photodiodes.

The reflected light from the main beam is intensity-modulated by the pits in the track that moves across the beam at a high velocity as the disc spins (Fig. 9.9). The secondary beams strike the disc in between tracks as shown. The angle between the secondary beams and the main beam is known as the *grating angle,* and it is set by the *grating-angle adjustment* (Sec. 9.10).

9.3.2 Focus

Figure 9.10 shows the photodiode arrangement and focus principle. There is some wobbling as the disc spins over the laser beam because a disc is never perfectly flat. As a result, there is a need to constantly adjust the focus of the beam on the reflective surface as the disc spins. The beam has an elliptical shape if the disc is too close to or too far from the objective lens. For example, if the disc is too far from the objective lens, photodiodes D_1 and D_2 receive more light than D_3 and D_4, as shown. The opposite occurs if the disc moves too close to the objective lens.

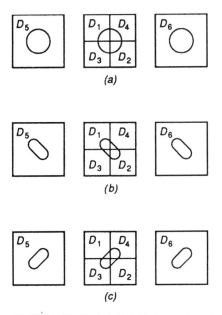

FIGURE 9.10   Typical photodiode arrangement and focus principles. (*a*) Disc perfectly focused; (*b*) disc too far from lens; (*c*) disc too near lens.

9.3.3 Photodiode Operation

All of the photodiodes operate in the photoconductive (reverse biased) mode. The arrangement of the photodiodes allows the three beams to fall, as shown in Fig. 9.10. The center beam falls on the four center photodiodes D_1 through D_4. The secondary beams fall on the two adjacent photodiodes D_5 and D_6. (The elliptical shape falling on D_5 and D_6 has no effect on player operation.)

If the beam is perfectly focused on the reflective surface of the disc, the center beam falls equally on each of the four center photodiodes D_1 through D_4. Thus, the sum current of D_1 and D_2 is equal to the sum current of D_3 and D_4.

If the beam is out of focus, the shape of the beam falling on the photodiodes becomes elliptical, and the total current of one pair of photodiodes is greater than the other pair. The difference in photodiode pairs creates the focus error (FE) signal, which is used by the focus servo to make the necessary focus corrections.

9.3.4 Radial Tracking

Figure 9.11 shows the method used to maintain proper radial tracking. When the main beam is directly centered over the track, the radial beams fall just outside the track. If the beams move either to the right or left of center, the intensity of one reflected beam is greater than the other. For example, when there is a shift to the right of the track, the reflection from radial beam R is greater than the reflection from radial beam L. (The reflection is less from L because much of the beam is falling on the track which contains pits.)

Because much of the light falling on a pit is diffused, less light intensity reaches the corresponding D_5, and the current from D_5 is less than from D_6. Photodiodes D_5 and D_6 are equally illuminated when the beam is perfectly centered on the track, and the current from D_5 is equal to that of D_6. The radial error (RE) signal is created by the difference in current between D_5 and D_6.

9.4 CDV TRACK FORMATS

The following is a summary of CDV tracking principles.

9.4.1 Converting Movie Film to CDV

Movie cameras use a frame rate of 24 Hz, whereas TV uses 30 Hz. To overcome this problem, movie film is converted to TV fields according to the format in Fig. 9.12a when recording on CDV. The film frames are alternately scanned for three video fields and then two video fields as shown. So five video frames are consumed while scanning four film frames.

Picture numbers are added only to the first video field per film frame. Because only four picture numbers are used for each five video frames, the picture numbers reach only about 43,200 (4/5 × 54,000) at the end of a 30-minute CDV.

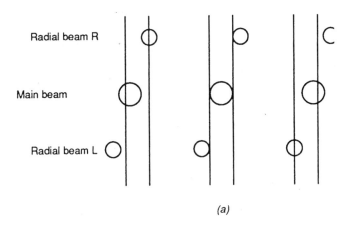

(a)

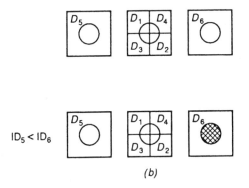

(b)

FIGURE 9.11 Typical photodiode diode arrangement and radial-error principles.

9.4.2 Special Operating Modes

CDV players are capable of many special operating modes, such as reverse play, fast forward, still picture, and slow motion (all under control of a microprocessor, as discussed in Sec. 9.6). These modes are possible because the beam can be made to "jump" from track to track at appropriate times by controlling the radial-tracking servo. To prevent the jump from being seen on the TV screen, all track jumping takes place only during the TV vertical-sync interval.

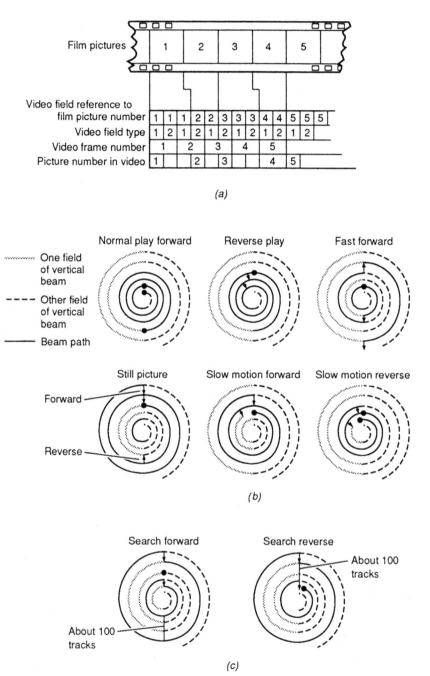

FIGURE 9.12 Converting movie film to video discs.

9.4.3 Track-Jumping Formats

Figure 9.12b shows the track-jumping formats required to create the various modes of operation. Each spiral drawing represents a few tracks of the disc. The left half (solid line) is one field of the TV vertical scan, and the right half (dotted line) is the other field of the vertical scan. The dark line represents the path of the beam. The path begins at the heavy dot in each illustration of Fig. 9.12b.

In the *normal-play* mode no track jumping takes place. The beam follows the spiral track from inside to outside. *Reverse play* requires the beam to jump back one track after each field. At the end of each revolution, the beam is on track inside of where the beam is started. The result is *normal-speed reverse play.*

A *still picture* occurs when one track is continuously repeated. One track is read and then the beam jumps back one track and repeats. The result is the same whether the forward or reverse Still-Picture button is pressed on the front panel. However, once the still-picture mode is in effect, additional actuations of the Still-Picture button makes the picture move forward or backward (on this particular player).

When the Still-Picture button is pressed, the beam is made to jump inward one track (shown as reverse in Fig. 9.12b). However, this jump takes place on the *field opposite* that of the previous jump. (If this were not so, the beam would have to jump two tracks to get to the original track.) Pressing the Still-Picture Forward button allows the beam to continue on to the next track. After one revolution, the beam simply begins repeating the next track.

Slow motion is a series of still pictures, each one lasting for a certain number of revolutions. The slow-motion control determines how long each still picture remains on until the beam continues to the next track and creates the next still picture. The effect is the same as rapid manual actuation of the Still-Picture Forward button. *Slow-motion reverse* uses the same principles, except the still picture advances in the reverse direction.

Fast forward is created by jumping forward one track after each field. After one revolution, the beam is three tracks further out than it was. As a result, the fast-forward speed is actually 3 times the normal forward-play speed.

9.4.4 Search Modes

Figure 9.12c shows the beam action during search modes. When search is actuated, the optical pickup begins to move rapidly across the bottom of the disc. Sections of the picture are momentarily displayed during the search process.

During *search forward*, the beam displays a series of pictures and then skips several hundred tracks, displays another series of pictures, and so on. These pictures are displayed very rapidly because the optical pickup takes only about 25 s to scan the entire disc. The *search-reverse* action is the same except that the beam moves inward when skipping the tracks.

9.5 LASER SAFETY

CDV players have a laser diode (or laser tube, in the case of older CDV players) which creates two possible service hazards. First, both the laser diode and tube produce a *potentially dangerous light beam.* As in the case of any other very intense light source, direct exposure to a laser beam can cause *permanent eye injury or skin burns.*

Second, the light beam produced by a solid-state laser diode is invisible. (This is in contrast to the red light beam produced by a laser tube.) Because the diode beam is invisible, you are never quite sure when the beam is present.

CDV players are designed to be operated without the operator being exposed to the beam. This is also essentially true for the servicer, with one major exception. If you gain access to the laser (by removing covers, opening the disc compartment, etc.) and keep power on the laser (by overriding interlocks, etc.), the beam may get you. None of this should frighten you, but the problems should keep you on your toes when servicing. It is the servicer's job to exercise all caution to avoid any direct exposure to the laser beam.

U.S. federal law requires (as do the laws of most countries) that servicers be advised of possible laser dangers. There is at least one warning label on all players, and often more than one (typically one on the outer cover and one in the disc compartment near the objective lens). This label reads something like "CLASS 1 LASER PRODUCT: Product complies with DHHS rules CRF subchapter J part 1040;10 at date of manufacture; DANGER: Invisible laser radiation when open and interlock failed or defeated. AVOID DIRECT EXPOSURE TO BEAM." On players designed for Canadian use, the warning label is strengthened by a *light-burst pattern within a triangle*.

Always be on the alert for these warning labels when servicing CDV players. Equally important, make sure all shields and covers are in place and that interlocks are working *before* you turn the player over to the customer.

In addition to producing a potentially dangerous beam, *lasers produce strong electromagnetic radiation*. This is usually not harmful to people but can be disastrous to magnetic tape, some wristwatches, and anything affected by magnetic fields (possibly pacemakers). Do not have magnetic tape, audio or video cassettes, or any other magnetic device nearby when servicing CDV players. To be on the safe side, keep all magnetic tape away, even when all player covers are in place.

9.5.1 Replacing the Laser

As with many electronic components, the laser diode can suffer *electrostatic breakdown* because of the potential difference generated by the charged electrostatic load on clothing and the human body. The laser diode is usually considered part of the optical system or pickup assembly, and most player manufacturers recommend replacement of the complete pickup assembly as a package. Most manufacturers do not supply individual parts for the optical system.

Figure 9.13 shows recommended procedures for handling the optical system of a typical CDV player when the entire optical system (pickup assembly) is to be replaced. The following notes supplement the procedures shown.

Place a conductive sheet on the workbench. In some players, the *black sheet used as a repair parts wrapping* is a conductive sheet.

Place the player on the conductive sheet so that the chassis touches the sheet. This makes the chassis (and pickup assembly) the same potential as the conductive sheet.

Place your hands on the conductive sheet. This makes your hands the same potential as the sheet.

Remove the optical-system block from the bag (which is made of conductive material).

Perform the necessary work on the conductive sheet. Be careful that clothing does not touch the optical-pickup block.

If practical, use a wrist strap with an impedance-to-ground of less than 1 megohm.

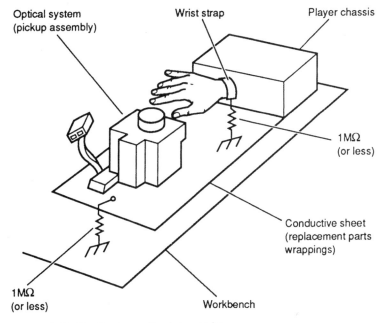

Optical system
(pickup assembly)

Wrist strap

Player chassis

1MΩ
(or less)

Conductive sheet
(replacement parts
wrappings)

1MΩ
(or less)

Workbench

FIGURE 9.13   Handling laser optics during service.

The workbench and/or conductive sheet should also have an impedance-to-ground of less than 1 megohm. Remember that static electricity builds up on clothing and is often not fully drained off, even with wrist straps and grounded workbenches.

Try to avoid servicing the player immediately after moving it from a cold to a warm place or soon after heating a room that was cold. Both of these conditions can cause moisture condensation. Excessive condensation (which is rare) can cause possible damage to circuits. More likely, *condensation can fog the lenses in the optical system.* You can clean the surface of the objective lens (with a clean, soft, dry cloth), but the remaining lenses are not accessible. You must wait until the condensation evaporates from the internal lenses. You can also try turning on the power (but not operating the player) and allowing heat from the transformers to remove the moisture.

9.6   CDV PLAYER CIRCUITS

Figure 9.14 shows the relationship of typical CDV player circuits. The following paragraphs describe the major functions of the circuits. Full circuit details are discussed in Secs. 9.7 through 9.21.

9.6.1   Start-Up and Loading

When power is first applied, the player goes into a start-up mode to check for the presence of a disc. Control of the start-up operation is provided by the system-control cir-

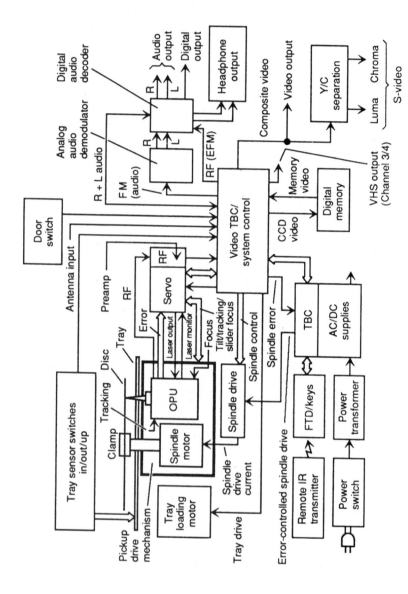

FIGURE 9.14 Relationship of CDV player circuits.

cuits (Sec. 9.9). First, the tray-loading motor is instructed to turn on and perform clamping, which includes moving the tray and clamper assembly to position a disc on the spindle. (This is similar to the clamping functions of a CD player.)

The clamping operation is determined by the position of the tray, indicated by the in, out, and up sensor switches. If the tray is in the up position, the loading tray and clamper assembly move down to clamp a disc (if present) onto the spindle. At the same time, the optical pickup unit (OPU) is moved by the slider drive from the park position to the LD-search position (a position to check for the presence of an 8- or 12-in disc, on this player).

The LD-search position is outside the diameter of a 5-in CD or CDV, about 2.5 in from the center of the spindle. A second clamping operation (double clamping) is performed as the OPU moves out to the correct LD-search position. Only one clamping operation is performed if the tray is in the down position when the player is turned on.

9.6.2 Focus

After the OPU reaches the LD-search position and the clamping sequence is completed, the laser diode turns on, and the focus-search sequence is executed (through the servo-control bus) by commands from system control. If focus is found (indicating that a CDV disc is in place), the laser turns off and system control receives data from the servo to indicate the presence of a disc. The OPU then returns to the park position, and the player is placed in the stop mode, awaiting instructions from the front-panel keyboard and/or IR remote.

If focus is not found (indicating that there is no disc in place), the OPU is instructed to move to a CD-search position (about 1.25 in from the center of the spindle). The focus sequence is then repeated to search for a 3- or 5-in CD or a 5-in CDV (CDV Single or Gold CD).

If a disc is present, the spindle drive receives a command from system control (through the spindle-control line) to turn the spindle motor on. The OPU then moves to the correct position to read the table of contents (TOC) on the disc. After reading the TOC, the laser diode turns off, and the OPU returns to the park position. The TOC is decoded by system control, and information regarding the disc (playing time, number of tracks, etc.) is displayed on the front-panel fluorescent tube display (FTD) and on the TV screen. The player remains in the stop mode until play is activated by the keyboard or remote unit.

If there is no disc present, the OPU returns to the park position and the player is placed in the stop mode, awaiting the command to open the tray so that a disc can be inserted. Keep this in mind when troubleshooting a "the player turns on and off again but will not play" symptom. Make certain that there is some kind of disc in place.

9.6.3 Keyboard and Remote Commands

Commands from the keyboard or remote control are applied to system control through circuits on the power-supply board. Likewise, control information to the FTD is applied through the power-supply circuits. Command information includes the RC_5 code (Phillips remote control transmitter code 5) sent to system control. The RC_5 code is also sent out to an external device (such as an amplifier) or received from an external device through an RC_5 in-out jack. The system-control circuit also receives signals from the power supply, the door switch, and the in, out, and up switches.

9.6.4   Spindle Drive

The system-control circuits provide the spindle drive with a spindle-control signal and provide the power supply with a spindle-error signal. This error signal is developed by the time-base correction (TBC) circuit. The error-controlled spindle-drive signal is then provided to the spindle-drive circuits.

The spindle drive provides spindle-drive current to the spindle-motor coils. When the disc spins, the optical pickup reads the disc and develops both RF and error signals. The error signals are sent to the servo to develop radial and focus-error signals. In turn, the radial and focus-error signals are fed back to the pickup-drive mechanism, keeping the pickup in focus and moving over the disc at the correct speed. The RF signals are processed by system control.

9.6.5   RF Signal Processing

The system-control circuits extract three components from the RF signals recorded on the disc: video FM, analog-audio FM, and digital-audio HF. The analog-audio FM is sent to the analog-audio circuit for demodulation. The demodulated right and left audio signals are sent to the digital audio circuits. The digital-audio HF (when present on the disc) is also sent to the digital-audio circuits.

If the disc is recorded in analog only, the analog audio is automatically selected and routed to the audio-output jacks. If the disc is encoded with digital audio, the digital audio is selected and routed to the audio-output jacks. If the disc contains both analog audio and digital audio, the audio-output source can be selected by the user.

The selected audio signals (right and left) are also sent to the headphone-output circuits. Likewise, the audio signals are combined and applied to the RF modulator.

9.6.6   Composite Video Signal

The video FM (extracted from the RF signals) is demodulated by system control to provide the composite-video signal. This signal is applied to digital-memory circuits to provide digital special effects (memory, still strobe, etc.). The digitally processed signal (when used) is then returned to the system-control circuits.

The composite-video signal (with or without digital processing for special effects) is sent to the video-output jack and to the Y/C separation circuits. When used, the Y/C circuits provide separate chroma and luma to the S-Video jack (for use with TV monitors capable of S-VHS operation).

The system-control circuits also provide video and audio (L + R) signals modulated on TV Channel 3 or 4 through the VHF output jack. As with most VCRs, the antenna or cable signal for this player may also be routed to the VHF output jack if selected (when the player is not in use).

9.7   MUTE CIRCUIT

Figure 9.15 shows the mute circuit and related timing charts. This circuit mutes the audio so that a "pop" sound is not heard in the speakers when the player is turned on or off. The voltages shown in Fig. 9.15 are present after the player is turned on. In this configuration (after initial turn-on) both Q_{909} and Q_{910} are off.

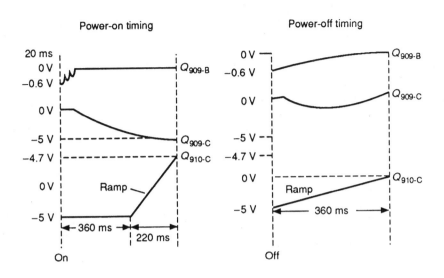

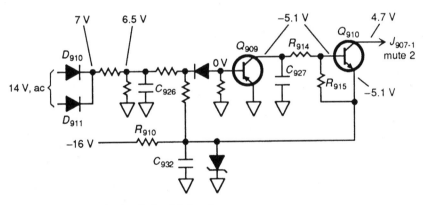

FIGURE 9.15   Mute circuit and related timing charts.

When power is initially applied, 14 V is applied across rectifiers D_{910} and D_{911}, producing about 7 V. At the same time, -16 V is applied to the mute circuit through R_{910}. When the player is turned on, Q_{909} is forward-biased for about 20 ms. During the time that Q_{909} is on, current flows through R_{915} and R_{914}, forward-biasing Q_{910}. After Q_{909} turns off (after 20 ms), C_{927} is allowed to charge, permitting current to continue flowing through R_{914} and R_{915}.

While C_{927} is charging, Q_{910} is in saturation for about 360 ms and applies about -5 V to pin 1 of J_{907}. This -5 V is the mute-2 signal. C_{927} continues to charge for another 200 ms until the -5 V is reached. The mute-2 signal rises up to about 4.7 V (forming a ramp signal, as shown), permitting the audio to have a soft turn on from mute.

When the player is turned off, the base of Q_{909} goes momentarily low (about -0.6

V). The positive 6.5 V across C_{926} discharges quickly, applying the negative potential from the charged C_{932} to the base of Q_{909}. When Q_{909} is on, Q_{910} is forward-biased to apply the mute-2 signal to pin 1 of J_{907}. It takes about 2.5 s for C_{932} to discharge and turn Q_{909} and Q_{910} off.

9.8 SPINDLE-MOTOR START-UP

Figure 9.16 shows the spindle-motor start-up circuits. Figure 9.17 shows the spindle-motor drive circuits and related timing charts. The spindle servo is discussed in Sec. 9.17.

To start the spindle motor, the spindle circuits receive two control signals from system control: the run (stop-start) signal and the forward-reverse (F/R) signal. A frequency generator (FG) signal, developed by the spindle-motor IC QD_{01}, is returned to system control (Sec. 9.9) through pin 3 of JD_{02}.

As shown in Fig. 9.16, QD_{01} receives motor-rotation information from the Hall sensors (HA, HB, and HC) located on the spindle-motor board (adjacent to the spindle). When QD_{01} receives a high (5 V) at the run input (pin 9), the spindle motor begins to rotate. As soon as the motor rotates, the FG signal developed by QD_{01} is sent to system control, thus monitoring the spindle-motor speed.

When play is activated, the F/R signal (pin 8) goes low (0 V) to start the clockwise, or forward, rotation of the spindle motor. When the motor reaches the correct speed

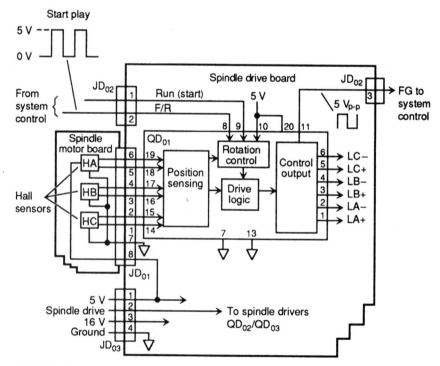

FIGURE 9.16 Spindle-motor start-up circuits.

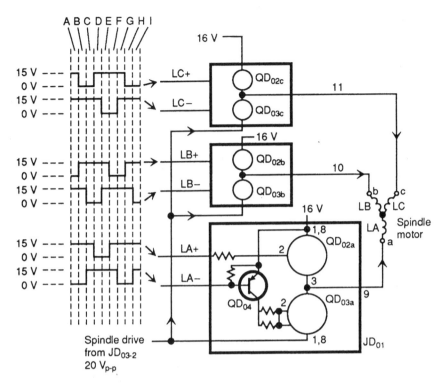

FIGURE 9.17 Spindle-motor drive circuits.

(spindle-lock condition), the F/R signal toggles from low to high several times at various intervals to prevent runaway of the spindle-motor speed. When the motor settles to the correct speed, the F/R signal goes low and remains low until the stop mode is activated.

When stop is activated, QD_{01} receives a high at the F/R input to stop the motor. This places QD_{01} in the reverse (CCW) mode to brake the spindle-motor rotation. (On our player, the brake input at pin 10 of QD_{01} is not used. Instead, pin 10 is connected to 5 V.) When the motor stops, the run signal goes low and the motor is in the stop mode. The F/R signal remains high until play is activated.

As shown in Fig. 9.17, QD_{01} controls the switching of the spindle-driver transistors QD_{02} and QD_{03}. Although QD_{02} contains three PNP Darlington pairs and QD_{03} contains three NPN Darlingtons, only one set of driver circuits is shown in any detail. The spindle-drive signal to QD_{02} and QD_{03} is discussed in Sec. 9.17.

The timing chart in Fig. 9.17 shows switching of the three transistors to control current through three coils of the three-phase spindle motor when playing a standard play (CAV) disc. *During period A*, QD_{03a} and QD_{03b} are forward-biased; current flows from the 16-V source through QD_{02b}, LB, LA, and QD_{03a}. *During period B*, current continues to flow through LA and QD_{03a}, but the current from the 16-V source now flows through QD_{02c} and LC. *During period C*, current flows through QD_{02c}, LC, LB, and QD_{03b}. This cycle continues as the spindle motor makes one complete revolution (time periods D through I) and continues to repeat the cycle as the motor spins.

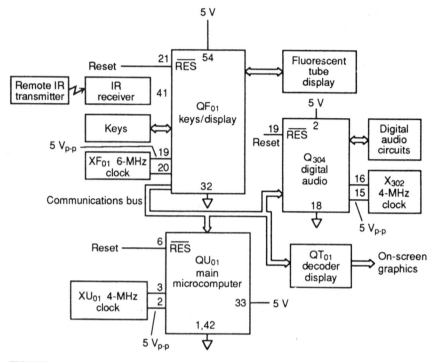

FIGURE 9.18   Relationship of main system-control microcomputer QU_{01} and peripheral devices.

9.9 SYSTEM CONTROL

Figure 9.18 shows the relationship of the main system-control microprocessor QU_{01} (called a microcomputer) and the peripheral devices, including the sub or slave microcomputers. Figure 9.19 shows the interface of QU_{01} and the spindle TBC, focus and tracking servo circuits, and input-output expander QU_{03}.

As shown in Fig. 9.18, QU_{01} controls all of the player circuits through a *communications bus*. This includes the keys and display microcomputer QF_{01}, the digital-audio microcomputer Q_{304}, and the display decoder QT_{01}. QF_{01} receives commands from either the remote unit or the front-panel functions keys and sends data to the fluorescent tube display (FTD). Q_{304} controls decoding of the digital audio. QT_{01} displays data on the TV or monitor screen.

As shown in Fig. 9.19, QU_{01} communicates with the input-output expander QU_{03} through *data and address buses*. With the expansion capability of QU_{03}, QU_{01} is able to control additional player functions such as digital memory, video demodulator, analog and audio and the loading motor. In addition, QU_{01} is able to receive data through QU_{03} from the door, up, in, and out switches and to make decisions based on these switch inputs.

The bidirectional data and address buses are also linked with two memories: a 512K ROM QU_{02} and a favorite-track selection (FTS) RAM QU_{08}. Addressing of the various devices is controlled by address controller QU_{05}.

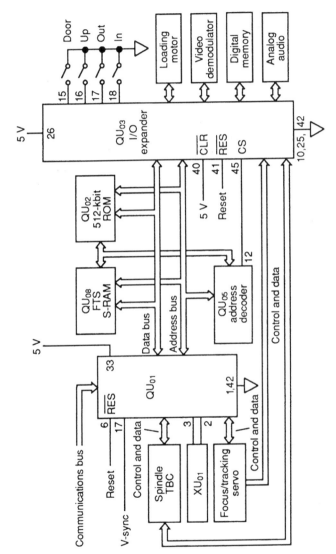

FIGURE 9.19 Interface of QU$_{01}$ and other control circuits.

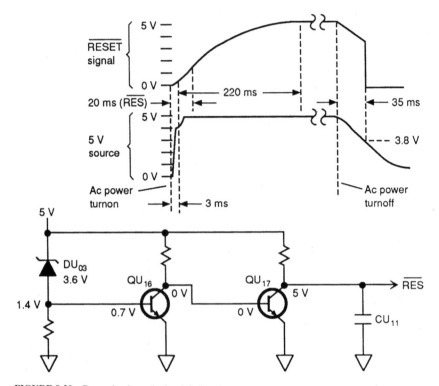

FIGURE 9.20 Reset circuits and related timing charts.

9.9.1 Reset, Power, Ground, and Clock Functions

All of the microcomputers shown in Figs. 9.18 and 9.19 require power and ground connections, as well as clock and reset signals (as do all microprocessors). As discussed in Sec. 4.6, if you suspect a malfunction in any microprocessor or digital IC, one of the first troubleshooting steps is to check all of these inputs. For example, the main microcomputer QU_{01} requires 5 V at pin 33, a ground at pins 1 and 42, a 4-MHz clock at pins 2 and 3, and a reset at pin 6. If any of these inputs are absent or abnormal, the microcomputer will not function properly (if at all).

The power, ground, and clock functions are obvious (Sec. 4.6) and need not be discussed here. However, all microprocessors in our player use a common reset signal (called $\overline{RES}$). Figure 9.20 shows the reset circuits and related timing charts.

Each microcomputer requires a low (0 to 0.8 V) at the $\overline{RES}$ input for about 20 ms after power is applied. When power is first applied to the player, 5 V is applied to the power-on reset circuits shown in Fig. 9.20. Capacitor CU_{11} charges up to 5 V in about 220 ms, applying a high to the $\overline{RES}$ line. However, $\overline{RES}$ is low (0 to 0.8 V) for about 20 ms, fulfilling the reset requirements for all microcomputers in the system.

If power is removed, the 5-V source dissipates slowly as the power-supply capacitors discharge. To pull the $\overline{RES}$ line low (to about 3.8 V) in a short time, QU_{16} becomes reverse-biased and QU_{17} turns on, pulling the $\overline{RES}$ line low. Thus, if power is momentarily lost, or if the power supply falls too low, all of the microcomputers are reset, preventing a fault or error from occurring.

Tray position	Door (15) PC$_4$	Up (16) PC$_5$	Out (17) PC$_6$	In (18) PC$_7$
Down	L	H	H	L
Up	L	L	H	H
Out	H	L	L	H

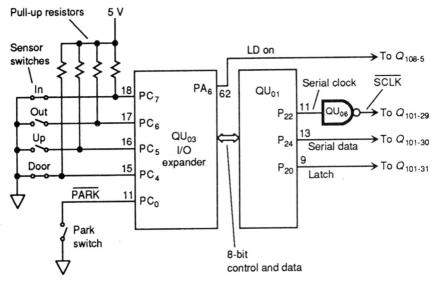

FIGURE 9.21 System-control functions during start-up.

9.10 START-UP AND LASER CONTROL

Figure 9.21 shows the system-control functions involved during start-up. Figure 9.22 shows the related focus and laser-control circuits. Figure 9.23 shows the disc-detection sequence during start-up.

9.10.1 Mechanical Sequence

When the loading tray is pulled into the player, QU_{10} instructs the circuit to look for a disc (disc detection). As the tray is pulled in, the pick-up drive mechanism scans (using a high-speed scan) from the park position to an outward direction (for about 1 s) to a position that allows for detection of an 8- or 12-in disc.

Next, the loading motor turns on to place the tray in the up position. When the tray reaches the up position, the up switch opens, pin 16 of QU_{03} goes high, and the tray

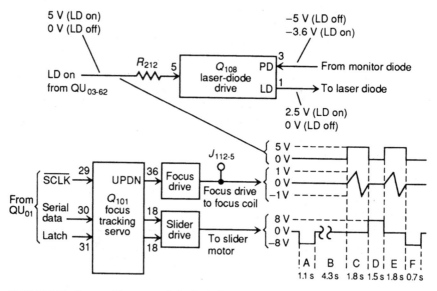

FIGURE 9.22 Focus and laser-control circuits during start-up.

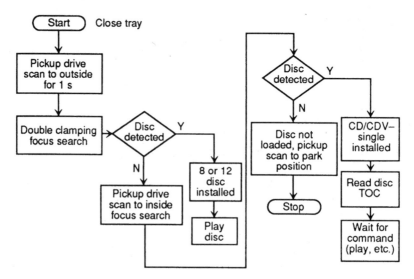

FIGURE 9.23 Disc-detection sequence during start-up.

motor reverses direction to place the tray in the play position. This action, known as *clamping*, allows the disc to be seated properly on the spindle.

9.10.2 Focus Search

After the tray goes into the play position (down), the laser turns on and the objective lens performs a focus-search operation to detect the presence of an 8- or 12-in disc. If focus is achieved, the presence of the large-size disc is indicated, and play can be activated.

If focus is not found, the pickup drive moves inward until the $\overline{\text{PARK}}$ line goes high (about 5 V) and continues to move inward for 300 ms after $\overline{\text{PARK}}$ goes high. (The $\overline{\text{PARK}}$ line at pin 11 of QU_{03} is controlled by a park switch, actuated by the pickup-drive mechanism.)

The laser again turns on, and the objective lens again performs a focus search to detect the presence of a 3- or 5-in disc (CD or CDV Single). If focus is still not found (no disc of any size installed), the pickup drive returns to the park position and play cannot be initiated.

9.10.3 Mechanical Sensor Switches

The tray-position truth table in Fig. 9.21 shows the logic of the expander QU_{03} inputs under various conditions. When a mechanical-sensor switch is closed, a low (ground) is applied to the QU_{03} input. When the switch is open, a high (5 V) is applied through the corresponding pull-up resistor.

With no disc installed, and with power applied, the sensor switches are as follows: in and door switches closed (low) and up and out switches open (high). As a result, pins 15 and 18 are low (ground) and pins 16 and 17 are high (about 5 V). Pin 11 receives a low from the park switch.

9.10.4 Focus and Tracking Control

When a command to read a disc is executed from QU_{01}, pin 62 (LD On) of expander QU_{03} goes high to turn the laser on. Control of focus and tracking is provided by serial data (8 bit) through the serial-clock ($\overline{\text{SCLK}}$), serial-data, and latch buses. The serial-clock signal is inverted by QU_{06} and coupled to focus-tracking servo Q_{101}.

9.10.5 Laser Control

The laser-diode on (LD On) signal is applied to laser drive IC Q_{108} through R_{212}, as shown in Fig. 9.22. When pin 5 of Q_{108} goes high, the laser is turned on (through pin 1 of Q_{108}). The intensity of the laser diode is monitored by a monitor diode, which develops a voltage that is proportional to the laser intensity. The laser light output is about 0.3 mW (with a maximum of about 0.8 mW).

The monitor-diode signal applied to pin 3 of Q_{108} regulates drive to the laser. When the laser is on, pin 1 of Q_{108} (LD) is about 2.5 V, with pin 3 (PD) at about -3.6 V. Pin 3 is -5 V when the laser is off.

9.10.6 Objective-Lens Control

The serial-data bits from QU_{01} are fed to Q_{101} to control focus-drive and slider-drive circuits during start-up. The signal at pin 36 of Q_{101} (UPDN) controls the objective lens during start-up focus, which is initiated (1) when the loading tray is retracted into the player, (2) when power is applied to the player, or (3) when play mode is activated (if a disc is loaded into the player).

9.10.7 Disc-Detection Sequence

The disc-detection timing chart of Fig. 9.22 shows what takes place as the loading tray is retracted (with no disc present). When the close-tray command is entered, the slider motor receives about -8 V for 1.1 s (period A) to move the pickup drive outward. After 4.3 s (period B), the laser turns on and the focus drive (to the focus coil) begins a focus search pattern (objective lens up and down) for 1.8 s (period C). Because there is no disc in the player, focus is not achieved.

After this first focus search (focus start-up), a second focus search is initiated. The pickup drive is moved inward by applying 8 V to the slider motor for 1.5 s (period D). During period E (1.8 s), the laser turns on and a focus start-up again takes place. Because there is no disc present (in our example), the pickup drive is returned to the park position by applying -8 V to the slider motor for 0.7 s (period F).

This sequence must take place to detect the presence of a disc before play can occur. Because no disc is detected (in our example), the player cannot be placed in the play condition.

9.11 SERVO SYSTEMS

Figure 9.24 shows the four basic servo systems used in our player. Note that the four photodiodes B_1 through B_4 provide error signals to the focus, tracking, and slider servos. These same photodiodes are used to extract video- and audio-signal information from the disc as described in Sec. 9.15. The tilt servo receives error signals from a separate set of diodes.

The focus-error signal developed by the focus circuits (Sec. 9.12) is fed to the focus-drive circuits. In turn, the focus-drive circuits provide the drive to the focus coil on the pick-up drive mechanism to keep the laser beam focused on the disc.

The tracking servo (Sec. 9.13) receives tracking-error (TR) signals from the photodiodes and develops the tracking-error signal. The TR error is fed to the tracking-drive circuit, which applies drive to the tracking coil on the pickup. The coil provides a TR return to the tracking drive and to the slider servo.

The slider servo (Sec. 9.13) provides low-frequency (LF) tracking corrections. The TR LF error signal is sent to the slider-drive circuit to drive the slider motor.

The tilt servo (Sec. 9.14) uses the error signal from the tilt diodes to make tilt corrections on the pickup drive mechanism.

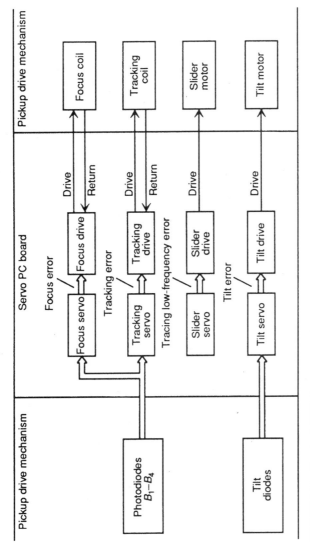

FIGURE 9.24 Four basic CDV player servo systems.

9.12 FOCUS SERVO

Figures 9.25 through 9.29 show the focus-servo circuits. The following paragraphs describe each of the circuits.

9.12.1 Focus Error

As shown in Fig. 9.25, two signals are developed in the focus-servo circuits to keep the laser beam focused on the disc. The $\overline{\text{DSUM}}$(diode sum) and FE (focus error) signals are derived from the outputs of photodiodes B_1 through B_4, which are part of the pickup assembly. All of the photodiode signals are summed, inverted, and doubled in op amp Q_{103b}.

The $\overline{\text{DSUM}}$ signal is applied to pin 5 of Q_{106}, to pin 2 of Q_{103a}, and to pin 3 of Q_{101}. Note that the signal to pin 3 of Q_{101} is set by focus-sum level adjustment R_{200}. In simple terms, R_{200} is adjusted (using a test disc) to provide 1.6 V at pin 6 of J_{111}. Full adjustment details are given in Sec. 9.25.

Two of the photodiode signals, B_2 and B_4, are applied to Q_{103}, along with the inverted DSUM signal. The resultant FE signal at the output of Q_{103} is applied to pin 12 of Q_{107} and pin 4 of Q_{101}. Note that the gain of Q_{103} is set by focus-gain control R_{168}. In simple terms, R_{168} is adjusted to provide correct focus-servo loop gain (as described in Sec. 9.25).

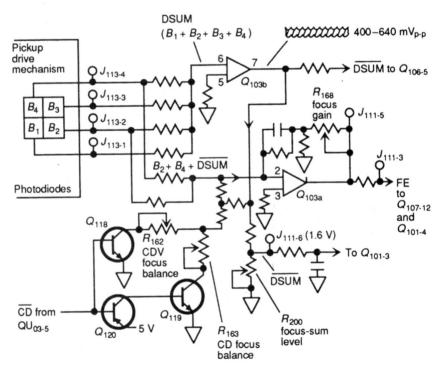

FIGURE 9.25 Focus-error circuits.

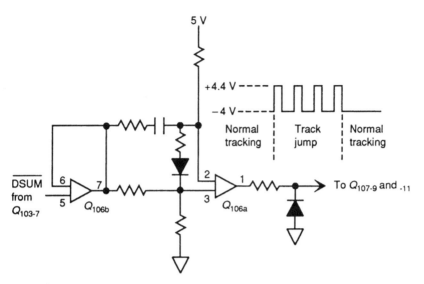

FIGURE 9.26   Focus-sum control circuits.

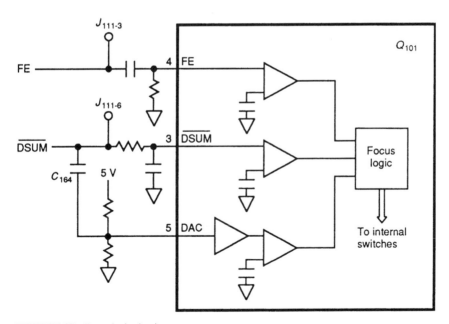

FIGURE 9.27   Focus-logic circuits.

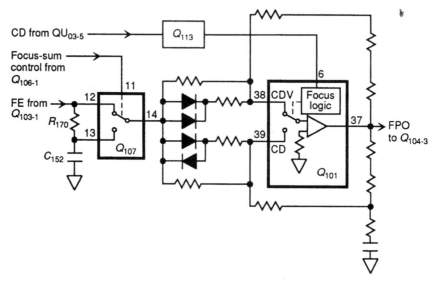

FIGURE 9.28 Focus-output circuits.

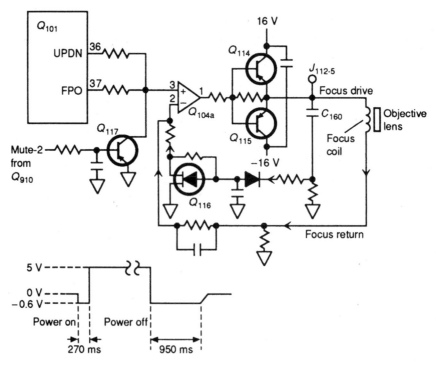

FIGURE 9.29 Focus-drive and mute circuits.

9.12.2 Focus Balance

The focus-balance circuits adjust the focus according to the type of disc being played (as determined by system control). As shown in Fig. 9.25, a CD signal from pin 5 of system-control expander QU_{03} is applied to switching transistors Q_{118} through Q_{120}. If a CD is played, the CD line at QU_{03-5} is low. If the disc is a CDV, QU_{03-5} is high.

With the $\overline{CD}$ line high, Q_{118} is on, with Q_{119} and Q_{120} off. This places the CDV focus balance control R_{162} in the circuit. When $\overline{QU}_{03-5}$ is low, Q_{119} and Q_{120} are on and Q_{118} is off, placing the CD focus-balance control R_{163} in the circuit.

The focus-balance controls R_{162} and R_{163} are adjusted to keep the objective lens at an optimum distance from the disc. If the proper balance is not maintained while playing a CD or the audio portion of a CDV, the audio is noisy. The video portion of a CDV will also show crosstalk. The adjustment procedures for C_{162} and C_{163} are given in Sec. 9.25.

9.12.3 Focus-Sum Control Signal

As shown in Fig. 9.25, the $\overline{DSUM}$ signal is applied to pin 5 of Q_{106b}. As shown in Fig. 9.26, Q_{106b} acts together with Q_{106a} as a unity-gain op amp to develop a control signal. This control signal is applied to pins 9 and 11 of Q_{107} and serves to change the focus level during normal and special tracking.

During normal tracking, pin 1 of Q_{106a} is a steady -4 V. During track jumping (search, fast forward, fast reverse, track loss, etc.), the signal toggles between -4 and $+4.4$ V. This is discussed further in Sec. 9.12.5.

9.12.4 Focus Logic

As shown in Fig. 9.27, the FE and $\overline{DSUM}$ signals are applied to the focus-logic circuit, which is part of the focus-tracking slider (FTS) servo IC Q_{101}. The focus-logic circuit turns the servo loop on when certain conditions (which indicate focus lock) are met during start-up (Sec. 9.10).

The focus lock condition is met when the $\overline{DSUM}$ input (pin 3) to Q_{101} reaches 0.4 V and when the FE input (pin 4) reaches 0.3 V. If the $\overline{DSUM}$ signal falls below 0.4 V because of a focus loss, Q_{101} instructs the focus servo to open, and a focus-search pattern is activated. (A damaged disc and excessive vibration are the most common causes for loss of focus.)

Note that the input at pin 5 of Q_{101} is derived from the $\overline{DSUM}$ signal (through C_{164}) and is applied to a comparator within Q_{101}. This signal provides for main-beam on-off track detection. If the signal at Q_{101-5} is below a threshold of 0.5 V, this indicates that the main beam is off track.

9.12.5 Focus Output

As shown in Fig. 9.28, the focus-sum control signal (Sec. 9.12.3) is applied at pin 11 of Q_{107} to control the input for the CD or CDV focus-output circuit. When the focus-sum signal is low, pin 12 of Q_{107} is connected to pin 14, and the FE signal is passed directly. When the focus-sum is high, pin 13 is connected to pin 14, and the FE signal is attenuated by R_{170} and C_{152}. Because the focus sum toggles between high and low during track jumping, FE is also attenuated during track jumps but remains constant during normal tracking.

The FE signal is applied to the FTS servo Q_{101} through feedback phase-compensation networks at pins 38 and 39. The correct network is selected according to the type of disc being played (as determined by system control). A CD signal from pin 5 of expander QU_{03} is applied to pin 6 of Q_{101} through Q_{113}.

If a CD is played, the $\overline{CD}$ line at QU_{03-5} is low, connecting the network at pin 39. If a CDV is played, QU_{03-5} is high, and the network at pin 38 is used. Either way, the FE signal is applied to the focus drive and mute circuit through pin 37 of Q_{101} as a focus output (FPO) signal.

9.12.6 Focus Drive and Mute

As shown in Fig. 9.29, both the FPO and UPDN (Sec. 9.10.6) signals are applied to the focus coil of the objective lens through Q_{104a}, Q_{114}, and Q_{115}. Note that both of the signals are muted during power-on and power-off by the mute-2 signal (Sec. 9.7), which is applied through Q_{117}. This prevents the objective lens from moving up and down excessively during power-on or power-off.

Feedback is provided to the inverting input of Q_{104a} (from a focus return loop) to stabilize focus-servo operation. The FPO signal at the focus coil is also returned to a high-frequency focus-limiter loop through C_{160}. This lowers the loop gain when the focus-error signals are high frequency (to further stabilize the focus servo operation).

9.13 TRACKING AND SLIDER SERVO

Figures 9.30 through 9.33 show the tracking-servo circuits. Figure 9.34 shows the slider servo. The following paragraphs describe the functions of these two interrelated servos.

9.13.1 Tracking Error

Two photodiodes are used to read the secondary (or radial) beams reflected from a CDV disc to provide radial tracking. As shown in Fig. 9.30, the two radial-tracking photodiodes are labeled A and C (whereas the four main photodiodes are labeled B_1 through B_4). The signals from photodiodes A and C are sent to a difference amplifier in Q_{102b} to produce a tracking-error signal at pin 7. The signals are also added through R_{101} and R_{102} at test point J_{111-7} to produce a tracking-sum signal.

The tracking-error signal is applied to the noninverting input (pin 3) of tracking and equalization (TR/EQ) amplifier Q_{102}. The feedback equalization (or gain) for Q_{102a} is set by the type of disc being played (Sec. 9.12.2). If a CD or the audio portion of a CDV is being played, the $\overline{CD}$ line is low, placing the switch in Q_{107a} in the CD condition, with C_{103} and C_{113} in the circuit. When the video portion of a CDV is being played, the $\overline{CD}$ line is high, placing the Q_{107} switch in the CDV position (with C_{103} and R_{113} out of the circuit, shorted by the Q_{107} switch).

The tracking-error signal can be checked at J_{111-1}. Tracking-balance control R_{105} is used to adjust the signal at J_{111-1} so that the signal is zero when tracking is properly balanced (Sec. 9.25). Note that the tracking-error signal is applied to both the tracking-logic circuit (Sec. 9.13.2) and the CD/CDV phase-compensation circuit (Sec. 9.13.3) before the error signal is applied to the objective lens (Sec. 9.13.4).

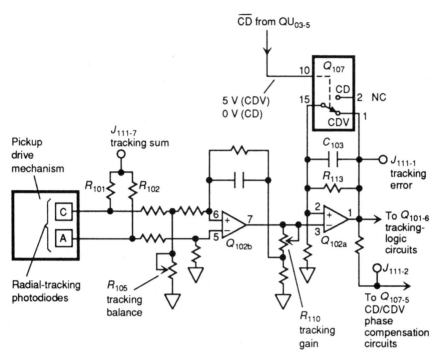

FIGURE 9.30   Tracking-error circuits.

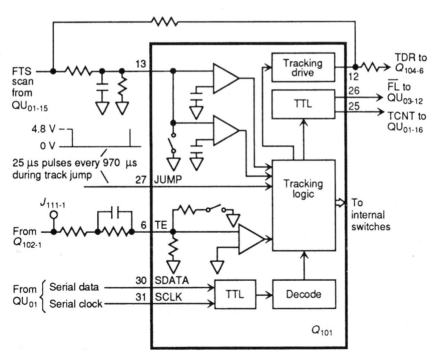

FIGURE 9.31   Tracking-logic circuits.

9.39

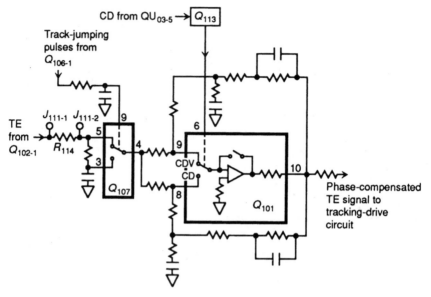

FIGURE 9.32   CD and CDV phase-compensation circuits.

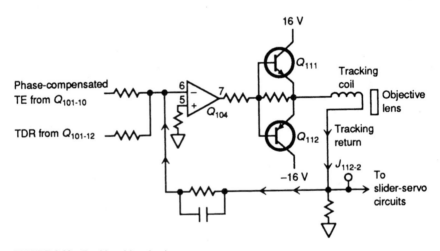

FIGURE 9.33   Tracking-drive circuits.

9.13.2 Tracking Logic

As shown in Fig. 9.31, the tracking-logic circuit monitors several signals that are used to control the tracking servo. The tracking-logic circuit also receives commands (forward, reverse, start, stop, etc.) from system control through the serial data-bus lines (Sec. 9.9).

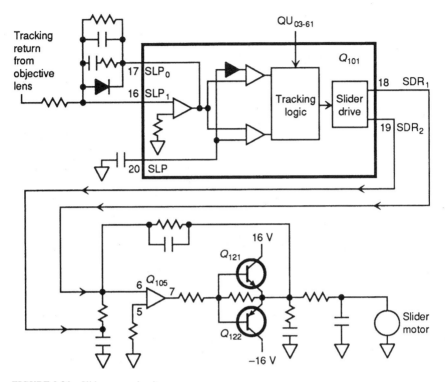

FIGURE 9.34 Slider-servo circuits.

Pin 6 of FTS servo Q_{101} receives the tracking-error (TE) signal to determine tracking conditions. Q_{101} monitors the frequency of the TE signal to open and close the tracking-servo loop or to change the servo-loop characteristics. When operations such as search, fast forward, or fast reverse are activated, pin 27 of Q_{101} receives pulses from system-control expander QU_{03} to move the pickup-drive mechanism in the proper direction. During such an operation, the tracking-servo loop must be open to allow jumping or skipping over tracks on the disc.

The tracking-logic circuit also provides tracking information to system control. The tracking-logic circuit provides a tracking count (TCNT) signal to pin 16 of system-control QU_{01}, allowing QU_{01} to determine the position of the laser beam on the disc.

The focus-lock ($\overline{FL}$) signal is applied to QU_{03} at pin 12, indicating the condition (locked or unlocked) of the focus-servo loop. The $\overline{FL}$ signal is low (0 V) when focus is achieved (locked) and high (5 V) when the beam is not in focus (unlocked). Note that the tracking circuit is not activated until the focus servo loop is locked.

QU_{01} provides an FTS scan signal at pin 12 of Q_{101} to open the tracking-logic circuit during slow and medium scan operation. When pin 13 of Q_{101} is at 0.2 V, the tracking loop opens until the frequency of the TE signal at pin 6 drops below a certain frequency (determined by the scan speed). While the tracking loop is open, a brake pulse is taken from the tracking-drive (TDR) output at pin 12. The TDR signal is applied to the tracking-drive circuits (Sec. 9.13.4).

9.13.3 CD and CDV Phase Compensation

As shown in Fig. 9.32, the TE signal is applied to pin 5 of Q_{107a}. The control signal at pin 9 is the same signal used in the focus-servo circuit, and it operates the Q_{107} switch as described in Sec. 9.12.3 and 9.12.5. That is, the TE signal is attenuated during track jumping but remains steady during normal tracking.

In either condition, the TE signal is applied to Q_{101} through feedback phase-compensation networks at pins 8 and 9. The correct network is selected according to the type of disc being played (as determined by system control). A CD signal from pin 5 of QU_{03} is applied to pin 6 of Q_{101} through Q_{113}. If a CD is played, the CD line at QU_{03-5} is low, connecting the network at Q_{101-8}. If a CDV is played, QU_{03-5} is high, and the network at Q_{101-9} is used. Either way, the TE signal is applied to the tracking-drive and slider circuits through pin 10 of Q_{101}.

9.13.4 Tracking Drive

As shown in Fig. 9.33, both the tracking-drive (TDR) and tracking-error (TE) signals are applied to the tracking coil of the objective lens through Q_{104}, Q_{111}, and Q_{112}. Feedback is provided to the inverting input of Q_{104} (from a tracking-return loop). This feedback signal is also applied to the slider-servo circuits.

9.13.5 Slider Servo

As shown in Fig. 9.34, the tracking-return signal (used as a stabilizing feedback control) is also applied to the slider motor through Q_{101}, Q_{105}, Q_{121}, and Q_{122}. As the laser beam moves over the disc, the tracking servo applies current to the objective lens in an attempt to keep the beam on track. Because there is a limit as to how far the objective lens can move (right or left), the slider servo is used to move the entire pickup drive mechanism (including the objective lens) across the disc (toward the outer tracks during play).

When a track-jumping operation (search, fast forward, fast reverse, track loss, etc.) is necessary, the tracking loop is open, and the tracking-logic circuit controls movement of the pickup mechanism.

9.14 TILT SERVO

Figure 9.35 shows the tilt-servo circuits. These circuits ensure that the laser beam tracks perpendicularly to the disc plane. This is necessary for the laser beam to be reflected back into the objective lens and be picked up by the signal photodiodes (B_1 through B_4, Fig. 9.25).

Note that the tilt servo is active only when 8- and 12-in discs are played (because large discs are more likely to be warped). When 3- or 5-in discs are played, the pickup drive remains in a neutral position as sensed by the tilt-neutral opto-transistors located on the drive mechanism.

An LED emits a beam that is reflected by the disc and picked up for the tilt photodiodes D_1 and D_2. Each diode produces voltage in proportion to the intensity of the

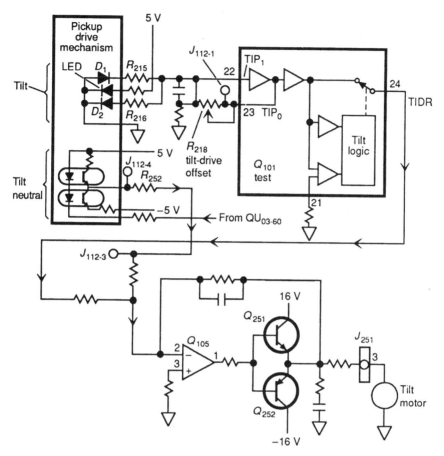

FIGURE 9.35   Tilt-servo circuits.

light received. If D_1 and D_2 receive an equal amount of light (disc flat), no current flows through R_{215} and R_{216}, because the polarity of the diodes is opposing. If the disc is not flat, the diodes receive unequal light and a difference signal is generated.

The difference signal is applied to pin 22 of Q_{101}, which develops a tilt-drive (TIDR) signal at pin 24. The TIDR signal is applied to the tilt motor through Q_{105}, Q_{251}, and Q_{252} as necessary to offset any warping of the disc. The tilt-drive signal offset is adjusted by R_{218} (Sec. 9.25). R_{218} is adjusted for zero offset when there is no tilt.

If a 3- or 5-in disc is played, pin 60 of QU_{03} goes low, turning on a pair of tilt-neutral opto-transistors. Under these conditions, -5 V is applied to pin 2 of Q_{105} through the tilt-neutral opto-transistors and R_{252}. This disables the tilt-drive circuit.

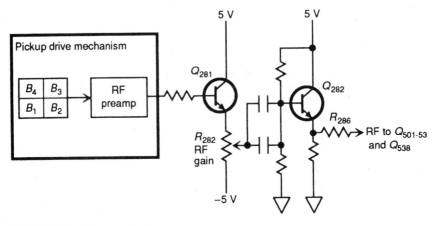

FIGURE 9.36   RF preamp circuits.

9.15   DISC SIGNAL PROCESSING

Figures 9.36 through 9.38 show the disc signal-processing circuits. The following paragraphs describe each of the signal-processing functions.

9.15.1   RF Preamp

As shown in Fig. 9.36, disc video and audio information picked up by the four photo-diodes B_1 through B_4 is amplified by an RF preamp. Both the photodiodes and preamp are part of the pickup mechanism. The amplified disc signal is applied to the RF signal-processing circuits through Q_{281} and Q_{282}. The signal gain is set by RF-gain control R_{282}.

9.15.2   RF Signal Processing

As shown in Fig. 9.37, the RF (with EFM, 8-to-14 modulation) signal is applied to both the video and audio circuits after amplification by the preamp. The term *EFM* refers to the modulation system used when information is placed on the disc during manufacture. An 8-bit data stream is used but is not placed directly on the disc on 8-bit format. Instead, each group of 8 data bits is converted, or "stretched," to 14 bits, as shown in Fig. 9.39. This prevents the laser beam from converting two adjacent transitions (from a pit to flat, and vice versa) at the same time. Note that CD, CDV, and CD-ROM use 8-to-14 modulation, whereas DVD and DVD-ROM use 8-to-16 modulation.

The main disc signal passes through an RF-correction circuit in Q_{501}, amplifier Q_{511}, buffer Q_{512}, a main bandpass filter (BPF) that passes signals between 3 and 13.5 MHz), a limiter in Q_{501}, and the video demodulator in Q_{501}. The demodulated video is passed through a drop-out control (DOC) circuit (Fig. 9.38) before the video is applied to the video-processing circuits (Sec. 9.16).

The disc signal also passes through a DOC BPF (that passes RF signals between 1

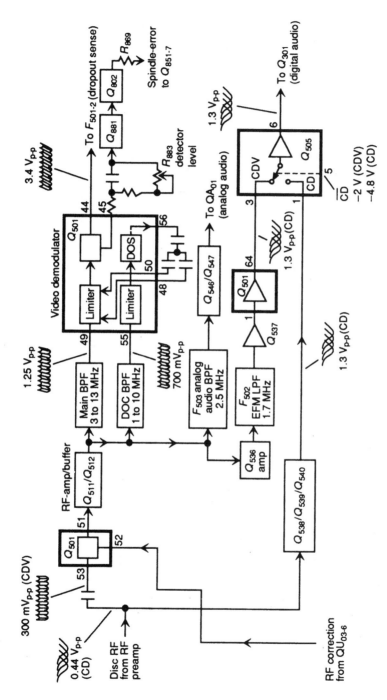

FIGURE 9.37 RF signal-processing circuits.

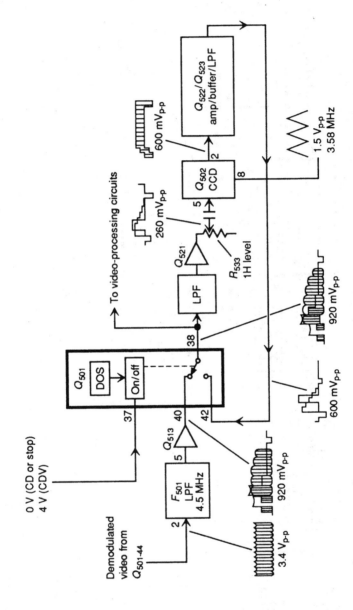

FIGURE 9.38 DOC circuits.

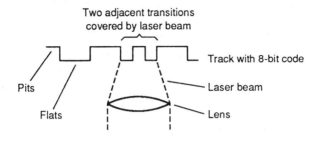

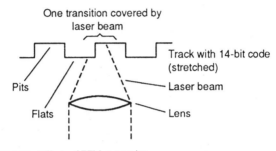

FIGURE 9.39   Effects of EFM conversion.

and 10 MHz), a limiter in Q_{501}, and a drop-out sense (DOS) circuit in Q_{501}. The resultant signal is added to the main disc signal in Q_{501}.

The analog-digital portion of the RF signal (2.5 MHz) is selected by BPF F_{503} and applied to the analog audio circuits (Sec. 9.20) through amplifier Q_{546} and buffer Q_{547}.

The EFM portion of the RF signal (below 1.7 MHz) is extracted by low-pass filter (LPF) F_{502} after amplification by Q_{536}. The EFM signal is then applied to the CD/CDV switch in Q_{505} after being buffered by Q_{537} and Q_{501}.

Q_{505} selects the EFM input to be processed by the digital-audio circuits (Sec. 9.21). If a CD is played, the CD line goes low (-4.8 V) to place the Q_{505} switch in the CD position. When a CD, or the audio portion of a CDV, is played, the EFM from the disc is routed directly from the RF preamp through Q_{538} and Q_{540} to the digital-audio circuit (without processing).

If a CDV (which contains digital audio) is played, the CD line is high (-2 V), placing the Q_{505} switch in the CDV position. Under these conditions, EFM from the disc is routed to the digital-audio circuit through the correction and processing functions shown in Fig. 9.37.

The main demodulated video is also used to provide a spindle-error signal that is applied to the spindle-drive and feedback circuits (Sec. 9.17) through R_{883}, Q_{881}, Q_{802}, and R_{869}. The spindle-drive and feedback system reduces the time-base error in reproduced video by detecting the reproduced composite-sync and then accelerating or decelerating (braking) the spindle motor as necessary to reduce the time-base error. Detector level R_{883} sets the level of the spindle-error signal.

9.15.3 DOC Circuit

As shown in Fig. 9.38, the main video signal is applied through a DOC circuit before the signal is applied to the video-processing circuits (Sec. 9.16). The demodulated video at pin 44 of Q_{501} is applied to the DOC circuits through 4.5-MHz LPF F_{501}, Q_{513}, and a switch in Q_{501}.

The Q_{501} switch normally connects pins 38 and 40 of Q_{501} to pass the video signal to the video-processing circuits. If a video dropout occurs (say, because of dirt or a scratch on the disc), the DOS circuits in Q_{501} sense the loss of video signals and move the Q_{501} switch so that pins 38 and 40 are connected. This applies the previous horizontal line (1H) of video to the video-processing circuits, thus making up for the loss of video.

The horizontal line of video is developed by the LPF at pin 38 of Q_{501} and at Q_{521}, Q_{522}, and Q_{523}. The LPF removes chroma from the video signal and produces a luma signal which is amplified by Q_{521}. The luma signal is then returned to pin 42 of Q_{501} through R_{533}, 1H-delay Q_{502}, and Q_{522} and Q_{523}. Q_{502} uses charge-coupled device (CCD) techniques to delay the video by 1H. Q_{502} is synchronized to the video signal by a 3.58-MHz clock at pin 8. R_{533} sets the level of the 1H signal (Sec. 9.25).

The DOC circuit is inhibited by a low at pin 37 of Q_{501} when a CD is being played or during stop. When a CDV is played, pin 37 is high (4 V), permitting the Q_{501} DOS circuit to control the Q_{501} DOC switch.

9.16 VIDEO PROCESSING

Figures 9.40 through 9.45 show the video-processing circuits. The following paragraphs describe each of the signal-processing functions.

9.16.1 Time-Base Correction (TBC)

As shown in Fig. 9.40, video from pin 38 of Q_{501} (Fig. 9.38) is applied to TBC circuit before being applied to the spindle-servo (Sec. 9.17) and video-distribution circuits (Sec. 9.18). The TBC circuits use CCD techniques for time-base correction. A TBC error loop (composed of sync and data separation, time-base error detection, and VCO blocks) develops a TBC error signal by comparing the playback-horizontal (PB-H) signal to a horizontal-reference (H-ref) signal.

The time-axis correction block is a coarse-correction circuit that removes horizontal jitter from the playback video. To provide fine correction of the chroma phase, the CCD video is fed to a phase-correction block. The phase-error detection block develops a video phase-shift error (VPS ERR) signal by comparing the 3.58-MHz playback burst (detected from the CCD video) to a 3.58-MHz reference signal. The VPS ERR signal is applied to the phase-correction block, along with the CCD video to provide correction of chroma-phase jitter.

9.16.2 Time-Axis Correction

As shown in Fig. 9.41, the time-axis correction circuit receives video from pin 38 of Q_{501}. The video is applied to Q_{504} through Q_{514}, and Q_{515}, C_{536}, and C_{537}. The TBC-error signal (Sec. 9.16.5) is applied to VCO Q_{503} to develop a variable clock signal (9 to 14 MHz), which is applied to Q_{504} to control the timing of the clock driver.

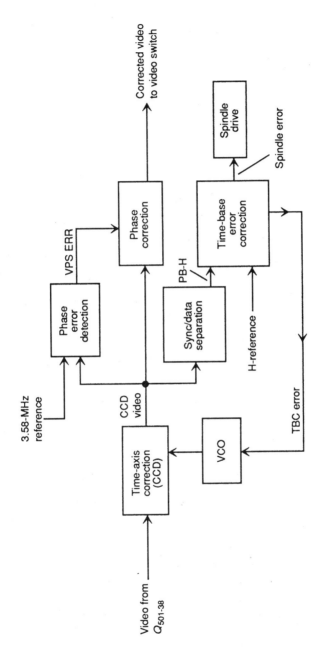

FIGURE 9.40 Relationship of time-base correction circuits.

9.49

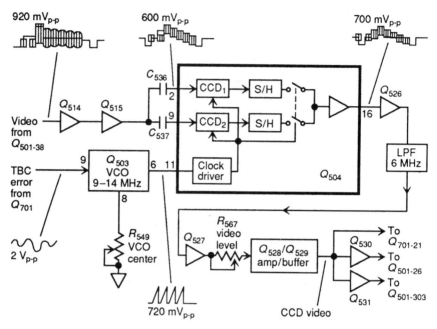

FIGURE 9.41 Time-axis correction circuits.

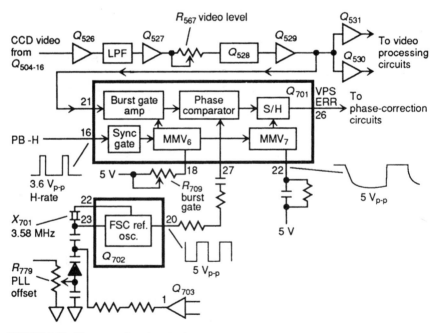

FIGURE 9.42 Phase-error detection circuits.

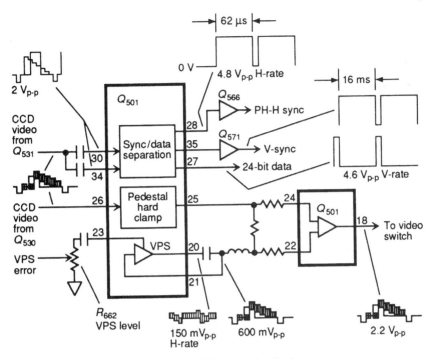

FIGURE 9.43   Phase-correction and sync and data-separation circuits.

The clock driver controls the timing of the CCDs to provide a variable delay of the video signal. The VCO center adjust R_{549} is used to provide an optimum average (or center) delay from the input video (at pin 9 of Q_{504}) to the CCD video (at the emitter of Q_{527}). Two CCDs are used in parallel to get sufficient bandwidth and delay. The average delay is adjusted to 70.7 μs (1H plus 7.2 μs) ±0.1 μs (Sec. 9.25). Note that a color-lock failure (or a slow color lock) after a search operation may indicate the need for adjustment of R_{549}.

The time-base corrected video (CCD video) at pin 16 of Q_{504} is applied to the phase-error detection circuits (Fig. 9.42) and the sync/data-separator and phase-correction circuits (Fig. 9.43) through Q_{526}, a 6-MHz LPF, Q_{527}, R_{567}, and Q_{523} and Q_{531}. Video-level adjustment R_{567} is adjusted to provide a video (pedestal to 100 percent white) of 0.71 V_{p-p} at the video-output jack (with 75-ohm termination), as described in Sec. 9.25.

9.16.3   Phase-Error Detection

As shown in Fig. 9.42, CCD video is applied to the burst-gate amplifier of Q_{701} at pin 21. The burst-gate amplifier supplies the burst extracted from the CCD video to the Q_{701} phase comparator where the burst is compared with a 3.58-MHz reference (FSC) from Q_{702}.

The phase-comparator output is applied to the phase-correction circuits (Fig. 9.43) through a sample and hold (S/H) circuit in Q_{701}, as the VPS error signal at pin 26 of

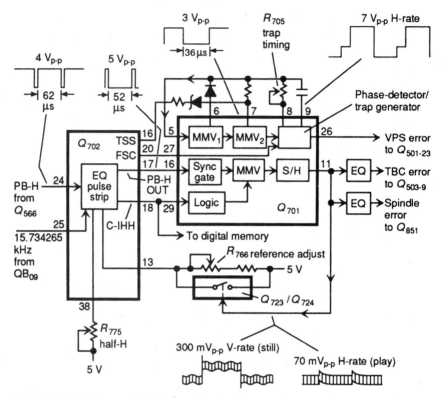

FIGURE 9.44　TBC-error and spindle-error detection circuits.

Q_{701}. The S/H circuit is clocked by MMV_7. The burst-gate adjust R_{709} is adjusted to provide a 1-μs delay from the leading edge of the video signal (at the emitter of Q_{527}).

9.16.4　Phase Correction and Sync and Data Separation

As shown in Fig. 9.43, CCD video is applied to a phase-correction circuit in Q_{501} through a pedestal hard-clamp circuit in Q_{501} (pins 24, 25, 26). The VPS error signal is also applied to the Q_{501} phase-correction circuit through VPS level adjustment R_{662} and an amplifier in Q_{501}. R_{662} is adjusted to provide proper color phase (Sec. 9.25). The output of the phase-correction circuit is applied to the video switch (Fig. 9.45).

The sync and data-separation circuit of Q_{501} separates PB-H sync (pin 28), vertical sync (pin 35), and 24-bit data (pin 27) from the luma signal at pins 30 and 34 (from the phase-error detector, Fig. 9.42).

9.16.5　TBC-Error and Spindle-Error Detection

As shown in Fig. 9.44, both TBC-error and spindle-error signals are developed by the TBC circuit. The PB-H signal from Q_{566} (Fig. 9.43) is applied to an equalization-pulse

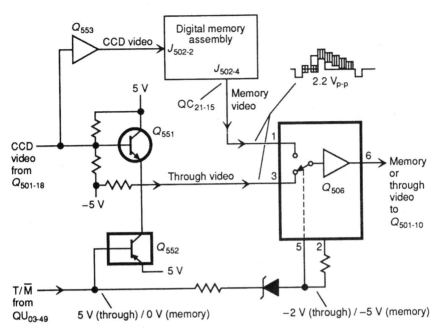

FIGURE 9.45   Video-switch circuits.

strip circuit in Q_{702} to remove equalization pulses from the PB-H signal. The half-H adjustment R_{775} is adjusted to provide 52 μs between positive-going pulses of the signal at pin 17 of Q_{702} (Sec. 9.25).

After the equalization pulses are removed from the PB-H signal, Q_{702} applies PB-H out and clock-inhibit (C-INH) signals to Q_{701} (pins 16 and 29). The reference shift adjust R_{766} sets the width of the C-INH pulse, which is then applied to the digital-memory circuit (Sec. 9.19) and to logic in Q_{701}.

A trapezoid signal is generated in Q_{701} as a reference to detect time-base errors. The signal at pin 11 of Q_{701} is developed by determining where the PB-H falls in respect to the slope of the trapezoid in the S/H circuit.

The horizontal-reference signal from pin 16 of Q_{702} is adjusted by trap-timing adjustment R_{705} to get 36 μs between positive-going pulses (Sec. 9.25). The signal at pin 11 of Q_{701} is applied through equalization circuits to both the time-axis correction circuits of Fig. 9.41 (as the TBC-error signal) and the spindle servo described in Sec. 9.17 (as the spindle-error signal).

9.16.6   Video Switch

As shown in Fig. 9.45, CCD video from pin 18 of Q_{501} (Fig. 9.43) is applied to a switch in Q_{506} through Q_{553} and the digital-memory circuits (Sec. 9.19) or directly through Q_{551}, depending on which mode is selected.

If memory video is selected (so that special effects can be used), the T/M line (from system control $QU_{03\text{-}49}$) is low (0 V). This biases Q_{552} on and applies 5 V to the emitter of Q_{551}, cutting Q_{551} off. At the same time, the voltage at pin 5 of Q_{506} goe‧

low, connecting the switch to pin 1 of Q_{506}. The CCD video is thus routed through the memory circuits (for special effects) to the video-distribution circuits (Sec. 9.18). If the memory circuits are not selected, the T/M line is high (5 V), Q_{552} is cut off, Q_{551} is on, the Q_{506} switch moves to pin 3, and CCD video (without special effects) is routed directly to the video-distribution circuits through Q_{551} and Q_{506}.

9.17 SPINDLE SERVO

Figures 9.46 through 9.49 show the spindle-servo circuits. The following paragraphs describe the functions of this servo system (used to drive the spindle motor, Sec. 9.8).

9.17.1 Spindle-Drive Development

As shown in Fig. 9.46, the spindle-error signal from the TBC circuit (Sec. 9.16.5) varies according to the type of disc being played (CD or CDV). The waveforms shown in Fig. 9.46 are for an audio CD and for a standard-play (CAV) disc. In either case, the spindle-error signal (combined with feedback, Sec. 9.17.4) is applied to pin 2 of integration amplifier Q_{851}. The integrated output at pin 1 of Q_{851} is applied to both CDV (Sec. 9.17.2) and CD (Sec. 9.17.3) spindle-error circuits.

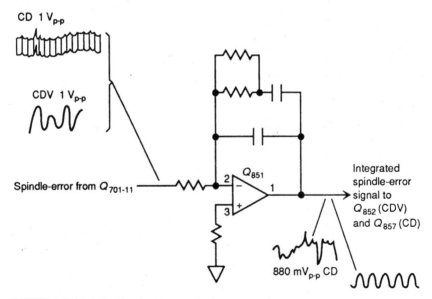

FIGURE 9.46 Spindle-drive development circuits.

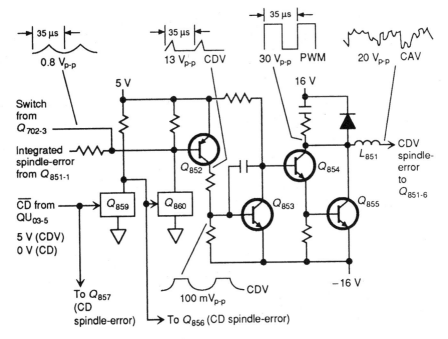

FIGURE 9.47   CDV spindle-error circuits.

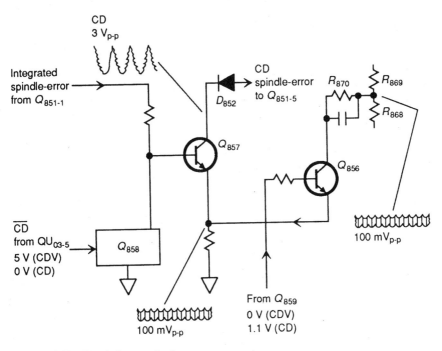

FIGURE 9.48   CD spindle-error circuits.

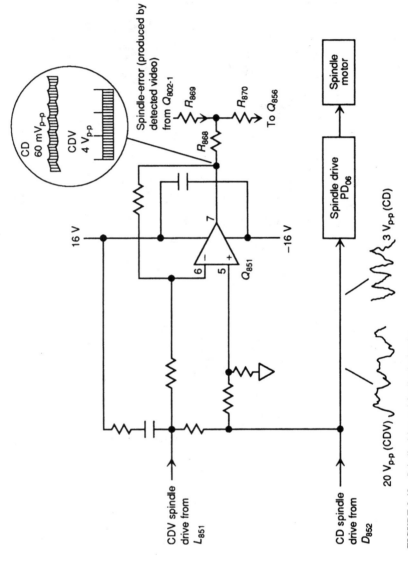

FIGURE 9.49 Spindle drive and feedback circuits.

9.56

9.17.2 CDV Spindle Error

As shown in Fig. 9.47, the integrated output from Q_{851} is combined with a switch signal from the TBC circuit and applied to the spindle-drive and feedback circuits (Sec. 9.17.4) through Q_{852} through Q_{855}. If CDV is selected, the $\overline{CD}$ line (from system-control QU_{03-5}) is high (5 V). This biases Q_{859} on and Q_{860} off. The spindle-error signal then passes through Q_{855} and L_{851} to the drive and feedback circuits. When a CD is played, the $\overline{CD}$ line is low, biasing Q_{859} off and Q_{860} on. (Actually, Q_{860} does not saturate but does conduct enough current to prevent the spindle signal from passing to Q_{852}.)

Both the spindle-error signal and the switch signal from the TBC circuits are applied to the base of Q_{852}. The collector signal (about 13 V_{p-p}) of Q_{852} is used to drive Q_{853} through Q_{855}, producing a pulse width modulation (PWM) signal of about 30 V_{p-p} at the collectors of Q_{854} and Q_{855}. The PWM signal (representing the spindle error or drive) is applied to the drive and feedback circuits through L_{851}.

9.17.3 CD Spindle Error

As shown in Fig. 9.48, the integrated output from Q_{851} is applied to the drive and feedback circuits through Q_{857} and Q_{856}. If CD is selected, the CD line (from system control QU_{03-5} is low (0 V). This biases Q_{858} off, allowing the spindle-error signal to be applied at the base of Q_{857}. The spindle-error signal is amplified to about 3 V_{p-p} by Q_{857} and coupled to the drive and feedback amplifier through D_{852}. Q_{856} is also biased on (by about 1.1 V from Q_{859}, Fig. 9.47), allowing the 100-mV signal at the emitter of Q_{856} to pass and be fed back.

9.17.4 Spindle Drive and Feedback

As shown in Fig. 9.49, both the CD and CDV spindle-drive signals are applied to the spindle motor (through spindle drive PD_{06}) and to the spindle-error circuits (through feedback amplifier Q_{851}). The feedback signals are combined with the spindle-error signal from Q_{701} (Fig. 9.44) and are applied to the spindle-drive development input at Q_{851} (Fig. 9.46). As discussed in Sec. 9.16.5, the spindle-error signals pass through equalization circuits before reaching Q_{851}. The spindle-error signal (produced by detecting the composite-sync signal, Sec. 9.15.2) is applied to the feedback circuits through R_{869}.

The spindle motor has three phases and has Hall-effect sensors. The motor is controlled by the usual Hall feedback circuits located on spindle-drive board PD_{06}, as described in Sec. 9.8.

9.18   VIDEO DISTRIBUTION, NR, AND OSD

Figure 9.50 shows the noise reduction (NR) and on-screen display (OSD) circuits for our CDV player. Figure 9.51 shows the video-distribution circuits.

9.18.1 Video Signal Processing

As shown in Fig. 9.50, the video signal at pin 6 of Q_{506} (Fig. 9.45) is applied to the video-distribution circuits (Sec. 9.18.3) through video processor Q_{501}.

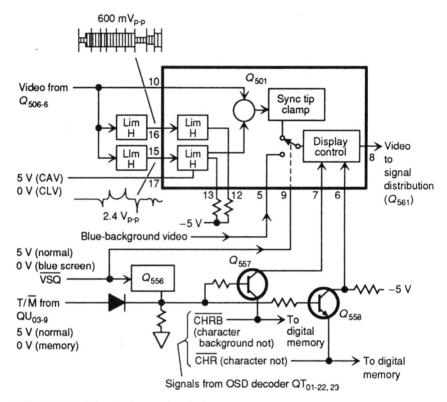

FIGURE 9.50   Video signal-processing circuits.

The video signal is subtracted from the low- and high-frequency signals at pins 15 and 16 (obtained from internal and external limiters controlled by the CAV/CLV line at pin 17 of Q_{501}). The resultant signal is applied to a sync-tip clamp in Q_{501}. The clamped video is applied to a video squelch (VSQ) switch, which is controlled by the VSO signal at pin 9.

When a video disc is playing, the VSQ signal is high (5 V) and the video signal from the clamp is connected to the display-control circuits. When the player is in the stop mode, the VSQ line is low (0 V), connecting the video to a blue-background video signal input at pin 5 (causing the TV or monitor screen to go blue until the player starts).

9.18.2 On-Screen Display

The OSD circuit is controlled by the display-control circuits in Q_{501} (Fig. 9.50) and by the T/M control signal from $QU_{03\text{-}9}$ (Sec. 9.16.6). Signals from the OSD decoder QT_{01} are applied to the display-control circuits through Q_{557} and Q_{558}. In turn, Q_{557} and Q_{558} are controlled by the T/M line.

When the player is not in the memory mode (no special effects), the T/M line is

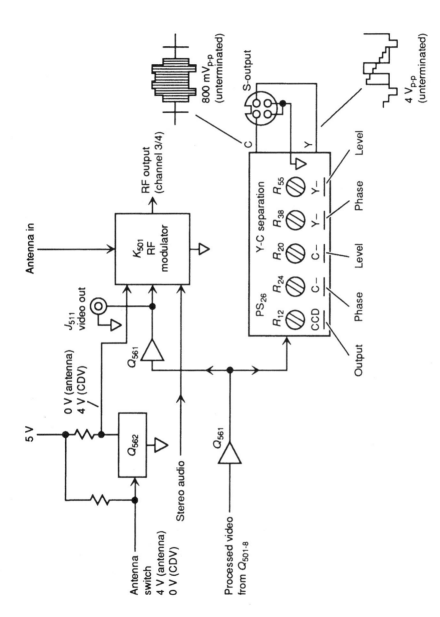

FIGURE 9.51 Video distribution circuits.

high. This biases Q_{557} and Q_{558} on and passes the CSD signals to the video distribution through the Q_{501} display-control circuits. When the player is in memory, the T/M line is low. This biases Q_{557} and Q_{558} off, preventing the OSD signals from passing to Q_{501}. Instead, the OSD signals are applied to the digital memory circuits (Sec. 9.19).

9.18.3 Video Distribution

As shown in Fig. 9.51, the processed video from pin 8 of Q_{501} is distributed to the RF modulator K_{501}, to the video out jack J_{511} and the S-output connector through Q_{561} and Q_{562}. The output from J_{511} is a standard composite video signal to be viewed on a monitor TV. The signal from the S-output connector requires a monitor capable of displaying separated Y (luma) and C (chroma) video (an S-VHS monitor).

RF modulator K_{501} combines the standard composite video signal with right- and left-channel stereo audio for conversion to RF at TV channels 3 and 4 in the usual manner (the same as for VCRs and camcorders). K_{501} also includes an antenna input to feed external RF (from an antenna or cable TV) to the TV set.

9.19 DIGITAL MEMORY

Figure 9.52 shows the digital-memory circuits used in our CDV player. These circuits provide for the special effects. The memory circuits can be bypassed if desired (Sec. 9.16.6). Although the circuits are shown in block form, all of the input-output test points and signal paths are given by pin number.

CCD video is applied to the digital-memory circuits at analog-digital (A/D) converter QB_{08} through an LPF. The CCD video is converted to 8-bit data (4-bit upper and 4-bit lower) by QB_{08} at a rate of 576 samples per horizontal line. The A/D clock input at QB_{08-12} is developed by memory control QB_{09} and is at a frequency of 9.06 MHz (which is 576 times the horizontal rate, or 576fH).

Memory control QB_{09} controls both the writing and reading operations to and from the memory IC Q_{810} (a 1-Mbit DRAM). The 18.12-MHz (1152fH) input clock at pin 10 of QB_{09} is required to write each sample into memory through the 4-bit parallel-serial bus.

Writing into memory is enabled by the write-enable (WE) inputs from system control and the WE input from pin 36 of QB_{09}. The 15-bit address bus A_0 through A_{14} is used to write the 4-bit-by-8-bit (four samples, or 32 bits) data to a four-sample RAM address in QB_{10}.

QB_{09} controls the reading of data from QB_{10} for special effects. The 8-bit memory video is converted by digital-to-analog converter (DAC) QC_{29} to an analog output at pin 6 of QC_{29}. Horizontal sync is added to the analog signal by QC_{31}, and the composite output is applied to QC_{14} where the OSD is inserted (Sec. 9.18.2).

The output from pin 1 of QC_{14} (including any OSD characters) is applied to a switch in QC_{21} through a 140-ns phase-shift circuit. The phase shift provides the correct burst-phase for the composite video signal and is operated by the switching pulse (CINV) from pin 60 of QB_{09}. The CINV pulse is developed by an edge-detect circuit and QB_{09}. The edge-detect circuit compares a 3.58-MHz reference with the burst signal of the composite video. The CINV pulse switches at the frame rate to provide a 140-ns phase-shift burst every other frame.

QC_{21} is operated by the same T/M control line used in the video switch and video

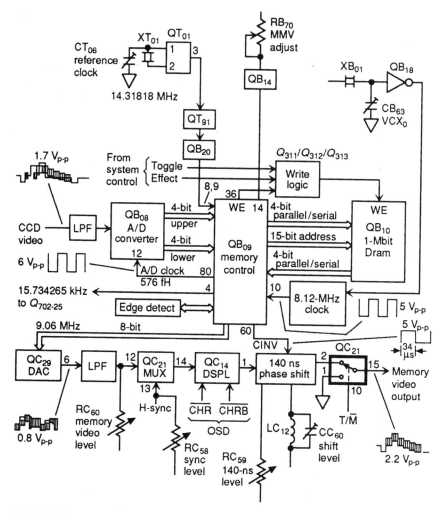

FIGURE 9.52 Digital-memory (special effects) circuits.

distribution circuits (Figs. 9.45 and 9.50). When the line is low, the memory circuit output is applied to video distribution, and any special effects selected by system control appear in the video at pin 8 of Q_{501} (Fig. 9.50). The memory circuits are completely bypassed when the T/M line is high, thus inhibiting any special effects.

9.20 ANALOG AUDIO

Figure 9.53 shows the analog audio circuits for our CDV player. Again, the circuits are shown in block form, but all significant input-output test points and signal paths are given by pin number.

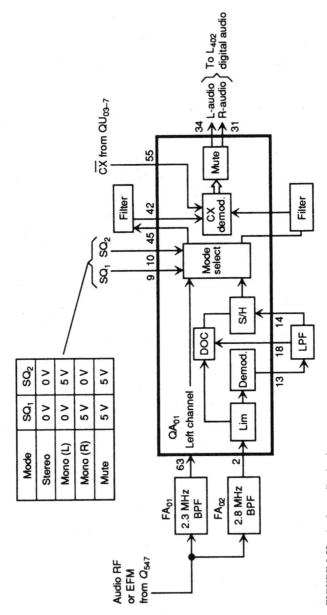

FIGURE 9.53 Analog audio circuits.

The audio RF or EFM signal (Fig. 9.37) is applied to the audio demodulation IC QA_{01} through two BPFs which extract FM left- and right-channel audio from the RF signal. The demodulators for the two channels are identical, so only the right channel is shown.

The audio FM signal is applied through the QA_{01} limiter to the FM-demodulator and DOC circuits. The DOC circuits sense signal dropouts and operate an S/H circuit to hold a sample of audio in the usual manner. The demodulated audio signal is applied through an external LPF and the S/H circuits to mode-select circuits.

The mode-select circuits select stereo, left-channel, right-channel, or mute according to the logic levels of the SQ_1 and SQ_2 signals (from system control QU_{03}) at pins 9 and 10 of QA_{01}. The selected audio is passed through external filters to CX noise-reduction circuits (Sec. 9.2.7). If the disc being played contains the CX noise-reduction code, the CX line (from pin 7 of QU_{03}) is low, and the CX circuits in QA_{01} are activated.

The processed audio is then passed through mute circuits to the digital audio circuits. The mute circuits mute all audio during power-on and power-off.

9.21 DIGITAL AUDIO, ANALOG- AND DIGITAL-SELECT, AND BILINGUAL

Figure 9.54 shows the digital-audio and analog- and digital-select circuits for our CDV player. Figure 9.55 shows the bilingual circuits.

9.21.1 Digital Audio

The digital-audio circuits are similar to those of an audio CD player and include demodulator Q_{301}, digital filter Q_{302}, and DAC Q_{401}. The analog output from Q_{401} is applied to the player digital and analog output jacks through analog-filter circuits Q_{403} and Q_{404} and analog-digital relay L_{402}.

9.21.2 Analog- and Digital-Select

When a disc containing digital audio is played, microprocessor Q_{304} applies a low to relay driver Q_{415} which, in turn, places relay L_{402} in the digital position (DL and DR). Under these conditions, the audio appearing at the player output jacks is "digital audio" (decoded CD digital audio) in stereo analog form.

When playing a CD or CDV Single, L_{402} cannot be switched to "analog audio" because only digitally encoded audio is recorded on the CD or CDV Single. However, when playing a CDV which contains both digital and analog audio, the output can be switched to either digital or analog by the user.

When a CDV containing only analog audio is played, Q_{304} applies a high to Q_{415} which, in turn, places L_{402} in the analog position (AR and AL). Under these conditions, the audio appearing at the player output jacks is stereo analog audio as decoded by QA_{01} (Fig. 9.53).

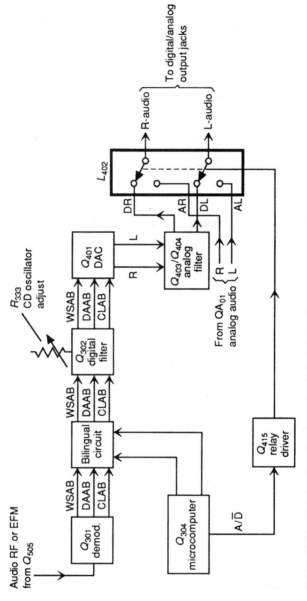

FIGURE 9.54 Digital audio and analog- and digital-select circuits.

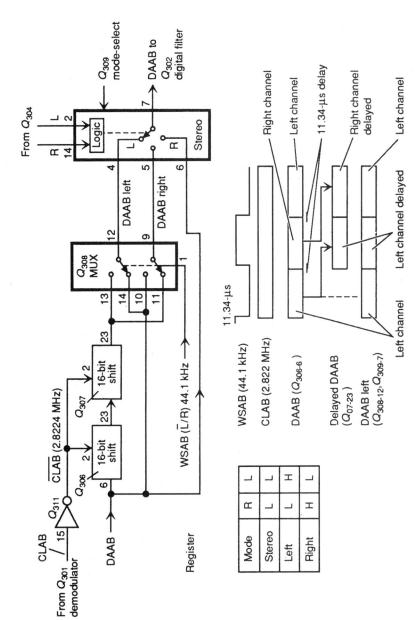

FIGURE 9.55 Bilingual circuits.

9.65

9.21.3 Bilingual Circuit

The bilingual circuit provides for play of bilingual discs as well as stereo discs. Bilingual discs are recorded with one language on the left channel and another language on the right channel. The bilingual circuit chooses one of three modes: stereo, left-channel only, or right-channel only.

As shown in Fig. 9.55, the bilingual circuit receives CLAB (bit clock, 2.8224 MHz), DAAB (data A-chip to B-chip), and WASB (word select A-chip to B-chip) demodulated signals from Q_{301}. (These same signals are applied to digital filter Q_{302}, as shown in Fig. 9.54).

The CLAB signal is inverted by Q_{311} and applied to 16-bit shift registers in Q_{306} and Q_{307}. CLAB clocks the DAAB signals in and out of Q_{306} and Q_{307} to delay the DAAB data bits by 11.34 μs (as shown by the timing chart in Fig. 9.55).

Both the delayed and undelayed DAAB signals are applied to digital filter Q_{302} through multiplexer Q_{308} and mode-select switch Q_{309}. Q_{308} is switched between left and right (delayed and undelayed) DAAB signals by the WASB signal at a rate of 44.1 kHz. Q_{309} is switched between left and right by select signals from Q_{304}.

When Q_{309} is in the left position (pin 4 connected to pin 7), only the left-channel DAAB samples (SL) are passed to Q_{302}, and SL is connected to pin 13 to receive the left-channel sample a second time. Thus, instead of delivering left and right samples to Q_{302}, only left-channel samples are sent (in sequence) to Q_{302} (as shown by the timing chart).

Q_{309} can also be switched to the right position (pin 5 connected to pin 7) so that only right-channel samples are sent to Q_{302}. Likewise, Q_{309} can be switched to stereo, where pin 6 is connected to pin 7, and both channels are passed to Q_{302} in the normal manner (normal stereo operation).

9.22 MECHANICAL SECTION

This section is devoted to the mechanical functions of our CDV player. Again, these descriptions must be compared with the player you are servicing (which, of course, will be altogether different).

Figure 9.56 shows major components of our player (with covers removed) as well as a close-up view of the optical pickup assembly. The remainder of this section shows the steps necessary to replace and adjust those mechanical components most likely to require service (which, it is hoped, will never happen).

Caution: If it becomes necessary to work with the pickup assembly (especially if you remove and reinstall the pickup), always follow all of the precautions described in Sec. 9.5 and illustrated in Fig. 9.13. The laser-diode and photodiodes in any CDV optical pickup assembly are electrostatically sensitive devices (ESD) and must be so treated (using wrist straps, working on a conductive sheet, etc.).

9.22.1 Replacing the Pickup Assembly

This section describes the steps necessary to remove and replace the entire pickup assembly. Refer to Fig. 9.57 for location of parts.

1. Remove the top cover and bottom plate by removing the screws, as shown in Fig. 9.57*a*.

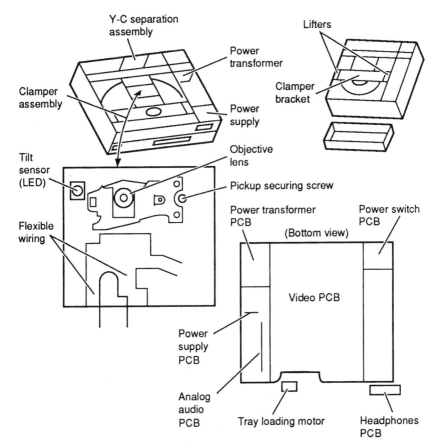

FIGURE 9.56 Major components of CDV player (with covers removed).

2. Switch the power on and press the Open/Close keys to eject the disc tray. Then switch the power off.

3. With the tray out, move the pickup assembly to the position shown in Fig. 9.57*b*. The pickup can be moved by rotating the slider motor manually. On most players, you can also operate the slider motor with a 1.5-V battery connected across the slider-motor terminals.

4. Stand the player on its side, as shown in Fig. 9.57*c* (with the power transformer at the top). Unfasten the three video-assembly screws and the three rear-panel screws. Open the video assembly and disconnect J_{402} from the digital-audio assembly.

5. Disengage the J_{101} lock in the servo assembly and carefully remove the flexible cable. Place a paper clip across the foil conductors at the end of the flexible cable after removal from J_{101}.

6. Unfasten the pickup-securing screw from the player top, as shown in Fig. 9.57*d*. Carefully remove the pickup assembly. Try to avoid touching the soldered sections of the pickup assembly because these connect to the laser diode and photodiodes (which are ESD).

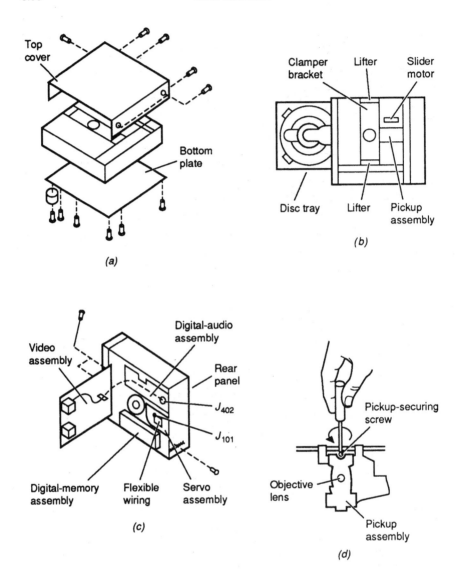

FIGURE 9.57   Removing and replacing CDV pickup assembly.

7. Using a wrist strap and conducting sheet, mount a new pickup assembly and tighten the securing screw (Fig. 9.57d). Carefully reconnect (and lock) the flexible cable to J_{101} in the servo assembly (Fig. 9.57c).

8. If a new pickup assembly is installed, or even if the old pickup is returned to the player, check the adjustments described in Sec. 9.25. Pay particular attention to the spindle-motor centering check (Sec. 9.25.3).

9.22.2 Dismantling the Tray Assembly

This section describes the steps necessary to remove and replace the tray assembly (which should be avoided). Refer to Fig. 9.58 for the location of the parts.

1. Remove the top cover (but not the bottom plate) by removing the screws, as shown in Fig. 9.58*a*.
2. Switch the power on and press the Open/Close key to eject the disc tray. Then switch the power off.
3. If the tray does not move to the full-out position (say, because of a defective loading motor or no microprocessor open signals to the motor, etc.), use the manual tray-opening procedures in Sec. 9.22.3.
4. With the tray out, remove the screw, stopper, and bushing and then remove the tray assembly (by pushing the center of the tray slightly) as shown in Fig. 9.58*b*.

9.22.3 Manual Tray-Open Procedure

This section describes the steps necessary to open (or eject) the disc tray manually (when the tray cannot be moved out with the Open/Close key and loading motor).

1. With the top cover removed, push on the white nylon movements to raise the disc tray, as shown in Fig. 9.58*c*.
2. Turn either gear in the direction of the arrows until the tray just starts to move toward the front.
3. Push gently at the rear to move the tray to the full-out position.

9.22.4 Installing the Tray-Loading Mechanism

This section describes the steps necessary to install (or reinstall) the tray-loading components after the tray has been removed from the player.

1. Attach the link and left and right movements so that the three parts engage as shown in Fig. 9.58*d*. Then turn the link fully counterclockwise.
2. Engage the lock assembly with the rack-gear portion of the movement and then attach the cam gear, as shown in Fig. 9.58*e*.
3. Attach the belt around the timing gears so that both timing gears face in the same direction, as shown in Fig. 9.58*f*.
4. Check that the Up and In switches are attached in the directions shown in Fig. 9.58*e*.

9.22.5 Attaching the Tray Assembly

This section describes the steps necessary to attach the tray assembly (if you should be so foolhardy as to remove it during service). Refer to Fig. 9.59 for the location of parts.

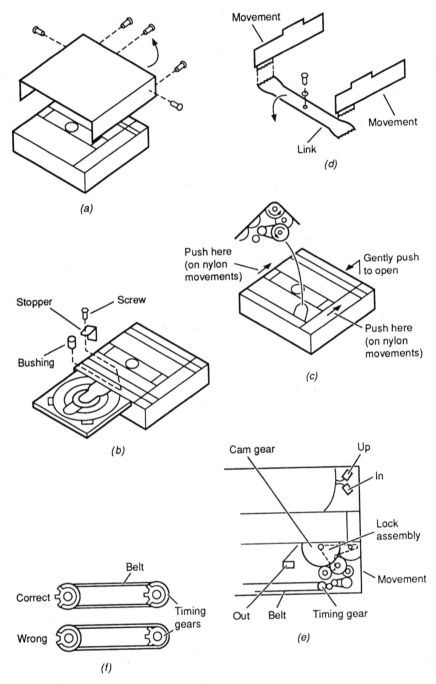

FIGURE 9.58 Removing and replacing CDV tray assembly.

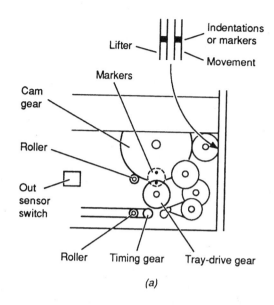

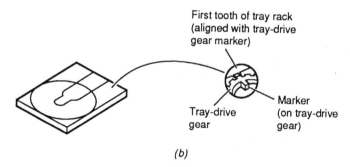

FIGURE 9.59 Attaching CDV tray assembly and aligning tray-loading gears.

1. Switch the power on and press the Open/Close key to move the tray out. (Because there is no tray, watch the cam gears rotate.) When the cam gear stops rotating, switch the power off while holding the Out sensor switch, as shown in Fig. 9.59a.

2. After making sure that the cam gear has rotated fully clockwise (as is the case when the tray is fully ejected, or out), check the alignment of the loading-gear markers as follows: (a) the markers on the cam gear and tray-drive gear must be aligned and (b) the indentations (or markers) on the movements and lifters must also be aligned. Do not check either alignment until you are certain that the cam gear is fully clockwise. If the markers are not aligned, use the alignment procedure of Sec. 9.22.6.

3. With all gears properly aligned, open the front-panel door and insert the tray with the timing gears on the left and right (Fig. 9.59a). When inserting the tray, make certain that the rollers are well seated in the grooves of the tray rack.

4. Continue pushing the tray until the first tooth of the tray rack engages the tray-drive gear teeth at the marker, as shown in Fig. 9.59b.

5. With the tray properly engaged, secure the stopper with the screw, and insert the bushing (Fig. 9.58b).

6. Switch the power on and press· the Open/Close key to move the tray in and out, making certain that the loading motor stops when the tray is fully in or out. There should be no problem if the in, out, and up switches are in the positions shown in Figs. 9.58e and 9.59a.

9.22.6 Aligning the Tray-Loading Gears

The tray-loading gears do not often go out of alignment during normal operation. Generally, you never need to check alignment unless the tray is removed from the player. However, if excessive force is applied to the tray, it is possible for the gear teeth to slip (particularly if the force is applied when the teeth tips are engaged). Examples would be if the tray is moving out and hits against a solid object or if someone tries to push the tray in manually.

Should the gears go out of alignment, as evidenced by failure of the tray to go fully in or fully out (or both), check the alignment as shown in Fig. 9.59a. Then, if absolutely necessary, align the gears as follows:

If the markers of the cam gear and tray-drive gear are not aligned, remove the plastic washer that retains the tray-drive gear, and reset the gear as necessary for proper alignment.

If the markers of the lifter and movement are not aligned, rotate the cam gear clockwise (trying desperately not to damage the gear teeth) until just before the lock assembly has locked the tray-drive gear. Then align the markers (indentations) manually.

With both sets of markers aligned, repeat the tray-attaching procedures in Sec. 9.22.5 from the start.

9.23 VIDEO-DISC PLAYER TROUBLESHOOTING BASICS

Because a video-disc player reproduces both audio and video, it is essential that you have both a shop-standard stereo system and a shop-standard TV or monitor of known quality. All players passing through the shop can be compared against the same standard. The monitor should be capable of displaying S-VHS (with separate Y and C signals). Many CDV players (including ours) can produce both conventional TV signals (composite video and/or RF) as well as S-Video and separate Y and C.

One practical way to confirm troubles in a CDV player is to monitor a TV broadcast with a TV set or monitor and compare the playback quality with that of the broadcast.. However, assuming a good test disc, the player picture is usually better than a broadcast picture (and considerably better than a VCR picture). Also, remember that an improperly adjusted monitor or TV can make a perfectly good player appear to be bad. This is always a problem when servicing CDV players (and VCRs) and is another good reason for a shop-standard monitor or TV.

9.24 CDV PLAYER TROUBLESHOOTING APPROACH

The troubleshooting approach for our player is based on functional circuit groups. The CDV players of all manufacturers have certain circuits in common (spindle-drive, system-control, video and TBC, audio, digital-memory, and special effects). The troubleshooting approaches for these circuit groups of our player are given in Secs. 9.27 through 9.32. Before you start any troubleshooting, read Secs. 9.8 through 9.21, where you will find (1) an introduction that describes the purpose or function of the circuit and (2) some typical circuit descriptions or circuit theory.

In many cases, the troubleshooting procedure requires adjustment. For that reason, the adjustment procedures for our player are given first, in Sec. 9.25. These adjustments are referred to in the troubleshooting procedures as necessary. Then Sec. 9.26 summarizes trouble symptoms that can be caused by improper adjustment.

9.25 CDV PLAYER ADJUSTMENTS

The following paragraphs describe adjustment procedures for our player. Each procedure is accompanied by diagrams that show the electrical locations for controls and test points, as well as waveforms or signals that should appear at the points. Remember that these procedures are for our player. Other players may require more (or less) adjustment. It is your job to use the correct procedures for each player you are servicing. Also, some disassembly and reassembly may be required to reach test and/or adjustment points (Sec. 9.22).

9.25.1 Rough-Grating and Tracking-Balance Adjustments

Figure 9.60 shows the adjustment diagram. Figure 9.30 shows additional circuit details. This adjustment sets the laser beam to the optimal position on the disc tracks and sets the tracking servo offset voltage to 0 V.

Rough-Grating Adjustment. Play a test disc and operate the controls to display the frame number on the monitor. Move the pickup to frame 16,000 by scanning or searching. Open (turn off) the tracking servo. Observe the scope display.

Insert a *special grating adjustment tool* into the grating adjustment hole and turn the grating so that the amplitude of the tracking-error signal varies from maximum to minimum. Note that any tool other than the manufacturer's special tool might (and probably will) damage the adjustment point. This will usually require replacement of the pickup assembly (at nerve-shattering expense to the customer).

Using the tool, find the position where the waveform amplitude reaches a minimum with a smooth waveform envelope, as shown in Fig. 9.60. This condition indicates that the three-way split laser beam is directed onto a single track, which is called the *on-track* position.

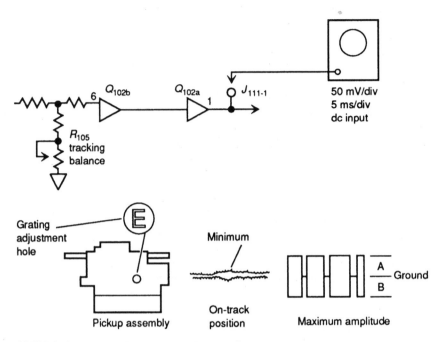

FIGURE 9.60   Rough-grating and tracking-balance adjustments.

Slowly turn the grating counterclockwise from the on-track position until the waveform reaches maximum amplitude. Close (turn on) the tracking servo and check that a normal picture is displayed on the monitor.

Tracking-Balance Adjustment.   Leave all connections in their present condition but align the scope trace with the center of the scope screen. Adjust R_{105} so that the positive and negative halves (A and B) of the tracking-error waveform are equal (Fig. 9.60).

9.25.2   RF Gain Adjustment

Figure 9.61 shows the adjustment diagram. Figure 9.36 shows additional circuit details. This adjustment sets the RF signal amplitude (from the main pickup photodiodes) to the optimum value.

Play a test disc and operate the controls to move the pickup to frame 16,000. Observe the scope display and adjust R_{282} for an RF signal amplitude of 300 mV ± 20 mV.

9.25.3   Spindle-Motor Centering Check

Figure 9.62 shows the test diagram. Figure 9.30 shows additional circuit details. This test checks that a single (imaginary) line (running parallel to the pickup slide) passes through the center of the pickup laser beam and the center of the spindle.

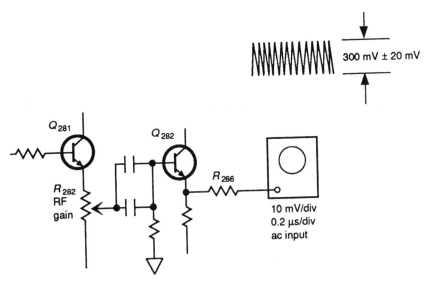

FIGURE 9.61   RF gain adjustment.

Play a test disc. Move the pickup to the inner tracks of the disc. Open (turn off) the tracking servo. Observe the scope display (scope set to X-Y, tracking error to Y, and tracking sum to X) and measure the Y-axis pattern amplitude.

Turn on the tracking servo and move the pickup to the outer tracks of the disc. Then open (turn off) the tracking servo and measure the Y-axis amplitude. If the amplitude of the Y axis is not substantially the same when the pickup is at the inner and outer tracks, the spindle-motor centering must be adjusted (Sec. 9.25.4). This check is required whenever the pickup is replaced.

9.25.4   Spindle-Motor Centering Adjustment

Figure 9.63 shows the test diagram. Figure 9.30 shows additional circuit details. This adjustment ensures that the single (imaginary) line (running parallel to the pickup) passes through the center of the pickup laser beam and the center of the spindle. This adjustment is necessary *only* when indicated by the spindle-motor centering check (Sec. 9.25.3).

1. Loosen the three spindle-motor set screws by turning each about half a turn.
2. Connect X and Y inputs of the scope (Fig. 9.63).
3. Play a test disc and move the pickup to the outer tracks.
4. Open (turn off) the tracking servo and observe the scope display.
5. Using the procedures of Sec. 9.25.1, fine-adjust the grating until the amplitude along the Y axis is minimum.
6. Close (turn on) the tracking servo and move the pickup to the inner tracks.
7. Open (turn off) the tracking servo again and observe the scope display. Record the amplitude of the Y axis.

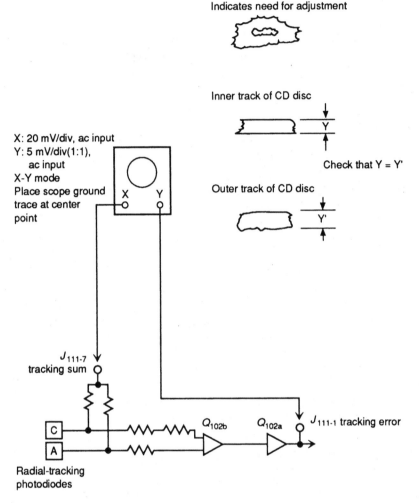

FIGURE 9.62 Spindle-motor centering check.

8. Insert an eccentric tool into the adjustment hole and slowly turn in the direction which *reduces* the Y-axis amplitude. (The eccentric tool can be fabricated, as shown in Fig. 9.64.) After reaching minimum Y-axis amplitude, continue turning the eccentric tool in the same direction until the Y-axis amplitude is the same as recorded in step 7.

9. Close (turn on) the tracking servo and move the pickup back to the outer tracks.

10. Repeat steps 4, 5, and 6.

11. Open (turn off) the tracking servo again and observe the scope display. Check that the amplitude of the Y axis has reached a minimum. If the Y axis appears to be greater than minimum, repeat steps 8 through 11. Tighten the spindle-motor set screws.

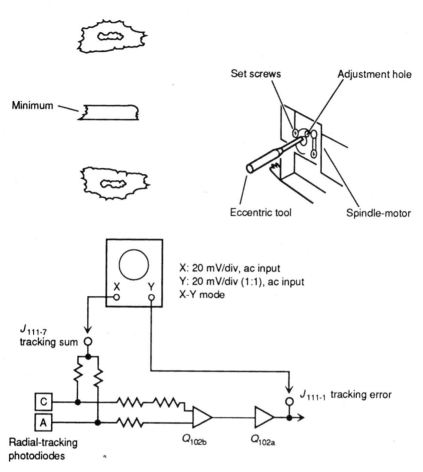

FIGURE 9.63 Spindle-motor centering adjustment.

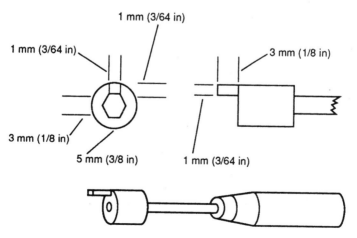

FIGURE 9.64 Eccentric tool for spindle-motor adjustments.

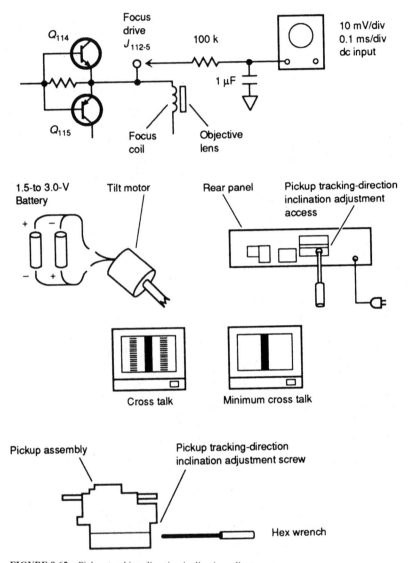

FIGURE 9.65   Pickup tracking-direction inclination adjustment.

9.25.5   Pickup Tracking-Direction Inclination Adjustment

Figure 9.65 shows the adjustment diagram. Figures 9.29 and 9.35 show additional circuit details. Adjustment of the slider shaft inclination ensures that the pickup assembly moves parallel to the disc surface. Adjustment of the pickup tracking-direction angle ensures that the laser beam is perpendicular to the disc.

1. Disconnect the tilt-motor connector J_{251} (Fig. 9.35). Do not reconnect J_{251} until the procedures of Secs. 9.25.5 through 9.25.11 are complete.

2. Play a test disc and search to frame 4760. This is the tilt fulcrum point (for our player).

3. Connect the scope to $J_{112\text{-}5}$ through a low-pass filter (100K, 1 μF) as shown in Fig. 9.65 and observe the focus-drive voltage.

4. Adjust the scope Y-axis position knob and move the focus-drive waveform to the center of the scope screen. Measure the focus-drive voltage again.

5. Continue to measure the focus-drive voltage while scanning to frame 46,135. If the voltage differs from that measured in step 4, connect a battery (1.5 to 3 V) to the tilt motor (Fig. 9.65) and turn the motor until the focus-drive voltage is within ±50 mV of that in step 4.

6. Remove the small lid on the rear panel and insert a hex wrench through the opening. Adjust the pickup tracking-direction inclination adjustment screw to minimize the crosstalk on the left and right sides of the monitor screen.

7. Search to frame 115 and check that crosstalk on the left and right sides of the monitor screen is minimized (and is substantially equal on both sides). Repeat steps 6 and 7 as necessary.

9.25.6 Focus-Error Balance Adjustment

Figure 9.66 shows the adjustment diagram. Figure 9.25 shows additional circuit details. This adjustment ensures that the focus servo maintains the objective lens at the optimum distance from the disc during playback.

Play a test disc and search to frame 115. Adjust the focus balance R_{162} to minimize crosstalk on both sides of the monitor screen. If this adjustment fails to reduce crosstalk down to an allowable level, go to the pickup tangential-direction angle adjustment (Sec. 9.25.7). Leave R_{162} at whatever setting that produces minimum crosstalk.

9.25.7 Tangential-Direction Angle Adjustment

Figure 9.67 shows the adjustment diagram. Figure 9.30 shows additional circuit details. This adjustment is to reduce crosstalk. However, the adjustment is usually necessary only if crosstalk remains conspicuous after completing the tracking-direction (Sec. 9.25.5) and the focus-error balance (Sec. 9.25.12) adjustments.

1. Play a test disc, search to frame 28,600, and open the tracking servo.

2. Connect the scope to $J_{111\text{-}1}$ (Fig. 9.30) and measure the tracking error signal.

3. Insert a hex wrench through the gap between the chassis and mechanical assembly (Fig. 9.67) to the tangential-direction inclination adjustment screw.

4. Adjust the screw until the tracking-error waveform is maximum.

5. Remove the hex wrench. Search to frame 115 and check that crosstalk on the left and right sides of the monitor is minimum (and is substantially equal on both sides). Repeat steps 4 and 5 as necessary.

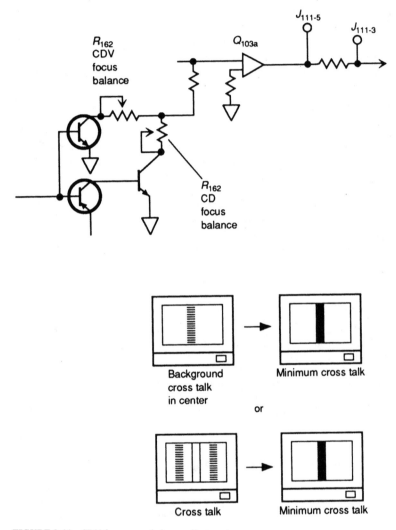

FIGURE 9.66 CDV focus-error balance adjustment.

9.25.8 Tilt-Sensor Inclination Adjustment

Figure 9.68 shows the adjustment diagram. Figure 9.35 shows additional circuit details. This adjustment sets the tilt-servo offset voltage to 0 V (by adjustment of the tilt-sensor inclination) when the pickup is replaced.

1. Check the color of the dot marked on the flexible cable next to the tilt sensor. There are two types of dots. Adjust tilt gain R_{218} accordingly. For a red dot, turn R_{218} fully clockwise. For a blue dot, turn R_{218} fully counterclockwise. If there is no dot or mark, set R_{218} to the center position.

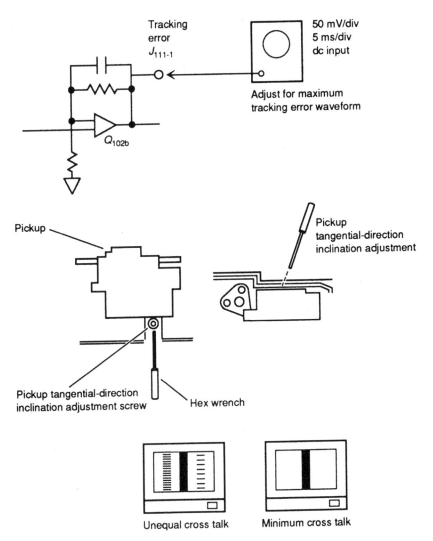

FIGURE 9.67 Pickup tangential-direction angle adjustment.

2. Play a test disc and search to frame 4760.

3. Connect the scope to J_{112-1} and measure the tilt-error voltage.

4. Insert a long-shaft Phillips screwdriver through the rear panel (Fig. 9.68) and adjust the tilt-sensor inclination adjustment screw until the tilt-error voltage is 0 V.

5. Connect the tilt-motor connector J_{251} that was disconnected during the pickup tracking-direction inclination adjustment (Sec. 9.25.5).

6. Search to frame 115 and check that the crosstalk on the left and right sides of the monitor has been minimized and is substantially equal on both sides.

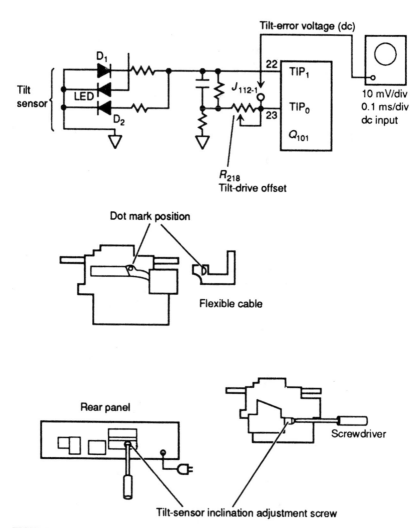

FIGURE 9.68   Tilt-sensor inclination adjustment.

9.25.9   Fine-Grating and Tracking-Balance Adjustments

Figure 9.69 shows the adjustment diagram. Figure 9.30 shows additional circuit details. This adjustment (or check) sets the laser beam so that the two tracking beams (on either side of the main beam, Fig. 9.9) are at the optimum position between the disc tracks and sets the tracking-servo offset voltage to 0 V.

1. Play a test disc, search to frame 16,000, and open the tracking servo.
2. Connect the scope as shown in Fig. 9.69 and observe the tracking-error and tracking-sum signals.

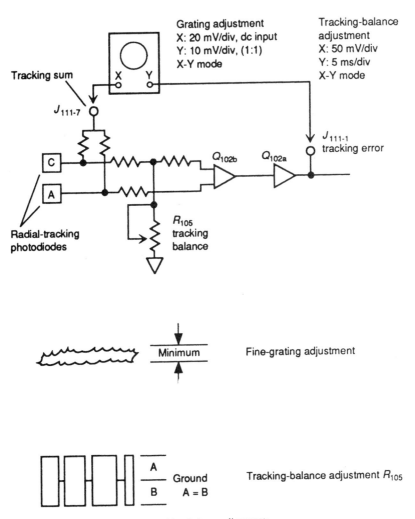

FIGURE 9.69   Fine-grating and tracking-balance adjustments.

3. Insert the special grating adjustment tool (Sec. 9.25.1) and fine-adjust the grating until the amplitude of the Y axis reaches minimum. If the grating is turned too far, and the optimum position can no longer be found, repeat the rough-grating adjustment (Sec. 9.25.1).

4. Leave all connections in their present condition but align the scope trace with the center of the scope screen. Check that the positive and negative halves (A and B) of the tracking-error waveform are equal, as shown in Fig. 9.69. If they are not, repeat the tracking-balance adjustment (Sec. 9.25.1).

5. Close the tracking servo and check that a normal picture is shown on the monitor.

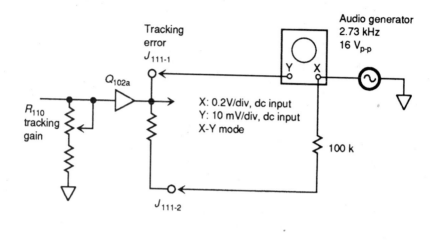

Out of adjustment

After adjustment

FIGURE 9.70 Tracking-servo loop-gain adjustment.

9.25.10 Tracking-Servo Loop-Gain Adjustment

Figure 9.70 shows the adjustment diagram. Figure 9.30 shows additional circuit details. This adjustment sets the tracking-servo loop-gain to the optimum value.

1. Play a test disc and search to frame 16,000.
2. Connect the resistor, audio generator, and scope to J_{111}, as shown in Fig. 9.70.
3. Set the audio generator to 16 V_{p-p} at 2.73 kHz.
4. Observe the scope pattern with the scope in the X-Y mode.
5. Adjust R_{110} until the scope pattern is symmetrical.

Note that if the audio-generator output is not capable of delivering 16 V, it might be necessary to reduce the value of the series resistor between J_{111-2} and the X input of the scope. Also, if the disc surface is scratched, the waveforms might not be read because of noise.

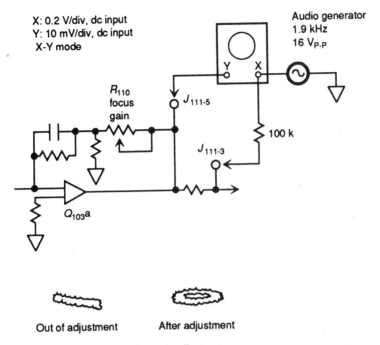

FIGURE 9.71 Focus-servo loop-gain adjustment.

9.25.11 Focus-Servo Loop-Gain Adjustment

Figure 9.71 shows the adjustment diagram. Figures 9.25 and 9.29 show additional circuit details. This adjustment sets the focus-error loop-gain to the optimum value.

1. Connect the gate of Q_{116} to ground (Fig. 9.29). This disables the high-frequency focus limiter circuit (Sec. 9.12.6).
2. Connect the resistor, audio generator, and scope to J_{111} as shown in Fig. 9.71.
3. Set the audio generator to 16 V_{p-p} at 1.9 kHz.
4. Observe the scope pattern with the scope in the X-Y mode.
5. Adjust R_{168} (Fig. 9.25) until the scope pattern is symmetrical.
6. Disconnect the gate of Q_{116} from ground.

Note that if the audio-generator is not capable of delivering 16 V, it might be necessary to reduce the value of the series resistor between J_{111-3} and the scope X input. Also, if the disc surface is scratched, the waveform might not be read because of noise.

9.25.12 CD Focus-Error Balance Adjustment

Figure 9.72 is the adjustment diagram. Figures 9.25 and 9.37 show additional circuit details. This adjustment ensures that the focus servo maintains the objective lens at the

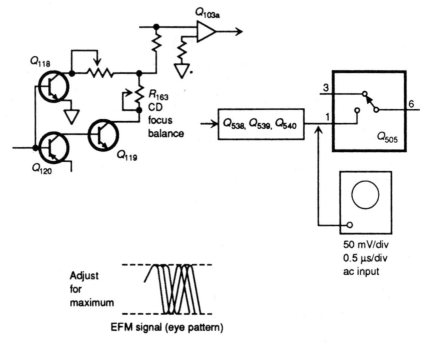

FIGURE 9.72   CD focus-error balance adjustment.

optimum distance from the disc during CD play (because our player is capable of playing both CDs and CDVs).

1. Play a CD test disc.

2. Connect the scope to $Q_{505\text{-}1}$ (Fig. 9.37), as shown in Fig. 9.72. Observe the EFM signal (so-called eye pattern).

3. Adjust the CD focus balance R_{163} until the EFM signal is at maximum.

9.25.13   Focus-Sum Level Adjustment

Figure 9.73 shows the adjustment diagram. Figure 9.25 shows additional circuit details. This adjustment sets the focus-sum level to the optimum value.

1. Play a test disc and search to frame 4760.

2. Measure the focus-sum voltage at $J_{111\text{-}6}$.

3. Adjust R_{200} until the voltage at $J_{111\text{-}6}$ is at 1.6 ±0.05 V.

Note that it is possible for the player to operate with a focus-sum voltage of less than 1.595 V. However, if the focus-sum voltage is substantially below 1.595 V, this indicates a possible failure of the focus-error circuits (such as Q_{103} and the photodiodes B_1 through B_4 in Fig. 9.25). If the focus-sum voltage is slightly low, set R_{200} for maximum voltage at $J_{111\text{-}6}$ but do not exceed 1.605 V.

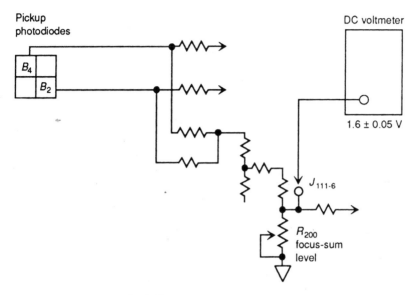

FIGURE 9.73 Focus-sum level adjustment.

9.25.14 Reference Shift Adjustment

Figure 9.74 shows the adjustment diagram. Figure 9.44 shows additional circuit details. This adjustment sets the amount of reference shift required for detecting a spindle-servo frequency error.

1. Connect the scope to $Q_{702\text{-}18}$, as shown in Fig. 9.74.
2. Adjust the scope trigger control to stabilize the pulse waveform (there may be jitter or pull on the scope trace).
3. Adjust reference shift R_{766} until the start of the high portion of the waveform to the start of the low portion is 28 μs ± 1 μs.

9.25.15 PLL Offset Adjustment

Figure 9.75 is the adjustment diagram. Figure 9.42 shows additional circuit details.

1. Play a test disc.
2. Connect the scope to Q_{703} as shown in Fig. 9.75.
3. Adjust PLL offset R_{779} until the offset voltage is 0 V ± 2 mV.

9.25.16 Half-H Rejection Adjustment

Figure 9.76 shows the adjustment diagram. Figure 9.44 shows additional circuit details. This adjustment sets the MMV pulse width for half-H rejection.

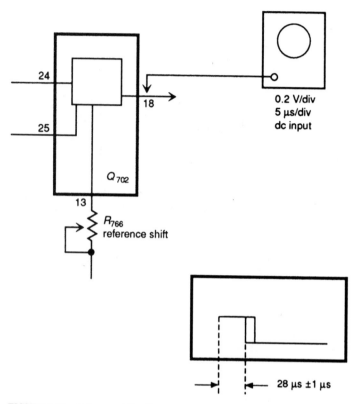

FIGURE 9.74   Reference shift adjustment.

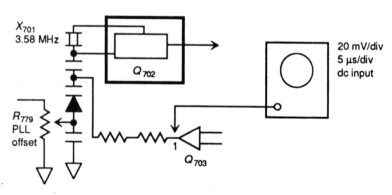

FIGURE 9.75   PLL offset adjustment.

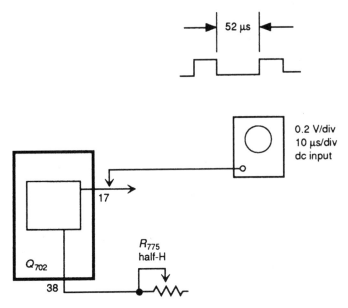

FIGURE 9.76   Half-H rejection adjustment.

1. Play a test disc.
2. Connect the scope to Q_{702-17}, as shown in Fig. 9.76.
3. Adjust half-H R_{775} until the width of the low interval of the pulse waveform is 52 μs ± 1 μs, as shown.

9.25.17   Trapezoid Inclination (TBC Error) Adjustment

Figure 9.77 shows the adjustment diagram. Figure 9.44 shows additional circuit details. This adjustment sets the slope of the trapezoid waveform used in TBC error-detection timing (the video phase shift, or VPS waveform).

1. Play a test disc. Put any frame into the still mode.
2. Switch the digital-memory function off.
3. Connect the scope to Q_{701-26}, as shown in Fig. 9.77.
4. Adjust trap timing R_{705} until the VPS error waveform is flat, as shown in Fig. 9.77.

9.25.18   Burst-Gate Position Adjustment

Figure 9.78 shows the adjustment diagram. Figure 9.42 shows additional circuit details. This adjustment sets the position of the burst gate.

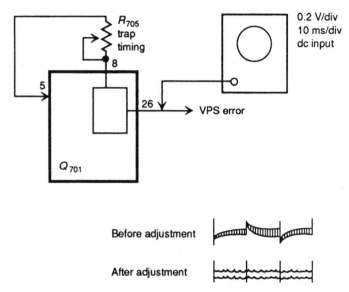

FIGURE 9.77 Trapezoid inclination (TBC error) adjustment.

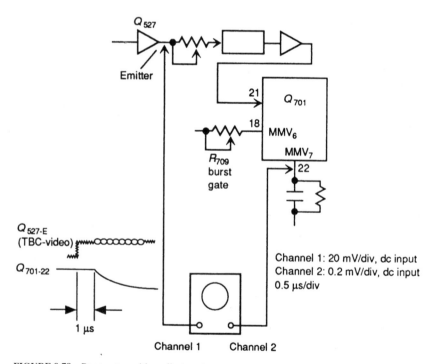

FIGURE 9.78 Burst-gate position adjustment.

1. Play a test disc.
2. Connect the scope to Q_{527} and $Q_{701\text{-}22}$ as shown in Fig. 9.78.
3. Adjust burst gate R_{708} until the trailing edge of the Memory Video (MMV) output is delayed by about 1 µs ± 0.1 µs with respect to the leading edge of the video signal, as shown in Fig. 9.78.

9.25.19 Reference Clock (14.31818 MHz) Frequency Adjustment

Figure 9.79 shows the adjustment diagram. Figures 9.44 and 9.52 show additional circuit details. This adjustment sets the reference clock frequency.

1. Connect a frequency counter to $QT_{01\text{-}3}$ as shown in Fig. 9.79.
2. Apply power to the player but do not play a disc.
3. Adjust reference clock CT_{06} until the counter reads 14.31818 MHz ± 10 Hz.
4. If this frequency is difficult to adjust (say, because the frequency counter does not provide sufficient resolution), play a test disc and adjust CT_{06} until the counter reads 15.734265 kHz when connected to $Q_{702\text{-}25}$ (Fig. 9.44).

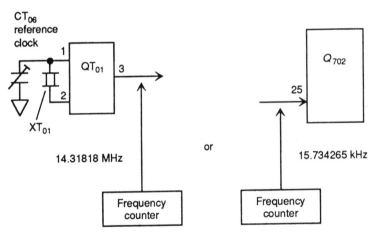

FIGURE 9.79   Reference clock (14.31818 MHz) frequency adjustment.

9.25.20   1H Delayed-Video Level Adjustment

Figure 9.80 shows the adjustment diagram. Figure 9.38 shows additional circuit details. This adjustment sets the amplitude of the 1H delayed-video signal (for dropout compensation) to the same level as the main video signal (so that any dropout is invisible).

1. Play a test disc and search to frame 19,801.
2. Connect the scope to Q_{501}, pins 40 and 42, as shown in Fig. 9.80.
3. Adjust 1H level R_{533} until the amplitude from the sync tip to the white level in the 1H delayed-video signal is the same as the amplitude of the main video signal.

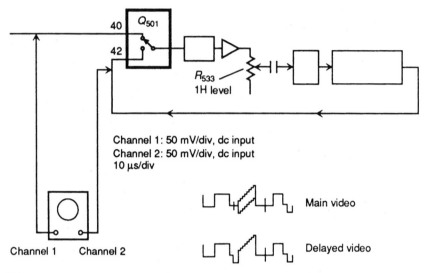

FIGURE 9.80   1H delayed-video level adjustment.

4. When properly adjusted, any dropout will be gray and will be almost invisible on the monitor. If R_{533} is not adjusted properly, the dropout will be visible (black if the delayed level is low or white if the level is high.)

9.25.21   VCO Center-Frequency Adjustment

Figure 9.81 shows the adjustment diagram. Figure 9.41 shows additional circuit details. This adjustment optimizes the CCD delay time for time-base error compensation.

1. Play a test disc and search to frame 19,801.
2. Connect the scope of Q_{515} and Q_{527} as shown in Fig. 9.81.
3. The video signal following the time-base error compensation in channel 1 (emitter of Q_{527}) contains jitter. Adjust VCO center R_{549} to delay the center of the jitter by 70.7 μs (1H +7.2 ±0.1 μs) from the trailing edge of the horizontal sync (H-sync) in the video signal prior to time-base error compensation at channel 2, as shown in Fig. 9.81.

9.25.22   Output-Video Level Adjustment

Figure 9.82 shows the adjustment diagram. Figure 9.41 shows additional circuit details. This adjustment sets the level of the player video output (from pedestal to 100 percent white) to 0.71 V_{p-p}.

1. Connect the player video output terminal to a monitor, and terminate the monitor internally with 75 ohms (if not already so terminated). If you are using a TV set with no video input terminal, terminate the player video output with 75 ohms.

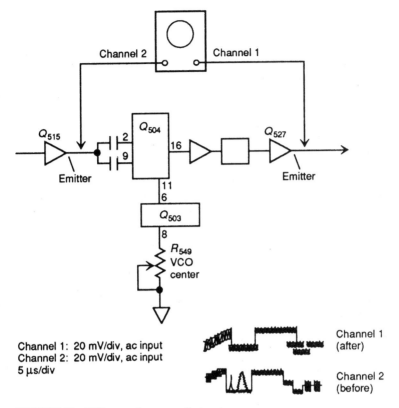

FIGURE 9.81   VCO center-frequency adjustment.

2. Play a test disc and search to frame 19,801.

3. Connect the scope to the player video output terminal (in parallel with the monitor) and observe the playback video waveform on the scope.

4. Adjust video level R_{567} until the amplitude from the pedestal level to the white level of the playback video waveform is 0.71 V ± 5 percent.

9.25.23   Color Phase-Correction Level Adjustment

Figure 9.83 shows the adjustment diagram. Figure 9.43 shows additional circuit details. This adjustment optimizes the amount of color phase correction.

1. Play a test disc and search to frame 7201.

2. Observe the monitor for color streaking, particularly any magenta color irregularities.

3. Adjust the VPS level R_{662} to minimize magenta streaking.

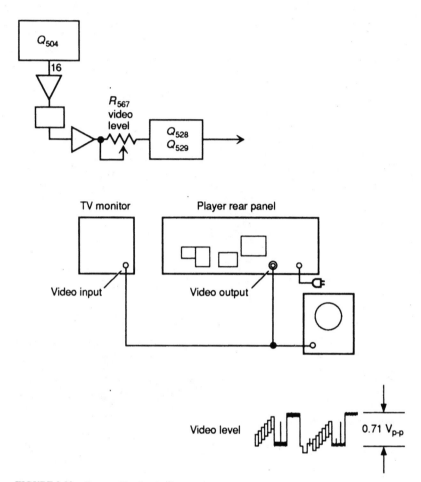

FIGURE 9.82   Output-video level adjustment.

9.25.24   Detector Level Adjustment

Figure 9.84 shows the adjustment diagram. Figure 9.37 shows additional circuit details. This adjustment optimizes the input voltage (spindle error) applied to the spindle-motor speed detector circuit.

1. Play a test disc and search to frame 4801.
2. Connect the DVM to Q_{881}, as shown in Fig. 9.84.
3. Adjust detector level R_{883} to obtain a difference voltage of 330 mV ± 5 mV.

9.25.25   Digital-Audio CD-Oscillator Adjustment

Figure 9.85 shows the adjustment diagram. Figure 9.54 shows additional circuit details. This adjustment sets the digital-audio CD clock to the correct frequency.

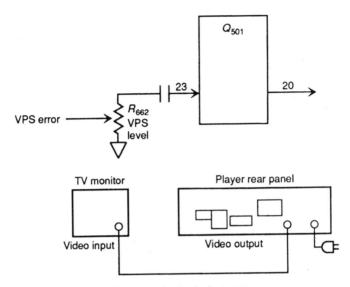

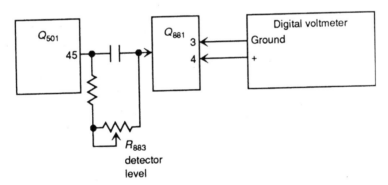

FIGURE 9.83 Color phase-correction level adjustment.

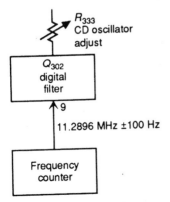

FIGURE 9.84 Detector level adjustment.

FIGURE 9.85 Digital-audio CD oscillator adjustment.

1. Connect a frequency counter to $Q_{302\text{-}9}$, as shown in Fig. 9.85.
2. Play a CD test disc.
3. Adjust R_{333} to set the clock to 11.2896 MHz ± 100 Hz.

9.25.26 Digital-Memory VCXO-Clock Adjustment

Figure 9.86 shows the adjustment diagram. Figure 9.52 shows additional circuit details.

1. Gain access to the digital-memory unit. On our player, this involves removing the unit from the shield case with the digital-memory assembly inside (by taking off five connecting cables and four screws that hold the shield case to the unit). Then remove the shields and reconnect the five cables.
2. Connect the scope to QB_{18} as shown in Fig. 9.86.
3. Adjust VCXO CB_{63} to set the voltage at pin 12 of QB_{18} to 3.0 V ± 0.2 V.

9.25.27 Sync-Level Adjustment

Figure 9.87 shows the adjustment diagram. Figures 9.45 and 9.52 show additional circuit details. This adjustment equalizes the amplitude of the H-sync signals of the input video signal applied to digital memory (the "through video" signal) and the H-sync signals of the output video signal from digital memory (the "memory video" signal).

1. Play a test disc and search to frame 4801.
2. Connect the scope to J_{502}, as shown in Fig. 9.87. This connection makes it possible to monitor CCD video and memory video simultaneously.
3. Press the player Digital Picture key.
4. Adjust sync-level RC_{58} until the amplitudes of the H-sync signals in digital-memory input and output video signals are the same.

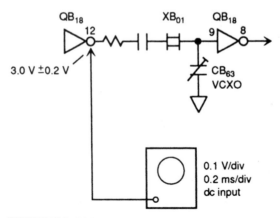

FIGURE 9.86 Digital-memory VCXO-clock adjustment.

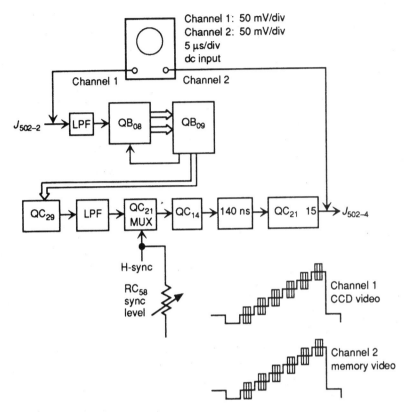

FIGURE 9.87   Sync-level adjustment.

9.25.28   Memory-Video Level Adjustment

Figure 9.88 shows the adjustment diagram. Figures 9.45 and 9.52 show additional circuit details. This adjustment equalizes the amplitude of the luma signals of the input video signal applied to digital memory (the through video signal) and the luma signals of the output video signal from digital memory (the memory video signal).

1. Play a test disc and search to frame 3900.
2. Connect the scope to J_{502} as shown in Fig. 9.88. This connection makes it possible to monitor TBC video and memory video simultaneously.
3. Press the player Digital Picture key.
4. Adjust memory-video level RC_{60} until the amplitude of the luma signals in the digital-memory input and output signals are the same.

9.25.29   Memory-Video (MMV) Adjustment·

Figure 9.89 shows the adjustment diagram. Figure 9.52 shows additional circuit details. This adjustment synchronizes the time base of the through-video and memory-video signals.

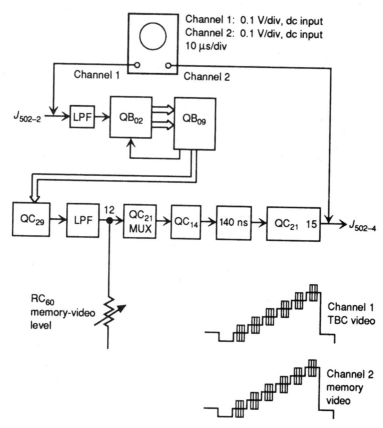

FIGURE 9.88 Memory-video level adjustment.

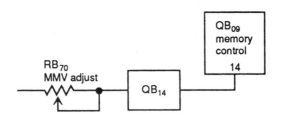

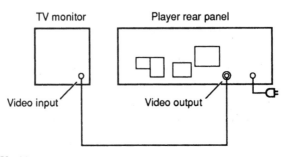

FIGURE 9.89 Memory-video adjustment.

1. Play a test disc.
2. Press the player Digital Picture key on and off while observing the TV or monitor.
3. Adjust MMV RB_{70} to minimize lateral displacement (horizontal shift) between direct pictures (through video) and pictures passed by the digital memory.

9.25.30 140-ns Shift-Level Adjustment

Figure 9.90 shows the adjustment diagram. Figure 9.52 shows additional circuit details. This adjustment sets the amount of phase shift in the memory-video 140-ns phase-shift circuit (to provide the correct burst phase to the composite video signal).

1. Play a test disc and search to frame 7201.
2. Connect the scope to QB_{18}, as shown in Fig. 9.90.
3. Operate the scope in the *X-Y* mode and observe the scope pattern.

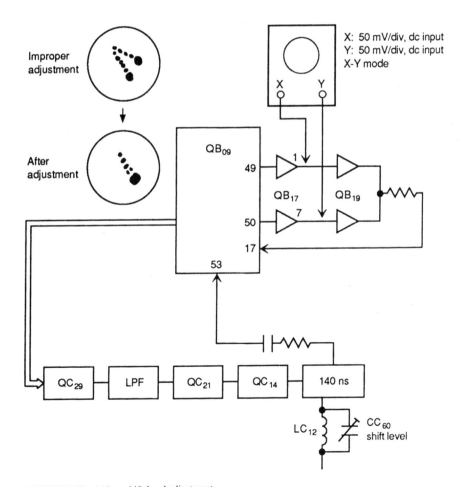

FIGURE 9.90 140-ns shift-level adjustment.

4. Adjust shift level CC_{60} until the two bright points in the scope pattern are as close to each other as possible (overlap is best), as shown in Fig. 9.90.

9.25.31 140-ns Level Adjustment

Figure 9.91 shows the adjustment diagram. Figures 9.44 and 9.52 show additional circuit details. This adjustment equalizes the amplitude of the chroma signal (in the composite video) before and after the 140-ns phase shift.

1. Play a test disc and search to frame 7201.

2. Connect the scope to QC_{21-15}.

3. Adjust 140-ns level RC_{59} to minimize vertical deviation of the chroma signal (as shown on the scope).

4. Observe the magenta image on the monitor screen to check for minimum flicker of the magenta image. If excessive magenta flicker persists, repeat the 140-ns shift-level adjustment (Sec. 9.25.30). If a vectorscope (Sec. 3.10) is available, adjust RC_{59} to align the magenta gain in the vectorscope screen to the prescribed position (typically $61°$ from B-Y, chroma 80 IEEE units, and luma 36 IEEE units, Figs. 3.3 and 3.10).

9.25.32 *Y/C* Separation Adjustments

Figure 9.92 shows the adjustment diagram. Figure 9.51 shows additional circuit details. These adjustments provide for the correct separation of Y and C signals from the composite video signal.

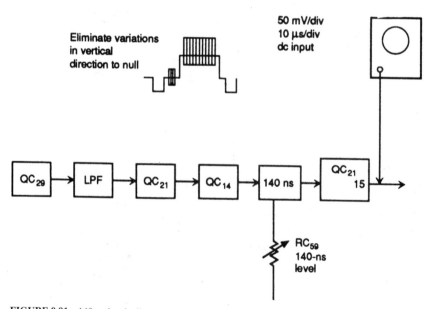

FIGURE 9.91 140-ns level adjustment.

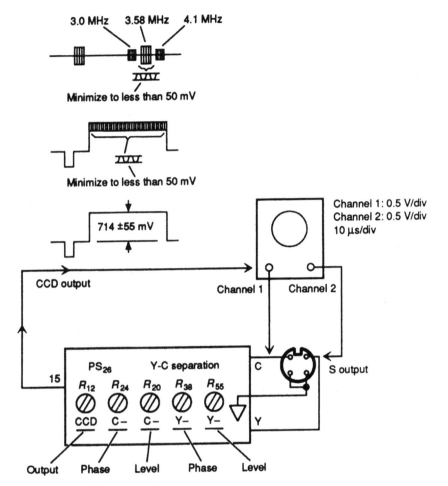

FIGURE 9.92   *Y/C* separation adjustments.

1. Remove the *Y/C* separation circuit board (this may not be necessary on all players).
2. Play a test disc and search to frame 9500.
3. Connect the scope to pin 15 of PS_{26} and observe the CCD out waveform (after passing through the LPF). Adjust CCD output R_{12} to minimize distortion and maximize the level simultaneously. Use a 0.1-μs/div sweep on the scope.
4. Search to frame 34,000 and observe the chroma-output (C-Out) waveform. Adjust C-Phase R_{24} and C-Level R_{20} alternately to minimize the residual *Y* component at 3.58 MHz. Use the luma-output (*Y*-Out) on channel 2 of the scope as a reference for the location of the 3.58-MHz burst.
5. Search to frame 9500 and observe the luma-output (*Y*-Out) waveform. Adjust *Y*-phase R_{38} and *Y*-Level R_{55} alternately to minimize the residual *C* component.
6. Search to frame 6500 and check that the 100 percent level is 714 ±55 mV. If the

level is not correct, repeat the video output-level adjustments (Sec. 9.25.22) as necessary.

9.26 CDV TROUBLE SYMPTOMS RELATED TO ADJUSTMENT

The following notes describe trouble symptoms that can be caused by improper adjustment. Remember that these same symptoms can also be caused by circuit failures. (We describe circuit failure as the basis of trouble symptoms in the remaining sections of this chapter.)

Improper tracking (track jumping or skipping):   Although they are not the only possible causes, these problems are often caused by improper grating and tracking-balance adjustments (Secs. 9.25.1 and 9.25.9) or the focus-sum level adjustment (Sec. 9.25.13).

Frequent dropouts: If there are frequent dropouts (with more than one disc), start by checking the RF-gain adjustment (Sec. 9.25.2). If the RF signal is weak, even a properly functioning DOC circuit cannot function properly.

Excessive dropouts: When you are certain that RF gain is good (Sec. 9.25.2), check the 1H delayed-video level adjustment (Sec. 9.25.20).

Track jumping during long search times: If tracking is good between nearby tracks but tends to jump when there is a long search time, check the spindle-motor centering (Secs. 9.25.3 and 9.25.4).

Crosstalk on TV or monitor display: If the TV or monitor display shows repeated images or blending of colors (left, right, or both), check the pickup tracking-direction inclination adjustment (Sec. 9.25.5), the CDV focus-error balance adjustment (Sec. 9.25.6), and the tilt-sensor inclination adjustment (Sec. 9.25.8). *If crosstalk is excessive,* check the pickup tangential-direction angle adjustment (Sec. 9.25.7).

Degraded performance: If more than one known-good disc (both CDV and CD show poor quality, both video and audio), check the tracking-servo loop-gain adjustment (Sec. 9.25.10) and the focus-servo loop-gain adjustment (Sec. 9.25.11).

Noise in CD sound: If there is noise *only* in the CD sound, check the focus-error balance adjustment (Sec. 9.25.12).

Spindle-servo locking failure: If the spindle-motor runs but fails to lock in any mode, check the reference shift adjustment (Sec. 9.25.14).

Color irregularities: If the color is poor (or incorrect) on more than one disc, check the PLL offset adjustment (Sec. 9.25.15). If there are color irregularities together with other symptoms, check as described in the following notes. *If color irregularities are excessive,* check the color phase-correction level adjustment (Sec. 9.25.23).

Picture irregularities and spindle-locking failure: If these symptoms occur simultaneously, check the half-H rejection adjustment (Sec. 9.25.16) and the detector level adjustment (Sec. 9.25.24).

Flickering in still mode and displacement in memory pictures: The first place to check for flickering or horizontal displacement in memory pictures is the trapezoid inclination (TBC error) adjustment (Sec. 9.25.17).

Playback does not start at the start of disc; missing or irregular color, fine stripes: Start by checking the burst-gate position adjustment (Sec. 9.25.18).

Color irregularities and spindle locking failure: If these symptoms occur simultaneously, check the reference clock (14.31818 MHz) frequency adjustment (Sec. 9.25.19).

Color-lock failure or slow color lock after search: Start by checking the VCO center-frequency adjustment (Sec. 9.25.21).

TV or monitor screen too dark or bright. Replay started at wrong position because of misread data: Start by checking the output video-level adjustment (Sec. 9.25.22).

Digital sound interrupted or irregular: Start by checking the digital-audio CD-oscillator adjustment (Sec. 9.25.25).

Irregular horizontal sync during digital memory mode: Start by checking the digital-memory VCXO clock adjustment (Sec. 9.25.26).

Unstable memory-video play but good nonmemory video: Start by checking the sync-level adjustment (Sec. 9.25.27).

Variations in brightness of the memory video play but good nonmemory video: Start by checking the memory-video level adjustment (Sec. 9.25.28).

Horizontal displacement of memory-video picture with respect to the nonmemory video: Start by checking the MMV adjustment (Sec. 9.25.29).

Flickering on TV or monitor display in all modes: Start by checking both the 140-ns shift-level (Sec. 9.25.30) and the 140-ns level (Sec. 9.25.31) adjustments.

S-Video monitor-display problems but good nonseparated video: If video from VHF out and video out is good, but S-Video is not good (striped noise, images not colored, etc.), check the *Y/C* separation adjustments (Sec. 9.25.32).

9.27 SPINDLE-DRIVE TROUBLESHOOTING

Figures 9.16 and 9.17 show the circuits involved in spindle-drive troubleshooting. Several symptoms can appear if a fault exists in the spindle-drive circuits.

If the spindle motor does not spin with any disc, check the voltage sources to the drive circuits, particularly the 16- and 5-V sources. If either voltage is absent, check the power-supply circuits. If both voltages are present, check at pin 9 of the spindle-motor control QD_{01} for a 5-V start signal. If it is absent, suspect the system-control circuits (Sec. 9.28). If the signal is present, suspect QD_{01}.

If the spindle motor spins when a CD is played but not when a CDV is played, the fault may be in the spindle motor or the motor drivers. If there is a fault in either circuit, the audio from a CD will be very poor because of dropouts.

To verify that a fault exists in the spindle motor, check for continuity through the three phases of the motor with an ohmmeter. If each phase of the motor is good (about 4 ohms), suspect the spindle-motor drivers QD_{02} and QD_{03}. The exact resistance of each winding is not critical, but all three windings should show substantially the same resistance.

It is also possible that the spindle drivers are not receiving signals from QD_{01} or a spindle-drive signal, as shown in Fig. 9.17. If any of these signals are absent, check back to the source (Figs. 9.46 through 9.49).

9.28 SYSTEM-CONTROL TROUBLESHOOTING

Figures 9.18 through 9.21 show the circuits involved in system-control troubleshooting. A malfunction in any of the systems might occur if there is a failure in system control. Because there are several microprocessors involved, the symptoms might point to only one of the microprocessors, but do not count on it. More likely, operation of one system-control microprocessor depends on bus signals from other microprocessors (as is typical for most present-day electronic devices).

Before you start looking for missing inputs to a system-control microprocessor or for outputs (in response to an input), there are some points that can be checked.

9.28.1 Power and Ground Connections

The first step in tracing problems on a suspected system-control IC is to check all power and ground connections. Figures 9.18 and 9.19 show the power and ground connections for the system-control ICs in our player. Make certain to check *all* power and ground connections to *each* IC because many ICs have *more than one* power and one ground connection. For example, as shown in Fig. 9.18, pins 1 and 42 of QU_{01} are both grounded. Likewise, as shown in Fig. 9.19, pins 10, 25, and 42 of QU_{03} are grounded.

9.28.2 Reset Signals

With all of the power and ground connections confirmed, check that all of the ICs are properly reset. The reset connections for the IC are shown in Figs. 9.18 and 9.19. Figure 9.20 shows the reset circuit details.

One simple way to check the reset functions is to check for reset pulses at the appropriate pins. For example, pin 6 of QU_{01} (and the entire reset, or $\overline{RES}$, line) should show pulses, as illustrated in Fig. 9.20. If they do not, suspect QU_{16}, QU_{17}, DU_{03}, and CU_{11}.

As in the case of any microprocessor-controlled device, if the reset line is open or shorted to ground or to power (5 or 16 V), the ICs are not reset (or remain locked in reset) no matter what control signals are applied. This brings the entire player operation to a halt. So if you find a reset line (or reset pin) that is always high, always low, or apparently connected to nothing (floating), check the line carefully.

9.28.3 Clock Signals

As shown in Fig. 9.18, there are three clock signals in the system-control circuits. The main system-control IC QU_{01} has a 4-MHz clock at pins 2 and 3, keys/display IC QF_{01} has a 6-MHz clock at pins 19 and 20, and the digital-audio IC Q_{304} has a separate 4-MHz clock at pins 15 and 16.

It is possible to measure the presence of a clock signal with a scope or probe. However, a frequency counter provides the most accurate measurement. Obviously, if any of the ICs do not receive the clock signal, the IC cannot function. On the other hand, if the clock is off frequency, all of the ICs might appear to have a clock signal, but the IC function can be impaired. (Note that crystal-controlled clocks do not usually drift off frequency but can go into some overtone frequency, typically a third overtone.) If the main system-control 4-MHz clock is absent or abnormal, suspect XU_{01}. If the digital-audio 4-MHz clock is absent or abnormal, suspect X_{302}. If the 6-MHz clock is absent or abnormal, suspect XF_{01}.

9.28.4 Bus Circuits

Generally, if there is pulse activity on the buses that interconnect system-control ICs (or any ICs), it is reasonable to assume that the clock, reset, power, and ground are all good. Although it is not practical to interpret the codes passing between ICs on the system buses, the presence of pulse activity *on all of the bus lines* indicates that the system-control functions are normal (but do not count on it).

When you have checked all of the bus lines for pulse activity, the next step is to measure the resistance to ground (with the power off) of each line in the bus (data bus, address bus, control bus, etc.). Pay particular attention to any lines that do not show pulse activity. If any one line differs substantially, suspect a problem on this line. If two lines show the same (higher or lower) resistance, they may be shorted together. In any case, before going further, check the schematic to see if circuits connected to suspect lines could explain the differences.

9.29 SERVO TROUBLESHOOTING APPROACH

Several symptoms can show up if there is a fault in one of the servo circuits (Fig. 9.24). Such symptoms may include *no start-up* or *poor play* (poor tracking or skipping), although these symptoms are not exclusively related to the servo.

The most practical approach is to try localizing the problem to only one of the servo systems. If you run through the test and adjustment procedures described in Sec. 9.25 (using test discs), specific problems can be isolated to a particular servo. The following paragraphs describe some techniques for isolating problems in the servo circuits.

9.29.1 Player Will Not Play Any Disc

The start-up circuits (Figs. 9.21 through 9.29) are likely suspects for this symptom. The start-up sequence can be checked by observing the pickup drive mechanism when power is applied.

With the top removed and without a disc, the drive mechanism should move outward, and the loading tray should move up and down to perform the disc-clamping sequence. After the loading tray moves to the down position, the objective lens should move up and down to perform a focus search.

Because the laser beam cannot focus without a disc, the pickup drive moves to check for the presence of a small disc (CD or CDV Single). The focus-search pattern is repeated after reaching the CD focus-search position. If this sequence is not followed, check the focus-servo circuits as follows.

With no disc installed, place the player in the test mode, and monitor the signal at J_{112-5} (Figs. 9.22 and 9.29) with a scope. In the test mode, the voltage should rise to 1 V, down to -1 V, and back up to 0 V (producing a voltage ramp) in about 1 s. This pattern should be repeated five times.

If the signal is present at J_{112-5}, but the objective lens does not move up and down five times, suspect the laser-diode assembly. If the focus start-up signal is not present, suspect Q_{101}, Q_{104}, Q_{114}, and Q_{115}. Note that the UPDN (start-up) signal at pin 36 of Q_{101} should be about 7 V$_{\text{p-p}}$.

If the start-up sequence is functioning, check the laser output next. Monitor the voltage at pin 5 of Q_{108} (Fig. 9.22) and go into a test mode. While the objective lens moves up and down, pin Q_{108-5} should read about 5 V.

If there is no voltage at Q_{108-5}, check the LD On signal from QU_{03-62}. If there is a signal at Q_{108-5}, stay in the test mode and monitor the output at pin 1 of Q_{108}, Q_{108-1} should go to about 2.5 V. If not, suspect Q_{108}.

If Q_{108-1} is 2.5 V during the test mode, check the monitor-diode input at pin 3 of Q_{108}. Q_{108-3} should be −5 V with the laser diode off. During the test, Q_{108-3} should go to about −3.6 V. If the voltage does not change when the test mode is initiated, suspect the laser-diode assembly.

9.29.2   Player Operates but Play Is Poor

If the start-up circuits are functioning properly, all of the drives for the servo circuits should be good. However, the adjustment may be off just enough to cause play problems. An effective troubleshooting procedure is to perform the adjustment checks for each of the servo circuits, as described in Sec. 9.25. Section 9.26 relates the procedures to trouble symptoms.

By simply going through the adjustment procedures (in the proper sequence), you can often pinpoint servo-system troubles. For example, if you are adjusting the focus-servo gain (Sec. 9.25.11) and find the signals at J_{111-5} and J_{111-3} (Fig. 9.25) absent or abnormal, you have localized trouble to the photodiodes B_1 through B_4 or to the Q_{103} circuits. Note that it might be difficult to monitor the outputs directly from the photodiodes. However, the voltage at all four photodiode test points J_{113-4} through J_{113-1} should be substantially the same.

Another aid in troubleshooting the servo circuits is to play different types of discs. For example, if a CD or the audio part of a CDV Single plays properly but a CDV shows poor play, the problem can be isolated to a circuit common to the CDV servo only (such as the CDV focus balance R_{162} or to Q_{118}, Fig. 9.25).

9.30   *VIDEO AND TBC TROUBLESHOOTING APPROACH*

Figures 9.40 through 9.45 and Fig. 9.51 show the circuits involved in video and TBC troubleshooting. There are many different symptoms that can be caused by a failure in the video and TBC circuits. These include *no picture, noisy picture,* and *picture jitter.* Start by playing a test disc and observing the display. This can often pinpoint the problem. Listening to the spindle motor can also give a clue as to the problem. The following techniques will aid in isolating problems in the video and TBC circuits.

9.30.1 No Picture

It is possible that a no-picture symptom can be caused by the spindle motor not being locked. If the spindle motor sounds like it is searching (increasing and decreasing spindle-motor velocity), the problem is probably in the TBC error-detection circuits (Fig. 9.44).

Another indication of TBC-circuit failure is a lack of audio. So, if you get no picture and no audio (or very poor audio) but the spindle motor is running, check the TBC error-correction circuits (Figs. 9.40 through 9.45).

One way to check the TBC circuits from input to output is to perform the adjustment checks for each of the circuits, as described in Sec. 9.25. By going through the procedures (in proper sequence) you can often pinpoint TBC circuit problems. Section 9.26 relates adjustment procedures to trouble symptoms. For example, Sec. 9.26 lists our TBC-related trouble symptom of "spindle-servo locking failure" as possible improper adjustment of the reference shift R_{766} (Sec. 9.25.14).

9.30.2 Poor Video

If the spindle motor appears to be locked, there is good audio, but the video is poor (jitter, noise, etc.) or the video is totally absent, the circuits following the TBC block are suspect. These circuits are shown in Figs. 9.50 and 9.51.

Play a test disc and monitor the circuit signals with a scope. As usual, use the waveforms shown in Figs. 9.50 and 9.51 for reference only. (This applies to any of the simplified schematics in this book). The service-literature waveforms should provide much greater accuracy (particularly if they are in the form of photos).

If poor video is in the form of flickering on the TV or monitor in all modes, check the 140-ns level adjustments described in Sec. 9.25.31.

If video is poor only when an S-Video (S-VHS) monitor is used, check the Y/C separation adjustments described in Sec. 9.25.31.

9.31 AUDIO TROUBLESHOOTING APPROACH

Figures 9.36, 9.37, and 9.53 through 9.55 show the circuits involved in audio troubleshooting. This involves straightforward signal tracing. There are no adjustments. Obviously, if the problem is with analog audio, but there is good digital audio, check the circuits of Fig. 9.53. On the other hand, check the circuits in Figs. 9.54 and 9.55 if the problem is in digital audio but there is good analog audio.

Remember that both audio outputs must pass through relay L_{402} (Fig. 9.54). If both digital and analog audio are good up to L_{402} but not at the output jacks, suspect L_{402}, Q_{415}, and Q_{304}. Also remember that both digital and analog audio originate from the same source as video (the B_1 through B_4 photodiodes, Fig. 9.36) but are separated in the RF signal-processing circuits, Fig. 9.37.

For example, if video is good, it is reasonable to assume that B_1 through B_4, Q_{281}, Q_{282}, Q_{501}, Q_{511}, and Q_{512} are good. If only analog audio is bad (good video and digital audio), suspect F_{503}, Q_{546}, Q_{547}, and the circuits of Fig. 9.53.

If only digital audio is bad (good video and analog audio), suspect Q_{538}, Q_{539}, Q_{540}, Q_{536}, F_{502}, Q_{537}, Q_{501}, Q_{505}, and the circuits in Figs. 9.54 and 9.55.

9.32 DIGITAL-MEMORY (SPECIAL EFFECTS) TROUBLESHOOTING APPROACH

Figure 9.52 shows the circuits involved in digital-memory troubleshooting. A failure in the digital-memory circuits usually occurs *only* when special effects are selected. Rarely will a problem in digital memory affect normal video.

If the symptom is no special effects, check the video signal input to A/D converter QB_{08}. If the input is absent or abnormal, trace back to the source. If there is a good input to QB_{08}, select the memory mode and play a test disc. Then trace through the circuits in Fig. 9.52.

When a picture is displayed (such as a color-bar test pattern), press the Memory key on the remote control. (On our player, MEM.P should be displayed on the screen for 1 s, indicating storage of the picture being displayed.) Now press the Stop key. The player should go into the stop mode, but the memory picture should still be displayed.

If the memory picture is not displayed, check the output from pin 6 of DAC QC_{29}. If the signal is present, trace through the circuits from QC_{29} to pin 15 of QC_{21}. Remember that the 140-ns phase-shift circuits require a 5-V CINV signal from pin 60 of QB_{09} to produce the correct output.

INDEX

ABOUT THE AUTHOR

John D. Lenk is a technical author specializing in practical electronic service and troubleshooting guides for more than 40 years. A long-time writer of international bestsellers in the electronics field, he is the author of more than 87 books on electronics, which together have sold well over 2 million copies in nine languages.

Lenk's guides, which regularly become classics in their fields, include *Lenk's Audio Handbook, Lenk's Laser Handbook, Lenk's RF Handbook, Lenk's Digital Handbook, Lenk's Television Handbook, McGraw-Hill Electronic Testing Handbook, McGraw-Hill Electronic Troubleshooting Handbook, Simplified Design of Linear Power Supplies, Simplified Design of Switching Power Supplies, Simplified Design of Micropower/Battery Circuits, Simplified Design of IC Amplifiers,* and *McGraw-Hill Circuit Encylopedia & Troubleshooting Guide,* Volumes 1, 2, and 3.